DEPARTMENT OF THE ENVIRONMENT

# Planning Control in Western Europe

LONDON: HER MAJESTY'S STATIONERY OFFICE

Crown copyright 1989
First published 1989

ISBN 0 11 752079 9

# FOREWORD

This report was commissioned from the Joint Centre for Land Development Studies (College of Estate Management/University of Reading) by the Planning Land Use Policy Directorate (as it was then designated) of the Department of the Environment. The purpose of the study was to investigate how planning control in four Western European countries operates in practice, and to make some broad comparisons with the UK systems. The research included an examination of the methods and procedures for development control in each country and of the relationship between these and the plan-making function. The findings and comparisons are those of the consultants and do not necessarily reflect the views of the Department.

# TERMS OF REFERENCE

The objectives of the project, in order of priority, are

(i) to provide, for each of the countries studied, a factual account of how development control operates; of the role of development plans; and of the relationships between them

(ii) to compare and contrast the operation of the overseas planning systems with the English planning system

(iii) to comment on the efficiency and effectiveness of the selected systems as seen by their operators and users.

In particular, the research is required to provide answers for each of the following questions.

(i) The scope of development control, ie the definition of development; how much is controlled and how much is exempted or given a general permission.

(ii) The basic system of control, ie whether an individual application for permission is required for each development; precisely how the application is made and a decision arrived at.

(iii) The planning considerations or factors taken into account in making a decision on a planning application, and where they are found in national or regional policy guidance, development plans or elsewhere.

(iv) The other permissions required before development can start, such as building regulations.

(v) The arrangements for preparing and approving or adopting development plans, their legal status and review.

(vi) The relationship between systems for preparing and approving plans, for the control of development, and the courts.

# PREFACE

We were appointed by the Department of the Environment to provide a factual description of the planning system for the control of development in four West European countries and to compare their practices with those in England. The terms of reference are given on the opposite page. The study was to be brief, lasting nine months, and to be based primarily on published sources, reports, plans and regulations, and interviews with planning officials and others during ten-day visits to the overseas countries.

This volume contains the results of the study. The bulk of it comprises the factual accounts of the control of development in the four overseas countries, preceded by a summary description of the English system. Each is written by a researcher, working to a common format based on the definitions, procedures and policies for the planning control of development. But despite certain similarities between the four, and the contrasts with England, each country's system of control is both complex and unique. Therefore each national account is an independent piece of work.

The comparative study follows, in the concluding section. It shows that in many respects the five countries are facing broadly similar issues about the control of development in terms of procedures and substantive content. But, despite the similarities, the chief impression from the national accounts is that each country's system is actually based most strongly in its history and constitution. It follows that any simplistic transfer of ideas and methods between countries, or even drawing lessons from other countries, is something to be treated with caution and understanding.

The study would not have been possible without the advice and cooperation of our four national correspondents who commented so readily on our draft reports and arranged the meetings with planning officials at every level of government and developers. They were

Dr Bo Grunland, School of Architecture, Royal College of Art, Copenhagen
Professor C.H. Chaline, Institute of Urbanism, University of Paris
Professor C.-H. David, University of Dortmund
Dr Johannes Blokland, Erasmus University, Rotterdam.

We are particularly grateful to Professor David for making available an early draft of his forthcoming study on the German planning system. We also acknowledge the advice of our colleague D.G. Hay, formerly Senior Research Fellow, Joint Centre for Land Development Studies, based on his wide knowledge of European urban systems and government, and his invaluable assistance in the visits to overseas countries. We are also very grateful for the exceptional assistance by our secretaries, Miss Sarah Colledge, Mrs Sheena Gordon, Miss Hilary Leckie and Mrs Jane Walker.

We acknowledge this assistance but the responsibility for any errors is ours. The views expressed in the report are those of the authors and do not necessarily reflect those of the Department of the Environment.

Finally, the one feature common to all five countries is that their planning systems for the control of development have recently changed in many detailed ways. The report describes the systems as they were early in 1987. They are continuing to evolve.

HWE Davies
D Edwards
AJ Hooper
JV Punter

*Department of Land
Management & Development
University of Reading
December 1987*

# ACKNOWLEDGEMENTS

Permission to reproduce plans and other documents has been received from the following organisations.

Department of Highways and Planning, Royal County of Berkshire
Reading Borough Council

Copenhagen Planning Department
Geodeotisk Institut, København
Høje-Taastrup Kommune
Plancenter Fyn A/S, Odense
Planstyrelsen, Miljøministeriet, København
Georg Gottschalk, Statens Byggeforskningsinstitut, Horsholm

Direction de l'Amenagement Urbain, Paris
Direction de l'Architecture et de l'Urbanisme, Ministère de l'Urbanisme, Paris
L'Urbanisme, Paris
Ville de Villeurbanne, Lyon
Directions Départmentales d'Equipement Loiret

Prof.Dr.Jur. C.-H. David, Fachgebietrechtsgrundlagen der Raumplanung, Universität Dortmund
Baubehörde, Freie und Hansestadt Hamburg
Presse-und Informationsamt der Bundesregierung, Bonn

Bouw-en Woningtoezicht, Gemeente Rotterdam
Dienst Ruimte en Groen, Provincie Zuid Holland
Gemeente Albrandswaard
Gemeentelijke Dienst Stadsontwikkeling-Grondzaken, s'Gravenhage
Rijksplanologische Dienst, Ministerie van Volkhuisvesting, Ruimtelijke Ordening en Milieubeheer, s'Gravenhage

# CONTENTS

# ENGLAND

by HWE DAVIES

## E7 BUILDING AND OTHER CONTROLS

# DENMARK

## by D EDWARDS

12

**Plates**

# FRANCE

by JV PUNTER

16

## Tables

## Figures

## Plates

# FEDERAL REPUBLIC OF GERMANY

by AJ HOOPER

# THE NETHERLANDS

by HWE DAVIES

# COMPARATIVE STUDY

## by HWE DAVIES, D EDWARDS, AJ HOOPER AND JV PUNTER

# INTRODUCTION

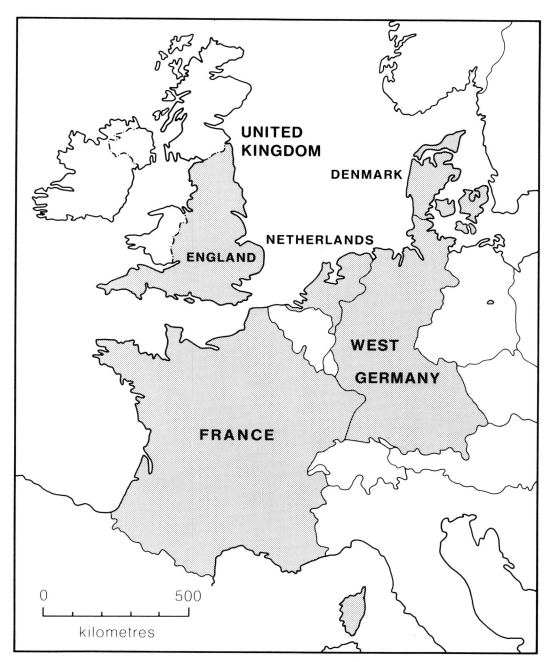

**Figure I1    Western Europe: The Study Countries**

# PLANNING AND THE CONTROL OF DEVELOPMENT

I.1   The use and development of land in England has been controlled under broadly the present system since 1947 although the basic legislation is now the Town and Country Planning Act 1971.

I.2   The legislation means that all development involving either new construction or a change of use has to have planning permission from a local authority to whom an application has to be made. The application is decided on its merits. That is, the local authority has to have regard to the development plan which it has prepared, but also to any other material considerations including national policies, other policies not in the development plan and planning judgement. Thus, the development plan does not give an automatic right to development and it can be overturned in any particular case.

I.3   The development plan itself is in two parts. Every county council has to prepare a structure plan which has to be approved by the Secretary of State for the Environment. The county or, more usually, the district or borough councils in a county may prepare and adopt more detailed local plans which have to be certified as being in general conformity with the structure plan. Local plans are not compulsory and, currently, only about a quarter of England is covered, chiefly the areas under pressure for new development or where development has to be restrained, for instance by a green belt.

I.4   If an applicant is refused planning permission, he or she may appeal to the Secretary of State who then appoints an inspector to conduct an inquiry into whether the original decision should be confirmed or reversed. In addition, applicants refused planning permission, objectors to a development plan and others concerned about a particular policy or decision have a limited opportunity for challenge in the High Court and thereafter in the Court of Appeal and the House of Lords. Such legal challenges however are on matters of law rather than policy, that is on whether procedures have been correctly followed, or whether the decision was within the powers of the Act. Finally, there are procedures for the enforcement of planning control, by taking action in the courts against those in breach of planning control.

I.5   The grant of permission for development under the Planning Act does not however complete the legislative requirements for the control of development. Permission will also be required under the public health legislation, including building regulations, and under other legislation on more particular matters.

I.6   The system of control in the four overseas countries of Denmark, France, West Germany and the Netherlands differs markedly from that in England. Briefly, the primary instrument of control in all four is a building permit issued by a local authority. It is required for virtually all building construction and most changes of use. The building permit however differs from an English planning permission in major respects. The permit covers both planning permission and that which in England is covered separately by building control.

I.7   More fundamentally, strong and clear criteria are laid down for the grant or refusal of a building permit. In general in the four countries if the proposed development is in accord with building regulations and the local plan, it cannot be refused permission; if not in accord, permission cannot be granted. That is, the local plan in each country is a legally binding document with detailed and precise regulations stating what is and is not permitted. It is therefore a very different type of document from the English local plan. It does in effect establish a legal right to development. The four plans, the *lokalplan* (Denmark), *plans d'occupation des sols* (France), *Bebauungsplan* (West Germany) and *bestemmingsplan* (the Netherlands) differ considerably among themselves. Nevertheless, within this apparently rigid, legal framework there is room for negotiation and discretion.

I.8   Local plans overseas do not necessarily cover the entire territory of local authorities any more than they do in England. However, in the absence of a local plan, the building permit is issued or withheld in accordance with the building regulations. But these regulations, unlike those in England, include many planning considerations as well as construction matters.

I.9   Furthermore, in each overseas country there is a hierarchy of higher order plans greater than that in England. Not only are there structure plans or their equivalent, as in England, but also regional plans, and national planning policies are prepared by different levels of government. In general, they are authoritative, and have a formal and explicit status in the planning systems for they are administratively binding on governments at every level, ultimately providing a strict framework for the local plans.

I.10   Many of these contrasts between England and the other four countries derive from their different legal systems and their written constitutions defining the relationships between citizens and government, and between central and local administrations within government. But it should also be remembered that the differences between the four can be as great as those between the four and England.

## The Approach to the Study

I.11 The aims of the study in summary were to present a detailed, factual account of the system for the control of development and the relationship with development plans in the four selected countries, Denmark, France, West Germany and the Netherlands (see Figure I1); and to compare their operation with that in the English system.

I.12 The four countries were chosen by the Department as, in the main, they were all highly developed, urbanised countries with long-established systems for planning and the control of development. They seemed to be an appropriate choice for comparison with England. Yet at the outset there were problems. A close examination of the terms of reference shows that they are written in language which strictly has meaning only in relation to the English system.

I.13 It is not simply a question of translation and understanding though that in itself is difficult. More fundamentally, the two crucial instruments in the English system, the development plan and development control, are unique to England, with a very precise legal and technical meaning. There are no comparable, equivalent terms in the other four countries. To take but one example, there is no Dutch word for 'development' in its technical, English sense. Conversely, there is no English term for the Dutch *bestemmingsplan* which is translated variously in the English literature as land use plan, local plan and development plan, all of which are both inaccurate and misleading. And the literal translation, destination plan, would be meaningless for the English reader.

I.14 Since the ultimate aim was to compare the overseas systems with the English, the starting point therefore was to describe the English system as closely as possible in terms that would be understood in relation to the overseas countries. This had the double advantage of defining the boundaries of what was meant by the control of development for the purposes of the study, and of describing the system and how it works. This was done by focusing on the four topics covered by the terms of reference, namely

(i) the definition of development

(ii) the system of development control by local authorities, its scope and procedures and the related systems of control such as that under building regulations

(iii) the preparation and role of development plans and their relationship with development control

(iv) the various challenges to the decisions in development control including administrative appeal to higher authority, and applications to the courts.

I.15 The second stage of the study was to review the available literature on the overseas systems, concentrating on the four topics already covered for the English system. This demonstrated the need for two further topics, if a comparable understanding was to be achieved, namely

(i) the historical background of planning control, especially, for instance, in the case of France where the system has been radically altered since 1983 following partial decentralisation to the *communes*

(ii) the political and administrative framework of control, important to an understanding of control in every country but especially, for instance, in Germany with its federal structure.

I.16 The next stage was to prepare draft national reports for each country. Together with the English report, they provided the basis for intensive discussions over a period of ten days in each country with the respective national correspondents and meetings with planning officials at every level of government and developers in the private sector. Copies were obtained of government reports, legislation, regulations and plans.

I.17 The national reports were then checked and revised, to cover as much as possible of the same ground as in the English report. Obviously this was not always possible. For instance, in none of the overseas countries were detailed statistics collected about the time taken to process applications for a building permit as this was not considered to be an important issue, as it is in England.

I.18 The final stage was to compare the different systems as far as possible from an English point of view, in compliance with the terms of reference. This was done by identifying the key topics and developing a comparative framework for the analysis of each topic, recognising the risk of oversimplification or even distortion that such a procedure might involve.

I.19 The report is laid out in separate, self-contained sections. The English system is briefly summarised, providing the framework to be followed in more detail in the four subsequent sections on the overseas countries, and the datum for the comparison between the English and the overseas countries in the final section.

I.20 The report comes to no general conclusions about the efficiency and effectiveness of the respective systems of control, and it makes no recommendations. The overseas accounts rely on published sources and a limited number of interviews in the countries. This means that the data for a rigorous comparison of the operation of control in practice in the overseas countries in general was not available on a consistent and comprehensive basis. Its collection would have required original surveys which were not possible in the time available for the study.

I.21 But there is a more fundamental reason for being wary of comparison. Each country's system is a response to its own perception of the issues and problems which it faces, and what it hopes to achieve through the control of development. Even more fundamentally, each system is constrained by its country's legal system and constitutional structure, and reflects the values of that society. And, finally, the control of development is only part, the regulatory part, of a wider system by which land and property is owned, managed and developed.

I.22 For all these reasons, the study has eschewed conclusions about the comparative efficiency and effectiveness of the different systems for the control of development. Its purpose rather has been to provide information and enhance understanding not only of the overseas systems but, through them, of the English planning system and its control of development.

# ENGLAND

HWE Davies

# E1 THE BACKGROUND TO CONTROL

## The 1947 Planning System

E1.1 The system for planning and controlling development currently used in England had its origins in the immediate post-war years and, although the original legislation has been largely repealed, the basic principles today are the same as those introduced through the Town and Country Planning Act 1947. Today, the primary instrument for planning and control is the Town and Country Planning Act 1971 although it too has been subsequently amended in a number of ways.

E1.2 The post-war decision to create a universal system of land use planning was a response to three main issues (Hall, 1982). Firstly, there had been a very rapid growth of suburban development around the main conurbations, especially London. It amounted to unchecked urban sprawl and unplanned ribbon development along highways. On the one hand, there was wide concern about the quality of the urban environment, the cost of providing utilities and the lack of planned provision of open spaces and community facilities. On the other hand, the loss of agricultural land and the sterilisation of mineral resources, especially sand and gravel, were serious as was the threat to the countryside.

E1.3 The pre-war town planning legislation had been based on legally binding zoning plans which allocated land for specific uses and densities. The first planning act had been the Housing, Town Planning, etc Act 1909. It had evolved from a long line of government reports and legislation during the nineteenth century about housing conditions and the problems of urbanisation (Cherry, 1974). Local authorities were given powers to prepare plans. However, few plans had actually been approved by 1939. No provision was made for limiting increases in land values, yet the system allowed compensation to be claimed if land values were reduced as a consequence of planning. The system was incapable of preventing the urban sprawl.

E1.4 The second issue was the economic depression of the interwar years. This had affected the older industrial districts and coal mining areas much more severely than the rest of the country. The resultant differences in unemployment levels reinforced the trends of migration from the depressed areas, intensifying the urban sprawl and congestion in the Midlands and south east England. The pre-war attempts to stem the migration and revive the economies of the depressed areas by a system of weak incentives for incoming industry were proving ineffective.

E1.5 Thirdly, the older, inner areas of the larger towns and conurbations were heavily congested, with a mixture of incompatible land uses, slum housing and obsolete road patterns. They required urgent planning and redevelopment, a situation made even worse by war-time damage and neglect. Yet town planning legislation was inadequate, in terms both of its financial provisions and its powers of control.

E1.6 Three major reports, on the geographical distribution of the industrial population, land utilisation in rural areas, and compensation and betterment, identified these problems and their interactions. Two solutions were proposed, within an overall framework of national planning. One was an effective system of land use planning which would not be impeded by problems of compensation and betterment for any loss or increase in land values as the result of planning controls. The other was to provide effective incentives and controls to guide the location of manufacturing industry and to reduce regional inequalities in unemployment.

E1.7 The 1947 Town and Country Planning Act enacted and consolidated all the instruments for a universal and comprehensive system of land use planning. The main executive responsibility for the new planning system was given to the upper tier of the elected local authorities, the so-called county councils, and for the larger towns, the county borough councils. These local planning authorities were given two main duties affecting the control of development, namely

(i) a duty to prepare a development plan for the whole of their area. Essentially, this was a land use map on an ordnance survey (topographic) base map. The development plan was for a period of twenty years, to be reviewed and rolled forward every five years

(ii) a duty to receive, and decide, applications for permission for every development within their area including changes of use of land or buildings. This requirement for planning permission was crucial. Failure to receive or comply with planning permission could result in enforcement action through the courts by the local planning authority against the guilty party.

E1.8 Central government was given two principal statutory duties and one statutory power, as well as a more general responsibility for the operation of the planning system. The duties were for the Secretary of State, as the cabinet minister responsible for the ministry or department responsible for town and country planning (variously the Ministry of Housing and Local Government, and, since 1970, the Department of the Environment)

(i) to approve the development plan and any subsequent alterations or amendments

(ii) to hear appeals against a refusal of planning permission or the imposition of conditions on its grant

(iii) to 'call in' planning applications for his decision where necessary.

E1.9 Four features of this system of development control were, and still are, particularly significant.

(i) The definition of development in the Act is extremely wide-ranging. However, the subordinate legislation, in particular the General Development Order and the Use Classes Order, limits the scope by defining various classes of development which are either exempted from planning control or are deemed to be permitted and therefore do not require formal planning permission. In particular, agriculture and forestry and many minor developments and changes of use were either exempted from the definition or classed as permitted development.

(ii) In making its decision about an application, the local authority has to have regard to the development plan and any other 'material' considerations relevant to the general purposes of the legislation and the specific circumstances of the application. The development plan by itself does not give an actual or implied right to develop land.

(iii) In general, there is no compensation payable for a refusal to grant planning permission, or indeed for an allocation of land for any particular purpose in a development plan.

(iv) Development by central government departments was formally exempted from development control through the planning system, although it is required to go through the procedures for consultation before being authorised.

E1.10 The 1947 Act contained many other provisions for the planning system. For instance, planning authorities were given powers for the compulsory purchase of land for planning purposes, and the preparation and implementation of plans for so-called areas of comprehensive development such as blitzed areas and, later, many town centres.

E1.11 The 1947 Act also contained financial provisions for compensation and betterment. The aim was that, in effect, land values should be frozen at existing use value on an appointed day in 1948. Landowners who thus lost development value should receive compensation and those who benefited by increased development value should pay a betterment charge. The intention was that all future land transactions, including those for compulsory purchase, should be at existing use value (Cullingworth, 1980).

E1.12 By, in effect, nationalising the right to develop land, the whole question of land use planning was separated from land policy about the ownership and value of land. Decisions about land use could be taken on planning grounds in the public interest rather than on the basis of land ownership and value. However, land policy since 1947 has been a political football, with major shifts in direction after every change of government. Labour governments introduced the original scheme for compensation and betterment in 1947, the Land Commission in 1967 and the Community Land Scheme in 1975. Each in turn was repealed by the incoming Conservative government, in order to restore market value as the basis of all land transactions including those for compulsory purchase.

E1.13 The effect of this instability on the material and distributional consequences of the system of development control has been little researched but is probably very substantial. But its procedural impact has been negligible. The system of control has remained unaltered and largely unchallenged for nearly forty years, whether land transactions are at market value as at the present time, or whether attempts are made for land transactions to be at something close to existing use value.

## Other Planning and Controls

E1.14 Three other acts are usually coupled with the 1947 Act as constituting the original English planning system, although they did not affect the basic methods of control. They were the New Towns Act 1946, the National Parks and Access to the Countryside Act 1949, and the Town Development Act 1952. The Distribution of Industry Act 1945, created the controls over the location of industry through industrial development certificates issued by what became the Department of Industry.

E1.15 Finally, planning is only one part of the control of development. Depending on the form of the proposed development, permission will also be required under other legislation, including

(i) building regulations, for any building works, from the district council. The system for building control was extensively modified in 1984

(ii) clean air regulations, fire precautions, waste disposal, highway connections, and development involving hazardous materials, from the county or district council, or other public authority

(iii) connection to gas, electricity and water supply, and sewerage, from the statutory undertakers or private organisations responsible for these services

(iv) the licensing of premises used for theatres and other entertainments including public houses and licensed restaurants, from the district council.

## The Planning System in 1987

E1.16    The 1947 planning system was designed primarily to deal with the problems of urban sprawl and congestion, and the loss of agricultural land. Development plans would coordinate the use of land by the private and public sectors. Private sector development would be controlled by the grant or refusal of planning permission.

E1.17    This still remains the system even though many of the circumstances which brought it into being have altered almost beyond recognition and today greater emphasis is given to the role of the planning system in encouraging development and, at the same time, conserving the urban environment and protecting the countryside. But the proportion and amount of development undertaken by the private sector has substantially increased, thus greatly intensifying the volume of work in, and the relative importance of, development control in the planning system. Local authorities still have to prepare development plans, but their form and content has changed, following a review of their purpose in the 1960s. Local government has been reorganised, in 1963 for London, in 1974 for the rest of England and again in 1986 for London and the metropolitan areas each time with consequences for the operation of the planning system.

E1.18    More recently, in the 1980s, new instruments have been introduced for selected areas. Development in accord with an approved planning scheme for an enterprise or simplified planning zone automatically has planning permission. The definition of development also has been revised, by amendments to the General Development and Use Classes Orders, in order to extend the scope of permitted development. Both are part of a wider trend towards deregulation (Great Britain, 1985).

E1.19    Today many aspects of the planning system are increasingly being questioned (Nuffield, 1986). During the 1970s concern focused largely on the administrative efficiency of the control of development, in particular the amount of time taken to make a decision on a planning application (Department of the Environment, 1975). More recently, the questions have been about the utility and effectiveness of development plans, especially structure plans. A consultation paper has been issued by the government, once again entitled **The Future of Development Plans** (Department of the Environment, 1986), foreshadowing possible future legislation, such as the abolition of structure plans and the further delegation of responsibility for planning and control to the lowest tier of elected local authorities, the district councils.

E1.20    Yet throughout the period since 1947 the basic definition of development, the basic system for the control of development, and the relationship between development plans and development control have remained unaltered. Even today, these principles are not fundamentally challenged.

E2.1   The United Kingdom is a unitary state. There is no written constitution defining the powers of central and local government or protecting the basic rights of citizens. All power derives from legislation enacted in Parliament, and is administered by central and local government. The courts provide a safeguard for the rights of citizens by ensuring that administrative decisions lie within the powers of the legislation.

## Central Government

E2.2   All parts of the United Kingdom have their own planning systems which differ in many of their legal and administrative arrangements even though all are based on the same fundamental principles. Thus Wales operates under the same legislation as for England but is administered in central government by the Welsh Office. Scotland and Northern Ireland have their own legislation as well as separate local and central government systems.

E2.3   The Department of the Environment, in the charge of a cabinet minister, the Secretary of State for the Environment, is the ministry responsible for planning in England (see Figure E1). Planning is only one of its functions. It also has housing policy; inner cities, now a separate function with its own legislation; and, most importantly, local government finance through which roughly half of local government expenditure is paid for by a block grant (the rate support grant) from central government and the other half through local property taxes (the rates) although the system is under review, and new legislation is promised for 1987/88 (Great Britain, 1986a). The Department's organisation is that

(i)   national legislation for enactment and major planning policies are formulated at headquarters in London, which is also responsible for technical advice on the operation of the system

(ii)  the day-to-day executive action, including the working relations with local authorities and the administration of policy including case work on development plans and planning applications, is in eight regional offices

(iii) planning inquiries and appeals are administered by the planning inspectorate

(iv)  the Secretary of State appoints the regional water authorities, responsible for water supply, land drainage and sewerage; the new town and urban

development corporations, which have planning control functions; and the Countryside Commission.

E2.4   The other government departments closely involved in land use planning, having to be consulted or providing various services, include

(i)   Ministry of Agriculture, Fisheries and Food, on loss of agricultural land

(ii)  Department of Transport, on the national road network and nationalised transport industries such as British Rail

(iii) Department of Industry, on industrial policy, including the location of industry

(iv)  other government departments as major land users either directly or indirectly through local government, and appointed bodies such as the National Health Service or the Central Electricity Generating Board.

All of these departments are separately accountable to Parliament.

## Local Government

E2.5   Local government within London and the six former metropolitan counties has, since 1 April 1986, been a unitary system with a single elected tier of boroughs or districts (see Table E2.1). They are responsible for all local government functions, principally education, housing, social services, leisure and recreation, highways, and planning. Thus planning in

**Table E2.1   Local Government in England, 1985**

| | Number of councils | Average population and approximate size range ('000s) |
|---|---|---|
| London and the metropolitan areas | | |
| county councils* | 7 | 1,949  (1,140–2,650)φ |
| London borough councils | 33 | 204  (132–321)+ |
| metropolitan district councils | 36 | 303  (158–1,003) |
| the remainder, 'shire' counties | | |
| county councils | 39 | 746  (121–1,511) |
| district councils | 296 | 98  (34–281) |

*Note:* *   abolished 1 April 1986
     φ   excluding Greater London, total 6,696,000
     +   excluding City of London, total 4,700

*Source:* Chartered Institute of Public Finance & Accountancy, 1986

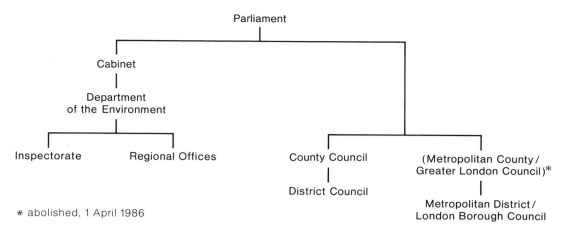

*abolished, 1 April 1986

**Figure E1  Structure of Government**

Figure E1 shows the structure of government for planning administration. Legislation is enacted by Parliament. The Department of the Environment is responsible at the national level, with the planning inspectorate responsible for appeals against refusal, etc. of planning permission by local authorities; and regional offices responsible for case work and liaison with local authorities.

London is the responsibility of 32 independent borough councils and the City of London, although a committee of the borough councils has been formed to act as an advisory body on London-wide planning issues, and the information and intelligence  system of the former Greater London Council continues in existence. A broadly similar situation exists in the other six areas, on a smaller scale. However, regional guidance for planning purposes in each of the areas is to be provided by the regional offices of the Department of the Environment.

E2.6   Local government elsewhere in the 39 so-called 'shire' counties has been since 1974 and is still a two-tier system of elected councils, with, in rural areas, a third tier of elected parish councils. Figure E2 shows the geographical pattern in and around Berkshire and its six district councils. It also shows the western areas of Greater London. The functions are that

(i) county councils are responsible for education, social services, waste disposal, highways and strategic planning

(ii) district councils are responsible for housing, leisure and recreation, refuse collection, and local planning and control of development

(iii) parish councils have a few, very limited responsibilities but have to be consulted on development control

(iv) in addition the national parks are designated by the Secretary of State, on the advice of the Countryside Commission, and administered by committees of the relevant county councils or national park authorities.

E2.7   County councils have an average population of about 750,000, and an area of about 310,000 hectares. They are directly elected bodies with a structure of committees and sub-committees which make the decisions on the advice of their permanent officials. County

Local government is given responsibility in the legislation for preparing development plans and development control. The responsibility is shared between elected county and district councils except in London and the larger conurbations where, since 1986, there has been just one elected tier of local government, the district or borough councils.

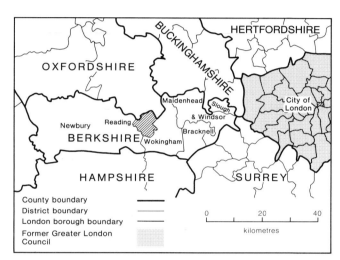

**Figure E2   Local Government Boundaries in Berkshire and Neighbouring Areas**

Figure E2 illustrates a part of the map of administrative boundaries in England. Berkshire County contains six districts within its boundaries. Neighbouring counties such as Oxfordshire also contain a number of districts. Since the abolition of the Greater London Council in 1986, local government in London has been the responsibility of 33 separate borough councils.

planning departments typically have a professional and administrative staff of about 60 to 70. They concentrate chiefly on the preparation and review of structure plans but also usually contain the county's research and intelligence unit and special sections dealing for instance with landscape, minerals and the countryside, and increasingly the promotion of economic development. In some counties, planning departments are combined with transportation or estates (land) departments.

E2.8   The average district council in England and Wales has a population of about 100,000 and a surface area of about 40,000 hectares. It is usually one of about seven districts within a single county. Like the county council, the district council is a directly elected body with committees and sub-committees which make decisions on the advice of their permanent officials. District councils thus have a planning committee and, in many, a sub-committee to deal with planning applications where these are not delegated to officers. They have a planning department with, on average, about 25 to 30 professional and administrative staff. It is usually organised into sections including a development control section and a forward planning (or plan-making) section.

E2.9    There is, of course, no such thing as an average district. Thus a district might be located entirely within a very large former metropolitan area in which case its population and its development control workload are twice as large as the average. At the other extreme a district might be a remote rural area with a very small population. Or it might be a medium-size town with its surrounding countryside.

E2.10    The variety of districts is matched only by the diversity of the policy stances of different districts. Some districts with historic towns have a strong emphasis on conservation. Suburban districts on the edge of the metropolitan areas often seek to prevent any further development, especially if it is for housing in a green belt. Inner city districts experiencing urban decay, or older industrial districts losing employment, seek to promote development.

## Other Planning Administration

E2.11    There is no level of regional government in England, other than the regional offices of central government departments. However, between 1964 and 1979, regional economic planning councils were appointed by the Secretary of State for the Environment to advise on regional issues and coordinate the policies of central government as they affected the regions. Regional strategies were prepared for each region. But the regional councils were abolished in 1979, and regional guidance is now provided by central government, working through its regional offices.

E2.12    Local authorities in some areas have formed *ad hoc* organisations to analyse information about regional trends on planning and development and as far as possible to coordinate their policies. The most highly developed example is SERPLAN, the South East Regional Planning Conference, which brings together planning authorities for London and south east England.

E2.13    Urban development corporations have been set up for the London and Merseyside docklands and others are promised. The corporations are appointed by, and responsible to, the Secretary of State for the Environment and are funded by government grant. They become the local planning authorities for development control in their areas, superseding the elected local authority. Their chief aim is to stimulate development in the private sector through the provision of serviced land and other means.

E2.14    New town development corporations were previously set up by government for each new town. The corporations were appointed by the Secretary of State to design and construct the new towns on their own and in partnership with the private sector. However, they never became local planning authorities. The majority have been wound up as the towns have been completed.

E2.15    Finally, public utilities such as gas, water and sewerage, and electricity are the responsibility of government appointed organisations such as the Central Electricity Generating Board or the regional water authorities. Current government policy is for some of these to be privatised. Thus British Gas has been privatised and others are to follow, including electricity and water.

# E3 THE DEFINITION OF DEVELOPMENT

E3.1 The definition of development is the crucial bedrock on which not only the control of development but the entire planning system is founded. The definition is found in the primary legislation (the Town and Country Planning Act 1971, as amended) and the subordinate legislation made by Parliament amplifying the Act.

E3.2 The following section gives a brief explanation of the definition. Inevitably, the subject is extremely complex and by now there is a large body of case law interpreting the definition. The procedures for obtaining permission are described later. But, if in doubt, a person can apply to a local authority to determine whether a particular proposal does constitute development, and can appeal to the Secretary of State against an adverse interpretation. The person can also challenge the interpretation on a point of law before a judge in the High Court and, ultimately, in the House of Lords sitting as the highest court in the land.

## The Basic Definition

E3.3 Section 22(1) of the 1971 Act defines development as

'the carrying out of building, engineering, mining and other operations in, on, over or under land'

or

'the making of any material change in the use of any buildings or other land'.

E3.4 The general definition however is narrowed down in the Act in several ways.

(i) Certain development is exempted from the definition, chiefly any improvements or alterations to a building which do not involve a change of use, or materially affect its external appearance; and, very importantly, the use of land for agriculture or forestry, although not agriculture and forestry operations as such.

(ii) Still other development is deemed to have received planning permission by having gone through various procedures under other legislation. This refers to development by central government departments, local authorities and statutory undertakers. In practice, this is a formal distinction as such development is nevertheless required to go through procedures similar to those for obtaining planning permission.

(iii) However the division of a single family dwelling into two or more flats or apartments explicitly does constitute development.

E3.5 It should be noted however that, with very few exceptions, demolition of buildings does not constitute development and therefore does not require planning permission. The chief exceptions are for listed buildings and all unlisted buildings in conservation areas (see below).

E3.6 The category of permitted development is still further narrowed down through subordinate legislation. The Secretary of State is empowered by the Act to make regulations and orders which have to be approved by Parliament using a quicker and more simple procedure than for amending an act. The two most important statutory instruments are the General Development Order, which sets down categories of so-called 'permitted' development as well as the detailed procedures for development control, and the Use Classes Order which still further refines the definition of development requiring planning permission.

E3.7 A large number of very minor developments involving building and other operations or changes of use are automatically granted planning permission by the General Development Order. Apart from matters such as temporary uses, and repairs, three are noteworthy, namely

(i) development within the curtilage (ie the grounds) of a dwelling, subject to conditions regarding height, location etc

(ii) construction of certain agricultural buildings, also subject to conditions

(iii) development for certain types of industrial processes.

E3.8 The current list of permitted development under the General Development Order is summarised in Table E3.1. The list has been fairly frequently revised since its first introduction following the 1947 Act. In nearly every case, the effect of the amendments has been to increase the scale of permitted development. The 1985 amendment for instance increased the volume of permitted development for industrial purposes from 10 per cent or not more than 500 square metres floorspace in the 1977 Order to 25 per cent or 1,000 square metres.

**Table E3.1 The General Development Order, Permitted Development: Summary**

Classes of Permitted Development specified in Schedule I of the General Development Order, as amended

| | |
|---|---|
| I | Development within the curtilage of a dwelling house (subject to qualifications relating to the height, set back, plot coverage and cubic volume) |
| II | Gates, fences, etc. external painting |
| III | Changes of use within use classes defined in the Use Classes Order |
| IV | Temporary buildings and uses |
| V | Use by members of specified recreational organisations |
| VI | Agricultural buildings, works and uses |
| VII | Forestry building and works |
| VIII | Development on land used for industrial purposes and processes |
| IX | Repairs to unadopted streets |
| X | Repairs to services (sewers etc) |
| XI | Replacement of war-damaged buildings, works and plant |
| XII | Development under local or private Acts or Orders |
| XIII–XXIII | Development by local authorities and statutory undertakers, such as repairs and improvements to highways |

*Note:* each class is defined precisely, with many qualifications and exceptions

*Source:* The Town and Country Planning General Development Orders 1977 to 1985, S.I. Numbers 1977/289, 1980/1946, 1981/245, 1569, 1983/1615, 1985/10011, 1981, 1987/702, London: HMSO

**Table E3.2 The Use Classes Order: Summary**

| | | | |
|---|---|---|---|
| A1 | Shops | B5 | Special industrial: bricks etc |
| A2 | Financial and Professional Services | B6 | Special Industrial: chemicals |
| | | B7 | etc |
| A3 | Food and Drink | | Special Industrial: animal products etc |
| B1 | Business, including offices, research and development, and light industrial | B8 | Storage and Distribution |
| | | C1 | Hotels and Hostels |
| | | C2 | Residential Institutions |
| B2 | General Industrial Building | C3 | Dwelling Houses |
| B3 | Special Industrial: alkalis etc | D1 | Non-Residential Institutions |
| B4 | Special Industrial: smelting etc | D2 | Assembly and Leisure |

*Note:* there is also a list of special uses, such as theatre, motor vehicle sales, etc which are not included in any class

*Source:* The Town and Country Planning (Use Classes) Order 1987, S.I. Number 1987/764, London: HMSO

E3.9 Another large number of changes of use are specified as not being development by the Use Classes Order. The current Use Classes Order was approved in 1987 after a long period of consultation. It is summarised in Table E3.2. Compared with the previous, 1972, Order, the number of classes has been reduced from 18 to 16 and a number of the classes have had their scope widened. As with the amendments to the General Development Order, the aim has been to reduce the amount of control in accord with the more general government policy for deregulation. But two of the changes are different. They are a response to changing patterns and forms of development (Department of the Environment, 1987c).

(i) A new financial and professional services class (A2) caters for business uses other than shops providing a direct service to the public and usually found in a shopping high street. Previously these were in an offices class which also had included all kinds of offices up to the very large headquarters of big organisations.

(ii) A new business class (B1) not only includes offices not necessarily providing a direct service to the public, but also research and development uses and even industrial uses which do not cause any environmental pollution incompatible with a residential area.

## Exceptions to the Definition of Development

E3.10 The basic definition and the general categories of exempted and permitted development apply universally. However

(i) certain types of permitted development under the General Development Order relating to the curtilage of dwellings and buildings do not apply in the ten national parks, the 34 areas of outstanding natural beauty or the hundreds of local conservation areas; nor do those relating to agricultural buildings apply in the national parks

(ii) a local authority or the Secretary of State can make an article 4 direction under the General Development Order by which specified categories of development in a defined area are no longer permitted development, and therefore require planning permission.

E3.11 In addition there are other circumstances where the definition of development as such is not altered but planning permission may be granted by special procedures. Four are relevant.

(i) The Secretary of State may make a Special Development Order for a particular site or area which has the effect of automatically granting planning permission for the development specified in the order.

(ii) A local or private Act of Parliament, for instance one introduced by an individual local authority, also has the effect of granting planning permission for the particular development. This is very rarely used: the most recent example was for the proposed Channel Tunnel.

(iii) Enterprise zones were introduced by the Local Government, Planning and Land Act 1980 and, so far, 18 have been designated. Their aim is to attract new development to industrial areas and inner cities.

They have a ten-year life and their chief attraction is to give exemption from certain national taxes and the local property tax. In planning terms, an enterprise zone scheme has to be adopted for each zone and the scheme, in most cases, has the effect of granting planning permission automatically for all kinds of development except those listed in the scheme.

(iv) Simplified planning zones were introduced by the Housing and Planning Act 1985 although, so far, none has been designated. The aim is to create a more relaxed planning regime, for instance in an industrial area designated by a local authority. In that case, the adoption of a planning scheme will have the effect of automatically granting planning permission for specified classes of development within the zone.

## Advertisements, Trees, Listed Buildings

E3.12   Finally, there are a number of special types of development concerned with amenity where regulations provide for special forms of control.

(i) Advertisements in outdoor locations, if they comply with a national code of practice, are deemed to be granted planning permission; otherwise 'express consent' is required. More detailed control is possible in areas of special control, mainly rural areas, but also in some non-rural areas, particularly conservation areas (Department of the Environment, 1984a).

(ii) Buildings of architectural or historic importance, the so-called listed buildings, require special 'listed building consent' for their alteration or demolition. The list of buildings is prepared and continuously brought up-to-date by the Secretary of State who has to be consulted before permission is given. In addition, all other non-listed buildings in conservation areas with a few minor exceptions require special 'conservation area consent' for their demolition (Department of the Environment, 1987b).

(iii) Trees and woodlands may have tree preservation orders placed on them, in the interests of amenity, in which case their felling requires consent. The control is further extended in conservation areas, even if not already protected by a tree preservation order.

## Conclusions

E3.13   Enough has been said to indicate the breadth and character of the definition of development and therefore the scope of development control. It is obviously important to stress that the definitions are given in much more precise terms than is possible in this summary.

E3.14   The definition of development is crucial as it establishes what is to be subject to development control. The basic principle is that the definition refers to changes in use, or development, taking place at a single moment in time; and that event has to be decided individually, on its merits. There has been no significant professional or political challenge to this basic principle since it was first introduced in 1947. Nevertheless, the definition of development does raise a number of points for consideration when comparing the English system with overseas systems.

E3.15   The first point concerns the limits of planning, that is, accepting the basic definition, what categories and scale of development should be exempted, permitted or deemed, and therefore not required to go through the formal systems of control. The main thrust of debate has been to extend the range of permitted development by amendment to the General Development and Use Classes Orders as part of a policy of deregulation, and to facilitate greater efficiency in the system by excluding very minor developments. But there have been counter arguments. Particular examples on which there is debate include arguments, for instance, that agricultural developments with an ecological impact, such as draining of wetlands, loss of hedgerows or afforestation, should be subject to general planning control and not just in nature reserves and sites of special scientific interest (Shoard, 1980).

E3.16   A second line of discussion is about the relevance of the definition, with its emphasis on physical change and land use, and whether other measures might be more appropriate either on their own, or as an extension to the current definition. Inevitably this raises a number of questions in the grey area about what precisely constitutes development in any particular case. But there have been arguments for example about the legitimacy of taking into account the ownership of a building, or the uses to which it is to be put in social terms, for instance the distinction between private and state schools, or between public and private sector housing. Another is the type and number of jobs created, or the amount and use of investment in plant and machinery.

E3.17   ) A third issue could be the effect of a cumulative build-up of a sequence of otherwise minor developments, establishing thresholds beyond which the developments have a qualitative or quantitative change in their impact. To some extent, in theory, this can be covered by policies in the development plan but it does raise questions about the importance of precedent in deciding even minor planning applications.

E3.18   Finally there is the continuing effect of a planning permission, in effect a question of the relevance of the continuing land and property management as distinct from that of land and property development. However, this can be taken into account by imposing conditions on the planning permission. For instance conditions have been imposed on the hours during which the construction, or subsequent operation, of say a shopping centre is permitted, or by giving a temporary planning permission, or by requiring the restoration of a site after mineral workings. But the conditions have to be relevant strictly in terms of the planning legislation.

E4.1   The primary control of development is exercised in the vast majority of applications by the district council (or in London the borough council) and in a very small minority of particular types of applications by the county council or the Secretary of State. Appeals against a refusal of planning permission are dealt with in a later section.

## The Development Control Workload

E4.2   The volume and variety of planning applications is very large. The total volume amounts to about 400,000 applications each year in England. Roughly 44 per cent are so-called 'householder' applications, that is, very small alterations to dwellings, such as building an extension or a garage; 12 per cent are changes of use not involving building works; 40 per cent are minor developments; and only 4 per cent are 'major' developments involving more than 10 dwellings, or 1,000 square metres of floorspace, or one hectare of land (Department of the Environment, 1987d).

E4.3   Table E4.1 shows the variation between the different categories of application for a small sample of districts in 1982. On the whole, the biggest difference is between the householder applications and the rest.

Table E4.1   Development Control: Workload, Time Taken and Decisions, 1982

|  | House-holder | Category of application change of use | minor | major | total |
|---|---|---|---|---|---|
| % of all applications | 47 | 13 | 37 | 3 | 100 |
| % of applications refused | 6 | 19 | 15 | 14 | 11 |
| time taken for decision % applications | | | | | |
| less than 8 weeks | 80 | 64 | 63 | 43 | 70 |
| more than 13 weeks | 5 | 12 | 13 | 29 | 10 |
| % of refused applications going to appeal | 17 | 15 | 37 | 75 | 28 |
| % of appeals dismissed | 48 | 71 | 72 | 67 | 68 |

*Note:*   statistics are indicative, for a sample of 12 districts
*Definitions:* householder: within curtilage, no increase in number of dwelling units
change of use: not involving building work
major: more than 10 units or 0.5 ha, residential; more than 1000 sq m or 1.0 ha other uses
minor: other
*Source:*   Davies, Edwards and Rowley, 1986

Householder applications are decided more quickly than the others; fewer are refused permission; and of those which go to appeal more are successful in their appeal. The only other noticeable distinction is that major applications take longer to decide, less than half being dealt with within the statutory eight week period.

E4.4   The workload varies widely, from perhaps 500 applications a year in a remote rural district to 2,000 or more applications in a large district in a metropolitan area. The rate of refusal of planning permission also varies widely, ranging on average from about 6 per cent of householder applications to as much as 25 per cent of applications for major residential development. The overall average refusal rate in the sample was about 11 per cent, although nationally it varied between 13 per cent and 15 per cent between 1981 and 1986.

E4.5   The average district in another small sample dealt with 1,285 applications in 1982. But the applications for planning permission made under section 25 of the 1971 Act are only part of the workload (see Table E4.2). As a consequence of the need for planning permission, there are many other types of administrative work, including

(i)   permitted developments for which planning permission is not required but on which informal enquiries are made to find this out

(ii)   formal applications under section 53 of the 1971 Act to determine whether planning permission is required

(iii)   formal applications under section 94 of the 1971 Act to confirm the established use of a site

(iv)   deemed consent for development by local authorities under section 40 of the 1971 Act in which the actual consent is given by the government department authorising the expenditure but on which the local authority comments as for an ordinary planning application

(v)   applications for consent to display advertisements

(vi)   applications for listed building or conservation area consent for the alteration or demolition of listed buildings of historic or architectural value, or the demolition of non-listed buildings in conservation areas

(vii)   consultations by other local authorities about planning applications in their areas

(viii) site visits and enforcement action for breaches of planning control.

**Table E4.2   The Planning Control of Development, 1981/82**

|  | Average number of applications/cases* |
|---|---|
| Development control |  |
| major | 75 |
| minor | 568 |
| change of use | 188 |
| householder | 454 |
| Total | 1,285 |
| Other applications |  |
| permitted development, no planning permission required, enquiry only | 215 |
| section 53 applications to determine whether planning permission required | 6 |
| deemed planning permission, proposals by local authorities/government departments | 55 |
| established use certificates, to confirm existing use rights | 7 |
| advertisement control | 165 |
| listed building applications | 53 |
| consultations on applications in neighbouring areas | 25 |
| others | 10 |
| Total | 536 |
| Appeals determined | 43 |
| Enforcement action |  |
| site visits | 1,993 |
| infringements recorded | 274 |
| notices issued | 20 |

*Note:* * statistics are averages for a sample of 11 districts
*Source:* Audit Inspectorate, 1983

E4.6   It follows from this workload that the size of a development control section in a district planning office can be substantial, varying between 3 and 24 full-time equivalent staff, or an average of 10.5, depending on the size of the workload. In addition, the equivalent of about on average nine staff in other sections of the planning department will be involved in development control, as well as their other duties. In more specific terms, the average planning application requires between 2.5 and 3.5 man-days for its processing, two-thirds of the time by professional staff (Audit Inspectorate, 1983).

## The Scope of Development Control

E4.7   The scope of development control is potentially very wide. The first question is, what considerations raised by an application are relevant in the sense that they lie within the powers of the planning acts as interpreted by development plans, professional expertise and local custom and practice. The second question is, which of the relevant considerations are actually material to the circumstances of the application and therefore need to be taken into account when arriving at a decision.

E4.8   In all cases, the relevance of planning considerations can be, and in many cases has been, tested in the courts when reviewing challenges to particular planning decisions whether on a planning application or a particular development plan. But, insofar as each planning application is dealt with 'on its merits', the test is not simply whether the considerations are relevant to the purposes of planning in general, but also whether they are material to the circumstances of the particular application. And it is the latter test which is crucial to the particular decision.

E4.9   Any one, single planning application can raise a host of planning considerations. At its widest, for instance in the application for full permission for a very large residential or commercial development such as a housing development of 500 dwellings on a green-field site or a mixed development for 20,000 square metres of shops, offices and other uses in an historic town centre, the potential range of planning considerations could cover everything from the proposed number of dwellings or amount of floorspace in relation to total demand in the county as a whole, to the detailed design and choice of materials for the building. Figure E3 illustrates a planning application for an office development in Reading, including a site plan, elevations and sections.

E4.10   Conversely, in a householder application the range of considerations will be much smaller, not much more than the design implications of any changes and matters such as sunlight, daylight or visual privacy affecting the development and its neighbours.

E4.11   There is no general list of planning considerations taken into account in development control in all districts as the list may vary in response to local policies and conditions and alter over time as public and professional attitudes shift. However, in general they include the following broad categories, further details being given in Table E4.3 (Davies, Edwards and Rowley, 1986a).

(i) Practical planning considerations relating to amenity and appearance (the design, external appearance, and physical impact of the development and its relationship to its surroundings); the arrangement of the development including layout and circulation on the site; access to the site and impact on the surrounding transport network; and its proximity to related or incompatible uses.

(ii) More general strategic planning considerations relating to the quantity and distribution or location of development; the phasing of development; and its coordination with other development including in some cases its financial viability.

(iii) Finally, precedent and the impact on existing occupants of the site or other, nearby sites, of the planning application, can also be taken into account.

## ROYAL COUNTY OF BERKSHIRE

**Application for permission to carry out Development—Town and Country Planning Acts**

Please read the accompanying notes before completing this Form and submit **four** copies of the Form and plans (Five are requested if the site adjoins a Trunk Road) to:

| For office use only | Trunk Road ☐ |
|---|---|
| Application or Plan No. | |
| Register or Site No. | |
| Date Received | |
| Date Acknowledged | |

**PART 1**

**1. Applicant** (block letters)
Name  Royal County of Berkshire
Address  Shire Hall
Reading, Berkshire.
Postcode  RG1 3XY   Tel. Number  55981

**Agent** (if any)
Name  Assistant County Property Officer
Address  Abbot's Walk,
Reading, Berkshire.
Postcode  RG1 3HN   Tel. Number  55981 Ext373

2. (a) State the full address or location of the land to which the application relates  (Outline the site in red on the accompanying plans, which must be based on an Ordnance Survey Map)
land at King's Road/Abbey Street,
Abbot's Walk, Reading, Berkshire.

| | (b) Ordnance Survey — Edition/Sheet Number | SU173NE 1970, SU717SSW 1967, SU72 73SW 1970 |
| Note 2 | | Parcel Number | Not shown on O.S. Sheet. |
| Note 3 | (c) Area of the site | Hectares or Acres | 1.6 hectares |
| | | | 3.9 acres |

3. (a) State the applicant's interest in the land which is the subject of the application
Freehold owner

(b) Does the applicant or owner of the land own or control any adjoining land?
YES/NO  (Delete whichever does not apply)
(verged blue on the site plan)

4. List all plans and drawings submitted with the application
1) Site plan - diagrammatic only,
2) (a) General Plan - reg. no. '1',
   (b) Car Parking - reg. no. '2'
   (c) Massing & Section - reg. no. '3'

5. Give brief particulars of the proposed development including the purpose(s) for which the land/buildings will be used
Offices (with associated car parking)

6. (a) Does the proposal involve any of the following?
1 New building  YES/NO
2 Alteration or extension  YES/NO
3 Change of use  YES/NO
4 Redevelopment  YES/NO
5 Demolition  YES/NO

(b) Which one of the following is the application for?
1 Outline planning permission  YES/NO
2 Full planning permission  YES/NO
3 Approval of reserved matters  YES/NO
4 Temporary permission or renewal of temporary permission  YES/NO
5 Relaxation of conditions of previous consent  YES/NO

(b) For the relevant previous planning decision state the Application Number and Date
Register No. 11547 File/DK/WH
TP 4347/RP 37638 26/3/65

Siting ☐  Design ☐  External Appearance ☐
Means of Access ☐  Landscaping ☐

(d) If the answer to 6(a) 3 or 5 is YES, state the particular conditions to which this application relates
N/A

---

**(Middle section)**

Offices/Car Parking/Temporary Offices/Retail/Ancillary/Residential

7. (a) State present use of land/buildings

(b) If vacant, state last previous use   N/A

8. (a) Are there any trees on the site?   YES/NO
(If YES, indicate positions and area covered on the site plans and show which, if any, are to be felled.)

(b) State material and colour of walls  )
(c) State material and colour of roof.  )   — — FOR SUBSEQUENT APPROVAL
(d) State means of enclosure of plot(s).  )
(e) How will surface water be disposed of?  )
(f) How will foul sewage be dealt with?  )

9. If the proposal involves access to a highway state the number of accesses and their widths, in each category
1 Existing vehicular accesses  King's Road - 2/Abbey Street - 4
2 Existing pedestrian accesses  King's Road - 6/Abbey St. 4/Abbots Walk 2
3 New vehicular accesses  )
4 New pedestrian accesses  )  — — FOR SUBSEQUENT APPROVAL
5 Altered vehicular accesses  )
6 Altered pedestrian accesses  )

10. (a) If the proposal involves residential development, state the number of new dwelling units and the total number of habitable rooms in each category

| | Number of new dwelling units | Total number of new habitable rooms |
|---|---|---|
| 1 House | | |
| 2 Bungalow | | |
| 3 Flat | N / A | |
| 4 Maisonette | | |
| 5 Bedsitter | | |
| 6 Other (specify) | | |

(b) If the proposal involves any less of residential accommodation (by demolition or change of use), state the number of dwelling units lost and the total number of habitable rooms in each category

| | Number of lost dwelling units | Total number of lost habitable rooms |
|---|---|---|
| 1 House | 4 | 24 |
| 2 Bungalow | None | None |
| 3 Flat | None | None |
| 4 Maisonette | None | None |
| 5 Bedsitter | None | None |
| 6 Other (specify) | None | None |

11. (a) Does this proposal involve any **non-residential** development? (If YES complete PART 2)   YES/NO

(b) Is the proposal for the Extraction of minerals? (If YES complete PART 3 which is a separate Form)   YES/NO

One copy of an appropriate certificate must accompany this application.

I/We hereby apply for permission/approval in respect of the particulars described above and in the attached plans and drawings.

Signed _____  on behalf of Royal County of Berkshire.

Date  22/1/1980

---

**PART 2**

**Additional Information required in respect of Applications for Non-residential Development**

If the occupier of the whole or major part of the proposed development is decided, please state this and complete all questions.
Otherwise complete all questions except 1, 4 and 6.   The occupier will be  Not Known

1. Give a description of the activity or the processes to be carried on, any end products, and the type of plant or machinery to be installed.
Office Function

2. If the proposal forms a stage of a larger scheme, for which planning permission is not at present sought, please give what information you can about the ultimate development   N/A

3. Is the proposal related to an existing use, on or near the site?   YES / NO  Unknown
If so, please explain the relationship?

4. (a) Is the proposal to replace existing premises in this area or elsewhere which have become obsolete, inadequate or otherwise unsatisfactory?
If so, give details including location and gross floor area, and state your intentions in respect of those premises.   YES / NO  Unknown

(b) Is the proposed occupier already established in this District Council area?
If not, please state present location or whether the firm is a new one   YES / NO  Unknown

5. Give the **Gross Floorspace** of all buildings to which this application relates, in total and subdivided by use, as shown.

| | Existing floorspace to be retained (if any) | Proposed additional floorspace (if any) | Existing floorspace to be lost (through demolition or change of use) |
|---|---|---|---|
| Industrial | None | None | 19500 s.f. |
| Office | 5,500 s.f. | 344,500s.f. | 4575 s.f. |
| Retail | None | None | 3650 s.f. |
| Warehousing | None | None | None |
| Storage | None | None | None |
| Ancillary - residential | None | None | 3800 s.f. |
| Residential | - | - | 31,525 |
| **TOTAL** | 5,500 | 344,500 | |

All figures are in square feet/square metres
(Delete whichever does not apply)

6. (a) Give the number of staff to be employed on the site, when the proposed development is complete, subdivided as shown

| | Office staff | Industrial staff | Other staff |
|---|---|---|---|
| New | Unknown | Unknown | Unknown |
| Transferred | | | |
| Existing | | | |
| **TOTAL** | | | |

(b) In the case of industrial or office development, if the premises affected

7. In the case of industrial or office development, if the application is accompanied by an Industrial Development Certificate or an Office Development Permit, give the certificate number and floorspace permitted
IDC / N/A   N/A  sq ft sq m
ODP N/A   N/A  sq ft sq m

8. What provisions have been made for parking, loading and unloading of vehicles on the site? For Subsequent Approval
(Please show their location on the plans and distinguish between parking for operational needs and other purposes)

9. What is the estimated vehicular traffic, flow to the site during a normal working day? Unknown
(Please include all vehicles except those used by individual employees driving to work.)

10. What is the nature, volume and proposed means of disposal of any trade effluents or trade refuse?   Unknown

11. Will the proposal involve the use or storage of any of the materials mentioned in the Note?   YES / NO
If so please state materials and approximate quantities

(a)

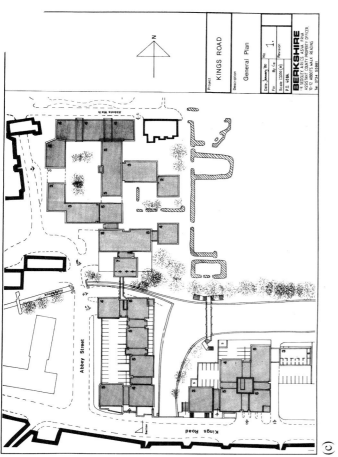

(c)

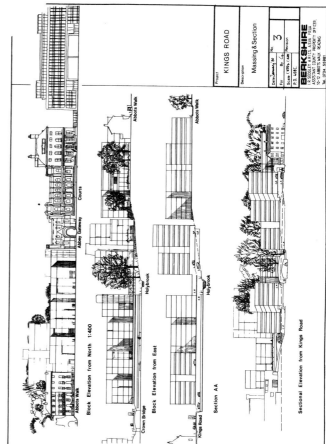

(d)

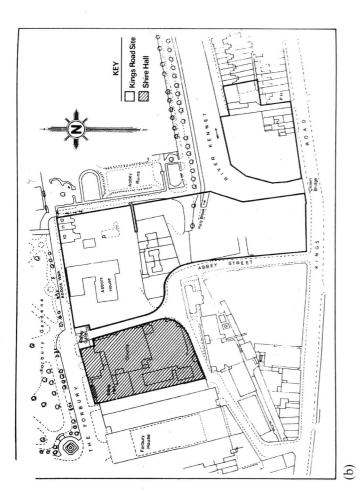

(b)

**Figure E3   A Planning Application for King's Road, Reading**

Figure E3 illustrates an outline application for an office development in Reading, submitted in 1980 (the application was eventually approved in outline, the reserved matters covering details of the design etc., approved and the building completed for occupation).

(a) shows the application form, with details of the proposed use, floorspace, etc.; (b) shows the site plan; (c) and (d) show for diagrammatic purposes only the general site plan, massing and section, and the relation to neighbouring buildings.

45

**Table E4.3  Planning Considerations in Development Control**

| General considerations | Examples of 3 detailed considerations$\phi$ |
|---|---|
| *Practical Planning Considerations* | |
| Amenity: | |
|   site characteristics (4) | topography |
| | *landscape features |
| | *archaeological features |
| design visual quality (5) | architectural style, merit |
| | *relationship with surroundings |
| | *treatment of external spaces |
| physical impact/quality (7) | daylight protection from noise |
| | visual privacy |
| operation, effect on amenity (4) | hours of operation |
| | effects of construction |
| | litter |
| relationship to surroundings (3) | *impact on historic buildings |
| | *impact on amenity landscapes etc |
| Arrangement: | |
|   on-site layout (21) | *roads, and |
| | parking-layout/capacity |
| | open space- layout/provision |
| | refuse collection |
| off-site relationships (9) | *incompatible/related |
| | uses-proximity/capacity |
| | *highway and public transport |
| | network-proximity/capacity |
| | *utilities-proximity/capacity |
| Efficiency: | condition of buildings |
|   use of resources, on site (6) | *vacant land/buildings |
| | *loss of agricultural land |
| *Strategic and Other Planning Considerations* | |
| Coordination: | linkages in mixed use schemes |
|   phasing, on site (2) | interim measures |
| phasing, off site (4) | other linked proposals, |
| | physical/socio-economic |
| | phasing, by quantity |
| | phasing, by area |
| operation/time (1) | temporary uses |
| Quantity/Distribution: | *loss of existing use |
|   quantity (9) | *addition/increase in use |
| | *employment generation |
| distribution location (2) | *by sub areas |
| | by group/category |
| Others: | planning gain |
|   others (a) | applicants' needs |
| | impact on existing occupiers |
| precedent (i) | precedent |

*Note:* based on a survey and interviews in 12 districts in 1982 to identify the factors potentially taken into account in considering planning applications

$\phi$   examples of the more detailed considerations taken into account. The figures in brackets in the second column show the total number identified. The weight given to any one consideration depends on the circumstances of the individual application.

\*   considerations for which there is usually authority/policy in the development plan

*Source*: Davies, Edwards and Rowley, 1986

E4.12   The authority for taking these considerations into account rests ultimately with the legislation and its interpretation by the courts. Broadly, the authority lies in the statutory development plan, where the considerations

may be described either as a specific target, such as a total office floorspace, or standard such as carparking space; or in more general terms as a criterion to be taken into account, such as amenity. Alternatively there will be no explicit statutory authority in a development plan in which case they become 'any other material considerations'.

E4.13   A special issue is that of aesthetic control, that is, the external appearance of development including the choice of materials, window and roofing details etc as well as the more general matters of bulk, mass and layout. Aesthetic considerations are firmly within the scope of development control but are much disputed. The present situation is that the Department of the Environment is seeking to limit the exercise of aesthetic control, whilst recognising its role in environmentally sensitive areas such as national parks, areas of outstanding natural beauty and conservation areas (Department of the Environment, 1985e).

## Planning Permission

E4.14   An application for planning permission may be granted in full or in outline, with or without conditions, or it may be refused. A full planning application includes all the information required by the local authority. An outline application includes only enough information to establish the principle of development, such as the use, size and location of a proposal. It would be sought, for instance, when the applicant does not wish to buy the land, or does not wish to incur the cost of preparing the final layout and design, until the principle of development has been established. This would typically be the case in a major, complex development. In such cases outline permission would be granted subject to certain, specified 'reserved matters' relating chiefly to the siting, design, appearance, landscaping and access of the final development. Once given, the outline permission cannot be reversed. But full approval for the development cannot be granted until the reserved matters have been approved, within three years of the outline permission.

E4.15   Planning conditions are a means of still further controlling a development as they form part of the grant of planning permission in full or in outline. They have been the source of dispute and raise many difficulties of interpretation, the subject of appeal to the Secretary of State, or challenge in the courts. Briefly, planning conditions must be necessary, relevant to planning and to the permitted development, precise and enforceable, and reasonable. Their scope can be demonstrated by listing some of the conditions which either have been ruled out by the courts or which the Secretary of State has discouraged. They include matters which could better be dealt with under other legislation; or matters limiting the tenure or occupancy of a development such as conditions preventing 'second homes' in holiday areas, or reserving a factory or office for 'local need'. On the other hand conditions relating, for instance, to carparking or access,

**Table E4.4 The Application for Planning Permission: Information Required**

Information usually required on the form when making a planning application

1. Name and address of applicant
2. Postal address of the application site
3. Brief description of the proposed development
4. Type of application:
   – whether change of use not involving building works
   – whether full/outline/reserved matters/removal of conditions
5. Outline applications:
   – matters reserved for future consideration: external appearance/siting/design/access/landscaping
   – site area (ha)
6. Reserved matters applications:
   – reference number of outline permission
   – which matters in this application
7. Access to roads (vehicular/pedestrian):
   – whether any new/altered access
   – name of road
8. Trees:
   – whether proposal does/does not involve felling of trees
9. Existing uses:
   – existing/last use of site
   – number of dwellings
10. Proposed uses, non residential:
   – alteration to existing gross floorspace (sq m)
   – proposed net floorspace, retail/office/industrial/warehousing/other
11. Drainage:
   – method for storm water
   – method for foul sewerage, mains sewer/cesspit/septic tank/other
12. Water supply:
   – whether required
   – if required, mains water supply/private water supply
13. Certificates, if required:
   – section 26, for bad neighbour development (e.g. coal yards, mineral workings, high buildings, some entertainment uses etc.)
   – section 27, that the applicant owns the site or, if not, that 20 days' notice has been given to the owner; and, in either case, any agricultural tenants on the site
14. Plans:
   – site plan, showing boundaries and relation to surrounding development (usually not less than 1/2500)
   – other drawings (except for outline applications) include layout plans (usually 1/1,250), block plans (1/500) and building plans (1/100) showing site features, proposed development, access, materials, elevation, fencing and landscaping etc.
15. Fees, depending on type of application and category of development.

Additional information which may be required for industrial, office, warehousing, storage or shops

1. Industrial processes and type of plant or machinery
2. Relation to the ultimate proposed development
3. Relation to existing uses on or near the site
4. Condition of existing premises, whether obsolete/inadequate etc
5. Number of employees (industrial, other) as a result of the proposal and whether this is a net increase/transfer from other premises
6. Arrangements for on-site parking, loading/unloading of vehicles
7. Estimated daily flow of vehicles to the site, excluding employees driving to work
8. Nature/volume/means of disposal of effluents/refuse
9. Storage of hazardous materials

*Note:* based on the application forms in three different districts

or to the restoration of mineral workings after use, or ensuring access for an archaeological investigation during the course of development, are acceptable (Department of the Environment, 1985a).

E4.16   A particular source of debate concerns so-called 'planning gain'. This is another means of, in effect, controlling the development but by means other than the granting of planning permission or the imposition of conditions. It takes the form of a legal agreement (usually under section 52 of the 1971 Act) between a developer and the local authority by which the developer agrees for instance to contribute to the cost of infrastructure; or to donate land, say, for a school, park or highway; or to confer some other benefit, such as to provide a public right of way; or to link the proposed development with some other proposed development, for instance the development of two separate sites in which activities are to be relocated from one site to the other (Department of the Environment, 1983).

E4.17   The whole question of planning gain is contentious and much disputed, the problem being what are the reasonable obligations which can and should be placed on a developer in the course of making an application for planning permission. However, entering into a section 52 agreement of itself cannot be made a formal condition to the grant of planning permission.

E4.18   Finally, detailed planning permission once granted has a life of five years in which it can be implemented, after which it lapses. It may be modified or revoked before the development is completed. If revoked or modified, it is open to challenge in the High Court; if confirmed, the revocation or modification may result in compensation being payable, or the applicant may serve a purchase notice on the local authority requiring it to acquire the property although this is rare.

## The Procedure for Planning Permission

E4.19   The procedure to be followed for submitting a planning application and receiving a decision on the application is to be found chiefly in the General Development Order and falls into five stages namely, the application; consultations; site visit and negotiations; the officer's report and recommendation; and the decision (see Figure E4). The various stages are illustrated in Figure E5 for a variety of different kinds of proposal.

E4.20   The application has to be made to the district council on a prescribed form, giving the details of the proposals together with a site plan and certificates indicating (a) whether or not the land is owned and occupied by the applicant and, if not, that the owner and certain tenants have been notified; (b) that site notices have been posted and the application advertised if it is a so-called 'bad neighbour' development involving noise, smell or other disturbances. The applicant also has to pay a fee, its size depending on the scale of the proposal. Further details may also be requested in the case of industrial or commercial proposals (see Table E4.4).

E4.21    The application is then registered in a form which is open to public inspection, and many district councils publish in a local newspaper weekly lists of planning applications received. However, there is no general requirement that planning applications be advertised. The only two exceptions are in the case of 'bad neighbour' development noted above, and applications in conservation areas.

E4.22    The second stage concerns consultations. There is a long list of organisations with which a local planning authority is obliged to consult, with sufficient time allowed for consultation, before making its decision. Depending on the type of development, the list includes regional water authorities; highway authorities and the Department of Transport; the Ministry of Agriculture, Fisheries and Food; the Nature Conservancy Council; British Coal, on matters of mining subsidence; and so on. Some of these organisations have the power to direct the district council as to its decision, for instance on access to a trunk road.

E4.23    Consultations also include the county council on applications which are thought likely to affect the structure plan. No clear criteria are laid down for such consultations although some county councils have prepared schedules of the types of development on which they wish to be consulted. Finally, the parish or town councils found in some parts of England have to be notified about all applications within their area.

E4.24    The third stage involves the development control case officer making a site visit, and if necessary having further discussions with the applicant in order to clarify points of detail. The discussions indeed might extend into negotiations about the proposal culminating in an amended application.

E4.25    The fourth stage concerns the report and recommendations which the planning officers make to their planning committee. This is the most mysterious part of the exercise, as very few guidelines are laid down in the legislation or the procedures. The only statutory guideline is that the local authority, in dealing with the application, 'shall have regard to the provisions of the development plan, so far as material to the application, and to any other material considerations' (Section 29(1) of the 1971 Act). This means that the planning officer may take into account any of the planning considerations within the scope of planning provided they are material to the application. The sources for such considerations therefore include

(i)   the statutory development plan which includes the structure plan, any local plans, and any still operative parts of the old-style development plan The plan may still be passing through the stages of its preparation. The nearer it is to being approved, the more likely it is to be a decisive factor

(ii)  the non-statutory written policies which the local authority (district or county) may have prepared

(iii) the national policy guidelines published by the Department of the Environment in circulars or otherwise

(iv)  the results of the various consultations, site visits and discussions with the applicant, amenity societies and other organisations or individuals wishing to comment

(v)   the professional planning judgement of the planning officers in the context of precedent and local unwritten custom and practice

(vi)  the unwritten policies, the so-called 'policy stance', of the local authority on which the planning officer should be well-informed from his knowledge of the planning committee (Healey, 1983).

E4.26    The final stage is the decision. This must be made within eight weeks of receiving the application unless the applicant agrees to an extension. If no such agreement is made, the applicant may assume that permission has been refused and accordingly appeal to the Secretary of State against a refusal of planning permission, though this is a comparatively rare occurrence. Most applicants agree to an extension as an appeal would take even longer and be more costly. The actual decision has to be formally taken in public by the full council of the local authority. Most councils however delegate most if not all decisions to a planning committee or development control sub-committee. Many councils delegate minor applications where the decision is to approve the development within specified guidelines to a senior planning officer. In those districts where this is done, about 60 per cent to 70 per cent of applications are delegated to officers.

E4.27    On average, the most recent figures for 1986 show that roughly 64 per cent of all applications are determined within eight weeks, though obviously the proportion of major applications so determined is smaller, at about 38 per cent. But the proportion determined within eight weeks varies considerably with 16 authorities deciding less than 35 per cent of applications within eight weeks, whereas another 16 decided more than 90 per cent in the same period (Department of the Environment, 1987d). The time taken has been the subject of much concern but research has failed to find any general, structural or procedural reason why the differences are of this order.

E4.28    A more detailed study on a small sample of districts shows how the time is taken in dealing with a typical planning application (see Table E4.5). It shows that the minimum time for a straightforward application

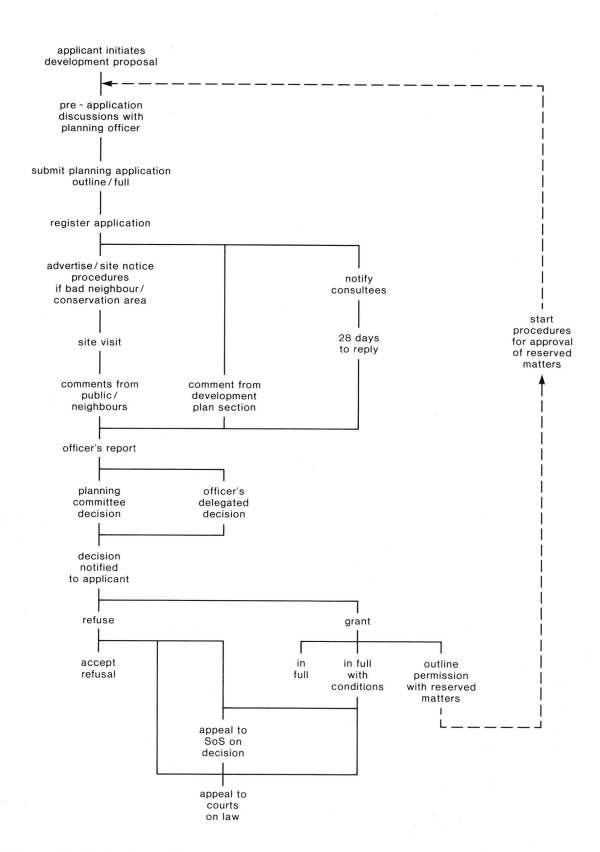

**Figure E4   Procedure for Development Control**

Figure E4 shows the development control process from the initial discussions between a possible applicant for planning permission and the planning officer, to the final decision or, in the case of a refusal, an appeal to the Secretary of State or the Courts.

The application itself may be initially in outline, covering only the basic principles of the proposed development such as the site, the use, and the amount of development. If so, the further details about design, landscaping etc. (the so-called reserved matters) will be the subject of a further approval.

**Figure E5  The Control of Development: Procedure and Examples**

Figure E5 gives five examples of development control in practice, identifying various consultations, the matters to be taken into consideration etc.

| Basic procedure | household application (ie within curtilage of dwelling) | change of use application (ie no building works) |
|---|---|---|
| **Proposed development** | eg  an extension of 20 sq m to provide an extra room at rear of house | eg  change of use of a shop to a casino |
| **Applicant** | eg  house owner, advised by builder | eg  professional planner, for client |

**Pre-application discussions/negotiations between applicant and planning officer**

| | | |
|---|---|---|
| - does it need planning permission<br>- what information is needed<br>- policies affecting the site<br>- planning considerations likely to be taken into account<br>- prospects of success | - perhaps 1 visit, to collect forms, check procedures | - perhaps 1 visit to collect forms, etc and discuss policy and requirements<br>- a casino is a "bad neighbour" development with special procedures for advertisement.  It is also likely to be contentious on grounds of noise, customers etc |

**Application, registration, advertisement**

| | | |
|---|---|---|
| - application form, site plan, certificate of ownership, fees<br>- application recorded in public register, possibly advertised in press<br>- advertisement compulsory in special cases, eg bad neighbour, conservation area | - application for full planning permission, perhaps on simplified form | - application for full planning permission<br>compulsory advertisement/site notice as a bad neighbour |

**Consultation with public bodies, replies within 28 days**

| | | |
|---|---|---|
| - parish council<br>- county council/highways/gas, electricity, water/agriculture etc as needed | - probably only parish council,<br>- if in rural area<br>maybe, neighbours notified | - as required |

**Site visit, discussions etc**

| | | |
|---|---|---|
| - site visit, clarification/ negotiation with applicant<br>- comment from neighbours, civic groups etc | - maybe, comment from neighbours | - maybe, comment from neighbours etc likely to be affected |

**Policy guidance**

| | | |
|---|---|---|
| - development plan<br>- other non-statutory guides/plans<br>- precedent/custom and practice<br>- national policies/appeal decisions | - probably, no mention in - development plan<br>- maybe, design guidance on materials, elevation etc | - indirectly, maybe in development plan, policy against loss of shops/ break in shopping frontage |

**Analysis, planning officer's report and recommendation**

| | | |
|---|---|---|
| - based on development plan (if relevant) and any other material considerations (ie consultations, comment, other policy guidance, etc) | - mainly based on practical con- siderations, eg relation to existing house and neighbours, views/privacy/daylight etc | - mainly based on practical consider- ations of conformity with neigh- bouring uses, effect on shopping |

**Committee decision, within 8 weeks of application unless extension agreed**

| | | |
|---|---|---|
| - grant, with/without conditions*<br>- refuse, with reasons*<br>- agreements re planning gain (not part of planning permission)<br>- issue decision notice | - most district, minor decisions to grant permission in accordance with guidelines, with no object- ions from neighbours etc, are delegated to the planning officer | - maybe delegated to planning officer for most changes of use but probably not for a bad neighbour development and for a potentially contentious issue<br>- conditions may be imposed eg about hours of opening |

\*     may be followed by appeal against conditions/refusal

50

| minor application | major application (ie > 10 houses, 0.5 ha residential; > 1000 sq m, 1.0 ha other users) | |
|---|---|---|
| eg redevelopment of site in shopping street for shop with office and flat, < 1000 sq m in a conservation area | eg outline application for 750 houses on 30 ha of farmland on edge of town | eg full application for reserved matters, for stage 1 of site granted outline permission |
| eg architect, for client | eg surveyor, for landowner | eg architect, for development company |
| - collect forms, discuss land use policies and requirements<br>- clarify design criteria for conservation area<br>- discuss/modify architect's schemes in light of comment by planning officer, to find a scheme likely to be acceptable to the planning committee<br>- discussions continue until agreed or applicant submits application<br>- at any time and risks later requests by planning officer for more information/alterations | - aim, likelihood of getting permission, and, if so, how much housing and what conditions-<br>- negotiations also involve other public agencies (water, etc) and cover need/demand for housing; supply of land; site quality and farmland; utilities and highways; need for other uses<br>a) if in green belt, etc, highly speculative attempt to get policy relaxed<br>b) if on land with no policy, lengthy negotiations to establish whether permission is likely<br>c) if allocated in development plan, negotiations on general principles how site to be developed | - protracted negotiations likely<br>a) to vary the original permission* (eg to get more houses)<br>b) to settle the reserved matters<br><br>* original permission cannot be revoked without compensation |
| - application for full planning permission with detailed drawings<br>- compulsory advertisement/site notice as in conservation area | - application for outline permission (ie site and amount of housing but planning officer may want more information)<br>- advertise, if departure from development plan<br>- advise Secretary of State who may call-in application | - application for reserved matters as specified in outline permission<br>- submit within 3 years of outline permission |
| - as needed, including civic amenity groups, conservation area committee and maybe Royal Fine Arts Committee | - as needed, full list, lengthy consultations unless agreed at pre-application stage<br>- consult county council on structure plan impact | - as needed, limited if matters settled at outline |
| - maybe, detailed discussions with applicant with alterations needing revised application | - consultations may identify problems (eg about water, highways) to be resolved by planning gain | - maybe further discussions with applicant about detail |
| - strong, general policy in development plan, plus conservation area policy, plus design guidance<br>- general land use policy etc | - strong policy in development plan (but maybe out-of-date) or land availability study<br>- national policy guidance | - probably, not much specific policy in development plan<br>- supplementary planning guidance on design, layout etc |
| - mainly based on<br>a) land use policy about shopping, car parking etc<br>b) need to retain residential use<br>c) aesthetic considerations for the conservation area | - mainly based on strategic planning considerations about<br>a) location, phasing, servicing of of development<br>b) amenity impact including green belt/landscape | - mainly on practical considerations of design and layout |
| - full committee decision, maybe on advice of independent architects' panel, architects' department of the district council<br>- conditions unlikely as application will be very specific and detailed | - full committee decision, if granted outline permission<br>a) conditions about access, landscaping, archaelogy, etc<br>b) subject to reserved matters siting, layout, design etc<br>- maybe section 52 (1971 Act) agreements for costs of services, donate land for highways etc (planning gain) | - full committee decision |

without any protracted consultations and delegated to a planning officer takes about 23 working days. The time taken increases to about nine or ten weeks if the application has to go to committee and if it involves allowing the full waiting time for completion of consultations.

Table E4.5    Time Taken for Development Control (Days)*

| | Applications decided by committee | Applications delegated to officers |
|---|---|---|
| 1. Receipt, registration and acknowledgements of the application, including checking and postage time | 5 | 5 |
| 2. Advise potential consultees (see #9) | 1 | 1 |
| 3. Site visit | 1 | 1 |
| 4. Consultation with neighbours and parish councils | 10 | 10 |
| 5. Report preparation including agenda cut-off period for prior circulation of report to committee if necessary | 9 | 5 |
| 6. Committee meeting | 1 | |
| 7. Issue of decision notices including postage time and receipt by applicant, or by alternative methods | 5 | 1 |
| 8. Weekends included in above | 12 | 8 |
| Total elapsed time | 44 (6–7 weeks) | 31 (4–5 weeks) |
| 9. Additional time not so far included if consultees wish to comment within the statutory period (28 days) | 14 | 14 |
| Overall elapsed time (8–9 weeks) | 58 (8–9 weeks) | 45 (6–7 weeks) |
| 10. possible additional time to fit in with the committee cycle, assuming meetings are every 2 weeks, say | 7 | |
| overall elapsed time | 65 (9–10 weeks) | = |

*Note:* * the actual time taken, including postage and circulation time etc, between the receipt of the application by the local authority and the receipt of the decision notice by the applicant. The figures are for a typical, straightforward application.

*Source*: Audit Inspectorate, 1983

E4.29    However, this would be the case for a fairly straightforward application. It makes no allowance for incomplete, or incorrect planning applications; or for modifications and alterations to the application during its processing, possibly requiring an amended application. Finally, it makes no allowance for the discussions and negotiations between the applicant and the planning officer before the application is ever submitted. This is usually a crucial, often time-consuming, stage in the exercise which demonstrates the real character of the process. The discussions can cover everything from a statement of what information is required from the applicant and clarification of relevant policies, to an indication of what is likely to be acceptable to the

planning committee. It will often take the form of a negotiation to find the extent of compromise that is possible by way of a planning permission and conditions, and the amount and form of planning gain which the local authority will also require.

## County Matters, Departures and Called-in Applications

E4.30    The usual procedure for dealing with planning applications is that the decision is made by the district council, albeit sometimes on the direction of some other authority such as the Department of Transport or the county highway authority if access to a trunk or main road is involved. But there are special instances where the application may be decided by the county council or directly by the Secretary of State. They are very few in number.

E4.31    The county council may be involved in two ways.

(i)    Applications for the working of minerals and the disposal of waste are decided by the county council, the last remnant where this is so. In addition, however, county councils also decide planning applications in national parks or straddling national park boundaries, unless there is a separate national park authority to do the work.

(ii)    In addition, however, a district council must consult the county council on applications which materially conflict with or prejudice the implementation of the structure plan, though it is not bound to accept the county's advice. However, such applications would also have to go through the procedure for departures from a development plan.

E4.32    The development plan does not automatically give a right to develop land, or to restrict development. But any application which would conflict with or prejudice the implementation of an approved structure plan, and which the district does not intend to refuse, has to go through special procedures. In such cases of a departure from the plan,

(i)    in all cases, the local authority must advertise the application, giving 21 days for objection

(ii)    in the more serious cases involving a material conflict, the local authority must also notify the Secretary of State who, within 21 days, may direct that the application be refused.

E4.33    The problem with the county consultations and the departures is that the policies and proposals in development plans are often expressed in very general terms or are qualified by words such as 'normally'. It is thus often difficult to be certain that a departure is actually involved, yet the choice is left to the district council. But the use of the departures procedure is further discouraged by the advice in recent government circulars to give low priority to old, or out-of-date development plans which are not relevant to current circumstances (Department of the Environment, 1985b). Thus the departures procedure is very little used.

E4.34 However, the Secretary of State also has a reserve power to call-in any application and make the decision for himself. The procedure in such cases is very similar to that for dealing with appeals against refusal of planning permission described in section E6. It is comparatively rarely invoked, in the case of departures from development plans; for new forms of development not anticipated when the development plans were prepared; and for major proposals involving a national interest. Examples include proposals for a nuclear power station.

## The Crown, Statutory Undertakers and Local Authorities

E4.35 The control of development by these three types of public sector agencies differs from the normal procedures.

(i) A local planning authority must apply to itself for planning permission for its own development by (a) resolving to seek planning permission; and (b) resolving to carry out the development, in which case planning permission is deemed to have been granted by the Secretary of State.

(ii) If a local authority or statutory undertaker is proposing to undertake development which will require authorisation by a government department it is deemed to receive planning permission when granted the authorisation if the government department so directs.

(iii) Development by the Crown (ie government departments etc) for its own use does not require planning permission. However, government has agreed on a non-statutory basis that departments should consult a local authority for development which otherwise would require planning permission. A notice of proposed development is submitted to the local authority which then goes through the normal procedures. If it objects to the proposal, the government department notifies the Secretary of State and the matter is treated in a similar way to an appeal against a refusal of planning permission.

(iv) Proposals for the future development of Crown land after its disposal to the private sector, or for the development of a privately-owned interest on Crown land, are both treated as normal applications for planning permission.

## Listed Buildings and Conservation Areas

E4.36 These have become a significant part of planning and control since the passage of the Civic Amenities Act 1967. In both instances, special procedures apply for development control (Department of the Environment, 1987b).

E4.37 Special listed building consent is required to demolish, alter or extend any building which has been listed by the Secretary of State as being of architectural or historic interest on the advice of the Historic Buildings and Monuments Commission for England. About 260,000 buildings were so listed in 1978 of which about 1 per cent were Grade I, being of exceptional interest; and the rest were grade II (or Grade II*, of particular interest) including all remaining pre-1700 buildings, most buildings from 1700-1840, those of definite quality from 1840-1914, and a few more recent buildings. The list is continually reviewed and brought up to date, with a special provision for 'spot' listing if a relevant unlisted building is threatened with demolition. The owner of a listed building is notified but may not object to the listing. There are various possibilities for financial assistance or tax relief for the repair and maintenance of the listed buildings.

E4.38 It is an offence to demolish a listed building, or to alter or extend it so as to affect its character, without obtaining listed building consent from the district council or Secretary of State. In summary

(i) the applicant fills in a special form, requesting listed building consent

(ii) the district council advertises the application, with site notices; and notifies the Historic Buildings and Monuments Commission (and certain national amenity societies) about all proposed demolitions, and all alterations or extensions to Grade I/II* buildings

(iii) the district concil notifies the Secretary of State if it is minded to give listed buildings consent, in which case he has 28 days to decide whether to determine the application himself, or leave it to the discretion of the council

(iv) the consent may be granted with or without conditions, and there is provision for an appeal against refusal.

E4.39 Conservation areas are designed by local authorities as being of special architectural or historic interest and, after due process, are notified to the Secretary of State. By 1979 there were over 4,000 such areas, ranging from large areas of towns such as Bath to small villages. In general, the normal processes of development control apply, possibly with a more strict regime following a direction under article 4 of the General Development Order to suspend specified categories of permitted development. However, special conservation area consent does have to be obtained from the local authority before non-listed buildings in the area can be demolished.

# Enforcement of Planning Control

E4.40 Clear procedures are laid down for the enforcement of planning control. In brief summary, they involve

(i) the issuing of an enforcement notice by the local planning authority that a person is in breach of planning control, meaning either that development has occurred without planning permission, or that a refusal of planning permission, or conditions, have not been complied with. The enforcement notice must specify precisely the action to be taken to remedy the breach. There is provision for appeal to the Secretary of State against an enforcement notice

(ii) once an enforcement notice has been issued, the local authority may then serve a stop notice to prevent the person from using delaying tactics such as continuing in breach whilst appealing against the enforcement notice

(iii) if the enforcement notice is not complied with, the local authority may then itself carry out the works (eg demolition of buildings) and recover the cost from the owner or developer; and the person concerned can be fined by the courts.

E4.41 Enforcement procedures are very complex and are used only when a local authority regards the breach of planning control as being sufficiently serious. Thus, in a small sample of districts, on average in each district there were nearly 2,000 site inspections in which about 270 infringements were recorded but only 20 notices were issued (see Table E4.2).

## Staffing and Costs

E4.42 The cost of processing planning applications is closely related to staffing requirements. Table E4.6 gives some details, bearing in mind that the average figures conceal wide differences between different districts, and the problems of allocating staff time and costs to the different activities in the planning function. Development control and enforcement is chiefly a function of district councils and requires a professional, administrative and support staff of about 11.4 persons per 100,000 population. The total cost of administering this in 1985/86, excluding central administration charges, was about £1,527 per 1,000 population of which about £679 was recovered in the form of a sliding scale of fees charged for making an application irrespective of the outcome (Chartered Institute of Public Finance and Accountancy, 1986).

E4.43 Another way of expressing the local cost of the planning control is to estimate the average cost to local government of processing planning applications including the average cost of preparing the development plans and the other forms of non-statutory planning policies whether by county or district councils. In summary, the figures for 1985/86, excluding London, show a net cost of £214 per application (see Table E4.7).

**Table E4.6  Staffing, Costs for Development Control, 1985/86**

|  | counties* | districts* |
|---|---|---|
| Budgetted employees/100,000 population | | |
| planning policy | 2.4 | 5.6 |
| control and enforcement | 1.2 | 11.5 |
| Average net expenditure, £/1000 population | | |
| planning policy | 410 | 785 |
| control and enforcement | 168 | 848 |
| conservation and enhancement | 374 | 592 |
| central establishment | 271 | 850 |
| Total, planning functions | 1223 | 3075 |
| Fee income, £/1000 population | | |
| planning applications† | 14 | 679 |

*Notes:* * weighted average, metropolitan and non-metropolitan areas excluding London

† this is included in the net expenditure for control and enforcement; counties receive fee income only for applications which they determine, not for those on which they are consulted by districts

*Source:* Chartered Institute of Public Finance and Accountancy, 1986

**Table E4.7  Cost of Planning Applications, 1985/86***

|  |  | (£) |
|---|---|---|
| Cost per application of development control and enforcement, including applications determined by county councils, and consultations on planning applications determined by other councils | | 164 |
| of which, fees recovered | | 66 |
| Net cost per application | | 98 |
| Cost per application of the preparation of structure and local development plans and other planning policies | | 116 |
| of which, county councils | £41 | |
| and district councils | £75 | |
| Total net cost per application | | 214 |

*Note:* * the figures exclude central administration charges

*Source:* from Chartered Institute of Public Finance and Accountancy, 1986

## Conclusions

E4.44 This chapter has described in very broad outline the system for controlling development, postponing to a later section the subject of appeals to the Secretary of State. More general points do need to be made, for there are a number of problematic aspects of control which need to be borne in mind when reviewing the overseas systems.

E4.45 Firstly, the Department of the Environment has stressed in Circular 14/85, repeating the advice of earlier years, that there should in general be a presumption in favour of development unless there are good grounds, involving matters of acknowledged importance, for refusing permission. It has emphasised that the planning system must respond positively and promptly to proposals for development, in effect to promote and encourage development. Nevertheless, the circular also recognises the strength of the circumstances where development may not always be acceptable. It mentions the preservation of the heritage, the quality of environment, the conservation of good agricultural land, and the protection of green belts. Thus development control is about achieving a balance broadly between development and conservation, and between national and local interests.

E4.46 Each of these polarities has many aspects. Thus the conflict between conservation and development is not simply a localised affair between, for instance, those wishing to protect a green belt and those wanting new house building. It can also reflect more substantial differences in policy stance between, for instance, districts under pressure but wishing to restrain the development, and those districts lacking any pressure for development, trying to attract new development by placing more emphasis on a relaxed planning regime.

E4.47 The potential conflict between national and local interests is fundamental, exemplified by the Department's exhortation that there be a presumption in favour of development, and endorsing this as national policy. But the policy is simultaneously very wide and extremely vague, and lacks any clear criteria. One consequence is that the Department gets involved in minute matters of detail in commenting on development plans and hearing appeals against refusals of planning permission. There are those who argue that this is to concern itself unnecessarily in local concerns. In giving the initial responsibility for planning and control to local authorities, it is argued, central government should then concern itself with national and regional issues of control.

E4.48 Secondly, the system of control is less of a regulatory system for implementing clear, explicit policies, than a discretionary system for achieving more general objectives, many of them far from explicit. It means that in addition to the balances to be struck between national and local interests, or conservation and development, there is a more specific bargain to be struck between the applicant and the local planning authority. The process of development control thus becomes a negotiation to discover what is acceptable to both parties in the light of the current circumstances of time and place. Much of that negotiation lies outside the framework of the formal, statutory procedures. It takes place before ever a planning application is submitted, or it may mean that an original application is withdrawn and replaced by a fresh, revised application more likely to be the basis of an agreement. It means, too, reliance on ever more sophisticated conditions attached to planning permission, or ever greater use of planning gain.

E4.49 Thus the system for controlling development is not a simple, legalistic or administrative, device. Nor is it one on which there is broad agreement as to its scope, content and methods. The procedures are sufficiently flexible to allow a wide degree of discretion in which the Department of the Environment is but one protagonist, and only the courts have the last word.

E4.50 The final issue about control which has aroused the most complaint from developers and applicants for planning permission, and has received the most attention from the Department, is that of delay. The argument is that development control procedures can be unduly long, and that delay costs money. The counter argument is that planning applications need time to sort out, with proper consultations. Research has failed to establish conclusively any causal relationships which explain the very considerable variations in the time taken by different councils. However, many suggestions have been made, and acted upon, including the delegation of decision-making to sub-committees and officers; improved procedures for consultations; and simplified application forms. One suggestion, by an expert report, has not been accepted, however. This was that at least for minor applications a failure to issue a decision within eight weeks should be deemed to mean a grant of planning permission (Department of the Environment, 1975).

## The Idea of a Development Plan

E5.1   The development plan is the name given to the plan which has gone through a rigorous process laid down in the legislation for its preparation and approval, with provision for public participation in its preparation and challenge to its legality in the courts (Department of the Environment, 1984b).

E5.2   The development plan is in two parts.

(i) The structure plan, which has to be prepared by the county council and formally approved by the Secretary of State.

(ii) Local plans, which are usually prepared by the district councils, but may be also be prepared by the county council. They are prepared only where a local plan is thought necessary. In all cases, the local plan is adopted by the local planning authority after being certified by the county as being in general conformity with the structure plan.

In addition, the development plan also still formally includes such old-style development plans from the pre-1968 system not yet replaced by a new local plan or revoked by the Secretary of State.

E5.3   The pattern can be illustrated by listing the documents comprising the statutory development plan for Reading borough (see Table E5.1). The key documents, the replacement structure plan, the Central Reading District Plan and the Kennet Valley Local Plan are illustrated in Plates 1, 2 and 3. In fact the two local plans cover only about half of the area of the district.

E5.4   From 1 April 1986, when the county councils for Greater London and the six metropolitan counties were abolished, the system of structure and local plans is to be replaced by unitary development plans. Each London borough and metropolitan district will have to prepare such a plan in which Part I will be similar to a structure plan for its area, working within strategic guidance by the regional offices of the Department of the Environment; and Part II, such local plans as may be necessary. The new unitary development plan will be introduced when directed by the Secretary of State. Until then the existing structure and local plans remain in force (Department of the Environment, 1985d).

E5.5   The old-style development plans were, in effect, land use plans on an ordnance survey base map. They

### Table E5.1   Statutory Development Plans, Reading Borough Council, 1986

*Prepared by Berkshire County Council*

(i) Central Berkshire Structure Plan
prepared mid-1970s, approved 1980
gives general targets for population, housing, employment (including office floorspace), retailing in the district, and policies and proposals for traffic and the environment

(ii) Berkshire Replacement Structure Plan
prepared in the early 1980s, submitted for approval by the Secretary of State, 1986
gives revised targets and policies, especially for housing, employment and retailing

(iii) Berkshire Minerals Subject Plan
prepared in the early 1980s, adopted 1984
identifies sites, gives policies and proposals for working of minerals and restoration after use, and protects other areas from any future mineral working

(iv) Berkshire Countryside Recreation Plan
prepared in the early 1980s, adopted, 1985
states policies and proposals for the development of countryside recreation, including the use of rivers and canals for leisure uses

*Prepared by Reading Borough Council*

(v) Central Reading District Plan
prepared early 1980s, adopted 1986
identifies sites for development and redevelopment in the town centre, with targets for dwellings and floorspace for offices, retail, industry and warehouse; and policies for land use, traffic and environment

(vi) Kennet Valley District Plan
prepared mid 1980s, not yet adopted
gives policies and proposals for a fringe area including recreation, minerals and urban uses

*Prepared by old Reading County Borough Council*

(vii) Reading Town Map and Written Statement
prepared in 1950s, approved 1957
a 20-year land use map for the entire Reading district, in principle remains in force except where revoked by (iii), (iv) and in future (v), that is for what is now largely the existing built up area outside the central area (In practice this was automatically revoked when the Berkshire Minerals Subject Plan, which covered the entire country, was adopted)

allocated areas for so-called primary use zones, roughly corresponding to mixtures of compatible, or conforming, uses. And, though they had inset maps at different scales, together they comprised the one and only development plan, prepared by a single, county planning authority.

E5.6   The new structure and local plans were introduced in 1968 following a critical review of the old-style development plans (Ministry of Housing and Local Government, 1965). They differ in a number of respects. The structure plans are not land use plans, but rather policy statements about strategic land use matters which go into more explicit and varied detail than the old-style plans but are not site-specific. The local plans show some site-specific land use proposals but they too are primarily detailed policy statements against which proposals for development can be evaluated. Structure and local plans were deliberately split into two distinct documents of which only the more generalised, the structure plan, needs to be approved by the Secretary of State. However, following local government reorganisation in 1974, the system was even further fragmented between the two autonomous tiers of local government in a way not foreseen when it was enacted in 1968. Public participation too was a new feature. It involves repeated consultations with other agencies, and opportunities for the expression of public opinion by anyone interested, built into the statutory procedures for preparing both structure and local plans.

E5.7 Local planning authorities, counties and districts, are required to survey their area and keep under review those matters affecting its development and future planning, including physical, economic, social and demographic matters and communications. The development plan must be based on the surveys, as must any subsequent alterations or replacements to the plan. Such alterations or replacement plans themselves have to go through the same statutory procedures as did the original development plan. They would be made either because of an altered time-scale, such as rolling the plan forward, or because of changes in the basic assumptions or other policies affecting the plan.

E5.8 The development plan, although statutory, does not give a right to development and therefore, other than in certain, specified cases, the allocation of land in a development plan does not give a right to compensation for assumed loss of development value. It is a guide, but no more, to the intentions of the local planning authority. But it is always potentially a material consideration to be taken into account in deciding a planning application although it may either not be relevant to the circumstances of that application or it might be set aside because of other, more important, material considerations. Thus an application might involve types of development not foreseen when the plan was prepared, or economic changes might mean different priorities.

E5.9 The development plan however is not the only source of written policy guidance taken into account in the control of development. Local authorities also make use of many, individual types of non-statutory plans and the Department of the Environment issues national policy guidance.

## Structure Plans

E5.10 Every county planning authority is required to prepare a structure plan, and to keep it under review. Structure plans now cover the whole of the country, although the last was approved only in 1985. They are usually for a period of about fifteen years. A number have been reviewed and altered or replaced. So far, by March 1986, 28 of the 46 counties in England had submitted, or had approved, alterations to their structure plans, usually on specific matters such as housing targets, and eight had submitted or had approved a complete replacement of their structure plan (Department of the Environment, 1987e). This means that the average structure plan is still based largely on surveys and analysis in the mid 1970s, using data from the 1971 Census of Population, and approved about 1980 though with later alterations on some matters.

E5.11 The aims of a structure plan are

(i) to define policies and general proposals for the use of land, including the scale of provision; the physical environment, including environmental improvement and protection; and traffic management and communications, taking into account policies at the national and regional levels

(ii) to provide a framework for local plans.

E5.12 The structure plan may be either a single plan for the whole of a county or a number of separate plans provided the whole county is covered. It comprises a written statement of the actual policies, together with a key diagram of the structure plan which explicitly is not a map or plan, but a sketch diagram of spatial relationships such as the distribution of settlements, communications and areas of restraint (see Plate 1). The structure plan is accompanied by an explanatory memorandum setting out the reasoned justification for the policies; their relationship to neighbouring areas, and to national and regional policies; and their resource implications.

E5.13 The procedure for preparing and approving a structure plan is lengthy, taking up to four or five years of which about half the time is taken by the Department of the Environment in conducting the examination-in-public and finally approving the plan (see Table E5.2). It involves

(i) a preliminary survey which usually includes an informal public participation in order to identify issues for the plan

(ii) preparation of a draft plan including alternative policies, and involving consultation with government departments, etc

(iii) publication of the draft plan and alternatives, for a statutory six-week period of public participation

(iv) revision of the draft plan and approval by the county council

(v) publication of the final, draft plan, and submission to the Secretary of State, with a further statutory six-week period for public comment and objection to the plan

(vi) the ordering, if necessary, by the Secretary of State of an examination-in-public conducted by an inspector and panel of experts, to seek further information and clarification on key issues

(vii) publication by the Secretary of State of his modifications (if any) to the plan with a further period for objection

(viii) approval by the Secretary of State of the structure plan, with a period of six weeks for challenge in the courts as to its legality.

**Table E5.2   Preparation and Approval of Structure Plans, Time Taken**

| | average time in months |
|---|---|
| Local authority | |
| – draft plan for public participation to submission to Department of the Environment | 11 |
| – initial preparation of draft plan, from initial survey to publication of draft | (say) 9–21 |
| Department of the Environment | |
| – submission to examination-in-public | 8 |
| – examination-in-public to publication of modifications | 13 |
| – publication of modifications to final approval | 7 |
| Total | 48–60 (4–5 years) |

*Source*: Department of the Environment, 1986

E5.14   Once the structure plan is approved, the county council is required to keep it under review, especially to ensure its continued relevance in the light of its underlying assumptions about economic and demographic change, national and regional policies and any other changes. Most county councils have created monitoring procedures, though in general not on a uniform basis (Davies, Edwards and Fielder, 1987). One exception concerns the release of land for housing development. The Department of the Environment requires county councils to ensure that there is always a five-year supply of land available for development and has issued guidance for monitoring this, involving consultation with private house building interests. Monitoring exercises are not only for review of the structure plan but for other purposes such as promoting local economic development.

E5.15   Structure plans vary widely in their form and content, and in their degree of detail and precision. Some policies are mandatory, in the sense of setting criteria which can be satisfied through the control of development, either by the county council or, more usually, the district council reinforced by its local plan. Other policies are statements of intention, that for instance the county council will give support to the policies of other agencies, such as the Countryside Commission, or regional water authorities.

E5.16   The typical structure plan however is likely to have policies covering the following topics

(i) housing, including the number of houses for which land should be made available and its broad distribution

(ii) industry and commerce, including the amount of land to be made available, and its distribution

(iii) shopping, a broad guide to floorspace provision, the hierarchy of centres and their location

(iv) settlement pattern, including constraints on future extensions to keep the separate identity of particular towns and villages; identifying towns and villages for expansion; and policies for isolated dwellings in the countryside

(v) transport and communications, including the principal road network, policies for encouraging public transport and for traffic management

(vi) agriculture and forestry, including policies for the retention, or reclamation, of land for agricultural use, and areas for afforestation

(vii) minerals, including policies for the conservation and planned exploitation of mineral resources, and the restoration of sites

(viii) waste disposal strategies and policies regarding hazardous development, and the control of pollution

(ix) environmental protection and conservation, including identifying broad areas for landscape preservation

(x) green belts and their general location

(xi) recreation and tourism including the general location of facilities such as country parks, and policies for access to the countryside

(xii) social amenities and facilities for education, health and welfare, social services etc.

E5.17   The list of possible topics is very long. But, in all cases, the policies and proposals in the structure plan are written policies. The key diagram does no more than show the broad, spatial structure of the county as is shown in the draft Berkshire Structure Plan in Plate 1. Furthermore, the written policies often are very detailed, spelling out for instance that housing provision should be for so-called local needs, geared to forecasts of the natural increase in population.

## Local Plans

E5.18   Local plans are not a statutory requirement although they are increasing in coverage with time. Where they are not required the old-style development plan will usually continue to exist. They are usually prepared only where necessary, for instance in areas under intense pressure for development or where areas of restraint, such as a green belt, have to be defined. The first local plans began to be prepared in the late 1970s and, by March

1986, 440 had been adopted and 250 deposited for adoption. However, this means that only about a quarter of England is covered, and only in a very few districts is the entire district covered. The more usual pattern is for a district to have one or more local plans covering only a small area. Their time scale is for a period of about ten years, shorter than a structure plan, but, so far, alterations have been deposited or adopted for sixteen local plans and in no case has a complete local plan been replaced (Department of the Environment, 1987e).

E5.19   Local plans have four official functions (Department of the Environment, 1984b), namely

(i)   to carry forward in more detail structure plan policies and general proposals

(ii)   to provide a framework for development control

(iii)   to coordinate proposals for the use of land

(iv)   to bring detailed planning issues before the public.

E5.20   Local plans come in three varieties, the vast majority being in the first category.

(i)   General, local, or district plans deal with the full range of policies in a particular area. This is usually an area for suburban development, a small town, or a central area of a larger town. In a very few counties and much of London, such a plan has been prepared for the whole of a district or borough, in some cases with inset plans in even more detail.

(ii)   Action area plans deal with small areas for comprehensive development, sometimes requiring compulsory purchase, in which development is to start within ten years and be completed fairly quickly.

(iii)   Subject plans deal with a specific topic over a fairly wide area some of which may also be covered by district plans. The most usual subjects are those for defining green belt boundaries; for the conservation and exploitation of mineral resources and their subsequent restoration after use; and for areas of special landscape protection, conservation and treatment.

E5.21   Most local plans currently are prepared by district or borough councils, the chief exceptions being the specialised subject plans extending over several districts within a county in which the county council does the job. The county council, in consultation with the districts, is required to prepare and keep up-to-date a development plan scheme. The scheme shows the programme of local plans to be prepared, and the responsibility for their preparation.

E5.22   Every local plan comprises a written statement and proposals map.

(i)   The written statement contains all the policies and proposals of the plan together with an explanation,

or reasoned justification of the policies. Like the original structure plans before 1980, and unlike their current form, the written statement of the local plan is a single document. But a distinction must be made clearly between policies and explanation, for instance by using a different type face.

(ii)   The proposals map is on an ordnance survey (topographic) base usually to a scale of 1/10,000 or 1/25,000, showing the sites for proposed development and the boundaries of policies such as a central area or office zone, or a conservation area.

E5.23   The procedure for preparing and adopting a local plan is broadly the same as for a structure plan, and takes a similar time, with the following crucial differences (see Table E5.3).

(i)   The formal procedures can be started only when the proposed plan has been included in a development plan scheme, or when directed by the Secretary of State. Furthermore, before the 1980 Act, work could start only after the structure plan for the area had been approved but this now has been changed.

(ii), (iii), (iv) are similar to the first three stages in preparing a structure plan, including the survey; the technical work which often includes rather more public participation and consultation with community groups and the like as the local plan is more specific, focused on a smaller area than for the structure plan; and the statutory requirement for publication of a draft plan with a formal opportunity for public participation.

(v)   At the stage corresponding to the submission of a draft structure plan to the Secretary of State, a draft local plan is placed on deposit for public inspection and objection, together with a certificate by the county council that it is in broad conformity either with the approved structure plan, or with the draft proposals of a structure plan currently being altered. The local plan however does not need to be precisely in conformity; and the county council or, indeed, the Secretary of State, can object to proposals in the local plan.

(vi)   The corresponding stage to the examination-in-public of a structure plan is that of the public local inquiry into objections to a local plan, conducted by an inspector appointed by the Secretary of State. However, the public local inquiry has crucial differences. First, it is held only if there are specific objections to the plan and they cannot be dealt with by written representations. Second, it is a quasi-judicial inquiry into the objections as such rather than the plan as a whole, conducted like an appeal against refusal of planning permission described in Section E6, rather than a more general inquiry into issues on which the Secretary of State requires more information. Third, the inspector makes his report and recommendations on the objections to the district council rather than the Secretary of State, and the district council need not accept the inspector's report although it has to publish the report and give reasons for not accepting it.

(vii) After the local council has published any modifications to its original draft plan, heard objections to the modifications, and made known its intention to adopt the local plan, there is an opportunity for the Secretary of State to call-in the plan or, under certain circumstances, for the Ministry of Agriculture, Fisheries and Food to object to the plan. In either case, in principle involving issues of regional or national policy or substantial controversy, the Secretary of State then takes on the job of considering the plan and either approving it, with or without modifications, or rejecting it. Call-in involves its own procedures.

(viii) The local plan is formally adopted by the council making it a part of the development plan, with a short six-week period for challenge in the courts on points of law.

(ix) Finally, new legislation provides for an expedited procedure which, in effect, cuts out the need for the first stage of public participation on the draft plan, so that the only formal requirement is for objection to the deposited plan.

**Table E5.3   Preparation and Adoption of Local Plans, Time Taken**

|  | average time in months* |
|---|---|
| Preparation/publicising local plan brief | 1.5 |
| Report of survey and public participation on issues | 18.0 |
| Draft local plan and public participation | 12.0 |
| Final draft local plan, certification and deposit | 4.5 |
| Public local inquiry and report | 9.5 |
| Modifications on deposit | 4.0 |
| Adoption of local plan | 2.5 |
| Total | 52.0 (4–5 years) |

*Note:* * figures relate to the early 1980s: time taken has since fallen
*Source*: Bruton and Nicholson, 1987

E5.24   Local plans of the more general, district plan variety in general cover much the same range of topics as for structure plans though obviously in much more detail, including the precise definition of areas and boundaries on ordnance survey maps. Nevertheless, the variety of local plans is very great, responding to the local circumstances. The local plan may cover the whole of a district though in the majority of districts it will be for only one or two areas, such as the central area or an area undergoing change on the edge of the built up area, as in the case of the two Reading plans (see Plates 2 and 3).

E5.25   The policies in the local plan can be expressed in a variety of ways including, for instance

(i) area-wide targets and policies relating to particular uses, or changes of uses, such as the total increase in

shopping floorspace, and the need to retain or replace existing shopping floorspace

(ii) definition of policy areas, backed by written policies, such as conservation areas, central shopping areas, landscape management areas, etc

(iii) allocation of areas of land for particular uses, such as for the planned and coordinated development of large areas for new housing

(iv) development briefs and criteria for particular sites, giving the mix of uses and amounts of development

(v) standards and criteria for matters such as carparking

(vi) traffic management and pedestrianisation schemes.

## Old-style Development Plans

E5.26   The old-style, pre-1968 development plans were, in effect, land use plans on a topographic base map for a period of 20 years into the future, with a requirement that they be reviewed and rolled forward every five years. An entire county was covered by a written statement and county map (scale 1/63,360) with inset town maps (scale 1/10,560) and their own written statements for most of the existing and proposed built-up areas. Every town and county map had to be approved centrally, by the Secretary of State, following an inquiry into objections to the map.

E5.27   The old development plans remain in force until they are revoked automatically for the area covered by a new-style local plan, or by order of the Secretary of State. The system was progressively dropped after 1968 because of criticisms that the maps were too rigid, and the procedures for their approval too centralised, cumbersome and long drawn out.

## Non-Statutory Planning Guidance

E5.28   Structure and local plans are general guides to policy. They do not necessarily, or usually, match the full scope of all the considerations likely to be raised in development control and noted in Table E4.3. In particular, they usually do not include very detailed and specific guidelines on practical considerations about the amenity and arrangement of development (Davies, Edwards and Rowley, 1986b).

E5.29   Partly for this reason, most local authorities have relied heavily on additional, non-statutory policy guidance (Bruton and Nicholson, 1987). Originally, such guidance started to be used extensively because of the delays in waiting for structure plans to be approved before adopting local plans. However, current use has been influenced by three other needs, namely

(i) to avoid the lengthy procedures for statutory adoption, in which case the procedures tend in effect to stop at a point equivalent to placing a local plan on deposit

(ii) to include within the local plan matters which relate to other legislation, since the local authority regards the linkages between, say, housing and planning as being indissoluble but they cannot be formally included in the statutory local plan

(iii) to avoid the necessity of the full process as policies might need to be reviewed very frequently, as in the development brief for a particular site, or need to be in very precise detail, as in parking standards.

E5.30 The non-statutory policy guidance falls broadly into two categories.

(i) Supplementary planning guidance, meaning development control policy notes on narrow, specific matters applying over the whole of a district such as for carparking standards; design guides on the general layout and design of new development or the conversion of older buildings; and development briefs for particular sites. These provide more detail to supplement the development plan.

(ii) Informal local plans similar in content to a statutory local plan, or the policy frameworks for particular issues such as shopping or employment. These are more usually a substitute for a development plan, going through many of the same technical stages and public participation but stopping short of a public inquiry and adoption.

E5.31 Non-statutory plans, especially the supplementary planning guidance, are a very important source of policy for the day-to-day control of development. The questions about their status are raised on appeal against a refusal of planning permission. The general attitude of the Department of the Environment has been that provided supplementary planning guidance has been through a form of public participation and adopted by resolution of the council, then it can be accepted as a material consideration (Department of the Environment, 1984b). However, the situation is less clear on the informal local plans and policies, even if they have been approved by the council, and Departmental policy has been to discourage their use. The real test is the weight which they are given by an inspector in relation to development plans, other material considerations and national policy guidance.

E5.32 The status of the non-statutory policies and guidance has recently been brought into question by a successful challenge in the courts to the legal validity of the City of Westminster borough plan. The courts, confirmed by the House of Lords, found that there is a clear duty on local planning authorities to include all of their policies and proposals affecting the use and development of land in their statutory development plan.

E5.33 The implications of this finding have still to be fully worked out. On the one hand, it suggests that any planning decisions based on non-statutory plans could be set aside by the courts. But on the other hand, if the matters covered by the plans are found to be material considerations in a particular application or appeal, they may properly be taken into account. Thus the future of the non-statutory local guidance is uncertain.

## National Planning Policy Guidance

E5.34 Finally, development plans, like development control, have to have regard to national policy guidance. This is found in

(i) the legislation which comprises the acts and the subordinate legislation, namely the statutory instruments, orders and regulations

(ii) ministerial statements and government white papers (ie command papers) which contain statements of government policy, laid before Parliament

(iii) circulars published by the Department giving substantive policy statements and mandatory directions to local authorities such as, recently, on green belts, land for housing, industrial development and aesthetic control; and providing information and advice about procedural matters. These have recently (January 1988) been reinforced by a series of Planning Policy Guidance, issued by the Department

(iv) appeal decisions and the letters of approval of structure plans

(v) development control policy notes on particular topics such as large retail stores, etc.

E5.35 Another source of national guidance, similar to circulars, are the letters which the Secretary of State publishes on regional policy. Thus, he recently published a letter to the South East Regional Planning Conference of local authorities setting out in very general terms his views on matters affecting the scale and distribution of development in the region (reproduced in Department of the Environment, 1986). He has promised similar letters and strategic guidance to the London boroughs and metropolitan districts for their unitary development plans. The letters are brief and in much less detail than the regional strategies prepared by teams appointed by the Secretary of State in the late 1960s and 1970s, but they all have the same, formal status. They are indications of national or regional policy. They are not formally binding on local authorities, but give a strong lead.

E5.36 Together, these add up to a considerable amount of material. But they do not add up to a single coherent and comprehensive statement of national policy for the use and development of land. There is no statutory requirement for such a statement and there is no regular, annual report on planning to Parliament. Furthermore the status, expecially of the circulars, in relation to the statutory development plan is uncertain. The problem is that the national policies are very broad and general statements and it can be argued that in any particular case the local policies should have precedence.

## Conclusions

E5.37 The preparation of development plans and, more generally, the role and status of policies and plans have been the biggest single source of criticism and debate about the planning system. The original 1947 system of development plans and their preparation was altered completely after 1968, and is once more now being reviewed with the prospect of fresh legislation (Department of the Environment, 1986). The chief points include

(i) the proper content of development plans. One view is that they should be confined to a narrow interpretation of land use. The contrary view is that they provide a vehicle for a wider, more comprehensive approach to planning and development

(ii) the procedural delays, and the costs of preparing and approving development plans, including the requirement for repeated public participation and the amount of checking by the Department of the Environment before approval, given the number of policy statements in structure plans and their level of detail.

E5.38 The argument is that, as a result, development plans risk becoming unwieldy, irrelevant and rapidly out-of-date. This further contributes to the uncertainty which characterises the relationship between development plans and development control. Section 29 of the 1971 Act states that although the plan may be taken into account, so may any other material consideration. Further uncertainty arises from the division of responsibility between structure and local plans, their sequential preparation and relative authority; and the widespread use of non-statutory plans. Finally, there is the question of whether national and regional guidance is required, in what form, and with what authority over structure and local plans.

E5.39 The current government view, now being widely discussed, is for a much simpler system with one district development plan for the entire district, prepared and approved more quickly by the district council itself; with policy guidance from the county council in a new, formally prepared, narrowly focused, county statement replacing the structure plan; and with regional guidance provided where necessary by government. The basic relationship with development control however would remain unchanged.

E6.1   The planning system is an administrative mechanism established by the legislation. The main sources of challenge are internal, within the system, chiefly through appeals to the Secretary of State against a refusal of planning permission. But the decisions of local authorities and the Secretary of State may be further challenged in the courts, on the grounds of their legality. And, finally, the planning system is subject to review through a number of other ways, notably in complaints to the Parliamentary and Local Ombudsmen, and oversight by the Audit Inspectorate, the National Development Control Forum and by Parliament.

## The Workload and Outcome of Appeals

E6.2   The main method by which the Secretary of State may control development directly is through appeals by an aggrieved applicant against refusal of planning permission under section 36 of the 1971 Act, together with a very small number of appeals against a failure by a local authority to issue a decision on a planning application. In 1984, 16,200 appeals of this kind were decided in England, or roughly 3 per cent of all planning applications, or 28 per cent of all applications refused planning permission (see Table E6.1). Of those appeals, roughly 32 per cent were allowed but in 68 per cent, the original decision of the local planning authority was confirmed and the appeal dismissed.

E6.3   Section 36 appeals cover the full spectrum of planning applications. In general the majority of major applications refused planning permission go to appeal, as the rewards of a successful appeal can be very large. Conversely only a small proportion of householder and change of use applications go to appeal. But, irrespective of the size of the application, only the householder applications vary significantly from the average rate of success. Thus in 1982 only 48 per cent of householder appeals were dismissed compared with between 67 and 72 per cent of the other classes of appeal (see Table E4.1).

E6.4   The figures do fluctuate considerably and there are growing tendencies both for the numbers of appeals and the proportion of successful appeals to increase (Great Britain, 1986b). One possible reason for this lies in the reiteration by the Department of the Environment of a presumption in favour of development, and their making this very clear in circular 22/80 and even more in circular 14/85.

**Table E6.1   Types of Planning Appeal, 1984**

| Act | Grounds of appeal | Number of appeals received in 1984 |
|---|---|---|
| *Development control* | | |
| Sections 36/37 | Refusal/conditions on planning permission and failure to determine within 8 weeks | 16,192 |
| Section 53 | Whether planning permission required | 75 |
| Section 56, etc | Relating to listed buildings | 261 |
| Regulations | Relating to advertisements | 1,730 |
| Regulations | Relating to tree preservation | 262 |
| Section 95 | Relating to established use certificates | 89 |
| Sub-total, development control | | 18,609 |
| *Enforcement* | | |
| Section 88 | Enforcement notice for breach of control | 4,058 |
| Section 97 | Enforcement notice for breach of listed building control | 138 |
| Sub-total, enforcement | | 4,196 |
| Total | | 22,805 |
| Applications, called-in by the Secretary of State | | 41 |

*Source*: House of Commons, 1986

E6.5   Section 36 appeals are only part of the full spectrum of cases brought to the Secretary. The full list for 1984 is shown in Table E6.1 and includes those relating to advertisements, listed buildings and tree preservation orders. They also include a large number of appeals against enforcement notice.

E6.6   Finally, the Secretary of State may also call in a planning application because it is a major departure from an approved structure plan which the local authority is minded to permit, or because it raises very major national issues such as a proposal for a nuclear power station.

## Appeal Procedures

E6.7   Although the appeal formally is to the Secretary of State, there is a choice of methods for dealing with it (see Table E6.2).

(i)   In all cases, the Secretary of State appoints an inspector to conduct the appeal. The inspector is usually a professionally qualified planner either on the staff of the Inspectorate or appointed for the purpose. On average, in 96 per cent of cases in 1985, the responsibility for making the final decision on the appeal was transferred, that is delegated, to the inspector. In about 5 per cent of appeals were the issues of principle, or the circumstances of the case, such that the inspector simply made a recommendation and the final decision was made by the Secretary of State, or his officials acting on his behalf.

(ii) In 86 per cent of the appeals, the matters were sufficiently clear for the appellant to agree that the decision be made on the basis of written representations by the various parties to the appeal. But in 12 per cent, including a third of the Secretary of State's cases, the decision followed a public local inquiry. Finally, a new procedure has recently been introduced for an informal hearing conducted by an inspector; it was used in 3 per cent of appeals.

**Table E6.2    Outcome of Planning Appeals, 1985**

| Procedure | Number of appeals decided | Number of appeals allowed | Allowed as % of decided |
|---|---|---|---|
| Transferred to Inspectors for decision | | | |
| Written representations | 12,068 | 4,297 | 35.6 |
| Public inquiry | 1,434 | 699 | 48..7 |
| Informal hearing | 382 | 161 | 42.1 |
| Recovered by Secretary of State for decision | | | |
| Written representations | 466 | 177 | 38.0 |
| Public inquiry | 289 | 143 | 49.5 |
| Total | 14,639 | 5,477 | 37.4 |

*Source*: Great Britain, 1986b

E6.8    Planning appeals take a relatively long time to process, ranging from an average of 20 weeks from the appeal to the decision for one decided by an inspector on written representations, to 56 weeks on one decided by the Secretary of State involving a public inquiry (see Table E6.3). The time taken however has been reduced by transferring more appeals to inspectors, by relying more heavily on written representations and by introducing the new, informal hearings. Also, a greater use is being made of the award of costs (see below).

**Table E6.3    Planning Appeals: Time Taken, 1985**

| Procedure | Median time in weeks |
|---|---|
| Transferred to Inspectors | |
| Written representations | 20 |
| Public inquiries | 32 |
| Informal hearing | 28 |
| Recovered by Secretary of State | |
| Written representations | 41 |
| Public inquiry | 56 |

*Source*: Great Britain, 1986

### The Public Inquiry

E6.9    The choice of procedure lies with the Inspectorate but if either party requests it, the appeal will be by public inquiry. Public inquiries are not courts of law, but they are governed by strict rules of procedure having their origins in the 1957 report of the Committee on Administrative Tribunals and Procedures. The rules relating to planning inquiries are laid down in regulations and their basic aim is to ensure natural justice, that is the procedures should be open, fair and impartial. The proceedings thus become quasi-judicial, conducted without bias, and in such a way that each party to the inquiry is given an adequate hearing, extended to mean that facts must be presented at the inquiry itself.

E6.10    The local planning authority has to publish in advance of the inquiry its written statement of the reasons for refusal, together with a list of the plans and documents which it will be presenting in support of its case; and the documents must be available for public inspection before the inquiry.

E6.11    Those entitled to appear at the inquiry, in each case with evidence, witnesses and cross-examination, include the appellant; the local planning authority; and third parties with a statutory right, or at the inspector's discretion. The inquiry, which may last anything between half-a-day and many weeks, is followed by an accompanied site visit.

E6.12    The inspector writes his report and recommendations to the Secretary of State, or makes his decision. In all cases, if any new evidence of fact emerges, or if any expert advice is sought after the inquiry, it has to be notified to everyone who had appeared and if necessary the inquiry may be reopened.

E6.13    Finally, the decision letter, together with the inspector's report and recommendations is published. The decision letter and report together must summarise all of the evidence given at the inquiry, decide on what is relevant and material to the particular application, and give the grounds for the decision, whether it be to uphold the appeal, to dismiss it, or to impose or relax the conditions.

E6.14    Costs may be awarded against either the appellant or the local planning authority if the appeal was thought to be vexatious or frivolous, for instance, repeating earlier appeals, or the grounds of refusal by the local planning authority were thought unreasonable. However, costs were applied for in less than 2 per cent of appeals in 1984, and in only one in three were they awarded, although recently the numbers have increased following Departmental advice (Department of the Environment, 1987a).

### Written Representations

E6.15    These follow much the same procedure, except that they are carried on by written correspondance between the inspector and the various parties to the appeal. Each piece of correspondence is shown to the other parties, giving them a chance to challenge, in writing, the facts or their interpretation.

*Informal Hearing*
E6.16   This method is comparatively new, and is offered by the Inspectorate when the circumstances warrant in an attempt to shift away from time-consuming legalistic procedures. In effect, the inspector acts as an examining magistrate, on the French model. He or she studies all the written evidence and then prepares an agenda for informal discussion between all the parties to the appeal. On the basis of that hearing, the inspector comes to a decision, and announces it, at the hearing.

## Appeals and the Control of Development

E6.17   The decision in an appeal in principle is made in precisely the same way as that on a planning application, from first principles. The inspector has to take into account the development plan and any other material considerations. These may include national policy guidance, on which the inspector must be fully informed. But the inspector's task is not to impose national policy. Rather it is to balance the relative weight of national and local policy in the specific circumstances of that particular appeal. Nor is the inspector bound by precedent although he or she will be aware of recent decisions in the locality of the appeal, or on similar types of development, through the Department's system of computerised records.

E6.18   Recent research showed that in one-third of all appeals in 1982 there was no mention of policy, whether national or local (Davies, Edwards and Rowley, 1986b). These were mainly appeals on householder applications, changes of use and other minor developments. They turned mainly on practical considerations such as amenity and design or the layout and arrangement of buildings and uses. In these circumstances, even in 1982, inspectors tended to find in favour of the appellant more often than when there was a strong local policy.

E6.19   In the two-thirds of appeals in which policy was mentioned, the most frequently mentioned policies were the structure plan, followed by the local plan, the old-style development plan and national policy guidance. Supplementary planning guidance was mentioned in a significant proportion of the minor developments, changes of use and householder applications. The problem for the inspector therefore was what relative weight to give the local and the national guidance.

E6.20   In general, structure plans were supported, and the appeals dismissed in most cases, especially those which turned on strategic matters such as the location and volume of development. But in cases where the practical matters were more important the national guidance was given greater weight, especially the presumption in favour of development, and the appeal was allowed and local authority's refusal of permission reversed.

## Major Public Inquiries

E6.21   Finally there is the problem posed by the really large developments such as the proposals for a nuclear power station at Sizewell in Suffolk, a new coal mine at the Vale of Belvoir in Nottinghamshire, or the extension of Stanstead airport near London. Fortunately, they are comparatively few in number but they are characterised by the extent to which they raise matters of national importance transcending the planning legislation. Thus the examples quoted raise national energy and transport policies. Yet in terms of procedure they are dealt with in precisely the same way as any small development under the 1971 Act. The only difference is that they are called-in by the Secretary of State, rather than being left to the local authority to decide (House of Commons, 1986).

E6.22   A number of criticisms have been levelled. For the developer such as the Central Electricity Generating Board it can mean an extremely long inquiry during which they can be challenged on every aspect of their proposal, from the need for, and safety of, nuclear power, to the number of jobs created and the landscaping of the site. In the case of Sizewell, this resulted in an inquiry which lasted more than two years, held in a remote area near the site of the proposal. The second problem is the demands of time and cost which it makes not only on the developer and local authority, but also on the public interest groups which usually are the most concerned, and carry the burden of objecting to the proposal. It also much more sharply raises questions of natural justice as these cases refer almost always to proposals by government which is thus acting as judge in its own cause.

E6.23   The 1971 Act does provide for a special procedure, the so-called Planning Inquiry Commission, to be appointed each time one of these major cases threatens. But the idea has never been used, chiefly because, although it widens the remit of the inquiry in theory, in practice the ordinary inquiries have covered a very wide range of considerations.

E6.24   The real problem is that these developments on the one hand raise questions of major national policy for which, it can be argued, a public inquiry under the planning legislation is not the proper forum. On the other hand, they raise the more local planning issues which are the proper scope of a local public inquiry.

E6.25   One answer, currently being used for the proposed channel tunnel to France, is by a special act of parliament, going through the full legislative procedure which, however, offers fewer opportunities and time for objections to be heard. But the problem remains.

## The Courts

E6.26   The main controlling processes of the town and country planning system over the use and development of land are administrative, rather than judicial. It is the

Secretary of State who usually has the last word. Yet town and country planning has become one of the most consistently controversial areas of administrative law in Britain.

E6.27 Two interacting factors create this situation. On the one hand has been the massive amount of administrative discretion rather than legislative regulation built into the planning system. Appeals and public local inquiries are quasi-judicial, but their decisions are based on policy rather than legal rules. On the other hand, there is no written constitution in Britain. Instead there is what amounts to a continuous process of piecemeal additions to the law, and a day-to-day flow of judicial decisions by the courts interpreting the law by reference to its intentions and precedent, thus establishing rules for the future.

E6.28 The chief safeguard of the rights of the citizen in this is that the administrative processes are in the last resort supervised by the courts. That is, application can be made by an aggrieved person to the High Court, the Court of Appeal and, ultimately, the House of Lords acting as the final court of appeal. The application to the courts can only be on points of law and the courts have no power to make a new decision, or substitute a new decision for the appealed decision. They can only suspend or quash the operation of the original decision, in whole or in part.

E6.29 The grounds of appeal in principle must be that the challenged decision was not within the powers of the Act, or that it had not complied with the statutory procedures. But these restrictions can be widely interpreted through the process of judicial review in which it is alleged that, in making his decision, the inspector or other authority disregarded proper considerations which were relevant and material to the case; or improperly took into account considerations not relevant to planning; or came to an unreasonable conclusion. It is the last of these that in effect opened the door to judicial review on grounds of policy. That is, by challenging the reasonableness of a decision, it opened the possibility of policy being made in the courts rather than by the executive.

E6.30 Judicial review, of course, is a constitutional matter potentially of great importance which applies across the whole field of administrative law, going well beyond the grounds of strict legality and natural justice. One aspect concerns who are the aggrieved persons who can apply to the courts. Recent decisions by the courts have accepted appeals not only from the parties with a direct interest (the original applicant, the local authority or Secretary of State etc) but also from concerned groups of citizens such as an amenity group.

E6.31 The majority of challenges have been to decisions and appeals on planning applications although, even so, only 83 decisions (0.6 per cent of the total) by inspectors on section 36 appeals were challenged in 1983, and 35 (4.6

per cent) of those by the Secretary of State. They have been mainly about the definition of development and the propriety of planning conditions on a grant of planning permission. Other reasons included a failure on the part of an inspector to have regard to national policy as laid down in circular 14/85 (Great Britain, 1986b).

E6.32 Section 245 of the 1971 Act provides a period of just six weeks after its approval by the Secretary of State or adoption by the local council in which the legality of a structure or local plan can be challenged in the courts. However, the challenges have been few in number, thus tending to create a situation in which the policy statements on plans cover potentially a much wider range than development control (Davies, Rowley and Edwards, 1986a).

E6.33 Finally, the courts are more directly involved in enforcement action. Any costs for works undertaken by the local authority have to be recovered from the person in breach of planning control through the courts, and only the courts have power to levy a fine.

## The Commission for Local Administration

E6.34 A further method of challenge to planning decisions is by making a complaint to a Parliamentary or Local Commissioner (Ombudsman), created by the Local Government Act 1974. Complaints can be on any injustice caused by maladministration. Examples include a failure to comply with local codes of practice; or giving misleading information, and misrepresentation of facts and arguments about a planning application or appeal. But the complaint can be considered only if no other course of action is open, by appeal to the Secretary of State or challenge in the courts.

E6.35 Complaints about planning, in particular development control, generate a large proportion of the work of the Local Commissioner. Nevertheless, the actual number of complaints is very small when compared with the total number of planning applications. Many of the complaints are by neighbours and third parties, that they were not correctly informed about, or given an opportunity to object to, a planning application. But in the end the Commission has no power to quash, or reverse, a decision or to impose any form of punishment for maladministration. Thus its direct power is very limited but nevertheless its influence is very great, especially on the rigour with which local authorities follow statutory and informal procedures.

## The Audit Inspectorate

E6.36 The Audit Inspectorate was appointed by the Secretary of State under the Local Government Finance Act 1982, replacing the previous system by which local authority accounts were audited by district auditors. The Inspectorate appoints auditors for individual local authorities. And from time to time an auditor may decide to review a particular service, such as planning, either at the request of the local authority, or as part of a national review of that service at the request of the Inspectorate.

E6.37 The original aim of the audit of accounts was to ensure the financial probity and legality of the local authority and, under this heading, the district auditors had considerable powers to refer matters to the courts. However, the purpose of setting up the Audit Inspectorate was to take matters a step further. Its remit is not only that of financial audit, but of establishing good practice in local government and management.

E6.38 Development control has, by this mechanism, been the subject of a national review, following the growing concern going back many years about alleged delays in the processing of planning applications. Their report (Audit Inspectorate, 1983) concentrated on the operational procedures within local authorities. As with previous studies, the Commission found no obvious correlation between the procedures and characteristics of local authorities and their efficiency whether measured in terms of costs, or speed of making decisions. Thus their general conclusion was simply to recommend good management practice and a commitment to an economic and speedy service.

## The National Development Control Forum

E6.39 This was an organisation set up in 1981 by the three associations of local authorities, the Association of County Councils, the Association of District Councils and the Association of Metropolitan Authorities. Its purpose is to review current practice in development control, hi-tech development, or land for housing. It is purely an advisory body, establishing codes of practice.

## The House of Commons Environment Committee

E6.40 This is one of a number of committees set up to oversee and review the work of government departments, in this case the Department of the Environment. The committees replace the former Select Committee on Expenditure, one of whose sub-committees had planning within its remit.

E6.41 The Committee sets its own agenda, calls for witnesses, commissions research, and holds public hearings. It has considered, for instance, green belts and the release of land for development. Its most recent, substantial report on planning (House of Commons, 1986) has been on the conduct of appeals against refusal of planning permission and major public inquiries. It amassed a large body of evidence and came to a number of recommendations about the conduct of appeals, including

(i) the greater use of costs against appellants and local authorities to deter frivolous appeals and unreasonable decisions by local authorities

(ii) improved procedures to speed the processes

(iii) efforts to reduce the number of appeals for instance by the Secretary of State spelling out the criteria likely to influence his consideration of appeals

(iv) clarification of the role of major public inquiries, by separating out where possible the questioning of national policy and the consideration of alternative locations from the site-specific issues. Where possible only the latter should be the subject of the usual appeals procedure.

E6.42 Reports by the Environment Committee do not carry any legislative authority. Its recommendations however could well influence future legislation, or the direction of national policy as expressed in white papers or circulars.

## Building Regulations, 1985

E7.1   A new system of building control was introduced by the Building Act 1984. The Building Regulations 1985, prepared under the Act, cover the construction, alteration and extension of buildings, changes of use and the provision of services and fittings. Certain small buildings not normally frequented by people are exempted.

E7.2.   The regulations list a schedule of requirements covering structure, fire, site preparation, noise, ventilation, hygiene, drainage and waste disposal, heat, stairways etc and conservation of fuel and power (see Table E7.1). The requirements are very general and are backed by twelve approved documents, not part of the regulations, which provide practical guidance on how to comply with the regulations. Other methods would be acceptable if it can be shown that they comply with the regulations. The only exception is the means of escape in fire, for which there are mandatory standards.

**Table E7.1   The Requirements for Building Control: Summary**

| | |
|---|---|
| A  Structure<br>  – loading<br>  – ground movement<br>  – disproportionate collapse | G  Hygiene<br>  – food storage<br>  – bathrooms<br>  – hot water storage<br>  – sanitary conveniences |
| B  Fire<br>  – means of escape<br>  – internal fire spread<br>  – external fire spread | H  Drainage and Waste Disposal<br>  – sanitary pipework and drainage<br>  – cess pools, septic tanks etc<br>  – rainwater drainage<br>  – solid waste storage |
| C  Site preparation and Resistance to Moisture<br>  – preparation of site<br>  – dangerous and offensive substances<br>  – subsoil drainage<br>  – resistance to weather and ground moisture | J  Heat Producing Appliances<br>  – air supply<br>  – discharge of products of combustion<br>  – protection of buildings |
| D  Toxic Substances<br>  – cavity insulation | K  Stairways, Ramps and Guards<br>  – stairways and ramps<br>  – protection from falling<br>  – vehicle barriers |
| E  Resistance to Passage of Sound<br>  – airborne sound | L  Conservation of Fuel and Power<br>  – resistance to the passage of heat<br>  – heating system controls insulation of heating services |
| F  Ventilation<br>  – means of ventilation<br>  – condensation | |

*Source:* Powell-Smith and Billington, 1986

E7.3   The procedures were altered in 1984, giving a choice of methods (see Figure E6).

(i)   Local authority control may be by either one of two methods

   (a)   full plans must be deposited with the local authority which must then approve or reject the plans within five weeks
   or

   (b)   a building notice must be issued, together with plans, for most buildings of a less complex kind, in which case the local authority does not have to approve or reject the plans but may call for further details.

   In either case, the local authority has to be given 48 hours' notice of the commencement of works, and 24 hours' of the completion of certain works such as foundations, to give an opportunity for inspection.

(ii)   Control by an Approved Inspector is by an inspector, approved by the Secretary of State, who takes responsibility for ensuring compliance by giving an initial notice to the local authority with a certificate of insurance cover. The local authority accepts the plans or rejects them as defective within ten days. If they are accepted, the inspector issues a plans certificate and, on completion, a final certificate.

E7.4   At each stage, a builder can appeal to the courts against rejection of any of the notices or certificates. Contravention of the building regulations is a criminal offence and a local authority may require the offending works to be pulled down etc. Finally, a local authority, approved inspector or builder can be liable in the courts for civil damages resulting from a breach of building control.

E7.5   Building control is exercised by officers of the district or borough council, often in a section of the planning department. Prior to 1986, and the introduction of the new system, the council was responsible for all building control, processing 444,000 permits in 1984/85 at a gross cost of £69.5 million, offset by a fee income of £33.4 million. Thus the net cost of each building permit was about £81, compared with about £98 for each planning application in development control. The levels of employment also were roughly the same as in development control, with roughly 11.1 staff per 100,000 population. More recent statistics under the new system are not yet available (Chartered Institute of Public Finance and Accountancy, 1986).

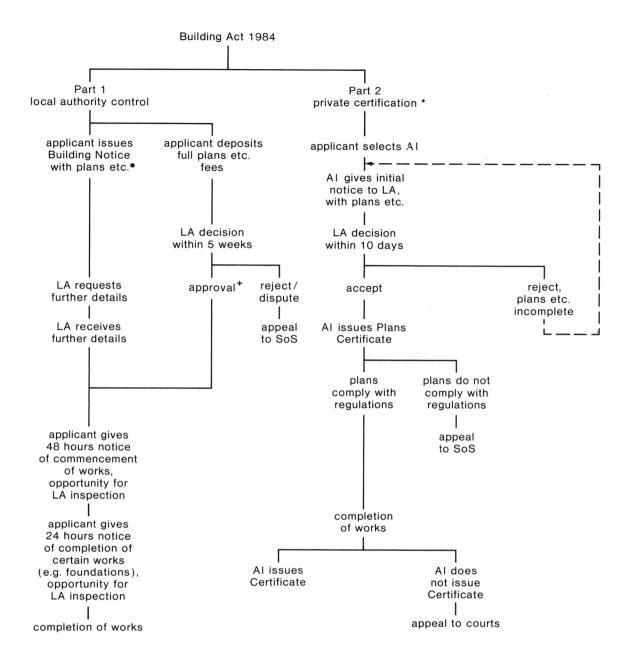

Building Act 1984

Part 1
local authority control

applicant issues
Building Notice
with plans etc.●

applicant deposits
full plans etc.
fees

LA decision
within 5 weeks

LA requests
further details

approval+      reject/
                      dispute

LA receives
further details

appeal
to SoS

applicant gives
48 hours notice
of commencement
of works,
opportunity for
LA inspection

applicant gives
24 hours notice
of completion of
certain works
(e.g. foundations),
opportunity for
LA inspection

completion of works

Part 2
private certification *

applicant selects AI

AI gives initial
notice to LA,
with plans etc.

LA decision
within 10 days

accept                          reject,
                                         plans etc.
                                         incomplete

AI issues Plans
Certificate

plans
comply with
regulations

plans do not
comply with
regulations

appeal
to SoS

completion
of works

AI issues
Certificate

AI does
not issue
Certificate

appeal to courts

notes :   LA   - local authority
          AI   - Approved Inspector
          SoS  - Secretary of State

   +   approval may be given with conditions or for the construction to be carried out in
       stages with inspection opportunities at each stage
   *   by either an Approved Inspector or approved public body doing work for its own use,
       with full insurance cover
   ●   for relatively straightforward buildings

**Figure E6   Building Control Procedure**

Figure E6 shows the alternative procedures for building
control introduced since the Building Act 1984. One method
is by inspection and control directly by the local authority,
with an expedited procedure for relatively straightforward
buildings. The other method involves selection of a surveyor
or inspector approved by the Secretary of State who then
takes responsibility for certifying that the construction is
satisfactory.

## Other Controls

E7.6   Other controls on development include

(i) additional, local building regulations covering matters specific to a particular area, especially in inner London, such as concerning basements, etc

(ii) controls over particular matters such as gas, water and electricity installation, access to highways, atmospheric pollution, noise abatement and licensing for use for entertainment etc

(iii) those affecting the initial construction and continuing requirements on owners to ensure the buildings remain safe, etc: the Factories Act 1961, the Offices, Shops and Railway Premises Act 1963 and the Health and Safety at Work Act 1974.

E7.7   All these other controls and building controls are additional to the planning control of development and have to be separately complied with.

# GLOSSARY

**appeal**
an application to the Secretary of State against a refusal, conditions or non-determination by a local planning authority on an application for planning permission

**area of outstanding natural beauty**
an area of countryside designated by the Countryside Commission where the intention is to protect its natural beauty

**Building Regulations**
the system under which the structural and other qualities of building, including health and safety, are controlled

**called-in application**
an application for planning permission which is called-in to be determined by the Secretary of State, usually as a major departure from an approved development plan

**circular**
a document issued and published by central government departments, containing statements of government policy, or advice on particular issues or procedures

**conservation area**
an area of particular amenity or historical character, usually in a town or village, within which planning control is more strictly enforced and for which the local planning authority has a duty to formulate and publish proposals for preservation and enhancement

**councillor**
the elected member of a county, district or borough council

**Countryside Commission**
the organisation appointed by the Secretary of State for the Environment with responsibility *inter alia* to oversee rural planning matters

**county map**
the old, pre-1968 part of a development plan showing the future pattern of land uses, etc, for an entire county

**deemed planning permission**
procedures by which government departments etc receive planning permission for their own development

**development**
the basic concept underlying all planning control, strictly and comprehensively defined in the primary legislation and statutory instruments, covering all building etc operations, and all changes of use of land and buildings between defined classes of use

**development control**
the system by which development is controlled through the grant or refusal of planning permission

**development plan**
the statutory statement of policy for the use and development of land, currently including structure plans, local plans and old-style development plans

**Department of the Environment**
the central government department responsible *inter alia* for the town and country planning system, housing and local government

**district council**
the lower tier of local government outside London and the metropolitan areas, with primary responsibility *inter alia* for local plans and development control; also known as borough council in some areas

**enterprise zone**
a small area designated by the Secretary of State for the Environment, usually in older, or inner city industrial areas, within which specified types of development are granted planning permission automatically, without having to submit an application and in which there are also tax advantages

**examination-in-public**
a special form of inquiry by a panel appointed by the Secretary of State to examine a submitted structure plan

**General Development Order**
the statutory instrument approved by Parliament, which sets out classes of 'permitted development', and procedures for development control

**Historic Buildings and Monuments Commission**
(otherwise known as English Heritage) the organisation appointed by the Secretary of State for the Environment with responsibility *inter alia* for listing buildings of architectural and historic interest

**informal local plan**
a plan which contains policies and proposals approved by resolution of a local planning authority but which has not gone through the formal statutory procedures for the preparation and adoption of local plans

**listed building**
a building of architectural or historic interest, listed by the Historic Buildings and Monuments Commission for England, and subject to strict planning control requiring 'listed building consent' to be obtained for any works affecting it

**local plan**
part of a development plan covering either a particular area, or an area for comprehensive development or redevelopment (an action area plan), or a particular issue or topic (a subject plan)

**local planning authority**
the elected local government council responsible for planning matters

**London**
since 1986 administered by the 33 pre-existing borough councils, following abolition of the Greater London Council, with a new form of development plan, the unitary development plan, to be prepared by each borough council

**metropolitan areas**
the six areas in England which, since the abolition of the metropolitan counties in 1986, have been administered by the pre-existing metropolitan district councils, with a new form of development plan, the unitary development plan, to be prepared by each district council

**Ministry of Housing & Local Government**
the central government department responsible for planning before its absorption into a new Department of the Environment in 1970

**national park**
an area of attractive scenery designated by the Countryside Commission within which planning control is exercised by either the county council or a specially appointed national park authority

**Nature Conservancy Council**
the organisation appointed by the Secretary of State for the Environment *inter alia* to designate nature reserves and sites of special scientific interest

**nature reserve**
an area of particular ecological interest designated by the Nature Conservancy Council and subject to special control

**new town development corporation**
an organisation appointed by the Secretary of State for the Environment for the planning and construction of a designated new town

**old-style development plan**
the style of development plan prepared under the Town and Country Planning Act 1947, and still operative until revoked or replaced by a local plan or unitary development plan

**outline planning permission**
a form of planning permission establishing the basic principles of a proposed development, but still requiring approval of reserved matters of detail

**permitted development**
a form of development, mainly minor in scale, for which planning permission is granted automatically by the General Development Order

**planning conditions**
the conditions which may be attached to the grant of planning permission and which run with the land

**planning gain**
a term applied whenever, in connection with a grant of planning permission, a developer enters into a legal agreement with the local planning authority to carry out works, or make some payment, or confer some other right etc when receiving the permission

**planning inspector**
an official of the Department of the Environment appointed to hear appeals against a refusal of planning permission by a local planning authority and to conduct public inquiries into objections to a deposited local plan

**planning permission**
the basic instrument of development control, following an application to a local planning authority for permission to develop land, including changes of use

**public inquiry**
a quasi-judicial administrative hearing conducted by a planning inspector into appeals against a refusal of planning permission, or objections to a deposited local plan

**public participation**
the arrangements in planning legislation for the public to be consulted on the preparation of development plans and other planning matters

**reserved matters**
the matters of detail which have to be approved following the grant of outline planning permission

**regional economic planning councils**
the committees appointed by the Secretary of State for the Environment for each of its administrative regions to provide advice on regional strategies, and abolished in 1980

**regional guidance**
the advice to be provided in future by the Department of the Environment, in particular to provide a framework for the preparation of unitary development plans in London and the metropolitan areas

**regional offices**
the eight offices of the Department of the Environment with responsibility for the day-to-day work of the Department in the regions, including liaison with local planning authorities

**regional strategies**
the planning policies for each of the administrative regions of central government prepared during the 1960s and 1970s by teams of central government officials *inter alia* to provide a framework for structure plans

**Secretary of State for the Environment**
the politician, a member of the Cabinet and usually an elected Member of Parliament, responsible for the Department of the Environment

**simplified planning zone**
a small area to be designated by a local planning authority under the Housing and Planning Act 1986, within which specified types of development will be granted planning permission automatically, without having to submit an application

**site of special scientific interest**
a small area of particular ecological or geological character, designated by the Nature Conservancy Council and subject to strict control

**Special Development Order**
a parliamentary procedure by which the Secretary of State for the Environment may grant planning permission under special circumstances, such as for a very major development probably not included in an approved development plan

**statutory instrument**
subordinate legislation produced by the Secretary of State and usually to be approved by Parliament, authorised by, and elaborating, the primary legislation in the Acts of Parliament

**structure plan**
the part of a development plan which every county council is required to prepare for its entire area, and which has to be approved by the Secretary of State

**supplementary planning guidance**
policies prepared, and adopted, by local planning authorities without going through the statutory procedures for the preparation and approval of a local plan, usually covering matters of detail such as design guidance, car parking standards or briefs for particular sites

**unitary development plan**
the type of development plan to be prepared by each district or borough council in the metropolitan areas and London, comprising part 1, corresponding roughly to a structure plan, and part 2, a local plan

**urban development corporation**
an organisation appointed by the Secretary of State for the Environment for selected urban areas with responsibility chiefly for the acquisition and preparation of land for development, and the promotion of development, by the private sector

**Use Classes Order**
the statutory instrument approved by Parliament which defines broad categories of the use of land and buildings within which changes of use do not constitute development and, therefore, do not require planning permission

# REFERENCES

There are several sources and references which are so fundamental that they are not individually referenced in the text. The basic administrative system and the history of planning are fully discussed with detailed bibliographies, in Cullingworth, J B (1988), **Town and Country Planning in Britain**, 10th edition, London: George Allen & Unwin. A commentary on planning law is found in Grant, M (1986), **Urban Planning Law**, 2nd edition, London: Sweet & Maxwell. The full text of all planning legislation (acts and statutory instruments) and government circulars, annotated by reference to case law, appeals and court decisions, is found in the **Encyclopaedia of Planning Law**, London: Sweet & Maxwell (4 volumes, continuous up-date). The system for building control is described in Powell-Smith, V & Bellington, M J (1986), **Whyte and Powell-Smith's The Building Regulations Explained and Illustrated**, 7th edition, London: Collins.

AUDIT INSPECTORATE (1983), **Local Planning: the Development Control Function**, London: HMSO

BRUTON, M and NICHOLSON, D (1987), **Local Planning in Practice**, London: Hutchinson

CHARTERED INSTITUTE OF PUBLIC FINANCE AND ACCOUNTANCY (1986), **Planning Statistics, Estimates, 1985-86**, London: CIPFA

CHERRY, G E (1974), **The Evolution of British Town Planning**, Leighton Buzzard: Leonard Hill

DAVIES H W E, EDWARDS, D and ROWLEY, A R (1986a), **The Relationship between Development Plans, Development Control and Appeals**, Department of Land Management Working Papers, Reading: University of Reading

DAVIES, H W E, EDWARDS, D and ROWLEY, A R (1986b), 'The Relationship between Development Plans, Development Control and Appeals', **The Planner**, vol 72, no 10

DAVIES, H W E, EDWARDS, D and FIELDER, S (1987), 'Monitoring the Planning System', **The Planner**, vol 73, no 5

DEPARTMENT OF THE ENVIRONMENT (1975), **Review of the Development Control System, a report by Sir George Dobry QC**, London: HMSO

DEPARTMENT OF THE ENVIRONMENT (1980), **Development Control, Policy and Practice**, Circular 22/80, London: HMSO

DEPARTMENT OF THE ENVIRONMENT (1983), **Planning Gain**, Circular 22/83, London: HMSO

DEPARTMENT OF THE ENVIRONMENT (1984a), **Town and Country Planning (Control of Advertisements) Regulations 1984**, Circular 11/84, London: HMSO

DEPARTMENT OF THE ENVIRONMENT (1984b), **Memorandum on Structure and Local Plans**, Circular 22/84, London: HMSO

DEPARTMENT OF THE ENVIRONMENT (1985a), **The Use of Conditions in Planning Permissions**, Circular 1/85, London: HMSO

DEPARTMENT OF THE ENVIRONMENT (1985b), **Development and Employment**, Circular 14/85, London: HMSO

DEPARTMENT OF THE ENVIRONMENT (1985c), **The Building Act 1984, The Building Regulations 1985, The Building (Approved Inspectors etc) Regulations 1985**, Circular 22/85, London: HMSO

DEPARTMENT OF THE ENVIRONMENT (1985d), **Local Government Act 1985, Town and Country Planning: Transitional Matters**, Circular 30/85, London: HMSO

DEPARTMENT OF THE ENVIRONMENT (1985e), **Aesthetic Control**, Circular 31/85, London: HMSO

DEPARTMENT OF THE ENVIRONMENT (1986), **The Future of Development Plans, a Consultation Paper**, London: DEPARTMENT OF THE ENVIRONMENT

DEPARTMENT OF THE ENVIRONMENT (1987a), **Award of Costs Incurred in Planning and Compulsory Purchase Proceedings**, Circular 2/87, London: HMSO

DEPARTMENT OF THE ENVIRONMENT (1987b), **Historic Buildings and Conservation Areas - Policy and Proceedings**, Circular 8/87, London: HMSO

DEPARTMENT OF THE ENVIRONMENT (1987c), **Change of Use of Buildings and Other Land: The Town and Country Planning (Use Classes) Order 1987**, Circular 13/87, London: HMSO

DEPARTMENT OF THE ENVIRONMENT (1987d), **Statistics of Planning Applications, April-September 1986**, Information Bulletin 335, 21 August 1987, London: Department of the Environment

DEPARTMENT OF THE ENVIRONMENT (1987e), **Structure and Local Plan Progress to 31 March 1986**, London: Department of the Environment (unpublished)

GREAT BRITAIN (1985), **Lifting the Burden**, Cmnd 9571, London: HMSO

GREAT BRITAIN (1986a), **Paying for Local Government**, Cmnd 9714, London: HMSO

GREAT BRITAIN (1986b), **Chief Planning Inspector's Report**, January 1985 to March 1986, London: Department of the Environment

HALL, P G (1982), **Urban and Regional Planning**, 2nd edition, Harmondsworth: Penguin Books

HEALEY, P (1983), **Local Plans in British Land Use Planning**, Oxford: Pergamon

HOUSE OF COMMONS (1986), **Fifth Report from the Environment Committee, Session 1985-86; Planning: Appeals, Call-in and Major Public Inquiries**, Volume 1, Report, HC 181-1, London: HMSO

MINISTRY OF HOUSING AND LOCAL GOVERNMENT (1965), **The Future of Development Plans, Report of the Planning Advisory Group**, London: HMSO

NUFFIELD FOUNDATION (1986), **Town and Country Planning**, London: Nuffield

SHOARD, M (1980), **The Theft of the Countryside**, London: Temple Smith

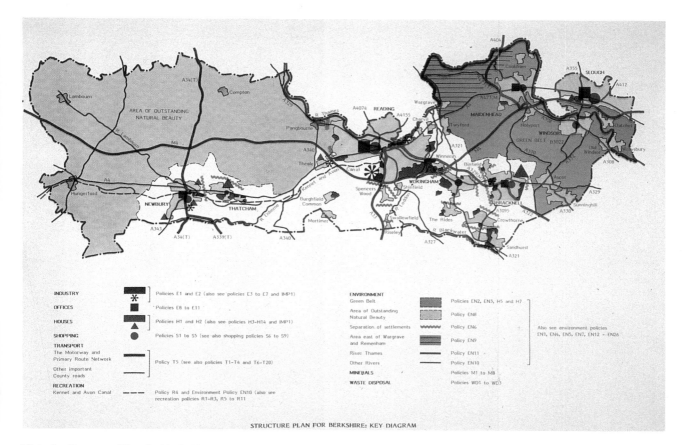

INDUSTRY    Policies E1 and E2 (also see policies E3 to E7 and IMP1)

OFFICES    Policies E8 to E11

HOUSES    Policies H1 and H2 (also see policies H3–H14 and IMP1)

SHOPPING    Policies S1 to S5 (see also shopping policies S6 to S9)

TRANSPORT
The Motorway and Primary Route Network    Policy T5 (see also policies T1–T4 and T6–T20)
Other important County roads

RECREATION
Kennet and Avon Canal    Policy R4 and Environment Policy EN10 (also see recreation policies R1–R3, R5 to R11)

ENVIRONMENT
Green Belt    Policies EN2, EN3, H5 and H7
Area of Outstanding Natural Beauty    Policy EN8
Separation of settlements    Policy EN6
Area east of Wargrave and Remenham    Policy EN9
River Thames    Policy EN11
Other Rivers    Policy EN10
MINERALS    Policies M1 to M8
WASTE DISPOSAL    Policies WD1 to WD3

Also see environment policies EN1, EN4, EN5, EN7, EN12 – EN26

STRUCTURE PLAN FOR BERKSHIRE: KEY DIAGRAM

## Plate 1    Structure Plan for Berkshire: Key Diagram

Plate 1 reproduces the key diagram from the draft replacement structure plan for Berkshire, revising the earlier structure plans approved by 1980. Every county council has to prepare a development plan, comprising a structure plan to provide a framework for the use and development of land in the county and any other local plans which may be necessary. The structure plan has to be approved by the Secretary of State. It is a written statement of policies illustrated by the key diagram. The chief features in this plan are the green belt in the east, part of the green belt surrounding London within which there is a strict limit on development; the area of outstanding natural beauty in the west, where development is also strictly limited; and the area around Reading and Bracknell where there is a very strong pressure for housing, offices and newer forms of high technology industry associated with the county's very good communications (rail, road and air) and attractive environment. County policy is to restrict growth as far as possible in order to preserve the character of the area.

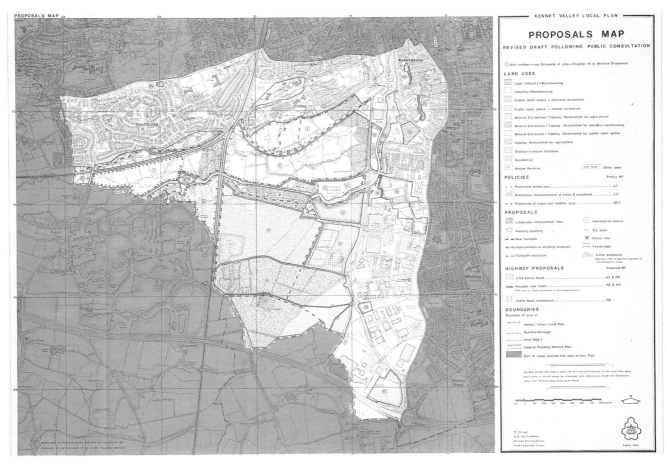

(a)

**Plate 2    Kennet Valley Local Plan: Proposals Map**

Plate 2 shows the statutory local plan for a part of Reading in the valley of the river Kennet, where there is a conflict between the release of land for housing and industry, mineral workings (gravel extraction) and recreational purposes. It overlaps with two other local plans prepared by the county council, for minerals and for countryside recreation. All of these plans have to conform to the Berkshire structure plan.

The main land uses, and the alignment of a new, relief road are shown in (a), much of it for reclamation for a variety of uses after mineral extraction. An inset map (b) shows in more detail the realignment of roads, giving access to the main road network for the town and to individual areas.

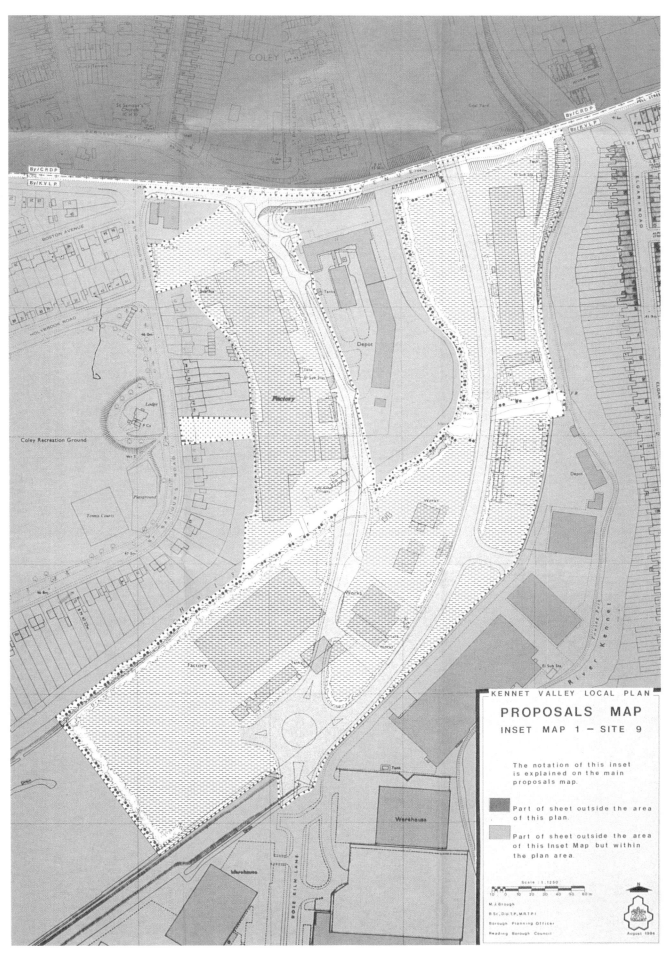

KENNET VALLEY LOCAL PLAN

# PROPOSALS MAP

INSET MAP 1 — SITE 9

The notation of this inset is explained on the main proposals map.

Part of sheet outside the area of this plan.

Part of sheet outside the area of this Inset Map but within the plan area.

Scale 1:1250

M.J.Brough

B.Sc.,Dip.T.P.,M.R.T.P.I

Borough Planning Officer

Reading Borough Council

August 1984

(b)

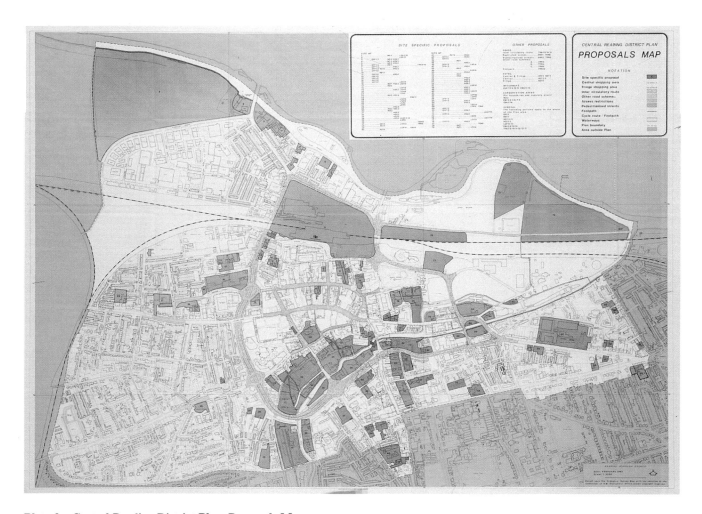

**Plate 3 Central Reading District Plan: Proposals Map**

Plate 3 illustrates the proposals map for the Central Reading District Plan. The plan is part of the development plan for Berkshire, being one of the statutory local plans which has been certified as being in accordance with the county council's structure plan. The district plan is prepared and adopted by the Reading borough council. Its chief function is to identify the main sites for redevelopment within the central area, chiefly for offices with some industry and shopping; and to strike a balance between the pressure for development; the traffic system including an inner distribution road; and the need for environmental conservation, with the boundaries of conservation areas shown on separate maps. The aim of policy in the rest of the area covered by the plan in general is to prevent large scale industrial or commercial development and to keep it in residential use.

# DENMARK

D Edwards

# D1 THE BACKGROUND TO PLANNING AND CONTROL

D1.1 Planning in Denmark embraces comprehensive physical planning and the sectoral planning of such aspects as housing, transport and education. In a report in 1977 on local democracy a working group under the Danish Ministry of the Interior stated that 'physical planning concerns the use and shaping of our physical environment, especially including the question of how to use the land and natural resources of the country' (Lemberg, 1982, p.239). Physical planning is under the jurisdiction of the National Agency for Physical Planning, within the Ministry of the Environment, while sectoral planning is the responsibility of a number of different agencies and ministries, depending on the subject. Physical planning includes national, regional and municipal planning. The legislation on building, housing, urban renewal and slum clearance that is under the jurisdiction of the Ministry of Housing is co-ordinated with the planning legislation.

D1.2 The legal basis for the present system was established in three stages.

(i) The Urban and Rural Zones Act of 1969 divides the entire country into urban zones, rural zones, and summer house districts. It provides the administrative authority for county councils to allow the sub-division of land and to determine the location of development in the countryside and in villages.

(ii) The National and Regional Planning Act of 1973 defines the national planning authority of the Minister of the Environment and the regulations governing the regional planning of the county councils and the metropolitan council of Greater Copenhagen.

(iii) The Municipal Planning Act of 1975 sets out the regulations governing planning at the municipal level including the preparation of local plans.

D1.3 The significant features of physical planning in Denmark are generally regarded as including the following aspects.

(i) The planning system has a high degree of decentralisation, so that to a great extent it is the responsibility of the municipal councils and the county councils to prepare and administer the plans.

(ii) Development in rural zones (outside the building zones allocated in approved plans) is highly regulated and requires special permission, as provided for in the Urban and Rural Zones Act. However, necessary building for agriculture, forestry and fishery is permitted.

(iii) The political component of planning, in terms of decisions on land use policies, is strongly emphasised. Decisions on development proposals generally are essentially administrative.

(iv) Public participation in the planning process is given a high degree of prominence and is assured through provisions for a public comment period in regional, municipal and local planning.

D1.4 Physical planning incorporates a framework management and control system known as *rammestyring* (see below), in which national planning provides the context for regional planning, which in turn constitutes the framework for municipal planning, which ultimately defines the parameters for the local plans.

| | |
|---|---|
| National planning | The State, the Government the Agencies |
| Regional plan | 11 county councils and the metropolitan council |
| Municipal plan | 275 municipal councils |
| Local plan | the citizens |
| the individual property | the landowners |

D1.5 National planning is managed by the Minister of the Environment on behalf of the Government. National planning is expressed first and foremost in binding directives (national planning directives) and in 'political' national plan reports that the Minister of the Environment submits every year to Parliament (*Folketing*). Thirteen national planning directives have been issued and a number are in preparation. They deal with, among other things, stopping the building of summer houses in areas near the coast, the placement of the main transmission network for natural gas, expansion of Copenhagen Airport, the location of plants for the treatment of power plant waste, the siting of antenna masts for television broadcasting etc. The annual national reports are partly a review of planning activity and a statement of likely future policy on particular issues.

D1.6  Regional plans are prepared by county councils and the metropolitan council and are approved by the Minister of the Environment. Regional plans have to contain the guidelines for development in the county for a 12 year time period. They are based on an evaluation of the technical and political questions related to the distribution of urban development, the designation and reservation of areas for specific purposes, and the resolution of conflicts between the interests of different sectors. In the period from 1979-1981 the first generation of regional plans was approved. During 1985 the county councils submitted to the Ministry the first regular supplements to the regional plans. The supplements to all the regional plans extend their time horizon by about four years and devote special attention to the planning of the countryside. All the supplements to the plans were expected to be approved by the Minister by the end of 1987.

D1.7  Municipal structure plans are prepared and ultimately adopted by the municipal councils, provided that the framework set out by the regional plan is respected. There is therefore no plan approval necessary from superior authorities. The plans have to determine the general structure for the municipality and contain a framework for the preparation of local plans. Nearly 100 municipalities have adopted a final municipal structure plan in accordance with the new legislation (see Plate 7). The remaining municipalities have presented proposals for municipal plans for public debate.

D1.8  Local plans are also prepared and adopted by the municipal councils. In the local plans more detailed regulations are established for individual properties. Larger or smaller areas can form the subject of local plans. The typical contents are provisions governing the use and development of properties included in the plan. The municipal councils have an obligation to prepare a local plan before larger projects, including the parcelling out of land, can be established. Since the Municipal Planning Act came into force in 1977 over 10,000 local plans have been prepared.

D1.9  Sectoral planning grew rapidly during the 1970's because the new legislation introduced at that time required specific sectoral plans. Three broad areas are covered by sectoral planning: land use and natural resources, infrastructure, and services. In terms of land use and natural resources the county councils, which have the major responsibility at local government level, are required to plan for

> (i)  landscape conservation
>
> (ii)  raw materials extraction
>
> (iii)  agricultural development
>
> (iv)  water catchment areas

(v)  water supply

(vi)  the siting of polluting enterprises.

D1.10  In relation to infrastructure and the provision of services the municipal councils and county councils must undertake the planning of such aspects as

> (i)  waste water
>
> (ii)  general waste
>
> (iii)  roads
>
> (iv)  energy supply
>
> (v)  urban renewal
>
> (vi)  schools
>
> (vii)  hospitals.

D1.11  In some areas there are already existing sectoral plans which have been adopted as a result of the recent legislation. For the sectors in the countryside (land and natural resources), the sectoral plans must be in accordance with the contents of the supplements to the regional plans that are now being approved by the Ministry of the Environment. When this process is completed the county councils will be able to finalise the sectoral plans for the countryside. Under the current legislation, most of these sectoral plans must be approved by the Minister of the Environment; the agricultural plans are however the responsibility of the Minister of Agriculture.

D1.12  The control of development is exercised through the simple one stage process of a building permit (byggetilladelse). The permit is granted if the proposal satisfies

(i)  the Building Act 1975 and the Building Regulations which have been subsequently promulgated

(ii)  the land use policies contained in the various plan documents but principally in the local plan which is a legally binding instrument, produced under the powers of the Municipal Planning Act.

The Building Regulations control the details of the built form in terms of size, siting, height, building materials etc, whilst the local plan is likely to be more concerned with the spatial distribution of functions. There is no concise definition of development, only regulations and requirements in relation to specific kinds of proposals in particular sets of circumstances.

## The Constitutional Context of Planning and Control

D1.13  Denmark has been since 1849 a constitutional monarchy with the legislative power vested jointly in the Queen and the Parliament, the executive power in the Queen, and the judicial power in the Courts of Justice.

The Queen is the Head of State but has no political power. The government is the actual holder of executive power with the 21 ministers making up the government, the administrative heads of the ministries. Since 1953, when the most recent revision to the Constitution was carried out, Parliament has consisted of only one assembly comprising 179 members, including two elected in the Faroe Islands and two in Greenland. The suffrage is universal and the voting age is 18. An election must be held every fourth year, but the Queen may at any time issue writs for another general election. There has been no majority government since 1909, and decisions reflect a bargaining process between the parties. The system of proportional representation that exists also produces a much greater advocacy of consensus politics than in Britain (Leslie, 1985). Legislation and policy tends therefore not to be reversed or changed dramatically (Skovsgaard, 1982). Much of Parliament's activity is concerned with discussing ministerial reports.

D1.14   According to the Constitution the judicial power is vested in the courts of justice. There are three levels

(i)   lower courts (*underretter*), with the country divided into just over 100 lower court districts

(ii)   two high courts (*landsretter*), one for Jutland and one for the rest of the country

(iii)   the Supreme Court (*højesteret*) located in Copenhagen.

Whereas the lower courts hear cases only in the first instance, the Supreme Court is a court of appeal only; the high courts are in some cases courts of first instance and in some courts of appeal. Generally a judgement given by a court can be the subject of only one appeal, but in exceptional cases involving principle, the Ministry of Justice may allow decisions which have started in a lower court to go to the Supreme Court (Christiansen, 1986). Administrative courts have not been introduced, although the Constitution expressly allows the establishment of such courts. The ordinary courts may, however, try all cases against the government and the administration, if there is not an express statutory provision to the contrary. There are also a number of commissions, boards, committees etc, such as the Environmental Appeal Board (see paragraph D6.12) which are important in relation to physical planning.

D1.15   There is a remarkable continuity in Danish law. It was never strongly influenced by any foreign system of law. Roman law was never accepted as the law of the land, although Roman concepts have filtered into Danish law, especially during the eighteenth century. Danish law cannot be classified as belonging to any of the well-known families of law. It is, however, closely related to the law of the other Scandinavian countries. The fact that these countries, as far as law is concerned, form a group of their own is due not only to the common historical and geographical background, but also to systematic endeavours to unify their law. These endeavours have taken place since the last decades of the nineteenth century. Committees with members from the different countries have drafted proposals for legislation within well defined fields, especially in family law and the law of contracts. Many of these proposals have later been enacted in all or at least in some of the countries (Knapp, 1972).

D1.16   Private ownership and freedom of contract may still be said to be the main principles of Danish private law. Section 73 of the Constitution states: 'The right of property shall be inviolable. No person shall be ordered to surrender his property, except where required by the public interest. It can be done only as provided by statute and against full compensation.' General restrictions, however, can be made without compensation and nowadays these are numerous.

D1.17   Danish law is pragmatic and empirical rather than theoretical and abstract. Legal decisions are usually based on case law, and not on statutes. An exception to this is the Land Registration Act 1926 (*Tinglysnings-loven*), which provides the basis for solving most of the problems in regard to real estate. This statute ensures that most rights to real estate (ownership, mortgages, etc) are registered by the courts, and contracts and executions can over-ride unregistered rights if they themselves are registered.

## Land Registration

D1.18   A notable feature of Denmark is the amount of information that is collected, and the range of records and statistics that is kept, most of which appear to be publicly accessible. A number of registers record property details but there are three which are most relevant.

(i)   Between 1790 and 1810 there was a land reform movement and in 1844 the old land register (*matrikel*) dating from 1688 was renewed with each plot of land being given a *matrikel* number, the only exceptions being some common land. The register has undergone some changes since but being used for tax purposes it is absolutely accurate. The register contains information about plot size, land use, whether for example there is a road, stream or protected wood on the plot, rateable value and other plots under the same ownership. Accompanying the *matrikel* are maps in scale from 1:800 to 1:4,000 showing the exact boundaries and size of each plot carrying a land register number. It is kept by the *matrikel direktorat*, an agency of the Ministry of Agriculture.

(ii)   There is also a Building and Housing Register which is a property register which was started in the 1970's. The register includes the *matrikel* number, ownership, plot ratio, number of floors of buildings, insulation and construction materials, and technical installations, but they are not always accurate because the municipalities which are responsible for them on behalf of the Ministry of Housing are not always informed of change.

(iii) The lower courts keep a register called the *tingbog* or local court book. This contains for each property information about ownership, mortages, loans outstanding, rights of way, *matrikel* number, local plan if present etc. It is information which relates to aspects protected by means of special public judicial registration and is the responsibility of the Ministry of Justice.

D2.1   There are three levels of government in Denmark: national, county and municipality. The country has a very long tradition of highly developed local self government. It clearly falls into the Scandinavian style of decentralised administrative decision-making, with legal effect, which is rather different from the practice in most European countries. Also, in the approach to decision-making, less importance is generally attached to sets of rules than elsewhere in Europe, with a correspondently greater emphasis on qualified judgement (Eilstrup, 1986).

D2.2   Local government is regarded by the Danes as a product, in one sense, of the law of necessity. A great number of common problems can be solved locally only if the central government is not to be overburdened with administrative work. This applies to fields such as social services and education, health and hospital services, local roads, water supply and sewage disposal. Local financing and independent local bodies relieve the central government of further administrative burdens. In a system of local self-government, central government's role may be reduced to that of supervision. In consequence of these practical needs and the ideas of representative government of the nineteenth century, the Danish Constitutional Act of 1849 explicitly laid down 'the right of local authorities to manage their own affairs independently, under State supervision'. Hence the responsibility for local affairs was gradually transferred from officers of the Crown to popularly elected representatives (Mathisen, 1983).

D2.3   In 1970 a major re-organisation of local government was carried out. Instead of the 1,388 small municipalities, most of which had a limited economic and administrative capacity, 275 new municipalities with a relatively large population averaging 15,000 and an improved administrative machinery were created. The aim was to ensure that by amalgamating an urban community with its surroundings, which in terms of population and housing, industry and trade could be regarded as an entity, the resultant municipality would be a viable local government unit. The number of counties was simultaneously reduced from 25 to 14, averaging 300,000 population (Svensson, 1981). Every municipality belongs to a county, except for the two ancient metropolitan municipalities of the capital, Copenhagen and Frederiksberg, each of which performs the dual function of county as well as municipality. Three of the counties make up the Greater Copenhagen Metropolitan Council (see Figure D1).

D2.4   The re-organisation of local government prepared the way for a changed distribution of responsibilities and obligations between the state, counties and municipalities. On the premise that there should be a merging of the responsibility for decisions and for the financial consequences of the decisions, a number of tasks were transferred from the state to county councils and municipal councils. The traditional detailed control of the administration of the municipalities was simultaneously reduced (Svensson, 1981). One basic principle of the local government reform of 1970 was the decentralisation of powers towards the lowest acceptable level of administration. Most tasks with a direct bearing on the daily life of the citizens have been made the responsibility of the municipalites.

## National Level

D2.5   The central government ministry responsible for physical planning has been since its establishment in October 1973 the Ministry of the Environment (*Miljøministeriet*). It took over planning functions from the Ministry of Housing and for the first time brought together a concern for physical planning with a wide range of responsibilities for the environment (Haywood, 1984). These comprise such aspects as nature conservation and the preservation of buildings, natural resources, environmental protection, re-cycling, foodstuffs, forest management and management of the areas under the jurisdiction of the Ministry of the Environment (approximately 4 per cent of the country). The Ministry consists of a main co-ordinating department and five agencies, namely the

National Forest and Nature Agency
National Agency for Environmental Protection
National Agency for Physical Planning
Geological Survey of Denmark
National Food Institute

Prior to January 1987 the National Forest and Nature Agency was divided into the National Forest Service and the National Agency for the Protection of Nature, Monuments and Sites. Whilst the responsibility for nature conservation has been combined with that for woodlands and forestry, the building preservation function has been taken over by the National Agency for Physical Planning.

**Figure D1    Denmark: Local Government Administrative Units**

D2.6    Under the National and Regional Planning Act 1973 the Minister for he Environment has responsibility for the carrying out of comprehensive physical national planning and for ensuring that the necessary investigations are undertaken in this connection (Section 2). To assist the Minister in this task, public authorities and similar organisations must provide information relating to the preparation or implementation of surveys, plans and major construction undertakings which might have significance for planning purposes (Section 3). The Minister may order the metropolitan council, the county councils and the municipal councils to provide informa-

tion for use in the carrying out of the national planning work.

D2.7    There is no national plan in the sense of a finite document. National planning is regarded more as a continuous process and the Minister has to submit an annual report (*Landsplanredegørelse*) on national planning activities to the Parliamentary Committee for Physical Planning. The other main responsibilities of, and instruments available to, the Ministry include

(i) the issuing of national planning directives (*lands-*

*plandirektiver)* (See Figure D9) and circulars
*(cirkulærer)*

(ii) a general obligation to ensure that the planning
system is operating efficiently and effectively and to
introduce any legislative changes that may be
necessary

(iii) the stimulation of national-local participation in
projects

(iv) the carrying out of investigations at a national level,
often using consultants

(v) giving advice and providing guidelines for counties
and municipalities

(vi) making decisions on certain appeal cases

(vii) approving the regional plans produced by the
county authorities. To ensure that regional plans
will not contain provisions which are contrary to
the national planning policies the Minister may
decide, after consultation with the ministers whose
field of responsibility will be especially affected, that
certain requirements shall be complied with, and
that specific conditions shall form the basis of,
regional planning (Section 4).

D2.8 Whilst physical planning is the responsibility of
the Minister of the Environment, as it deals with many
subjects which come under the responsibility of other
ministers, the facility exists for a special committee of
ministers to be established. But this has occurred only
very rarely. Normally, in response to the obligation
placed on the Minister to consult with his fellow
ministers, a committee of civil servants is set up, chaired
by an officer from the National Agency for Physical
Planning. Negotiations with ministers are carried out
only if necessary, for example, when a disagreement that
cannot be resolved has arisen in the committee.

D2.9 Within the Ministry it is the National Agency for
Physical Planning (*Planstyrelsen*) that is charged with the
task of physical planning, including the use of Denmark's
natural resources, and administering the Urban and
Rural Zones Act and the National and Regional Planning
Act. The Agency gives instructions and directions for the
planning work of county and municipal councils and also
advises other local and central authorities. It prepares
proposals for the national planning framework and may
lay down guidelines for the content of regional plans.

## County Level

D2.10 The reasoning behind the function of the
counties (*amter*) within the administrative system stems
from a widespread wish for decentralised solutions to
those tasks which are beyond the capability of the
municipalities. The main purpose for conferring tasks on
the counties is twofold.

(i) The counties have responsibilities which are

delegated to them for geographical reasons.
Typically, these are tasks which affect all municipalities within a county, such as the responsibility
for hospital services, or tasks that come up so rarely
that it would be pointless to equip the individual
municipalities with the necessary resources.

(ii) The counties are given tasks which are complicated
and technically demanding and therefore require an
administrative machinery which the county can
supply but which is not available in the average
municipality. This is the case with a number of
functions in the sphere of environmental protection
and conservation.

D2.11 The general rule is that the county is not superior
to the municipality, as counties and municipalities each
have their own responsibilities. This notion was adopted
in the reform of the legislation on physical planning which
was carried through in association with local government
re-organisation. Where previously a centralised
procedure of approval was required even for minor
modifications to the physical planning of a municipality,
what now applies is an all-encompassing guiding
principle: each administrative level looks after its own
interests and only its own, national interests in the case of
the State, regional in the case of the counties and local
interests in the case of the municipalities. This is intended
to give the lower administrative levels as free a hand as
possible, but whilst it is practical to follow this principle in
relation to most aspects, in certain situations the county
has the controlling influence in matters within the
individual municipality. For example, Parliament has
decided that control over nature protection cannot be
decentralized further than to the regional level (Eilstrup,
1986).

D2.12 The largest single field of responsibility for the
county authorities is the administration of hospitals, for
which the counties are practically solely responsible. In
education the counties are in principle responsible for the
16-19 age group, ie the upper secondary schools and
higher preparatory schools. The counties run a number of
round-the-clock services, including institutions for child
and youth welfare and for the handicapped. The
management of regional theatres, orchestras and
museums is carried out by the counties. More importantly
for planning is the counties' role in the building and
maintenance of major roads and the co-ordination and
operation of public transport. A task recently facing the
counties in the field of public transport has been to
establish regional transport authorities (Eilstrup, 1986).
Also they have significant functions in national resource
development and environmental protection. But it is in
regional planning that they are most relevant in the
overall planning system. Under Section 6 of the National
and Regional Planning Act each county council has to
produce a regional plan in accordance with the provisions
of the Act.

D2.13 As far as the countryside is concerned, it is the

counties which have the major responsibility for land use. The main guidelines, and thus the most essential considerations of interest in relation to each other, are to be incorporated in the regional plan. In relation to the utilisation of natural resources and environmental protection it is the responsibility of the counties to undertake the mapping work and surveying required to ensure that the necessary considerations are taken into account, usually in a situation where a number of often conflicting wishes and requirements have to be balanced together.

D2.14 The counties have responsibility for designating areas where the landscape and ecological features should be protected and which then become subject to special controls under the Conservation of Nature Act. The counties also have to map and survey the drift geology, and raw materials may not be exploited unless the county has granted its permission. The future of the area once extraction has stopped is usually identified and implemented through the conditions placed on the permission. Special water abstraction plans for water sources in the county are prepared and any taking of water requires the approval of the county unless it involves very small amounts for household use (Eilstrup, 1986).

D2.15 Anyone who has reached the age of 18 has the right to vote in county council elections and can be elected a member. County councils are directly elected for a four year period. They must have at least 13 and at most 31 members. The county mayor is elected by the county council from amongst its members and is chairman of the council and the paid full-time head of administration. Planning is the responsibility of the technical and environmental committee, one of the four standing committees, the others being the educational and cultural, the social security and health, and the hospital committees. As well as the standing committees there is also a finance committee.

D2.16 The counties established in 1913 a national body, the Association of County Councils (*Amtsrådsforeningen*). Both this and the Association of Local Authorities (*Kommunernes Landsforening*) at the municipal level are of particular importance in Denmark because, it is suggested, the single chamber of parliament does not give local government the same close contact with the legislature as a two chamber system (Eilstrup, 1986). The two associations form an important link in the collaboration between the tiers of administration. They are especially important in representing their members' interests in negotiations with central government and parliament, particularly when new laws and regulations are being proposed which have implications for local government. The associations collectively run a data processing centre and a training college with 10,000 people attending courses annually. Both associations are funded by their constituent members.

## Municipal Level

D2.17 The municipalities (*kommuner*) occupy a particularly important position in Danish society. The municipality is the public authority closest to the ordinary citizen. In most spheres it is the municipality which provides public services. It is also the municipality with which the individual citizen most frequently comes into contact, and it is here that the citizens have the greatest chance of making an impact on their community by engaging in local affairs and by exercising their votes in local government elections.

D2.18 It has been suggested that historically there have been four basic ideas underlying the system of local government and emphasising the municipal level (Mathisen, 1983).

(i) All local government builds on the concept of community. Traditionally everybody shares a responsibility for those with whom they live and work. Help to people in distress and the construction of local roads, for example, have always called for joint, supportive action.

(ii) Local government should be a system of self-government which entitles local authorities to manage their own affairs without, for example, the central government bothering about details. Since Denmark is not a federal state, however, certain limitations are imposed on local authority autonomy. Central and local politics must be coherent and integrated, but the system is based on the principle that within a broad, general framework the local authorities should be free to decide and control local policy.

(iii) Local government is a representative form of government. Every four years the citizens elect their representatives to the municipal council, which then manages local affairs until the next election, not without taking into account the interests of the population but with a fair amount of independence and freedom of action. Thus the municipal council has the authority to determine the level of local services and to select the sectors to be upgraded, while making allowance, if possible, for everybody's free selection of services. Among the powers vested in the councillors, those related to the levying of taxes and making of grants are of decisive importance. It is this right of the individual municipality to levy taxes which lends special strength to the Danish local authorities, although this now is within a framework laid down by central government.

(iv) Local government is organised as an all-purpose, unitary administration, where broadly speaking all local functions are administered by the municipal council. In no other Danish authority are vested such wide administrative powers as in the municipal councils. In principle they may deal with all domestic issues for which the public sector is responsible, from tax assessments to water and air purification. But the unitary administration is under constant pressure. From outside, grass-roots and local residents' movements are asserting themselves, and from within, different groupings of local authority employees are trying to exert influence. Increasingly, there are more people and groups to listen to and to consider when decisions are to be made, but that has not jeopardised the sovereign rights of the councils to make the final decisions.

D2.19 Since the local government reform of 1970 municipalities have gradually gained in importance, both in relation to the individual citizen and to the society as a whole. A successive transfer of functions away from the central government has concentrated the major part of local and regional functions in the municipalities and the new county authorities. New public functions are almost invariably performed by municipal and, to a lesser extent, county authorities. The municipalities increasingly provide the focus and co-ordinate the other parts of the public sector to ensure maximum utilisation of resources. The scale of the municipal operations is, however, likely to be very variable because of their range in size. In 1987 the population of municipalities had the following pattern

    37 over 30,000
    16 over 20-30,000
    82 over 10-20,000
    140 under 10,000.

D2.20 Whatever their size, however, all municipalities have the same range of functions. In the context of physical planning their main responsibilities are to

(i) prepare the municipal structure plan covering the authority's entire area

(ii) produce local plans where and when required

(iii) issue building permits

(iv) give permission in certain circumstances for proposals in the Rural and Summer House zones

(v) designate valued environments

(vi) provide the counties with advice on various environmental, natural resource, planning etc matters.

D2.21 As in the counties, anyone over the age of 18 can stand and vote for membership of the council. Each council must have a minimum of five and a maximum of 25 members, although Copenhagen City Council must have 55 members. Again as in counties, the members elect a mayor from their members who becomes the paid full-time chairman and chief executive. The technical and environmental committee is likely to be responsible for planning matters. Administration is usually organised on a departmental basis with, in the larger municipalities, a separate planning department concerned with plan making and a building administration department which

is mainly concerned with building control and so has responsibility for the issuing of building permits.

D2.22 In the municipality of Høje-Taastrup, with a population of 43,000 and also a higher than average amount of development activity, as an example, there is as part of the Technical Services Directorate a planning department comprising nine staff made up of

    a chief planner who is an engineer
    two architects
    one surveyor
    one building technician
    two technical assistants
    two secretaries.

This establishment would be regarded as small in terms of the municipality's population and development programme. The average number in a planning department is about five but the smallest municipalities may have only one person responsible for both planning and building aspects. In Høje-Taastrup the committee is made up of seven members. There is some delegation of decisions to officers, but whether, and the extent to which, delegation occurs is a local matter for each municipality to decide. The committee meets once a month and the public is not allowed to attend, but at full council meetings the public can not only be present, but can put questions at the beginning of the meeting, although no subsequent public participation is allowed. The building administration department in Høje-Taastrup is also part of the Technical Services Directorate.

D2.23 The municipalities, like the counties, possess a central organisation, the National Association of Local Authorities. Its purpose is to

(i) look after and represent the interests of the municipalities

(ii) promote co-operation between them

(iii) act as an intermediary between the municipalities and central government, eg in the consideration of bills and participation in government working groups

(iv) provide assistance for the municipalities in terms of advice, an information service, training programmes etc

(v) function as the employers' association of the municipalities' staff.

The Association has a staff of 350 people.

## Greater Copenhagen

D2.24 The local government system in the Copenhagen area differs from the rest of the country. The city has had a special position in the municipal administration of Denmark since 1660. The present system dates from 1973 and incorporates four levels of government

(i) national

(ii) regional, through the Greater Copenhagen Council which was established by the Copenhagen Metropolitan Council Act, 1973, although it was not until 1984 that the Council was granted permanent status

(iii) county, including Copenhagen, Frederisksborg and Roskilde

(iv) municipal, numbering 50 municipalities.

Under the Regional Planning Act for the Metropolitan Region it is the responsibility of the metropolitan council to contribute by means of regional planning towards a development of the metropolitan region that will be favourable to the well-being of the population and considers the interests of the national economy (Section 1). For the purposes of the Act the metropolitan region includes Copenhagen and Frederiksberg, which have a special status in that they are both counties and municipalities, and the counties of Copenhagen, Frederiksborg and Roskilde with their constituent municipalities. This area makes up 7 per cent of the country and comprises a third of the national population.

D2.25 The Greater Copenhagen Council comprises 37 members who are indirectly elected from those members of the local political parties who have already been elected to the constituent municipal and county councils. They thus have a dual loyalty and the strategic objectives and policies of the Council may be undermined because of local allegiances (Leslie, 1985). The Council was established on a temporary basis in 1974 primarily to advance the state of strategic planning in the metropolitan region at a time of economic growth. It was also intended to overcome the difficulties presented by the proliferation of joint committees and similar groups. It was thus required to undertake certain strategic functions, which needed a greater co-ordination than was possible within the region's existing local government structure. Its responsibilities now are for

(i) overall regional planning

(ii) co-ordination, development and the operation of the public transport system

(iii) protection of the environment

(iv) planning the water supply system

(v) overall hospital planning.

Ninety three per cent of the Council's expenditure is on public transport which employs 97 per cent of the Council's workforce. Forty five per cent of the cost of public transport is subsidised. The Council obtains some of its revenue from the public transport fares and the rest from indirect taxation and government subsidies. In 1984 the Danish parliament approved legislation which not only re-affirmed the Council's functions but also gave it a permanent status as a local government body.

D2.26 The municipality of Copenhagen is governed by a City Council and an Executive which is the administrative authority. The City Council has 55 members, elected for a four year term by citizens over the age of 18. The Executive comprises the Lord Mayor and six mayors elected by the City Council from its own members. They too serve a four year term. The Executive is elected under a system of proportional representation which reflects each party's strength on the City Council. No member of the Executive can sit on the City Council and mayors automatically resign from the City Council. Each mayor administers his own Executive Department and has full responsibility for all its affairs. Planning is in four Departments and also is under the responsibility of the Lord Mayor's Department. The system of a city council and an executive on the Copenhagen model also exists in Aarhus, Odense and Aalborg, the three other large towns in Denmark.

## Government Finances

D2.27 As with the rest of Scandinavia, Denmark is among those countries in the world where the total tax burden is heaviest. Total taxes and duties make up about half of the aggregate national income. Over half of the total revenue from tax and duty is income tax. In contrast to the situation in most other western nations the statutory contribution to social security schemes does not constitute a major source of revenue. Of the total taxes and duties, the state levies a little less than 70 per cent, the counties 8 per cent and the municipalities 22 per cent. Only the state levies indirect taxes - value added tax, other general dues and excise duties – in addition to which the state levies income tax. The counties and municipalities levy income and real property tax but the latter, compared to England, is a relatively minor source of income and both local government authorities are able to charge fees for some of their services. As a result of the full-scale delegation in recent years of tasks from the state, the counties and municipalities account for something approaching two-thirds of all public expenditure.

D2.28 In Denmark it is traditional for the local authorities themselves to establish the standard of services and independently to impose taxes to finance expenditure. In many important sectors, there are no hard and fast rules laid down, therefore, as to how tasks are to be solved. This is the case, for example, with the hospital sector, where the counties themselves determine the level of activity. There is also a time-honoured principle that all operational and planning outlay is financed from the total income. The allocation of priority to expenditure is thus not restricted by provisions that some incomes may be used only for certain purposes. The counties' and municipalities' revenue in the form of income taxes, land taxes, grants from the state, raising of loans and so on must therefore be used to finance all expenditure. Their freedom to raise loans is however quite restricted and generally is possible only in relation to long term capital investment and not for current expenditure. Even though the counties and municipalities cover about two-thirds of the total public expenditure they levy only something like one-third of the total taxes and duties. The state therefore transfers tax resources to the counties and municipalities.

D2.29 Two different kinds of transfer exist today: reimbursements or refunds, and block grants. Before local government re-organisation, by far the greater proportion of central government transfers to the local authorities were made in the form of full or partial refunding of certain types of expenditure. In order to achieve a more satisfactory ordering of priorities between the individual areas of expenditure, these reimbursements were to a large extent converted into block grants in the 1970s and 1980s. Those reimbursements remaining, such as pension benefits and expenditure on social security, concern exclusively the municipalities and there is likewise a move towards phasing these out. The block grants from the state are calculated annually. Allocation is based on the principle of equal distribution in relation to the tax base of the individual counties and municipalities. In April 1985 the government proposed a bill to abolish all government grants to the counties and municipalities. According to this proposal, counties and municipalities would in due course have to finance their entire expenditure out of their own income. It is suggested that this would strengthen the relationship between decision-making and economic responsibility. Besides the state grant, inter-municipal equalisation arrangements based on the economic capacity of the individual local authority are made by law. According to these settlements, parts of the differences in the counties' and municipalities' tax base and spending needs are equalised. Eighty per cent of the counties' tax base and spending needs are equalised in this way (Eilstrup, 1986).

D2.30 The right to levy taxes has made it easy for the legislature to impose new duties on the counties and municipalities. It has also made it financially possible for the municipalities, in particular, to undertake more functions, but there have been problems. The economic foundations of the individual local authorities differ widely; there are central government limits imposed on the magnitude of local taxes and there are, particularly in times of crisis, limits to public expenditure as a whole (Mathisen, 1983).

## Comments and Conclusions

D2.31 The local government reform of 1970 simplified and created a more effective basis for administration than had previously existed. It also brought about a decentralisation of responsibilities with the municipalities being given a number of additional tasks, but it has been commented, not without some truth, that 'you cannot decentralise without centralising - trouble is decentralised and the state strengthens its control' (Allpass, personal communication). Furthermore, through the creation of larger municipalities decision-making was perceived to be removed from the local level and in these larger local authorities there is some pressure for arrangements to be made to help overcome this situation. The effectiveness of the decentralised approach is very largely dictated by central government policy in relation to the allocation of finance, and at the present time local government spending is being severely restricted by cut-backs at state level. In order to attempt to reduce public expenditure there have recently been proposals to privatise some local government services, such as the fire service.

D2.32 It appears that generally the Ministry of the Environment, at least at civil servant level, has a good and productive relationship with the counties and districts. There is no obvious 'us and them' attitude because of the presence of a regular dialogue through the Ministry's organising of conferences and meetings and being generally accessible. They do not see themselves as having a policing duty. The counties probably suffered in the decentralisation process and this erosion of power is still continuing, for example in relation to the operation of the Urban and Rural Zones legislation. The increasing interest, however, in rural and environmental aspects seems to be focusing on the counties. The metropolitan authority is flourishing but one major problem in its organisation is that its members are indirectly elected and so there is often a conflict of loyalties between their municipality and the metropolitan council. The municipalities themselves vary greatly in their capacity to carry out their obligations but in planning matters it is common practice for consultants to be employed.

# D3   THE DEFINITION OF DEVELOPMENT

D3.1   There is no comprehensive general definition of development which provides the basis for exercising planning control. In its place are regulations, plans and legislation which precisely specify what may or may not be done in particular circumstances. Thus, the need for planning permission in the English sense is not required, only the need to satisfy building regulations and plans, because the prescribed function, and so land use in an area, are established through the plans and policies that are required to satisfy the legislative framework. The building regulations are clear and detailed and relate to the entire country, whereas the delimitation in a plan of what is an acceptable function in a particular location or on a specified site may be broad or narrow, depending on the degree of flexibility that is regarded as desirable.

D3.2.   Definitions of development, or perhaps more accurately of what building and uses are permitted and not permitted in specified circumstances, are found in the

(i)   Building Act 1975

(ii)   Building Regulations (*Bygningsreglement*), 1982

(iii)   Building Regulations for Small Buildings (*Bygningsreglement for Småhuse*), 1985

(iv)   legislation related to physical planning, especially the Municipal Planning Act 1975 and the Urban and Rural Zones Act 1969

(v)   local plans (*lokalplans*) produced under the Municipal Planning Act 1975

(vi)   other, mainly environmental, legislation.

## Building Act

D3.3   Five purposes of the Building Act are listed in the legislation, and not one is related to what might be regarded as planning objectives. The Act and the associated Building Regulations are concerned with the control of construction and the achievement of building and open space standards. They are both under the overall authority of the Ministry of Housing, and not the Ministry of the Environment. The purposes of the Act are

(i)   to ensure that a building is planned and executed in such a manner that it complies with adequate standards of fire safety, general safety and health

(ii)   to ensure that a building and any undeveloped areas belonging to the property are of an adequate

standard with due reference to the intended use and are adequately maintained

(iii)   to encourage measures designed to increase productivity in the building industry

(iv)   to encourage such measures as can minimise unnecessary consumption of energy in buildings

(v)   to encourage such measures as can minimise unnecessary consumption of resources in buildings (Section 1).

Although no mention is made of the wider planning objectives it is through this Act and the related Building Regulations which affect the granting of a building permit (*byggetilladelse*) that planning control is generally operated (See Figure D2).

D3.4   In the Act 'building' means buildings and walls and other permanent structures and installations provided that the application of the provisions of the Act to such structures etc, is justified by such considerations as the legislation seeks to accommodate. The Act applies to such moveable structures as are intended for use as buildings, provided such use is not of a purely temporary nature. This does not, however, include moveable structures erected on camp sites and permitted in pursuance of legislation pertaining to holiday homes and camping. The Act may also apply to existing building notwithstanding the date of erection of the building, provided this is explicitly stated in individual provisions (Section 2).

D3.5   The objectives of the legislation are mainly achieved through the building regulations for which the Act provides the necessary legal sanction. The Act itself does not establish precise requirements except in relation to the siting of buildings where it is stated that they must not be erected closer to a boundary with another site or path than 2.5 metres (or 5.0 metres in areas reserved exclusively for construction of holiday homes). Notwithstanding this provision, semi-detached houses can be erected on such a boundary. The building regulations can stipulate rules governing the siting of minor building components, garages, sheds, outhouses and other small buildings closer than stated above and governing the location of balconies, roof terraces, attics, gable windows, etc, at a greater distance than that stated above. Where an area is mainly occupied by unbroken rows of buildings the municipality can stipulate that any new building has to be erected on the road boundary or building line and

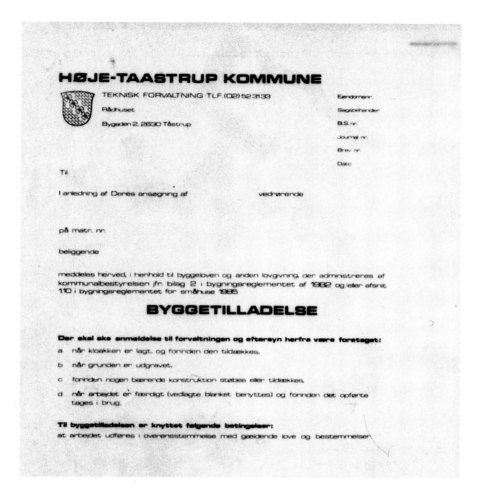

**Figure D2  A Building Permit**

Figure D2 is a building permit (*byggetilladelse*) issued by the municipal council. It is usually subject to conditions but it is common, as in this particular case, that there is only one general condition imposed requiring that the work is carried out according to the acts and regulations currently in force. It is also usual for the building permit to require the applicant to report to the municipality when various stages of the work have been reached to allow inspections to take place. Four stages have been identified in this building permit: when the sewerage system is in place but before it is covered, when the foundations have been excavated, prior to the construction of the building itself, and when the work is finished before it is occupied.

extended to the neighbouring boundary on each side (Section 8). Those provisions do not apply where a particular circumstance is subject to the provisions of a local plan.

D3.6   In foundation work, excavations, alterations of site level or other site alterations, irrespective of whether the work is otherwise subject to the provisions of the Act, all necessary measures must be taken to secure adjoining sites, buildings and pipes and cables of every description (Section 12).

D3.7   In built-up areas the natural contours and nature of a site cannot be modified by removal or addition of filling or by any means which can cause a nuisance to the adjoining sites. This requirement applies irrespective of whether the provisions of the Act are otherwise applicable to the work in question. The building regulations can stipulate detailed provisions relating to such modifications in the contours and nature of a site. These provisions

again do not apply where a particular circumstance is subject to the provisions of a local plan (Section 13).

D3.8   A building, the undeveloped areas belonging to the property, and any installations that have been erected must be maintained in a good state of repair in order that they present no danger to the tenants of the property or other persons and are not defective in any other manner. Moreover, with due reference to its location, the property must be maintained in a decent state. These provisions apply notwith-standing the date of construction of the property. In built-up areas the provisions also apply apply to undeveloped sites (Section 14).

## Building Regulations

D3.9   Under Section 5 of the Building Act the Minister of Housing has to draw up building regulations which relate to

(i) the erection of a building and to the extension of a building

(ii) structural and other alterations to a building which are substantial in relation to the provisions of the Act or to such regulations as may be promulgated in pursuance of the Act

(iii) changes of use of a building which are substantial in relation to the provisions of the Act or to such regulations as may be promulgated in pursuance of the Act

(iv) demolition of a building.

D3.10 The building regulations can stipulate rules governing execution of work and installation of fittings with regard to

(i) maintenance of safety, fire prevention and health standards

(ii) installation of technical fittings, including provisions permitting the subsequent installation of sanitary and technical fittings

(iii) adequate planning of residential buildings with due regard to their intended use

(iv) installation of fittings in residential buildings and buildings to which the public has access, in such a manner that the building can be used by persons under the handicap of reduced powers of mobility and orientation

(v) the use of standard dimensions to encourage the employment of standardised building components, service components and fittings

(vi) measures to minimise the unnecessary consumption of energy in both existing and new building work

(vii) calculation of building heights and areas and determination of levels for the building and the adjacent land

(viii) measures to minimise the unnecessary consumption of raw materials in both existing and new building work, including provisions concerning use of specified materials or structures and concerning re-use or recycling of materials (Section 6).

D3.11 The Building Act states that the building regulations can govern the size, planning and equipping of undeveloped areas belonging to the property to secure adequate outdoor living space for children and adults, entry and access facilities for the fire service, and parking facilities (Section 7). Rules issued in accordance with these aspects do not apply in cases where circumstances are covered by the provisions of a local plan.

D3.12 The building regulations can establish the relationship between the height of the building and the distance from a road, neighbouring boundary and other building on the same site in order to secure adequate

building distances and adequate daylight (Section 9). As previously, these rules do not apply in cases where circumstances are covered by the provisions of a local plan.

D3.13 The building regulations can control the location of projecting parts of the building, signs, display cases, etc beyond the road or building line. Rules can moreover be stipulated concerning the siting of garages and upward or downward sloping driveways to such garages in relation to the road line or building line. When a building is erected behind a building line fixed in order to secure construction of a roadway, buildings on the adjacent sites shall not without the approval of the municipality utilise such a set-back by installing doors, windows, signs, advertising devices etc. The building regulations can also include provisions relating to fences on boundaries shared with roadways. In built-up areas this can also apply to undeveloped land (Section 10).

D3.14 The 1982 Building Regulations in fact include provisions under the following headings

(i) building percentage, which is the percentage ratio of total floor space to the area of the site

(ii) unoccupied part of the site including parking and access

(iii) height of building and distance from boundaries

(iv) interior planning of buildings

(v) structural requirements

(vi) fire precautions, which is an extremely detailed section

(vii) resistance to moisture

(viii) thermal insulation

(ix) resistance to the transmission of sound

(x) fire-producing appliances and chimneys

(xi) ventilation

(xii) services.

The Danish Building Regulations for Small Buildings cover the same aspects but in different degrees of detail.

D3.15 Some buildings, unless otherwise stipulated in the building regulations, can be erected in accordance with such standards as the municipality may impose in each individual case with due regard to the aims of the Building Act. They are

(i) buildings and large premises for commercial enterprises, shops, offices, industrial enterprises, workshops and storage

(ii) churches, theatres, hotels, hospitals, prisons, barracks, buildings used for education, exhibition or entertainment and other buildings and large premises used as places of assembly (Section 11).

## Building Regulations for Small Buildings

D3.16 The National Building Agency (*Byggestyrelsen*) within the Ministry of Housing, as well as producing comprehensive Building Regulations has published the Building Regulations for Small Buildings. It came into force on 1 April 1985 and supersedes those provisions of the Building Regulations 1982 relating to small buildings. In addition to stating the statutory requirements it has a guide containing helpful sketches and comments to assist in interpretation. In addition the Agency makes freely available a number of pamphlets about different aspects of the Building Regulations for the public in general. The Building Regulations for Small Buildings apply to

(i) houses containing one dwelling for permanent occupation in the form of either detached single family houses or single family houses built partly or completely together with other such houses (semi-detached houses, terraced houses, chain houses, cluster houses, etc.)

(ii) summer houses in recreational areas, allotment-garden houses and camping huts

(iii) garages, carports, outhouses, greenhouses and similar small buildings and structures built in conjunction with detached or semi-detached single family houses or summer houses.

D3.17 The regulations cover the following building work

(i) the construction of new houses

(ii) extensions to existing houses

(iii) conversions and other alterations which are major in relation to the provisions of the Building Act or the provisions of these regulations

(iv) changes in use of a building which are major in relation to the provisions of the Building Act or the provisions of these regulations

(v) demolition of detached and semi-detached single-family houses.

## Municipal Planning Act

D3.18 When the Ministry of the Environment was established in 1973 the division of responsibilities was based on the planning function being transferred to the new ministry and building control being retained by the Ministry of Housing. In this division the building regulations included in the Building Act that were most closely related to the achievement of planning objectives were incorporated in the planning legislation. Thus, in Part 6 of the Municipal Planning Act there are included sections related to such aspects as plot size, space around buildings, plot ratio, building heights, external appearance, and mix of uses. The relevant regulations are discussed in detail in paragraphs D4.23-29.

## Urban and Rural Zones Act

D3.19 Through the provisions of the Urban and Rural Zones Act the whole of Denmark is divided into three zones : urban, summer house and rural (Section 2). In each of the zones the legislation specifies what building and type of function is acceptable and what requires permission, but in all three the Building Regulations generally apply.

D3.20 In rural zones permission has to be obtained from the county council outside the metropolitan region and within the region by the Metropolitan Council for the following activities.

(i) To make a parcelling out or a division of land unless the area parcelled out is added to an existing farm (Section 6).

(ii) To construct new buildings unless the buildings are necessary for the cultivation of the property concerned as farm or forestry land or for the carrying on of fishing. The same rule applies to the conversion of and construction of extensions to existing buildings. Permission is required, however, as regards the location and design of dwelling houses used by farmers, foresters or fishermen unless the buildings are built in connection with existing building areas (Section 7).

(iii) To take existing buildings and areas not built on into use in rural zones for purposes other than agriculture, forestry or fishery. Exploitation of raw materials in the ground can, however, take place without such permission but this is likely to be affected by other legislation. If it is proposed to use buildings which are no longer necessary for farming for industrial or commercial purposes, a permission has to be given unless the intended activity is of such a character or such an extent that it will present a flagrant contradiction of the interests which the Act aims to protect (Section 8). Recently there has been a further relaxation whereby it is now possible to use without permission redundant farm buildings for light industrial purposes as long as no more than five people are employed.

The permission that is being sought in each of the categories is for what, in effect, is zone dispensation (*zonedispensation*) because the Urban and Rural Zones Act has brought about a general regional zoning of the entire country.

D3.21 In areas covered by an adopted local plan the permissions are given by the municipal council. However, the Minister for the Environment may give directions requiring applications for permission to be referred to him in specific cases in which the matter at stake is of more than local importance or interest. The Minister may further decide to call in for his decision an application submitted to the county council or the Metropolitan Council if he deems the matter at stake to be of more than local importance or interest. The Minister is also able to

**Figure D3    A Local Plan in the Municipality of Høje-Taastrup**

Figure D3 is taken from the local plan and represents the illustrative material that is included. The whole document is 14 pages long and is unusually detailed, reflecting the control that the municipality seeks to exert over this major development project. Numerous local plans have been prepared covering the project area and all are very precise and specific in their requirements.

(a) General Location: is taken from the cover of the local plan and shows the general location of the area to which the plan relates. The total area is 7,000 square metres and has been allocated in the municipal structure plan for residential purposes.

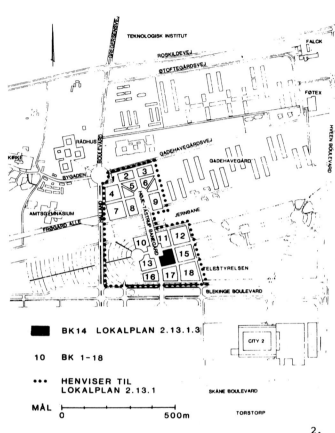

(b) Site Plan: shows the site covered by the local plan within the context of the wider area allocated for development.

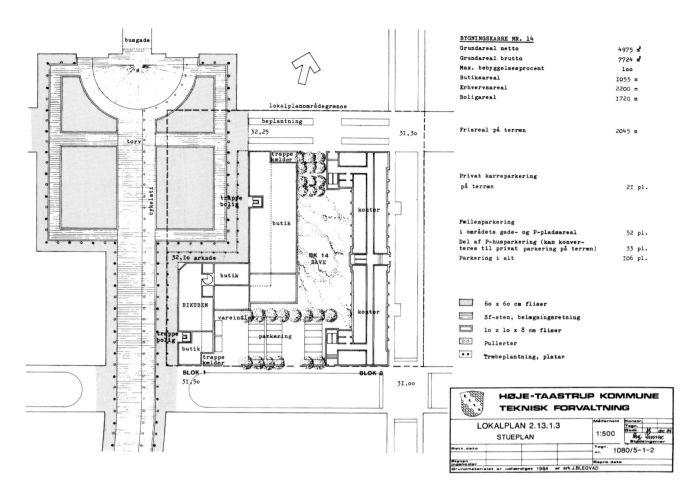

BYGNINGSKARRE NR. 14

| | |
|---|---|
| Grundareal netto | 4975 m² |
| Grundareal brutto | 7724 m² |
| Max. bebyggelsesprocent | 100 |
| Butiksareal | 1055 m |
| Erhvervsareal | 2200 m |
| Boligareal | 1720 m |
| Friareal på terræn | 2045 m |
| Privat karreparkering | |
| på terræn | 21 pl. |
| Fællesparkering | |
| i områdets gade- og P-pladsareal | 52 pl. |
| Del af P-husparkering (kan konver-<br>teres til privat parkering på terræn) | 33 pl. |
| Parkering i alt | 106 pl. |

60 x 60 cm fliser

Sf-sten, belægningsretning

10 x 10 x 8 cm fliser

Pullerter

Træbeplantning, platan

**HØJE-TAASTRUP KOMMUNE**
**TEKNISK FORVALTNING**

LOKALPLAN 2.13.1.3
STUEPLAN 1:500

Tegn. nr. 1080/5-1-2

Grundmaterialet er udfærdiget 1984 af ark.J.BLEGVAD

---

(c) Relationship to Adjacent Local Plans: shows the boundaries of the local plan and its relationship to the adjacent local plan areas. Each plan covers a residential block within which a particular architectural approach will be adopted.

Jernbane

BK 12
BK 11
BK 10
BK 13
BK 14
BK 15
BK 16
BK 17
BK 18

Blekinge Boulevard

•••••• Lokalplanens områdegrænse

Del af Høje-Tåstrup by, Høje-Tåstrup

**HØJE-TAASTRUP KOMMUNE**
**TEKNISK FORVALTNING**

LOKALPLAN 2.13.1.3
MATRIKELPLAN 1:2000

1080/5-1-1

Matrikelbetegnelserne er
ájour pr.25.10.1984

(d) Site Layout: provides the detailed site layout indicating the disposition of buildings, access, footpaths, parking and landscaping.

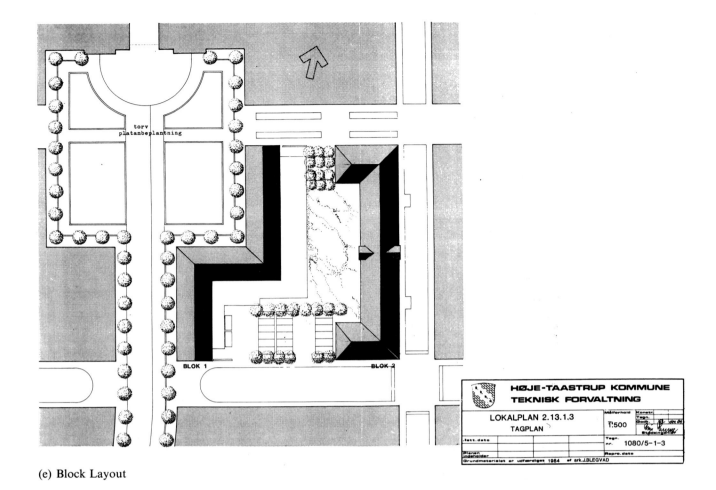

(e) Block Layout

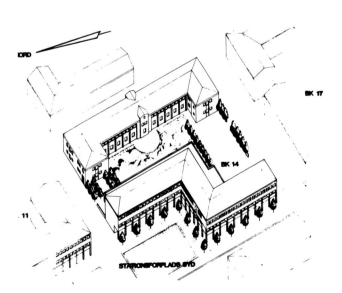

(f) Sketch Perspective

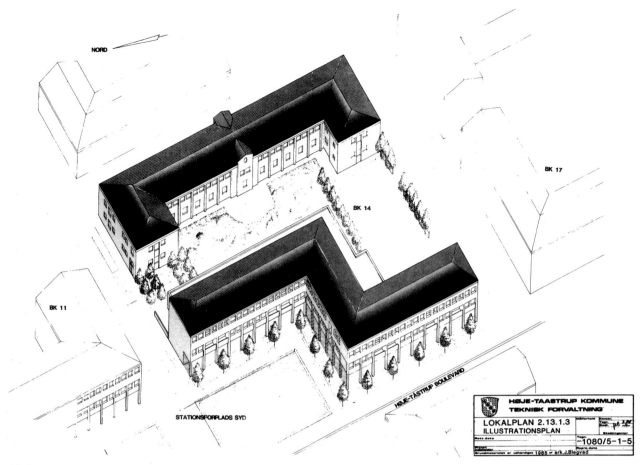

(g) Detailed Perspective

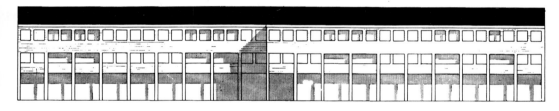

BLOK 1 FACADE MOD VEST

BLOK 2 FACADE MOD VEST

(h) Elevational Treatment.

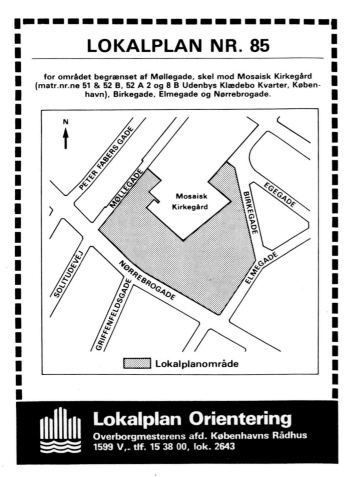

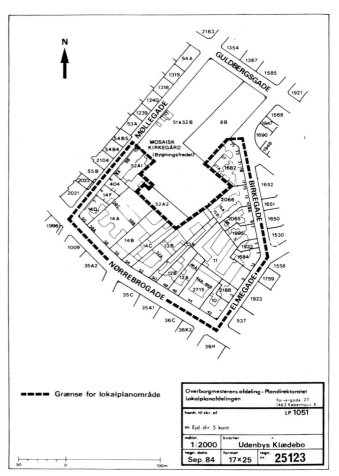

(a) Site Plan: shows the site that is covered by the local plan.

(b) Proposals: shows the boundaries of the area to which the proposals relate and the properties that will be affected.

**Figure D4    A Local Plan in the City of Copenhagen**

This Figure illustrates a very general local plan of the kind that is becoming increasingly common, especially in Copenhagen. The plan consists of the one map and just over one page of text.

The area currently consists of old tenement blocks and the proposals are to continue to use the area for residential purposes but also ancillary uses. The very general nature of the proposals clearly enables a wide range of developments to be legally acceptable and a high degree of discretion to be exercised by the council when building permission is being sought.

lay down rules to the effect that the permissions shall be given by the municipal council concerned when there is an approved regional plan and a finally adopted municipal structure plan for the municipality concerned, with general guidelines for the use of the areas in the rural zone (Section 9). Recently, it has been agreed that lines may be drawn around villages and within the areas defined, and in the villages themselves, the municipalities become responsible for making decisions on applications.

D3.22    A permission lapses if it has not been used within three years after the date when it was given (Section 9). Conditions attached to the permission are binding on owners and other holders of rights in the property irrespective of when such rights were created (Section 10).

## Local Plans

D3.23    Under Section 16 of the Municipal Planning Act a local plan has to be produced before the sub-division of

larger areas or before major building and investment works, including demolition of buildings, can be carried out (see Figures D3 and D4). Permission under the Building Act cannot normally be given until the local plan has been finally adopted. If there is a proposal to undertake major building or demolition work and it is in accordance with the municipal structure plan, the municipal council has to prepare as soon as possible a proposal for a local plan and then promote its implementation as expeditiously as possible. A building permit can usually be issued at this stage when the municipality has made the decision to prepare a local plan.

D3.24    Local plans can contain provisions about a wide range of subjects such as existing buildings, parks, gardens and planting, construction of new buildings, urban renewal and preservation. They deal with physical, legal and administrative matters, such as the relationship to other planning aspects and the establishment of landowners' associations. A local plan can provide that certain conditions shall be complied with, such as connection to communal facilities, establishment of a landscape belt etc before new buildings can be taken into use (Svensson, 1981).

D3.25 When a local plan has been adopted, and an announcement to this effect has to be made by the municipality, and owners of properties within the area sent a copy, there must not legally or actually be established conditions contrary to the provisions of the local plan. The local plan has to be entered in the land register of the local court (*the tingbog*) for the properties covered by the plan.

D3.26 The duty to provide a local plan can be considered one of the corner-stones of the planning system, as the duty ensures that major building and investment works are brought within the regulation of the planning system and within the rules about public participation which are so important in Denmark. The actual decision of the council as to whether there is a duty to provide a local plan in connection with specific applications for building permits often gives rise to debate in the local community. The decision of the council may be brought before the National Agency for Physical Planning for final decision. Since the coming into force of the Municipal Planning Act, hundreds of appeals as to the duty to provide a local plan have been decided. Such decisions have meant that local plans must be provided for the building of holiday hotels, schools and town halls (Christiansen, 1986).

D3.27 A local plan regulates only future transactions. The plan is binding on everybody in the sense that future transactions, factual as well as legal, must be in accordance with the plan. A local plan cannot in itself affect existing conditions. Thus, as Christiansen has pointed out, if a local plan reserves an area for the building of detached one-family houses, this is the only kind of development that will be permitted. But the owners cannot be ordered, simply by reference to the local plan, to build detached one-family houses; and if there is, for example, a lawful industrial building in the area at the time when the local plan is finally adopted, the owner of the building can continue to use the building for industrial

purposes. Consequently, the local plan does not give the council power to order the landowners to bring their properties into line with the plan (Christiansen, 1986, p.92). However, the local plan does enable the council to acquire by compulsory purchase any property that is of material importance for its implementation.

## Other Legislation

D3.28 The Building Act provides the statutory basis for the Building Regulations and so controls new construction. The planning legislation and the resulting local plans determine the future function and land use of particular areas. These are therefore fundamental to a definition, and the control, of development. However, other legislation, such as that relating to urban renewal, conservation of nature, preservation of buildings, identifies proposed changes which also become the subject of control (see section D4).

## Comments and Conclusions

D3.29 Development is a term not found in Danish and it implies a level of generality which is out of tune with the country's desire for precision and specificity. Thus where there is need for control to be exercised to achieve physical or more general sectoral planning objectives, those elements which have to be controlled are precisely defined in the Building Regulations and the statutes, together where necessary with the standards that are to be enforced. The key word seems to be precision although in reality there is greater opportunity for departures from the norms through the political process than might initially be apparent. Thus, as will be discussed later, local plans can be expressed in a general way allowing a high degree of flexibility as to what can be allowed, whilst there also are opportunities for dispensations and exemptions to be granted from the Building Regulations. The system therefore is one which contains apparent rigidity and precision but which at the operational level offers major opportunities for political influences to be brought to bear.

D4.1   It has been estimated that there are in Denmark about 50 planning systems which have an influence on land use and in consequence there is a large number of different legal means of control (Christiansen, 1986). Emphasis in this section will be placed on the controls that can be exercised in relation to physical planning, although reference will be made to some of the other more relevant controls under associated legislation.

## Building Permits

D4.2   Control of building and demolition work and change of use is exercised through the instrument of building permits. There is not a separate application for planning permission. In rural zones, however, there may also be the need to apply for zone dispensation if the proposal is not one that is an acceptable rural development (see paragraph D3.20). Some areas and some types of development proposals are subject to additional controls where other permissions may be required. Conformity with planning policies as expressed in municipal structure plans and local plans is checked as part of the process of deciding whether or not a building permit should be granted (see Figure D5).

D4.3   The Building Act and the associated Building Regulations 1982, the Building Regulations for Small Buildings 1985 and the Municipal Planning Act specify the circumstances where building permits are required and the procedures involved. In general terms, before developments falling into the categories listed in paragraph D3.9 (as specified in Section 2 of the Building Act) can be carried out, an application for a building permit has to be made, indicating clearly the work which it is intended to execute. Applications must be in writing and in duplicate, but usually the municipality which is responsible for making the decisions has its own application forms (see Figure D6). Most municipalities have a Technical Services Directorate within which a Building Department is responsible for building permits. They may take from two to eight weeks to be issued but the average time is two to four weeks. Building permits lapse if work is not commenced within a period of 12 months from the date of issue. They can be granted on a temporary basis or for a specified period (Section 26). It is rare for building permits not to be granted: less than 1 per cent of applications are in fact refused.

D4.4   Notice normally has to be given to the municipality before commencement of work for which a building

permit has been granted, whilst the permit can also include provisions to the effect that notice has to be given to the municipality at various stages of the work (see Figure D7). When the work has been completed, the municipality must be informed and the work must not be brought into use without the consent of the municipality. A permit to use the building (*ibrugtagningstilladelse*) often has to be issued (Section 16). However, as will become apparent, there are variations on this conventional process.

D4.5   The municipality can determine that a fee be charged for the issue of permits and that except for certain standard and minimum amounts fees are payable either as a fixed basic amount per square metre of floor area or per cubic metre of the building's volume or as a proportion of the building costs. The municipality chooses one of these as its standard method but it can also decide that in certain, specified types of building projects a fee will not be charged or will be in accordance with another method. The fee is payable when the building permit is about to be issued and the municipality can withold the permit until such time as the fee is paid.

D4.6   As well as the variable fees for building permits covered by the general Building Regulations 1982, a standard fee can be charged under the Regulations for Small Buildings for building permits for the construction of detached single family houses, semi-detached houses, terraced houses, chain houses, cluster houses etc. On 1 January 1985 the fee per housing unit was adjusted to 567 kroner (10.5 kroner = £1). A standard fee can be charged also for building permits granted for the construction of summer homes and this was fixed with effect from 1 January 1985 at 283 kroner per house. In the case of reports, the fee is payable at the time of submission of the report (see paragraph D4.15).

D4.7   The Building Regulations can stipulate that the municipal authority shall not grant permits until such time as it has ascertained that the work for which application has been made can be approved in accordance with other legislation administered by the municipality. It can moreover be stipulated that the municipality can postpone consideration of an application for such a permit until such time as the requirements of other legislation not administered by the municipality have been fulfilled.

D4.8   The general procedure relating to building permits can therefore be summarised as follows

(1) Material required before the case is treated:
Power of attorney
Application form
Drawings
Area of the ground
Built up area
Gross storeyed area
Building Percentage
Application to Environmental Ministry
Application to Food Quality Control
Application to Work Superintendent
Application for exemption.

(2) During the treatment of the case:
Local Plan
The Urban plan convention
Declaration (on building)
Other Declarations/Enforcements
B.B.R. register
Drainage proposals
Water proposals
Heating proposals
Supporting constructions
Fire treatment
Road proposals
Authorisation
Sewage treatment

(3) Organisations to be consulted – before the final decision is taken:
The committees on Planning and Environment
Fire Commission
The county council – in this case the metropolitan council
The Preservation Board
Work Superintendent Committee
Police
The Civil Defence Board
Food Quality Control Board
Department of Urban Planning
Road Department
Infrastructure Department
'Neighbour hearing'
Housing Commission
Environmental Section

**Figure D5    Checklist of Considerations in Deciding Building Permit Applications**

Figure D5 provides the checklist for the processing of an application in the municipality of Høje-Taastrup. The first section is to ensure that the application has been completed satisfactorily and is acceptable for determination. The second section identifies the checks to be carried out by the municipal Directorate of Technical Services and the third section those that are the responsibility of other organisations. In all three sections not all the items listed will be required for every application.

TEKNISK FORVALTNING
RÅDHUSET . BYGADEN 2
2630 TAASTRUP
(02) 52 31 33

| BS | / |

# Ansøgning om byggetilladelse

i Høje-Taastrup kommune.

**Denne ansøgning + bilag i 3 eksemplarer indsendes til Høje-Taastrup kommunes tekniske forvaltning**

Denne side skal udfyldes af ansøgeren.

| Matr. nr. | Ejerlav | | Sogn | | Gade/vej nr. | | | |
|---|---|---|---|---|---|---|---|---|
| Bygherrens navn | | | | | Stilling | | | |
| Adresse | | | | | Telefon nr. | | | |
| Ansøgerens navn | | | | | Stilling | | | |
| Adresse | | | | | Telefon nr. | | | |

| Grundens areal | Heraf vej | Indtil nu bebygg. areal | Indtil nu etageareal | Heraf beboelsesar. | og erhvervsareal | og garage/udhus |
|---|---|---|---|---|---|---|
| m² | m² | m² | m² | m² | m² | m² |
| Byggeriets art | | Project. bebyg. areal | Project. etageareal | Heraf beboelsesar. | og erhvervsareal | og garage/udhus |
| | | m² | m² | m² | m² | m² |
| Afstand fra skel mod nord | mod syd | | mod øst | | mod vest | |

| Kælderetage | Udnyttet til | Tag-etage | Udnyttet til | Antal butikker |
|---|---|---|---|---|
| m² | | | m² | |
| Antal etager | Antal WC | Antal bade | Bygningen opvarmes ved | Højde til tagfladens skæring med facade |
| | | | | m |

| Antal lejligheder | heraf | | med | og | | med | og | | med | og | | med | og | | med |
|---|---|---|---|---|---|---|---|---|---|---|---|---|---|---|---|
| i alt | vær. | | kam. | | vær. | kam. | | vær. | kam. | | vær. | kam. | | vær. | kam. |

De følgende 2 rubrikker udfyldes **kun** ved bebyggelse i værksteds- og industriområder.

| Eksisterende bebyggelses rumfang | Projekteret bebyggelses rumfang |
|---|---|
| m³ | m³ |

Runfang beregnes af hele den del af bygningen, der er over færdigt terræn, incl. tag-etage med evt. kvist, fremspring, skorsten m. v. udvendig målt.

**Fuldmagt.**
Undertegnede, der ifølge tingbogen er ejer af ovennævnte ejendom, bemyndiger herved nævnte ansøger
til på mine/vore vegne at ansøge om tilladelse til ovenstående byggeri.

_____
Ejers underskrift.

| Bilag | Stk. | | |
|---|---|---|---|
| Bygningstegning(er) | | | |
| Beskrivelse(r) | | | |
| Statiske beregninger | | | |
| Konstruktionstegninger | | | |
| | | | |
| | | | |
| I alt | | | |

_____ , den _____

_____
Underskrift/ansøger

HEDEHUSENE BOGTRYKKEN

---

**Figure D6  Application Form for Building Permission**

Figure D6 is an example for the municipality of Høje-Taastrup of an application form for building permission. Three copies, together with drawings showing the design and construction details and descriptive material, have to be sent to the Director of Technical Services. Detailed questions are asked about any existing buildings and the site itself, as well as about the proposed development. The applicant need not be the owner of the property but the application has always to be signed by the owner.

HØJE-TAASTRUP KOMMUNE

TEKNISK FORVALTNING
RÅDHUSET BYGADEN 2
2630 TAASTRUP
(02) 52 31 33

BS /

# Anmeldelse af byggearbejder

i Høje-Taastrup kommune

**Denne ansøgning + bilag i 3 eksemplarer indsendes til Høje-Taastrup kommunes tekniske forvaltning**

**Denne side skal udfyldes af ansøgeren.**

| | | | |
|---|---|---|---|
| Matr. nr. | Ejerlav | Sogn | Gade/vej nr. |
| Bygherrens navn | | | Stilling |
| Adresse | | | Telefon nr. |
| Ansøgerens navn | | | Stilling |
| Adresse | | | Telefon nr. |
| Grundens areal $m^2$ | Indtil nu bebygget areal $m^2$ | | Heraf garage/udhus $m^2$ |
| Byggeriets art | Projekteret bebygget areal $m^2$ | | Bygningen opvarmes ved |
| Afstand fra skel mod nord | mod syd | mod øst | mod vest |

**Fuldmagt.**
Undertegnede, der ifølge tingbogen er ejer af ovennævnte ejendom, bemyndiger herved nævnte ansøger til på mine/vore vegne at ansøge om tilladelse til ovenstående byggeri.

Ejers underskrift

| Bilag | Stk. | | |
|---|---|---|---|
| Bygningstegning(er) | | | |
| Beskrivelse(r) | | | |
| | | | |
| | | | |
| | | | |
| | | | |
| I alt | | | |

, den

Underskrift/ansøger

HEDEHUSENE BOGTRYKKERI

**Figure D7 Announcement Form for Proposed Building Work**

Figure D7 is the form that has to be completed in relation to those development proposals that according to the legislation and regulations require only to be reported to the municipality. These proposals relate to garages, greenhouses etc. and some demolition work. The questions and drawings required are less detailed than in an application for a building permit but again the announcement of the proposed work has to be signed by both the applicant and owner, if a different person.

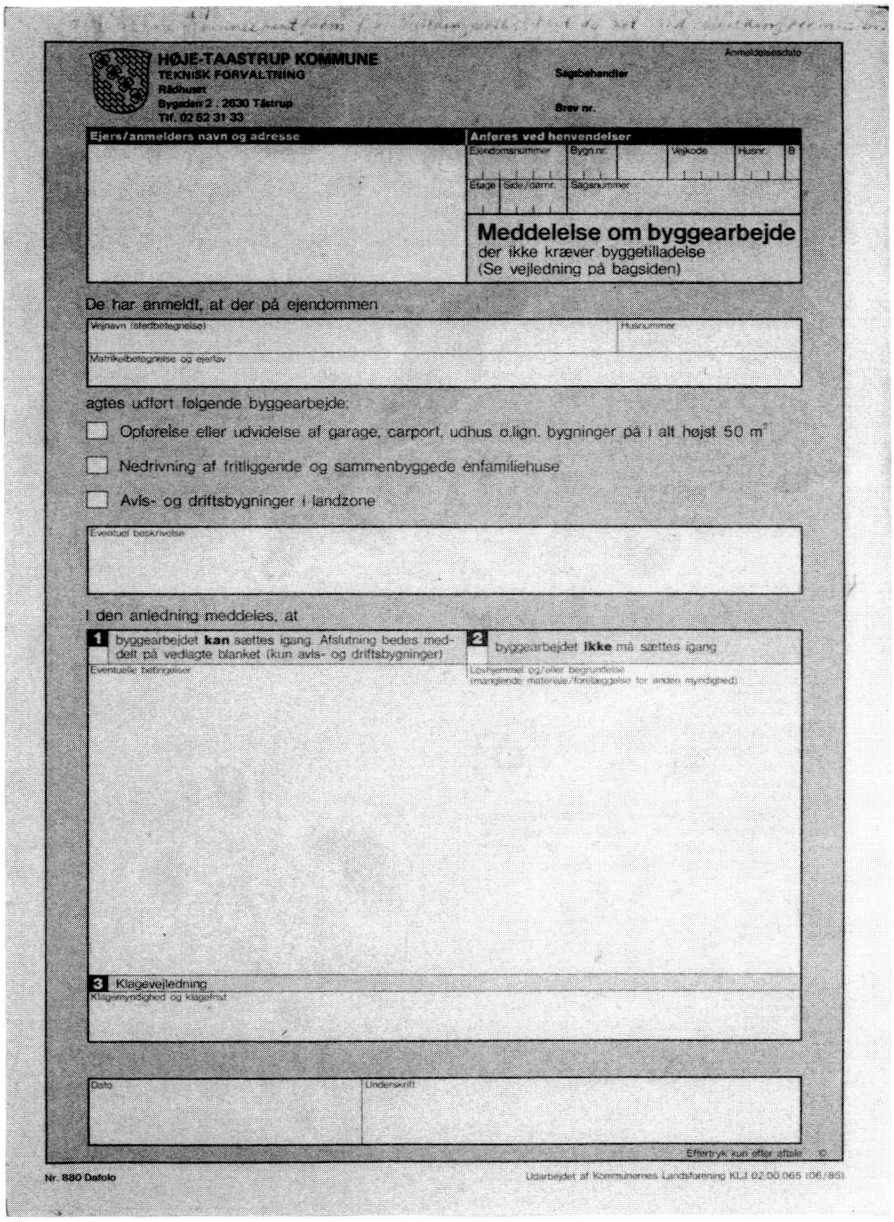

HØJE-TAASTRUP KOMMUNE
TEKNISK FORVALTNING
Rådhuset
Bygaden 2 . 2630 Tåstrup
Tlf. 02 52 31 33

Anmeldelsesdato

Sagsbehandler

Brev nr.

Ejers/anmelders navn og adresse

Anføres ved henvendelser

Ejendomsnummer | Bygn.nr. | Vejkode | Husnr. | B

Etage | Side/dørnr. | Sagsnummer

**Meddelelse om byggearbejde**
der ikke kræver byggetilladelse
(Se vejledning på bagsiden)

De har anmeldt, at der på ejendommen

Vejnavn (stedbetegnelse) | Husnummer

Matrikelbetegnelse og ejerlav

agtes udført følgende byggearbejde:

☐ Opførelse eller udvidelse af garage, carport, udhus o.lign. bygninger på i alt højst 50 m²

☐ Nedrivning af fritliggende og sammenbyggede enfamiliehuse

☐ Avls- og driftsbygninger i landzone

Eventuel beskrivelse

I den anledning meddeles, at

1 byggearbejdet **kan** sættes igang. Afslutning bedes meddelt på vedlagte blanket (kun avls- og driftsbygninger)

Eventuelle betingelser

2 byggearbejdet **ikke** må sættes igang

Lovhjemmel og/eller begrundelse
(manglende materiale/forelæggelse for anden myndighed)

3 Klagevejledning
Klagemyndighed og klagefrist

Dato | Underskrift

Eftertryk kun efter aftale ©

Nr. 880 Dafolo

Udarbejdet af Kommunernes Landsforening KL.J 02.00.065 (06/85)

**Figure D8 Announcement Form for Building Works that do not need Building Permission**

Figure D8 is the form used by the municipality of Høje-Taastrup for those building and demolition works that do not require permission. It is therefore primarily intended to ensure that records of any change in the environment are kept up to date and registered.

(i) application for permit for proposed work

(ii) notice given before commencement of work

(iii) notice given at various stages of the work

(iv) report completion of work

(v) application for permit to use.

All these procedures are, however, likely to be relevant only to the more complex proposals, and the need for the municipality to be informed when various stages of the work have been completed in particular is not a common occurrence. Section 3 of the Building Act indicates that the Building Regulations can stipulate that the legislation and such regulations as are promulgated under the legislation do not apply, or only partly apply to buildings of a specified type. It is also stated under Section 16 that the Building Regulations can incorporate provisions relating to relaxation of the requirements for building permits including provisions wholly or partly exempting certain categories of buildings from the requirements. Thus, some proposals may necessitate only one of the above stages being followed, whilst some minor works may be undertaken without any of the building permit procedures being required. Furthermore, in certain cases it may be possible for dispensations and exemptions from the legislation and Building Regulations to be applied for and obtained.

D4.9 The various departures from the conventional model identified above can be listed in terms of the following circumstances

(i) proposals where the Building Act and Building Regulations do not apply

(ii) work exempt from building permit procedures (see Figure D8)

(iii) exemptions from some requirements of the Building Regulations.

(iv) proposals where only the requirement of reporting to the municipality is present. The terms 'announcing' and 'giving notice' are also sometimes used in English translation (see Figure D7)

(v) work needing a permit and also the completion of the work to be reported

(vi) exemptions and dispensations.

Each of the above circumstances is now considered.

*Proposals where the Building Act and Building Regulations do not Apply*

D4.10 The Act and Building Regulations do not apply to

(i) road and railway bridges, road and pedestrian tunnels and electricity pylons

(ii) minor structures for electricity, gas, water, telephone and public transport systems

(iii) certain types of roof top aerials

(iv) minor buildings etc which fall into the following categories.

(a) A maximum of two small buildings, each occupying no more than 10 square metres and erected in connection with residential buildings but not used for habitation or commercial purposes. Small buildings of this type can be executed flush with the party boundary or closer than 2.5 metres to the party boundary, provided that the distance from other buildings (including small buildings) on the same site is not less than 2.5 metres. The height must not exceed 2.4 metres. Buildings executed flush with the party boundary or closer than 2.5 metres to the party boundary must not, however, be of a height exceeding an inclined height plane extending from 1.8 metres on the boundary to a height of 2.4 metres at a distance of 2.5 metres from the boundary. In areas reserved for summer houses these buildings must not be sited closer than 2.5 metres to the party boundary.

(b) Party walls on or adjacent to the party boundary, provided such a party wall is not higher than 1.8 metres.

(c) Camping chalets erected on campsites for which permission has been granted under the Summer Cottages, and Camping etc Act.

This means that the above buildings and structures can be executed, used and dismantled without the necessity of a permit under the Act or the Building Regulations. No special dispensations and exemptions are however available to particular categories of applicant such as government ministries or statutory undertakers.

*Work Exempt from Building Permit Procedures*
D4.11 In the Building Regulations for Small Buildings certain categories of building work are identified which can be executed without the necessity of a building permit and a permit to use the building. They are

(i) conversions of and other alterations to the building which do not increase its area

(ii) building work in connection with small buildings not exceeding 10 square metres in area and which satisfy specified height and distance standards

(iii) building work in connection with open swimming pools, garden fireplaces, boundary walls, terraces, and roof aerials which do not extend more than 5.5 metres above the roof

(iv) building work in connection with camping huts and allotment-garden houses. Similarly, no report has to be submitted to the municipality on completion of such building work. Notwithstanding the fact that the above building work can be executed without a building permit and without a duty to report the work, it has to comply with the provisions of the regulations insofar as they apply to the work in question. If they are unable to comply, an application for exemption from the relevant provisions must be submitted to the municipality and building work shall not commence until such exemption is granted. These general exemptions from the need for a building permit and permission to use the building may be expressly withdrawn through the stipulations of a local plan.

D4.12 In the notes for guidance in the Regulations it is stated that exemption from the need for a building permit also applies to such installations as fireplaces, woodburning stoves, boilers, etc. In the event that conversion work, etc results in an extension to the area of the building, a building permit has still to be obtained. Examples of such work would be all extensions, conservatories, garden rooms, winter rooms and the inclusion of all or part of an unused roof space in the dwelling area.

D4.13 In the guidance notes reference also is made to the categories of demolition work that need not be reported. They are

(i) summer homes in areas designated for summer cottage development

(ii) garages, carports, outhouses, greenhouses and covered terraces whose area does not exceed 50 square metres in conjunction with single family houses

(iii) small buildings not exceeding 10 square metres

(iv) allotment garden houses

However, each of these categories may have to be reported if stipulated in a local plan (see paragraph D4.31).

*Exemptions from Some Requirements of the Building Regulations*
D4.14 Under the provisions of Sections 3 and 16 of the Building Act the Building Regulations state that some works are subject only to a limited number of requirements. These works are

(i) outdoor swimming pools installed in connection with detached, single family houses

(ii) free-standing outdoor fireplaces

(iii) allotment-garden houses

(iv) minor heating and fuel storage facilities and grain drying plant

(v) agricultural buildings used for breeding etc which have to conform only to

(a) height of buildings and distance from boundary stipulations in Section 43 of the Municipal Planning

Act, ie that a building must not be built in more than two storeys and no part of the outer walls or the roof of a building must rise more than 8.5 metres above the surrounding ground

(b) some aspects of the design and execution of structural work

(c) fire precautions

(d) soil and waste drains

(vi) areas designated for summer cottages. Summer cottage development means buildings which may be used for habitation only during the period April 1 to September 30 and, outside this period, only during short holiday periods, weekends, etc.

*Proposals that Require only to be Reported*
D4.15 The Building Regulations for Small Buildings state that the following categories of building work do not require a permit and only have to be reported to the municipality

(i) construction of garages, carports, outhouses, greenhouses, covered terraces and similar buildings whose area does not exceed 50 square metres in conjunction with single family houses and summer cottages

(ii) extensions to garages, carports, outhouses, greenhouses, covered terraces and similar buildings, provided that their area on completion of the work does not exceed 50 square metres

(iii) demolition of detached and semi-detached single-family houses.

In the absence of any response from the municipality within two weeks from the date on which the report was received, building work may be commenced. Completion of the work need not be reported to the municipality.

D4.16 This relaxation of control may be over-ruled by the contents of a local plan, as may other provisions in the Building Regulations. The municipality is also obliged to investigate whether the building work described in the application infringes the provisions of the following legislation

Urban and Rural Zones Act
Municipal Planning Act
Environmental Protection Act
Conservation of Nature Act
Preservation of Buildings Act
Forestry Act
Sand Drift Prevention Act
Public Highways Act
Private Roads Act

In the event of any such infringement the municipality must respond to the person submitting the report within two weeks of the date of receipt of the report.

D4.17 The building permit procedures in rural zones in respect of agricultural buildings for crops, livestock etc on agricultural and forestry properties have also been relaxed. Instead of an application for a building permit, the municipality shall be furnished, as above, with written notification of the work it is intended to execute. Provided that the municipality has not raised objections within a period of four weeks after submission of the notification, work may be commenced. Where building work requires dispensation from the Building Regulations, explicit application to this effect has to be made in the notification, and work shall not commence until dispensation has been granted. The finished building work can be used without a permit to use.

*Work Requiring a Permit and Report of Completion*
D4.18 The Regulations for Small Buildings identify the categories of building work which require a permit. They are

(i) construction of detached single family houses, semi-detached houses, terraced houses, chain houses, cluster houses etc and summer cottages

(ii) extensions to detached single family houses, semi-detached houses, terraced houses, chain houses, cluster houses, etc, and summer cottages

(iii) conversions of and other alterations to detached single family houses, semi-detached houses, terraced houses, chain houses, cluster houses etc and summer cottages which increase their area

(iv) a change in the use of a summer cottage from temporary dwelling to permanent dwelling

(v) construction of garages, carports, outhouses, greenhouses, covered terraces and similar buildings whose area exceeds 50 square metres

(vi) extensions to garages, carports, outhouses, greenhouses, covered terraces and similar buildings, whose area on completion of the work exceeds 50 square metres.

Whilst completion of the work has to be reported to the municipality, the building can be used without a permit to use the building.

D4.19 The municipality must not however grant a building permit until it has ascertained that the building work does not infringe the provisions of the legislation listed in paragraph D4.16 and also the following

Summer Cottages and Camping, etc Act
Water Supply Act
Urban Renewal and Improvement of Housing Act
Housing Regulation Act
Heating Supply Act

In the event that the building permit stipulates requirements from other legislation special attention should be drawn to this fact.

*Exemptions and Dispensations*
D4.20 The Building Act states that dispensation from the provisions of the Act and regulations issued in pursuance of it can be granted when deemed compatible with the considerations on which the said provisions and regulations are based. The following statements are included in Section 22.

(i) Dispensation from regulations governing the interests of neighbours in terms of distances of buildings from boundaries and general siting requirements can be granted but not sooner than two weeks from the date on which the municipal authority has informed neighbours of the circumstances from which dispensation is requested to allow for their comments.

(ii) Dispensation lapses if a permit for the building work to start to which dispensation applies is not issued within a period of two years. If a building project for which dispensation has been granted is not implemented, dispensation lapses together with municipal authority permission to commence the work.

(iii) Dispensation is granted by the municipality. The Minister of Housing can lay down rules governing the exercise of the right of dispensation by the municipality and can provide that dispensation from specific regulations may be granted only by the county authority or in the local government areas of Copenhagen and Frederiksberg by the Minister of Housing.

(iv) The Minister of Housing can lay down rules governing permission for the municipal authority to make a relaxation from the regulations concerning provision of parking space on the property owner's property conditional upon payment being made to a municipal parking fund.

D4.21 In the Building Regulations it is stated that applications for dispensation, including an application to retain an unlawful structure, etc., are to be made to, and granted by, the municipality unless otherwise stipulated in the respective parts and sections of the Regulations. Dispensation for work in which the municipal authority is the building owner must, however, be granted by the county council. Dispensation granted under the Building Act lapses unless building permission is granted within two years for the building work to which the dispensation relates. Where a building project for which dispensation has been granted is not executed, dispensation lapses simultaneously with permission to commence building.

D4.22 The guidance notes in the Building Regulations for Small Buildings state that the municipal authority can grant exemption from the material provisions of the Building Act and the Building Regulations but that no exemption can be granted from the formal provisions, such as those determining when a project should be subject to a building permit, when neighbouring owners should be informed prior to the granting of exemption, appeal provisions, etc. The notes further state that the municipality can grant exemption when it considers that exemption is compatible with the considerations behind the provision from which exemption is requested. Further reference to exemptions and dispensations appears in the Municipal Planning Act.

## Municipal Planning Act and Local Plans

*Municipal Planning Act*
D4.23    The control of development through the Building Regulations based on the Building Act is further reinforced and amplified by provisions in the Municipal Planning Act. Under Section 40, unless otherwise stipulated in a local plan, no sub-division of land may be carried out which would create sites with an area smaller than 700 square metres, not including road areas which must not be built on as a consequence of corner splays or the imposition of building lines to ensure new roads or widening of existing roads. In the case of sub-division of land in summer cottage districts the area must not be smaller than 1,200 square metres. In the case of the sub-division of land for building purposes the municipal council must approve the frontage length of the site. Within urban zones or summer cottage districts the municipal council may oppose a sub-division of land which in the opinion of the municipal council will prevent a suitable building development or a suitable use of the area in question. Within urban zones the municipal council may also oppose a sub-division of land, which according to the timetable of the municipal council or until such a timetable has been adopted, in the opinion of the municipal council would be contrary to a proper and suitable urban development at the time in question. The positioning of new buildings on site must provide for open areas which can be laid out to ensure satisfactory leisure areas for children and grown-ups, access areas, the possibility for the fire brigade to fight fire, and parking areas (Section 41).

D4.24    Under Section 42 where a site is to be built on, including conversion or additional building, the building percentage of the site, unless otherwise stated in a local plan, must not exceed

(i)   25 per cent for detached houses for permanent residential purposes

(ii)  10 per cent for summer cottages and other buildings for holiday and leisure time purposes

(iii) 50 per cent for multi-storeyed buildings in an area allocated for this purpose in a municipal structure plan

(iv)  40 per cent for other buildings. However, the Minister for the Environment may lay down detailed rules as to the extent and the conditions on which the building percentage may be increased for sites with a special situation, always provided that the building percentage must not exceed 50 per cent.

D4.25    A building generally must not be built in more than two storeys, and no part of the outer walls or the roof of a building must rise more than 8.5 metres above the surrounding ground. Adjacent to a party boundary or path the height of the building must not exceed 1.4 times the distance from the boundary. Adjacent to a road the height must not exceed 0.4 times the distance from the opposite side of the road. In relation to the distance from a boundary, single-family houses must not be sited closer than 2.5 metres. The Minister for the Environment may lay down other restrictions on the height of buildings in summer cottage districts. Currently they must be single storey with a maximum height of 5 metres and they must be sited at least 5 metres from a party boundary or path. The Minister for the Environment may also decide that special kinds of buildings, structures or plants shall be exempted from these provisions (Section 43). Again, as with plot size and building percentage, a local plan can change the regulations on height of buildings.

D4.26    Aesthetic aspects are covered by the Municipal Planning Act. Under Sections 44 and 45 the municipal council may attach conditions to a permission under the Building Act that the building is given such an external form that in relation to its surroundings a satisfactory overall effect is attained. Advertisements, light installations and similar parts of a building must not cause inconvenience or have a disfiguring effect on their surroundings. The municipal council may by an order or a prohibition ensure the observance of this provision. If commenced building operations are given up or postponed for a longer period and the partially constructed building has a disfiguring effect in relation to the surroundings, the municipal council may order the owner of the property to remove the building or to bring it in such a condition that it will no longer have a disfiguring effect. The same rules apply if a building which is partially demolished or ruined remains for a longer period in such a condition that it has an adverse effect on its surroundings. These requirements do not apply to farm buildings. If the use of a site for storage, refuse disposal or similar purposes has a disfiguring effect the municipal council may order the owner of the site to undertake fencing, planting or other measures which may relieve the condition.

D4.27    These provisions apply irrespective of when the building was constructed, the advertisements etc placed or the site taken into use for storage and similar purposes. However, orders under this section of the Act may be appealed against to the Minister for the Environment by the owner concerned. An appeal has to be brought within four weeks from the date when the order was received (Section 44).

D4.28    Under Section 45, if in an area there are buildings for residential purposes the municipal council may forbid that industrial or commercial buildings detrimental to the environment are built in the area or that existing residential buildings or unbuilt sites in an area are taken into use for industrial or commercial purposes detrimental to the environment. If in an area there are industrial or commercial buildings the municipal council may forbid residential buildings being built in the area, or that existing buildings are taken into use for residential purposes if such buildings in the area for environmental considerations are deemed to be irreconcilable with the carrying on of industry and commerce.

D4.29 A prohibition or an order under this Section may also be appealed against to the Minister for the Environment by the owner concerned. Such an appeal again must be brought within four weeks from the date when the decision was received. The provisions under Section 45 do not apply when the area in question is covered by provisions about the building use in a local plan.

*Local Plans*
D4.30 Local Plans produced under the Municipal Planning legislation may because they are legally binding documents change and over-rule the Building Regulations, both those in accordance with the Building Act and those reflecting the provisions of the Municipal Planning Act. Where a local plan proposes modifications to the requirements relating to siting in Section 8 of the Building Act and to the minimum size of sites, building percentages and height of buildings in Sections 40, 42 and 43 respectively of the Municipal Planning Act, the Minister of the Environment's approval may need to be obtained depending on the circumstances.

D4.31 Local plans can be said to supplant and supplement the Building Regulations, because not only can they change the provisions in terms of the standards established but they can also introduce requirements beyond those covered in the Regulations. They can thus change the scope of control as well as its content. The dominance of the local plan as a control on development is apparent from the frequency to which attention is drawn in the Building Regulations to the local plan's possibility of stipulating different requirements.

D4.32 As well as affecting the scope and content of control local plans can also change the procedures that have to be followed. Thus the usual circumstances where the various building permit procedures (listed in paragraph D4.9) apply may be modified in a way which reflects the degree of control over development which the municipality may wish to exert. Some of the proposals which normally only need to be reported, for example, may become the subject of a building permit application.

D4.33 When a building permit application is received in an area covered by an adopted local plan, the decision will depend on the conformity of the proposal with the contents of the plan. If the plan's policies are precise and rigid the scope for flexibility is likely to be small and the decision straightforward. Recently, especially in Copenhagen, there has been a tendency for policies to be more general thus enabling more political imput to be made.

*Exemptions and Dispensations*
D4.34 There is scope under the Municipal Planning Act for certain exemptions against its provisions and those in a local plan. In Section 39 of the Act it is stated that in the case of sub-division of land or building on a site the rules relating to minimum plot area, maximum building percentage and height of buildings must be observed unless

(i) another decision has been made in an adopted local plan

(ii) an exemption is granted under the rules of Section 47. Section 47 specifies that exemptions may be granted in accordance with the following rules.

(a) Exemptions from minimum plot area, maximum building percentage and height of buildings may be granted by the municipal council within a framework laid down by the Minister for the Environment. More far reaching exemptions may be granted by direct application to the Minister. An exemption cannot be granted in the cases in which a local plan has to be prepared before the sub-division of large areas or before major building and investment works, including demolition, are carried out.

(b) Exemptions of minor importance from the provisions of a local plan which will not change the special character which the plan tries to create or maintain may be granted by the municipal council. The municipal council may authorise an association of landowners or with consent of the landowners concerned an association of tenants to grant such exemptions.

(c) More far reaching deviations than those described in (b) above may be made only by the production of a new local plan.

(d) From provisions protecting the interests of neighbours or other persons living in the neighbourhood, exemptions as mentioned above cannot be granted before the expiry of two weeks after the municipal council or the association of landowners or tenants respectively has notified the owners and tenants of the adjacent sites of the purpose for which exemption is applied. Comments, if any, may be submitted within two weeks. The municipal council may make it a condition for its consideration of an application for an exemption that the applicant gives this information to a wider circle of owners or tenants who in the opinion of the municipal council are deemed to have an interest in how the case is decided. The Minister for the Environment may lay down detailed provisions in these circumstances.

(e) Conditions attached to an exemption are binding on owners and other holders of rights in relation to the property irrespective of when the right was created. The municipal council shall at the owner's expense ensure that the conditions which relate to the use or other disposal of a property or parts of it are registered in the Land Register.

*Infringements and Enforcement*

D4.35 Infringement of the Building Regulations renders the offender liable to fines in accordance with Section 30 of the Building Act. This Section states that a fine or fines shall be imposed on any person who

(i) commences building work, uses completed building work or implements any other measures without obtaining a permit, as stipulated in the Act or in regulations issued in pursuance of the Act, or implements building work or other measures requiring the prior issue of a permit in any manner other than permitted by the relevant authority

(ii) disregards conditions stipulated in a permit pursuant to the Act or to regulations issued under the Act

(iii) fails to comply with an order or ban issued pursuant to the Act or to regulations issued under the Act

(iv) fails to carry out maintenance work necessary to avoid any danger for the tenants of the property or other persons

(v) contravenes the provisions of Section 7 of the Act. This Section states that

(a) the Building Regulations can stipulate rules governing the size, planning and equipping of undeveloped areas belonging to the property to secure adequate outdoor living space for children and adults, entry and access facilities, rescue facilities for the fire service and parking facilities

(b) undeveloped areas laid out and approved in accordance with the rules drawn up in pursuance of this section must not be used for any purpose contrary to that for which they were laid out. Where a building erected before commencement of the Act has undeveloped land suitable for fulfilling the aims of (a) above, the municipality can oppose any building upon or use of such land in a manner which would be contrary to such aims

(c) rules issued in accordance with this Section do not apply in cases where circumstances are covered by the provisions of a local plan.

D4.36 Section 30 of the Building Act further states that

(i) regulations issued by the Minister of Housing in pursuance of the Act can stipulate penalties of a fine or fines for contravention of provisions of the regulations

(ii) in the event that building work has been executed in an unauthorised manner, liability to penalty shall be incumbent upon the person or persons executing the work or responsible for its execution or, depending on the circumstances, both such parties. The person who ordered the work to be executed shall be held liable only when he is unable to nominate any other person upon whom liability can be placed or when he has participated in the violation fully aware or strongly suspecting that the matter was unauthorised

(iii) a fine or fines shall moreover be imposed on any person manufacturing or selling for use in the building industry building materials not authorised for such use under current Building Regulations

(iv) in the event of a contravention by a public joint-stock company, private company, co-operative or similar corporation, liability to a fine or fines shall be imposed on the company etc.

D4.37 Infringement of the Building Regulations for Small Buildings is moreover subject to the following.

(i) A fine or fines shall be imposed on any person who, contrary to the provisions of this regulation fails to report a building project or fails to report the completion of a building project.

(ii) A fine or fines shall be imposed on any person violating the provisions of the regulations.

D4.38 The Building Regulations state that any person commencing building work contrary to the provisions relating to applications in respect of agricultural buildings in rural areas shall be liable to a fine. This is linked to the legislation under the Urban and Rural Zones Act which requires the owner of a property to restore the building to its former lawful state if contravention has occurred. Similarly if the contravention relates to the unlawful use of the property the user of the property also has the same obligation. Furthermore, if an order made by a Court to restore a property to its former lawful state is not complied with within the time fixed by the Court, and the collection of enforcement fines cannot be supposed to result in the order being complied with, the authorities shall be entitled to do at the owner's expense what is necessary to restore the property to its former lawful state. For this purpose the authority concerned shall be entitled to procure admittance to the property. The police shall assist the authority in accordance with rules laid down after consultation between the Minister for the Environment and the Minister of Justice.

D4.39 Section 24 of the Urban and Rural Zones Act states that any person or company shall be liable to a fine if there is

(i) contravention of the necessity in rural zones to obtain permission for the parcelling out or division of land (Section 6), the construction of new buildings, other than those requiring conversion or extension of those existing, for agriculture, forestry and fishing (Section 7) or for the taking of existing buildings and areas not built on into use for purposes other than agriculture, forestry or fishing (Section 8)

(ii) violation of conditions attached to a permission given under the provisions of the Act

(iii) failure to comply with an order to restore a property to its former lawful state.

D4.40 Enforcement action can also be taken against infringements of the provisions of the Municipal

Planning Act. The municipal council is obliged to ensure the observance of the rules relating to the size of sub-divisions, building percentages and height of buildings and of the provisions in local plans (Section 46). Under Section 54 it is the duty of the owner of a property for the time being to remove the conditions resulting from an infringement of the Act. If the infringement consists of an unlawful use of the property, the use shall be subject to the same duty. If the owner or the user does not comply with an order given by the municipal council to remove the conditions resulting from an infringement he may by a judgement be ordered, subject to continuous fines, to remove the unlawful conditions within a period fixed by the court. If an order given in a judgement to remove unlawful conditions is not complied with in the time allowed by the court and collection of coercive fines cannot be expected to have the effect that the order is complied with, the municipal council may take the steps to have the unlawful conditions removed at the owner's expense.

D4.41   A person who

(i) infringes Section 42 relating to building percentages and Section 43 concerned with the height of buildings

(ii) infringes the provisions of a local plan

(iii) does not comply with conditions attached to a permission under the Act or the directions or plans issued under the Act

(iv) does not comply with conditions attached to an exemption under Section 47

(v) fails to comply with an order or a prohibition issued under the Act or the directions given under the Act, including an order to remove the conditions resulting from an infringement of the Act

shall be liable to a fine. If the infringement has been committed by a joint-stock company, another company with limited liability, a co-operative society or similar associations, the company or the society as such may be made liable to a fine (Section 55).

## Other Environmental Controls

D4.42   Whilst the building permit procedures apply over the whole country other, additional, permissions may be required depending on the local circumstances. Reference has already been made to applications for zone dispensation under the Urban and Rural Zones Act (paragraph D3.20) but the need for permission to carry out various proposals is specified under several other statutes. The most important acts in relation to physical planning are

Conservation of Nature Act
Agriculture Act
Summer Cottages and Camping etc Act

Preservation of Buildings Act
Forestry Act
Raw Materials Act.

*Conservation of Nature Act*
D4.43   This act statutorily controls and gives protection to certain areas and elements in nature and landscape. The statutory procedures for the protection of nature and landscapes include

(i) conservation regulations, designed among other things to preserve the existing condition of, or to lay down criteria for, the future use of the conserved objects or areas, including public access to particular areas. The conservation regulations may also include directives for construction within and maintenance of the area

(ii) fixed protection zones along coast, lakes, streams, woods, and ancient monuments

(iii) public purchase of land

(iv) strict nature conservation approval procedures in connection with public structures and works (eg roads and electrical installations) and structures, works etc in connection with wetlands

(v) preparation of Nature Conservation Plans (Kristiansen, 1984).

D4.44   Thus under the legislation it is forbidden to erect buildings or to undertake planting or to fence in areas within a distance of 150 metres from lakes and ancient monuments (Christiansen, 1986). It is also not permitted to construct buildings within a distance of 300 metres from forests and woods. These limits and the areas to which they apply can be changed by the National Forest and Nature Agency of the Ministry of the Environment in the context of local planning policy. Exemptions from the rules may be granted by a conservation board consisting of a chairman, appointed by the Minister for the Environment, and two other members elected by the county and municipality. There is a total of 26 conservation districts covering the whole country, each having its own board. Existing natural features are also directly protected through the necessity of obtaining permission from the county authority to make changes to the existing state of watercourses, lakes, bogs, marshes and heaths.

*Agriculture Act*
D4.45   The Ministry of Agriculture has to be notified of any proposed changes to the boundaries of individual properties in rural areas and permission has to be given before these can be implemented. An application for permission for such changes will be considered on the basis of such factors as whether the changes applied for are in accordance with other relevant land use legislation (Christiansen, 1986). Approval will not be granted until all other necessary permissions and exemptions have been granted. This procedure is similar to that relating to building permits.

D4.46 Under the Agriculture Act, specially defined agricultural holdings may be scheduled as land that must be reserved for agricultural purposes, with the result that the land must be used in a proper agricultural way to the extent to which it is suitable (Christiansen, 1986). This obligation may also be fulfilled by using the land for market gardening, orchards or similar purposes. The duty to use land for agricultural purposes may be removed in whole or in part by the Minister of Agriculture and normally this will be agreed if the land, because of planning policies, can be expected to be used in the near future for purposes other than agriculture.

*Summer Cottages and Camping etc Act*
D4.47 Under this legislation it is permitted to build one dwelling on a site in summer house districts which may be used only in the period from 1 April to 1 October, unless a special exemption is granted by the council. The decision of the council may be appealed against to the Environmental Appeal Board. Camping huts erected on campsites authorised under this Act do not have to comply with the Building Regulations. The erection of camping huts is subject to the permission of the county council.

*Preservation of Buildings Act*
D4.48 Since 1918 legislation has existed to list buildings of great architectural, cultural or historical value which normally are more than 100 years old. The building need not be intrinsically valuable; it may be associated with an historical event or an historically important figure (Saaby, 1984). The current Preservation of Buildings Act was passed in 1986 and its provisions also apply to structures, parts of buildings and the immediate surroundings if they are considered part of the total entity worthy of listing. The decision whether a building should be listed is made by the National Agency for Physical Planning and it then becomes subject to the building preservation legislation. Buildings constructed before 1536, the year which marks the Reformation and which is normally considered as the end of the Middle Ages in Denmark, are deemed listed buildings without a building preservation order and therefore fall under the provisions of the Act. Before the National Agency makes a decision about listing, the Historic Buildings Council has to be consulted and its recommendation obtained. The owner of the building and the municipal council concerned must also be notified and a period of not less than three months has to be allowed for the submission of comments on the proposal. During this period nothing may be done which could affect the possible listing. At the end of this period the National Agency decides whether the building should be included in the schedule. The fact that a building has been listed must be entered in the land register of the local court. The number of listed buildings is estimated to be between 26,000 and 30,000 with 100 to 200 added annually.

D4.49 A listed building must be maintained in a proper condition by its owner or user. Nothing may be done to the building which would change its condition and if the owner wishes to alter or demolish the building, application must be made to the National Agency for permission, and the Agency consults with the Historic Buildings Council. If the Agency has not decided on the application within three months from its receipt the owner may carry out the work or demolition in question. In the case of a demolition application if permission is refused the owner may demand that the State takes over the building subject to the payment of compensation. This obligation of the State only exists, however, if there is a considerable disparity between the rate of return from the property and that from properties with a similar location and function to which the prohibition does not apply. The National Agency for Physical Planning can provide financial assistance for the maintenance of listed buildings and also technical advice. Loans may also be obtained from the State Preservation of Buildings Fund.

D4.50 The Building Regulations for Small Buildings in the section relating to building work not subject to a permit or duty to report draws attention to the fact that all work on listed buildings beyond general exterior and interior maintenance requires a permit from the National Agency for Physical Planning under Section 10 of the Preservation of Buildings Act. 'Buildings worthy of preservation which are subject to a local plan, an urban preservation plan, a district preservation plan or a registered preservation easement, shall remain subject to the provisions stated therein concerning conversion and alteration work, etc'. Attention is also drawn to the fact that demolition of listed buildings requires a permit from the National Agency under Section 11 of the Preservation of Buildings Act.

*Forestry Act*
D4.51 The Forestry Act ensures the maintenance and proper management of Denmark's afforested area. Under the Act many of the country's forests are designated as forest reserves, which means that they must be kept covered with trees of such a kind and nature and in such a number that they form, or by continued growth within a reasonable time can form, a close stand of trees (Christansen, 1986). The owners are obliged to keep their woodland in good condition. This duty to keep forest areas covered with trees may be removed by the National Forest and Nature Agency in relation to the requirements of local planning policy, but it will normally be made a condition of the clearance of the forest that new, and often larger, areas are reserved for afforestation.

*Raw Materials Act*
D4.52 The planning and management of raw materials (excluding hydrocarbons) is based on the provision of a Raw Materials Act whose aim is to ensure economical utilisation of raw materials such as pebbles, gravel, clay, limestone and chalk. The extraction of raw materials is based on an integrated plan which takes account of other interests. A permit to recover is needed and this is granted by the National Forest and Nature Agency if the proposal conforms with the plan. The application for the extraction permit should normally be accompanied by a

plan showing the quantities, type of raw material and extraction rate. A permit will usually contain a stipulation requiring the applicant to restore the landscape once the recovery process is completed.

## Comments and Conclusions

D4.53 The systems of control over physical and environmental change are extremely comprehensive in their scope and content. In practice, over the last four years there has been some relaxation in the controls themselves and in the way they have been applied. For example, for some proposals the previous requirement for permission to be given to use the completed building has been replaced recently by the need simply to report the completion to the municipality. The one exception to this general relaxation has been in environmentally sensitive areas and in relation to environmental protection. Thus there was issued a national planning directive to prohibit the erection of new summer houses within three kilometres of the coast. The difficulty of achieving a balance between environmental concern and economic regeneration has recently been highlighted by the debate about whether redundant farm buildings in Rural Zones should be given permission for general employment purposes.

D4.54 What has also occurred is an increased flexibility in the operation of control so that local plans are being prepared in a way that gives greater discretion to the politicians. This can be illustrated by considering different scenarios relating to the application for a building permit.

(i) Where the proposal is in agreement with the local plan there is no problem and permission is given.

(ii) If the proposal is not in agreement with the plan and there is no political will to support the proposal, permission is refused.

(iii) But if the proposal is not in agreement with the plan and there is political pressure to support the plan, then problems arise. The plan may be changed in this situation, but this is unlikely to be achieved without conflict and consequent delay.

With what seems to be the increasing politicisation of planning, the system of control based on the nesting of plans with the local plan providing the ultimate authority can work successfully only if there is strong political support for the plan policies or if those policies are flexible enough for discretion to be exercised at the political level.

D4.55 In terms of the relationship between local plans and the Building Regulations as instruments of control, it has been suggested that the Building Regulations because of their restrictive nature might be considered the stick to force municipalities to produce local plans compared with the carrot offered in the freedom of the council to decide the precise content of the local plan, albeit within the given national and regional framework (Lemberg, personal communication).

# D5 DEVELOPMENT PLANS AND SOURCES OF POLICY

D5.1 Policy documents are produced at the three governmental levels: national, county and municipal. The guiding principle in the formulation of policy is framework management and control (*rammestyring*), whereby planning at any level has to be in agreement with the framework laid down at a higher level. Thus, the local plan which provides the basis for the implementation of development is produced in the context of the municipal structure plan, the county regional plan and national policy. Framework management and control is also the principle adopted to integrate the physical planning operation with that of the other sectors of interest such as agriculture, nature conservation, natural resource development etc, all within the context of economic planning and budgetary control.

## Objectives of National and Regional Planning

D5.2 Section 1 of the National and Regional Planning Act 1973 which came into force on 1st April 1974 states that through comprehensive physical national and regional planning it shall be the aim to provide for

(i) the utilisation of the land and natural resources of the country on the basis of an overall assessment of the interests of the community, and also to promote a uniform development in the country

(ii) the utilisation of the land to be determined in such a way that air, water and soil pollution, as well as noise nuisance, be forestalled, as for instance by the best possible separation of activities detrimental to the environment from housing developments

(iii) the co-ordination of individual measures taken within the framework of economic planning for the community.

D5.3 As a result of the Urban and Rural Zones Act of 1969 which came into force on 1st January 1970 the whole country is allocated into three zones: urban, summer house and rural (Section 2). Whilst the designations are the responsibility of the counties and municipalities their collective decisions amount to a broad national spatial planning framework. Furthermore the purpose of the Act is national in character in that it aims

(i) to ensure a planned, economically suitable development in accordance with national, regional and municipal planning; also in respect of ensuring such a development of the smaller urban communities

(ii) to be instrumental in protecting the recreational interests of the population and in preserving scenic values

(iii) to be instrumental in protecting interests in a suitable utilisation of the raw materials resources and the agricultural land resources (Section 1)

D5.4 Urban zones are defined as

(i) areas allocated for urban development by an urban development plan

(ii) areas allocated as building zones for urban development by a building by-law

(iii) areas allocated for urban development or public purposes by town planning by-law

(iv) areas transferred to an urban zone by a local plan.

The first three above are being superseded through the adoption of a local plans and are not discussed in this document.

D5.5 Summer house zones are

(i) areas allocated for summer house development by a building law or a town planning by-law. (This is not relevant after a local plan has been adopted)

(ii) areas transferred to a summer house zone by a local plan.

Rural zones are the areas not included in the urban and summer house zones.

## National Planning

D5.6 There is no national plan but the Ministry of the Environment has to submit an annual report to Parliament on national planning aspects. This is termed the Statement on National Planning (*Landsplanredegørelse*). The aim of the report is to

(i) provide guidance in a persuasive way to the counties and municipalities

(ii) cover special issues or particular topics

(iii) inform parliament and the country generally of what has been taking place during the year and what is foreseen in the future.

The first report in 1975 contained an outline of an urban

settlement strategy. Subsequent annual reports have included, in

1976: Land use planning, including urban development and protection of natural resources.

1977: The future of the villages, urban improvement issues. The relationship of physical planning to energy policy, to agriculture, to preservation of nature, etc.

1978: The Government's views on which national planning interests should be specially considered in connection with the approval of the regional plans.

1979: The relationship of physical planning to the building industry, traffic and energy supply and the relationship between economic and physical planning.

1980: The Government's principles for the approval of the proposed regional plans (Svensson, 1981).

D5.7 The 1982 report was short but particularly significant in that it discussed future spatial strategy. The most recent 1986 report was divided into two parts, the first being concerned with policy for undeveloped land in rural areas and the second, as a separate volume, with a review of physical planning in the previous year which included an appendix on all the regional planning documents currently approved and relevant in each of the counties. The report emphasised that national, regional and municipal planning should be an integrated activity with the problems being solved in a holistic way. The volume concerned with rural areas discussed the implications of recent trends in development in terms of their consequences for nature, the environment and socio-economic structure. Current policies were evaluated and possible changes recommended. It was suggested that a higher priority should be given to environmental aspects and their conservation and protection. 8,000 copies of the 1986 annual report were printed.

D5.8 The Minister of the Environment may, from time to time, issue national planning directives (*landsplan-direktiver*) which are mandatory on the counties and municipalities and thus have the force of law (see Figure D9). Thirteen had been published by 1986, covering the following topics

(i) first stage in the reservation of areas for main pipeline for natural gas distribution 1978

(ii) second stage in the reservation of areas for main pipeline for natural gas distribution 1979

(iii) areas reserved for the proposed bridge between Zealand and Fyn 1981

(iv) third stage in the reservation of areas for main pipeline for natural gas distribution 1981

(v) reservations of land for the extension of Copenhagen airport, including areas affected by noise, pollution etc 1981

(vi) prohibition of new summer houses within three kilometres of the coast 1981

(vii) reservation of areas for oil pipeline from North Sea to Fredericia 1982

(viii) reservation of sites for coastal natural gas terminals 1984

(ix) safeguarding an area for a new ferry terminal 1985

(x) reserving an area for defence purposes 1985

(xi) reservation of a site for a TV sender aerial 1986

(xii) reservation of other sites for TV sender aerials 1986

(xiii) area in the metropolitan region where disposal sites for power station waste will be located 1986.

D5.9 The Ministry may also issue circulars which generally refer to paragraphs in the legislation and provide a commentary on the legislation. They are thus generally more concerned with procedures and providing guidance to the counties and municipalities on how to fulfil their obligations than with policy. They complement the law and have the status of law but do not require before being issued the lengthy procedures necessary to change the statutes. Not surprisingly they are frequently used and follow a standard format with the formal text being accompanied by a commentary.

D5.10 Central government carries out economic planning and sector planning (*sektor-planlaegning*) whereby each of the ministries responsible for internal affairs is obliged to produce plans for its particular function. State sector planning is significant not only for the physical planning activities of the Ministry of the Environment but also for those of the counties and municipalities. The main spheres of sector planning are

(i) the sectors dealing with the planning of land use and natural resource, eg conservation of nature planning or agricultural land use planning

(ii) the infrastructure sectors, eg road planning, energy planning

(iii) the service sectors, eg education, health.

The sector plans are integrated with each other and co-ordinated in the physical planning activity which is the most highly developed form of planning. Sector plans are also co-ordinated with the economic planning, but as yet economic long term planning is not very well developed. Whilst physical planning is clearly the responsibility of the Minister of the Environment, because of its co-ordinating function a committee of civil servants from a number of different ministries has been established. All major issues arising in relation to physical planning and which have wider implications are considered by this committee. Thus the relationship between physical, economic and sector planning is as follows

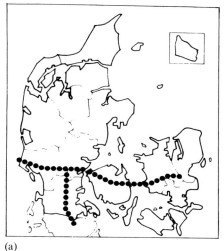

(a)

(b)

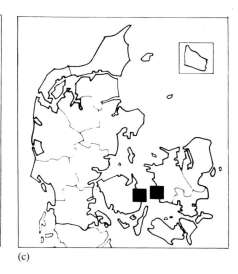

(c)

(d)

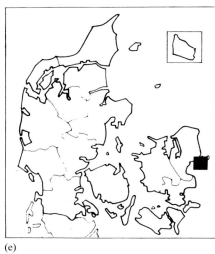

(e)

(f)

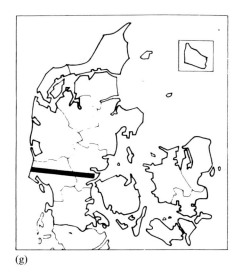

(g)

## Figure D9   National Planning Directives

Figure D9 shows the complete set of national planning directives that have been issued up to the beginning of 1987. Their main purpose is to safeguard areas for particular reasons and they are accompanied by more precise detail on the boundaries of the area that are subject to the directives.

(a) First stage in the reservation of areas for main pipeline for natural gas distribution 1978
(b) Second stage in the reservation of areas for main pipeline for natural gas distribution 1979
(c) Areas reserved for the proposed bridge between Zealand and Fyn 1981
(d) Third stage in the reservation of areas for main pipeline for natural gas distribution 1981
(e) Reservation of land for the extension of Copenhagen airport, including areas affected by noise, pollution etc. 1981
(f) Prohibition of new summer houses within three kilometres of the coast 1981
(g) Reservation of areas for oil pipeline from North Sea to Fredericia 1982

(h)

(i)

(j)

(k)

(l)

(m)

(h) Reservation of sites for coastal natural gas terminals 1984
(i) Safeguarding an area for a new ferry terminal 1985
(j) Reserving an area for defence purposes 1985
(k) Reservation of a site for a TV sender aerial 1986
(l) Reservation of other sites for TV sender aerials 1986
(m) Area in metropolitan region where disposal sites for
     power station waste will be located 1986.

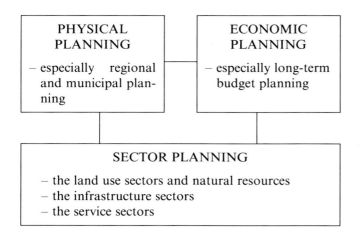

| PHYSICAL PLANNING<br><br>– especially regional and municipal planning | ECONOMIC PLANNING<br><br>– especially long-term budget planning |
|---|---|

| SECTOR PLANNING<br><br>– the land use sectors and natural resources<br>– the infrastructure sectors<br>– the service sectors |
|---|

## Regional Planning

D5.11   Each county council has to produce a regional plan, the general purpose of which is to utilise the land and natural resources of the county on the basis of an overall assessment of the interests of the community and to co-ordinate the individual measures within the framework of national economic planning (see Plate 4). Under Section 7 of the National and Regional Planning Act, the regional plan has to contain general guidelines for

(i) the distribution of future urban growth in the various parts of the county council area, including the demarcation of the urban zones

(ii) the extent and the location of major centres, major traffic facilities and other technical installations, and major public institutions

(iii) the location of special heat producing and heat consuming plants and establishments, major pipelines for heating purposes, and the areas in which specific methods of heat supply should be given a preferential use

(iv) the location of establishments etc for which special requirements must be advanced with regard to their location for the purposes of preventing pollution

(v) the size and the location of areas reserved for agricultural purposes

(vi) the utilisation of land for the exploitation of stone, gravel, and other natural resources in the ground

(vii) the preservation of nature including the designation and protection of nature reserves specially worthy of preservation

(viii) the size and the location of summer house districts and other areas for recreational purposes

(ix) the quality of water in watercourses, lakes and the parts of the territorial sea near the coasts

(x) the urban renewal in existing urban communities, including the chronological order of the urban renewal.

The regional plan may also contain general guidelines for other matters of vital importance for the promotion of a suitable urban and building development etc within the county council area.

D5.12   The National and Regional Planning Act lays down the process by which the regional plan should be prepared and approved. Regional planning has to be carried out in compliance with national planning directions under Section 4 and in all other respects on the basis of an overall assessment of the desirable development within the county council area. After negotiations with the local government associations, the Minister for the Environment may lay down rules for the carrying out of regional planning (Section 9).

(i) Within a time-limit fixed by the Minister for the Environment each municipal council must submit to the county council a proposal for the general guidelines for the land use in the municipality which the regional plan should contain, in the opinion of the municipal council. For the purpose of preparing this proposal, the county council is obliged to give the municipal councils information about the sector planning of the county council (Section 10).

(ii) The county council must collect information about existing plans and planning in progress for the county council area. This includes information about the municipal planning and the sector planning carried out by the state and the municipalities. At the request of the county council, the municipal councils and other public authorities must provide the information that is deemed necessary for regional planning. The county council is obliged to place the information obtained at the disposal of the municipalities concerned (Section 11).

(iii) After the expiry of the time-limit referred to above (Section 10) the county council negotiates with the municipal councils with a view to preparing alternative outlines of a proposed regional plan and an account of the conditions on which the outlines are based. The account should include the anticipated time-table for the regional plan's implementation and also state how it fits in with the planning hitherto undertaken for the county council area and its adjoining areas (Section 8). The outlines of the proposed regional plan and the account have to be published by the county council and sent to the municipal councils, the Minister for the Environment and the authorities and establishments from which the county council has received information under Section 11. The county council and the municipal councils must conduct information campaigns for the purpose of provoking a public debate on the objectives of the regional planning and its specific content (Section 12).

(iv) After the expiry of six months from the publication of the outlines, the county council negotiates with the municipal councils with a view to the preparation of a proposal for a regional plan, and when a county council has adopted a proposal for a regional plan, it publishes the proposal with the accompanying account and a summary of the comments received on the previously published outlines. The material is then forwarded to the authorities etc mentioned in Section 12, informing them that objections or comments, if any, can be sent to the county council within four months. After the expiry of the time-limit the county council sends the objections and comments together with an assessment of them to the Minister for the Environment (Section 13).

(v) The regional plan is approved by the Minister for the Environment after a review of the proposal in collaboration with the Ministers whose fields of competence are generally affected by physical planning. The Minister for the Environment may make such amendments as are deemed necessary in the interest of national planning, after the intended amendments have been reviewed in collaboration with the other Ministers. Before any amendments are made the county council and the municipal councils are given an opportunity to express their views. The approved regional plan is then published by the county council (see Plate 4).

D5.13   Following the approval of the regional plan, the planning activities of the county council and the municipal councils and their carrying through of investment work must not be contrary to the plan's context. In fact the county council and the municipal councils are obliged to promote the implementation of the regional plan in connection with the exercise of their statutory powers (Section 15). Every second year after the approval of a regional plan the county council must submit a report to the Minister for the Environment on the development that has taken place, including the planning which has been done in respect of the county council area or parts thereof. The report contains an assessment by the county council on the conformity of the development and the aforementioned planning with, or influence on, the regional planning (Section 16).

D5.14   The county council may prepare supplements to an approved regional plan until it is found necessary to produce a new overall regional plan, and the Minister is able to order the county council to prepare such supplements. The preparation of the supplements follows the same procedures as the original regional plan although the Minister can exempt the county council from some of them.

D5.15   These procedures introduced important innovations in that the regional planning work is to start with the introductory proposals of the municipalities, and that the county councils then prepare draft regional plans which have to be published for public debate. Only after the public debate may the proposed regional plan be prepared. By 1980 all county councils had prepared proposals for regional plans.

D5.16   A general purpose of the proposals is to encourage development in the most depressed parts of each county and of the country more generally. By means of guidelines for employment in the municipalities an attempt is being made to maintain or increase the number of jobs in these areas. Through guidelines for the location of grammar schools, hospitals and social institutions it is hoped to improve the public service. Several proposals also contain guidelines aimed at securing the shopping facilities in the minor urban communities by counteracting the concentration of shops in or close to the larger centres but owing to the economic situation and the rising energy prices the basis of these proposals has often suffered subsequent modification.

D5.17   Clearly regional planning does not end with the approval of the proposals, but is seen as a continuous task. Regional plan supplements have to be prepared because the assumptions relating to population growth that were made in the original plans have changed and are constantly changing. At the same time new legislation has made it necessary within a number of sectors to prepare plans which have to be made part of the regional plans. The Ministry of the Environment and the county councils have agreed that regional plan supplements should be prepared to cope with this situation. The Ministry demanded a complete supplement in 1984-85, and it has now been laid down that the regional plans have to be adjusted every four years by a complete supplement.

D5.18   In terms of content, regional planning deals not only with land use, but is intended to have the character of corporate planning. In the regional plans the various interests in natural resources, eg preservation of nature interests, and urban development interests are balanced against each other. A wide circle of politicians, organisations, authorities and citizens take part in the preparation of the plans. The subjects dealt with in the proposed regional plans are the urban pattern, housing and industrial and commercial building and the open country. They also deal with the county sectoral responsibilities and traffic.

D5.19   The county councils have used different planning methods in preparing the plans. In Viborg for example the county council has attached great importance to a debate on objectives between a broad section of the parties to the regional planning process. At Funen the county council has chosen a more technical, time-efficient method for the preparation of the plan (Svensson, 1981).

## Copenhagen Metropolitan Region

D5.20   Regional planning within the metropolitan region differs slightly from that in the rest of Denmark and has its own special act. Under this Regional Planning Act for the Metropolitan Region which came into force on 1 April 1974 the regional plan has to contain general guidelines for the same aspects that the counties have to cover in their regional planning activity (see paragraph D5.11). Section 2 of the Act states that the metropolitan regional plan also has to include guidelines and a general framework for

(i) the main features in the distribution of the various types of building development in the urban zones

(ii) the traffic services in the metropolitan region

(iii) the water supply and the waste water disposal in the region

(iv) the provision of hospitals etc in the region

(v) the provision of facilities etc for recreational purposes in the region.

The regional plan in addition has to contain a phasing programme for

(a) the development of the urban zones

(b) the urban renewal in the existing urban communities

(c) the provision of the necessary plants and facilities etc.

The plan can include guidelines for other matters of vital importance for the promotion of a suitable urban and building development etc within the metropolitan region. The procedures involved in the preparation and approval of the metropolitan regional plan are the same as those for the regional plans produced by the counties.

## Municipal Planning

D5.21 The Municipal Planning Act 1975, which came into operation on 1 February 1977, states that the municipal councils must contribute to the development of the municipality, within the framework of national and regional planning, in a way which promotes the welfare and prosperity of the population (Section 1). The aim of municipal planning is to ensure

(i) that the inter-relationship between the location of dwellings and employment, the provision of roads and traffic and public and private servies, open spaces and other urban facilities contribute to the creation of good working and residential surroundings in existing and future urban communities

(ii) that the land and other natural resources of the municipality are used on the basis of an overall assessment, inclusive of economic and ecological considerations

(iii) that pollution of air, water and soil as well as noise nuisance are prevented, eg by the best possible separation of activities detrimental to the environment and residential areas.

Two types of statutory policy documents are produced at the municipal level: the municipal structure plan (*kommuneplan*) and the local plan (*lokalplan*).

*Municipal Structure Plans*

D5.22 The structure plan is intended to provide a framework for the physical, economic and sectoral planning of the entire municipality (see Plates 5 and 6). It should be based on an overall assessment of the land and the natural and economic resources of the municipality, the objectives for the demographic and industrial development in the municipality and the objectives of the municipal sector planning (Section 2). It consists of one part which lays down the general structure of the municipality and another part which, in relation to the individual parts of the whole municipality, lays down the framework for the content of the local plans. The municipal structure plan must not be contrary to the regional plan or to any directions from the Ministry of the Environment. A proposal for a municipal structure plan has to be published within two years after the approval of a regional plan.

D5.23 The first part of the municipal structure plan relating to the general structure of the municipality, embracing the pattern of urban development and the distribution of service centres, has to be laid down by the comprehensive planning of

(i) the development of the urban communities of the municipality with dwellings and jobs, including the phasing of this development

(ii) the size and the location of summer house areas, again with the phasing proposals

(iii) traffic facilities and public transport services

(iv) institutions

(v) public supply services and other technical facilities

(vi) the location of special heat producing and heat consuming plants and establishments and the location of pipelines for heating purposes

(vii) the location of activities etc which in order to prevent pollution require that special regard is paid to them

(viii) the size and the location of areas reserved for agricultural or similar purposes

(ix) the use of areas in accordance with the purpose of the Nature Conservation Act

(x) other uses of land, including the exploitation of stone, gravel and other natural resources in the ground (Section 3).

D5.24 The second part of the municipal structure plan relating to the framework for the content of the local plans has to be determined for the individual parts of the municipality as regards

(i) the distribution of buildings according to their kind and uses

(ii) the areas in which specific methods of heat supply should be given a preferential use

(iii) institutions and technical plants

(iv) the supply of public and private services

(v) the traffic system

(vi) green areas, allotment garden areas etc

(vii) the density and height of residential and other buildings and other building conditions within areas delimited in accordance with (i) above, including the framework for the conservation of buildings and the urban environment

(viii) the urban renewal of existing urban communities

(ix) the transfer of areas to urban zones or summer house districts

(x) the chronological order of the development of the areas for urban purposes and summer house districts.

D5.25 The municipal structure plan may also contain guidelines for other matters of importance to the land use and the building development in the municipality. It has to be accompanied in the metropolitan area of Greater Copenhagen by a timetable for the construction of buildings and plants in the municipality in accordance with the guidelines approved in the regional plan and, outside the metropolitan region, by a plan for the chronological order of urban renewal in the existing urban communities in accordance with the approved guidelines in the regional plan (Section 4).

D5.26 The structure plan has to be accompanied by information on the assumptions of the plan, including information on the relationship of the plan to

(i) decisions under Section 4 of the Act on National and Regional Planning Act ie the specific conditions that the Minister lays down to form the basis of regional planning

(ii) the regional planning for the area

(iii) the goals of the demographic and industrial development in the municipality

(iv) the economic planning of the municipality, including the planned timetable for investments in operations aimed at making land ripe for building and urban development and for investments in urban renewal

(v) the municipal sector planning

(vi) the planning so far undertaken for the area of the municipality and

(vii) existing conditions, including the demographic and industrial structure and the landscape and building conditions of the municipality (Section 5).

The report containing this information is regarded by planners as being of great importance as it provides the context within which the plan has to be prepared.

D5.27 The municipal structure planning process consists of the following stages (see Figure D10).

(i) Before the preparation of a proposal for a municipal structure plan the municipal council publishes a brief account of the principal problems and planning possibilities in the coming planning work. Simultaneously with publication the account is sent to the Minister for the Environment and the county council and within the metropolitan area of Greater Copenhagen to the metropolitan council (Section 6).

(ii) The municipal council has to conduct an information campaign to arouse public debate on the objectives and the detailed content of the planning.

(iii) When the council finds that the public debate has thrown sufficient light on the principal problems and planning possibilities identified in the account it may adopt a proposal for a structure plan. The proposal can only be adopted at an earlier time than 12 weeks after its publication subject to the consent of the Minister for the Environment. After the adoption of the proposal the municipal council publishes it together with the information on the assumptions of the plan and an extract of the comments received on the previously published account and the assessment by the municipal council of the comments. The Minister for the Environment may lay down detailed rules as to the publication.

(iv) Any members of the council who have demanded to have their diverging opinions as to the adopted proposal for a structure plan entered in the minutes of the municipal council may demand that the council simultaneously publishes their diverging opinions and a brief statement of the reasons for their objections.

(v) The municipal council simultaneously with the publication of the proposal states the time allowed for the submission of comments on the proposal. The time, which must not be less than 16 weeks from the publication, is fixed by the municipal council.

(vi) Also simultaneously with publication, the proposal has to be sent to the Minister for the Environment and the county council and within the metropolitan area of Greater Copenhagen to the metropolitan council.

D5.28 The county council is obliged to ensure that a proposal for a municipal structure plan is in accordance with the regional plan and any specific conditions that the Minister has laid down for the regional plan (Section 8). The county council was also responsible for checking that the proposal had been given a form which would allow it to form the framework for local plans but this duty was removed in 1985 when an amendment to the Municipal Planning Act came into effect. The municipal council thereby was given more discretion and independence in its local planning activity. The Minister for the Environment may decide that the county council also controls other matters, for example that the proposal complies with directions under Section 51 which allow the Minister to assist the municipalities with guidance and to lay down detailed rules about their planning. A municipal structure plan which lays down a building percentage above 110 per cent for an area of the municipality cannot be finally adopted by the municipal council unless the Minister for the Environment has in advance approved the higher building percentage.

D5.29 Immediately after the final approval of a structure plan the municipal council makes a public announcement of the approval in accordance with rules laid down by the Minister for the Environment. The plan shall be accessible to the public (Section 11). The council is then obliged to work for the implementation of the structure plan by exercising the powers conferred on it by law. The building and investment activity of the council must not be contrary to a municipal structure plan (Section 13).

D5.30 Within the first half of the election period the council has to review the structure plan with a view to deciding whether it is necessary to amend the plan or to prepare a new one. The county council or the metropolitan council respectively may decide that a municipality amend a structure plan in order that the plan can be made to agree with a later approved regional plan or decisions made under Section 4 of the Act on National and Regional Planning relating to specific conditions that must form the basis of planning laid down by the Minister. Proposals for amendments to a municipal structure plan must be published (Section 14).

D5.31 As a result of increasing decentralisation the municipalities have become responsible for a large number of public services and it is the role of the structure plan to include the principal parts of the sector plans for each public service and to balance and co-ordinate them. The municipal sector plans comprise the development plan for social welfare and health, council schools, further education, and sewage discharge. Besides these sector plans required by law many municipalities prepare other sector plans, eg road and footpath plans. The sector plans are prepared on the basis of population forecasts divided into years. Normally the plans consist of three parts.

(i) The stocktaking part contains an account of the local and regional conditions for the solution of the problems in the subject in question.

(ii) The programme part contains the projects and activities which the municipal council has decided to carry through in the immediate future, normally the next four years.

(iii) The perspective part contains a description of the goals of the municipal council and the measures likely to be adopted over a longer period, normally twelve years, in order to realise the goals.

The sector planning is of great economic importance to the municipalities. The social welfare and the education sectors alone often account for more than 75 per cent of the gross expenditure of the municipality.

D5.32 The integration of the sector plans and their co-ordination with other types of planning take place as far as possible in the municipal structure planning process. The structure plan provides a vehicle for co-ordination and it has been used in this way to a varying extent and with various levels of success. A particular difficulty arises because the content and programming requirements of the different sector plans which are laid down in the relevant statutes are so dissimilar. Thus, more important than the municipal structure plans as integrating instruments are the annual budgets and the discussions and negotiations which precede their adoption.

*Local Plans*
D5.33 Local plans have to be produced when it is necessary to ensure the implementation of the municipal structure plan (Section 16). Their introduction in 1977 when the Municipal Planning Act came into operation has resulted in a simplification both for the public authorities and the citizens. Formerly there was a great number of different types of local plans, but now there is only the one plan. The local plan also replaces the detailed town planning by-laws which were to be submitted to central government for approval. As well as to ensure the implementation of the municipal structure plan, a local plan also has to be produced before the sub-division of larger areas or before major building and investment works, including demolition of buildings, are carried out. They can be produced therefore to prevent development proposals being realised as well as to provide the legal basis for proposals to be allowed.

D5.34 Local plans do not have to be submitted to the Ministry of the Environment for approval. The municipal council itself adopts the local plan and the county council ensures that the proposal is in accordance with the regional plan and the municipal structure plan. The Minister for the Environment may decide, however, that the municipal council cannot adopt a local plan without his approval.

D5.35 As already pointed out, if a major proposal is put forward, permission under the Building Act cannot normally be given until the local plan has been finally adopted (see paragraph D3.23). If it is desired to undertake major building or demolition work and it is in accordance with the municipal structure plan the municipal council must as soon as possible prepare a proposal for a local plan and then promote its implementation. The municipal council may in such a case demand that the person who is interested in the carrying out of the operation shall assist the municipality in the preparation of the plan (Section 16).

D5.36 Section 18 of the Municipal Planning Act states that a local plan may contain provisions as to

(i) the transfer of areas comprised by the plan to an urban zone or summer house zone

(ii) the use of the area, including provisions to the effect that specified areas shall be reserved for public purposes

(iii) the size and delimitation of properties

(iv) roads and paths and other matters of importance to traffic, including the rights of access to traffic areas with the intention of separating the various kinds of traffic

(v) the location of tracks, pipes and transmission lines, including power supply lines

(vi) the location of buildings on the sites, including the ground level at which a building shall be constructed

(vii) the extent and external appearance of buildings, including provisions for the regulation of the density of dwellings

(viii) the use of the individual buildings

(ix) layout, use and maintenance of unbuilt areas, including provisions as to the regulation of the ground, fences, conservation of plants and other matters pertaining to plants, and the lighting of roads and other traffic areas

(x) preservation of landscape features in connection with the development of an area allocated for urban purposes or a summer house district

(xi) provision of or connection to communal facilities situated within or without the area comprised by the plan as a condition for the taking into use of new buildings

(xii) provision of screening facilities such as a planting belt, a baffle, a wall or similar construction as a condition for the taking into use of new buildings or a changed use of an unbuilt area

(xiii) establishment of landowners' associations for new single family house areas, industrial areas or areas for leisure-time houses, including the landowners' compulsory membership and as to the right and duty of the association to be in charge of the establishment, operation and maintenance of communal areas and facilities

(xiv) preservation of existing buildings, so that buildings may be demolished, converted or in other ways altered only subject to the permission of the municipal council

(xv) keeping an area free from new building if buildings may be exposed to collapse, flood or other damage which may involve danger to the user's life, health or property

(xvi) cessation of the validity of expressly mentioned negative easements if the continued validity of the easement will be contrary to the purpose of the local plan, and if the easement will not cease to be valid under the provisions of the local plan

(xvii) joining of flats in existing housing

(xviii) noise insulation of existing housing

(xix) prohibition of major building works in existing buildings, so that such works can only be carried out subject to the permission of the municipal council or if they are required by a public authority in accordance with the law (Section 18).

The Minister for the Environment may decide that local plans can contain provisions about matters other than those mentioned above. A local plan must not contain provisions contrary to a municipal structure plan (Section 19).

D5.37 Local plans can thus contain provisions about virtually any aspect of development, eg land use change, site layout, open space and landscaping, the constructional details of new buildings, urban renewal and preservation. As well as being concerned with physical matters, local plans can also include provisions relating to legal and administrative aspects such as the relationship to other planned sectors and the establishment of landowners' associations. A local plan can provide that certain conditions are complied with before new buildings can be used, eg connection to communal facilities, establishment of a tree belt etc. The local plans are legally binding on the individual landowner (Svensson, 1981).

D5.38 Under Section 20 the local plan has to be accompanied by an account of the relationship of the plan to the municipal structure plan and the general planning for the area. The purpose of the local plan must be stated in the plan. The plan also has to state whether its implementation is dependent on permissions or exemptions granted by authorities other than the municipal council.

D5.39 The local planning process contains the following procedures (see Figure D10).

(i) After the adoption of a proposal for a local plan the municipal council publishes the proposal together with a brief description of its content and the legal effects of the plan. The council may publish alternative proposals (Section 21).

(ii) Any member of the municipal council who has demanded to have his divergent opinion as to the adopted proposal for a local plan entered in the minutes of the municipal council may demand that the municipal council simultaneously publishes the divergent opinion of the member concerned and a brief statement of his grounds drafted by him.

(iii) The municipal council simultaneously with the publication of the proposal gives information about the rules in Section 22 and about the time allowed for the submission of objections and proposed amend-

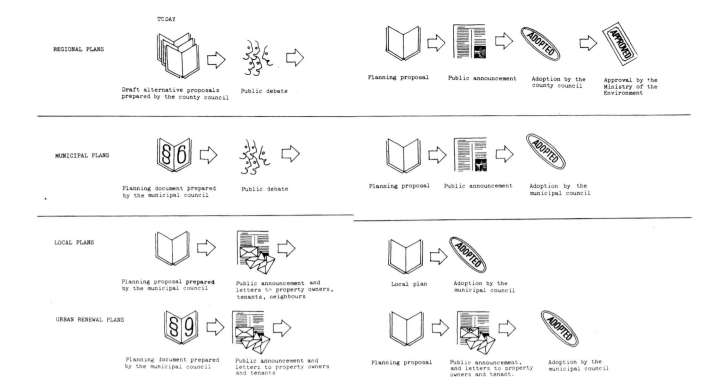

**Figure D10    The Planning Process (after Gottschalk)**

Figure D10 shows graphically the planning process as it applies in relation to the procedures involved in the production of the main plans. It identifies in particular the significance of public participation in the process.

ments. Section 22 relates to the fact that when a proposal for a local plan has been published, properties comprised by the proposal must not be built on or used in other ways which create a risk of prejudice to the content of the final plan.

(iv) At the same time as the publication of the proposal the municipal council gives notice in writing to

(a) the owners of properties comprised by the plan and the tenants and users of these properties

(b) the owners of properties outside the area of the plan and the tenants and users of these properties to the extent to which in the opinion of the municipal council the plan is of material importance to them

(c) the societies and similar associations domiciled or working in the municipality which have asked the municipal council in writing to be informed of proposals for local plans

(d) the Minister for the Environment, the county council and, in the metropolitan area of Greater Copenhagen, the metropolitan council

(e) the state authorities to which the plan is of special interest.

The notice states the time allowed for the submission of objections and proposed amendments.

D5.40    The procedure prescribed for the preparation and adoption of local plans aims at involving all interests in the planning process. The result of this is that the procedure has become time-consuming. The proposal has to be submitted for public inspection for a period of not less than eight weeks. If during that period the state authorities to which the plan is of special interest veto the proposal in relation to these special interests which they must protect, the council cannot finally adopt the proposal until agreement has been arrived at between the council and the authority in question about the content of the plan. If such agreement is not arrived at, the dispute may be brought before the Minister through the National Agency for Physical Planning. Some vetoes of this kind have been issued, but the majority of the disputes have been settled by negotiations without the involvement of the Minister. The Minister has, however, upheld a veto issued by the former National Agency for the Protection of Nature, Monuments and Sites against the building of a hotel in the neighbourhood of Frederiksberg Castle in North Zealand, so that the hotel project had to be abandoned (Christiansen, 1986).

D5.41 The Minister for the Environment assists the municipal councils with guidance and may lay down detailed rules about the planning of the municipalities (Section 51). The county council has to ensure that a proposal for a local plan is in agreement with the regional plan. The Minister may decide that the county council also controls other matters, such as that the proposal complies with directions issued under Section 51 above.

## Urban Renewal

D5.42 The current legislative base for urban renewal is provided by the 1982 Urban Renewal and Improvement of Housing Act. This enables larger urban areas and a wider span of activities to be covered, and offers greater opportunities for public participation than its predecessors the 1980 Urban Renewal and Housing Act and the Slum Clearance Act of 1969. A distinction is made in the legislation between urban renewal and rehabilitation. The former can involve a larger area with many properties and might embrace redevelopment, reorganisation of traffic, removal of incompatible uses, provision of community services and open space, whilst the latter is concerned with the improvement of individual houses in those areas where an acceptable standard can be achieved solely by the rehabilitation of individual properties (Svensson, 1986). The overall aim in the most recent act is to improve standards through modernisation rather than by demolition and new construction which were given priority in the earlier legislation.

The urban renewal process consists of the following stages.

(i) Preparatory phase. When an urban renewal project is being prepared, the municipality co-ordinates its content with other local planning, for the renewal is intended not only to provide better housing and properties but also, as mentioned earlier, to ensure proper traffic and environmental conditions and to include schools, institutions for children and the elderly, etc. An urban renewal decision should only comprise an area that can be renewed within 3-4 years.

(ii) Report phase. Before a renewal proposal is prepared, the municipal authority must issue a report on the problems of the district concerned so that the owners and tenants affected and other interested parties can discuss the basis for the planned measures. The report must include a record of the conditions in the various properties and an account of the authority's plans. The report must be open for debate for at least two months, and the inhabitants of the district affected are entitled to raise objections to the plans.

(iii) Proposal phase. On the basis of this debate, the municipal authority prepares a renewal proposal. This can contain minority views, from members of the council, for example, or views presented by an association of owners and tenants. The authority must also send the proposal to the property owners and tenants. The proposal must state what measures are planned for each property and which properties the local authority intends to take over and which properties the owners themselves can renovate. The municipality must make known any plans it may have for demolition of properties and for establishing communal areas and communal installations, eg yards, tenants' facilities, boiler-house and communal laundry. A local plan for the district may include criteria for the development of the district, environmental and traffic conditions, etc and these aspects can be included in the authority's proposal. When issuing its proposal, the authority must at the same time notify owners and tenants of the rules for public grants, rehousing, rent subsidies, and financial assistance toward rent deposits. The proposal must be open for debate for at least two months, and the tenants are entitled to raise objections or propose changes.

D5.43 Section 3 of the Urban Renewal and Improvement of Housing Act lays down standards for modernised properties. Some of the standards relate to the entire property.

(i) The building must be of sound construction. Reinstatement and exterior renovation works must include sealing roofs, windows, and doors and façades, and putting these into proper condition.

(ii) The building must be secured against fire.

(iii) The building must have a modern heating system. It must be prepared for the district heat supply planned for the area in question, possibly via a piped system. In areas in which a district heat supply is not planned, the property must be equipped with a modern, economical heating system.

(iv) Energy-saving measures must be introduced. The building must be insulated, windows must be tightened and fitted with two layers of glass, and the heating system must operate economically.

(v) The building's façade must be kept in good repair, taking due account of the general architectural qualities of the district.

(vi) The property's open spaces must be suitable for recreation. The tenants in each property must have access to open spaces suitable for the needs of the different age groups.

D5.44 Other standards relate to individual housing units.

(i) Each housing unit must have at least one living-room of 18 square metres or two living-rooms of reasonable size.

(ii) The housing unit must have a kitchen, normally 6 square metres minimum.

(iii) The housing unit must have at least one toilet with a washbasin in it.

(iv) The housing unit must have a hot water supply and a bathroom or be piped for this.

(v) The housing unit must be properly ventilated and sound-insulated.

D5.45 The municipality must notify the owners and tenants affected as soon as an urban renewal proposal has been adopted. The decision must be registered on the properties affected. Thereafter, no changes that are at variance with the decision (for example, major structural changes) may be made and no new loans be raised in the properties without the consent of the municipal authority. The authority then draws up a time schedule for the works to be carried out so that the owners and tenants know the order in which the various properties are to be improved etc. When a renewal process is planned, a distinction is made between properties which the municipal council may have to acquire (ie properties requiring so much renovation that this is deemed to be beyond the capacity of the owner) and properties where the owner himself must take care of the renovation (owner-renovated properties). The former must be acquired by the council if this is demanded by the owner. Besides properties where demolition is the only solution, they are typically properties requiring so much renovation that the occupants must be rehoused. The tenants of such an acquisition property may continue their tenancies until the renewal project is implemented.

D5.46 It is also the municipal authority that takes the initiative in the case of properties to be renovated by the owners - by requesting the owner to prepare a renovation project. The owner must discuss the project with the tenants before submitting it for official approval. The tenants must be informed about the content of the project, the cost, and the effect on their rent. The tenants than have a period of six weeks in which to propose changes or to veto the entire project or parts of it. In such cases, the redevelopment project can be implemented only by order of the municipal authority. Such an order can comprise only specific works, such as

(i) a modern heating system

(ii) measures to reduce fire and health hazards

(iii) measures generally improving the district or the building

(iv) measures accepted by a majority of the tenants.

D5.47 A central approval of urban renewal plans by the Ministry of Housing is not required. It is acknowledged that urban renewal, like municipal planning, is chiefly a local political question, so it is the municipal councils which are responsible for urban renewal and rehabilitation. However, central government, through its allocation of loans and subsidies, exerts significant control. Special non-profit associations for urban renewal and rehabilitation have been established to carry through the planning and implementation of projects. These urban renewal companies operate in partnership with the local authority but it is the latter which is legally responsible.

D5.48 Owners and tenants in renewal area can influence the municipal authority's decisions, and it is very important for them to exercise this influence at as early a stage of the process as possible. The local authority must consider the wishes of owners and tenants concerning the future standard of the properties and their wishes concerning phased implementation and/or the order in which the works are to be carried out. While the authority is in process of issuing its report and proposals, owners and tenants may raise objections and submit proposals for changes. Once the authority has reached decisions or adopted a redevelopment proposal owners and tenants are entitled to appeal against these (see paragraphs D6.18-D6.19).

D5.49 In districts where there is no need for a renewal decision, the municipality can adopt a housing improvement proposal, but only in respect of residential properties that are expected to remain in use for at least 15 years. The rules for housing improvement are less extensive than those for renewal. A housing improvement proposal must be publicised and sent to the affected owners and tenants. They must also be informed about registrations of the properties and receive the general information about public grants, rehousing, rent subsidies, etc. When the proposal has been published, the owners and tenants have a period of two months in which to object to the proposal. The municipality can then adopt the proposal. The procedure for implementation of a housing improvement decision corresponds to the rules for a renewal decision in the case in which the owner himself can carry out the works. If an owner and at least half the tenants of a given property so request, the municipality can adopt a housing improvement proposal in respect of that one property. The proposal can be adopted without prior publication. The owner and the tenants must be notified of the decision, which must also be registered on the property. If a request to a local authority for a housing improvement proposal is accompanied by an improvement project with information on financing etc the municipality need not obtain statements from the tenants. However, the individual tenants retain their right to veto works in their own homes.

D5.50 The municipal authority must inspect housing and living-rooms to ascertain whether these constitute a health or fire hazard. This can be done through a housing commission. If shortcomings of this nature in a property can be remedied in a reasonable way, the authority may order the owner to remedy them The owner is entitled to a public grant for such works under the same rules as for redevelopment works. If the shortcomings are so extensive that it does not pay to remedy them, the municipality can condemn the property, ie prohibit occupation and order the premises to be cleared. It can follow a condemnation order up with a request for the property to be closed, and a demolition order. An owner can try to get such a condemnation order lifted by proposing improvements to the property. However, if the municipality approves such proposals and lifts the condemnation order, the owner cannot obtain a public grant for implementation of the proposals. Condemnation does not in itself entitle the owner to compensation. The tenants of a condemned property must be informed about the municipality's decision but may not veto this as in the case of redevelopment decisions. However, both the owner and the tenants can appeal the authority's decision to the redevelopment board.

## Conservation

D5.51   There are two principal acts relating to conservation: the Conservation of Nature Act, which includes the protection of ancient monuments, and the Preservation of Buildings Act. The central government body responsible for the administration of the former is the National Forest and Nature Agency within the Ministry of the Environment which in addition to its forestry role also operates the Raw Materials Act so bringing it into close collaboration with the Geological Survey of Denmark. The Preservation of Building Act operates under the National Agency for Physical Planning with the active involvement of the national Historic Buildings Council.

*Nature Conservation*
D5.52   Nature conservation is related both to the landscape and to the safeguarding of wildlife and plants and their habitats, but the legislation also includes the aim of making areas accessible to the general public and there is an Acquisition of Rural Property for Open-air Activities Act which can be evoked if necessary to achieve this. By 1984 about 5 per cent of Denmark had been scheduled for nature conservation or had been acquired by public authorities. As well as the statutory procedures for the protection of nature and landscapes under the Conservation of Nature Act (see paragraph D4.43) more general conservation planning for the protection of the countryside can be carried out. It comprises the formulation of both a planning policy and a Nature Conservation Plan for each county and it is these conservation plans, which must be compatible with the regional plans, which constitute the framework for the administration of the legislation (Kristiansen, 1984).

D5.53   The plans must indicate overall aims for the protection, management, and use of the region's nature areas and cultural landscapes in accordance with the guidelines laid down in regional plans for the use of open land, and taking national conservation interests into consideration. They must identify the areas in the region which, within the framework of overall physical planning, ought to fulfil the purposes of the Act. The nature conservation plans must further indicate suitable methods and means for ensuring the enforcement of proposals for the protection, management, and use of the areas.

They thus represent the general framework of protection, care, and interpretation of monuments and sites in the context of the cultural landscape (Kristiansen, 1984).

D5.54   The plans are prepared by the counties but approved by the conservation boards that are present in each of the 26 conservation districts that cover the country (see paragraph D4.44). There is also a Superior Conservation Board for the whole country and this is composed of a chairman, two supreme court judges and at present eight members elected by the political parties represented in Parliament. The chairman is head of the secretariat of the Superior Conservation Board.

D5.55   Following formalised and comprehensive public participation, the conservation boards and the Superior Conservation Board decide on the implementation and sometimes annulment of conservation areas and in this connection determine the amount of the compensation which has to be paid out of public funds to the owners of land in the protected areas.

D5.56   A completed nature conservation plan entails binding precepts, notably on how the area in question may be utilised, and thus the boards decide on any conflict of interests. The Superior Conservation Board is the court of appeal for decisions made by the conservation boards in this respect and for compensation determined by the boards and also any major conservation programme must necessarily be scrutinised by the Superior Conservation Board in every respect.

D5.57   In addition to determining compensation, the conservation board decides applications for exemption from already implemented conservation programmes and from several generally applicable provisions in the Conservation of Nature Act, eg the ban on changing conditions in a zone along the coast or around lakes. These decisions made by a conservation board may be the subject of an appeal to the Superior Conservation Board.

D5.58   As well as the provisions relating to nature conservation in general the Conservation of Nature Act makes it possible to schedule areas as nature reserves and at the same time to determine how such areas shall be used. The procedure has much in common with the local planning procedure. The process may be initiated by the Minister, the county authority, the municipal councils or the private organisation, The Danish Society for the Preservation of Natural Amenities (which has 200,000 members). However, it is the conservation board which decides whether an area should be scheduled as a nature reserve. While the matter is being considered, it is forbidden to use the area in question in any way which is contrary to its proposed conservation.

D5.59   In contrast to local planning, but as in the case of restrictions on use imposed by conservation plans, the scheduling of areas as nature reserves will entitle the landowner to claim compensation, which will also be determined by the conservation board. The reason underlying the difference between local planning and the scheduling of areas as nature reserves is largely historical; but there is also the real difference that scheduling of areas as nature reserves, unlike planning policies, will often affect lawfully existing conditions or positively impose duties to act in a certain way. The state normally pays 75 per cent of the compensation and the relevant county authority 25 per cent.

D5.60   The decisions of the conservation board as to scheduling of areas as nature reserves may be appealed against to the Superior Conservation Board; and all scheduling of areas as nature reserves must be brought before the Superior Conservation Board for re-examination if the compensation awarded to the landowners exceeds 100,000 kroner. The decision of the Superior Conservation Board concerning compensation may be brought before a special valuation commission.

### Building Conservation

D5.61   The Preservation of Buildings Act 1986 provides for the listing of buildings in order to ensure their protection. The decision to list a building is taken centrally by the National Agency for Physical Planning and once protection has been established the building may neither be demolished nor modified without the permission of the Agency, and protection requires the owner to maintain the building to prevent it from falling into disrepair. As this maintenance may place a heavy financial burden on the owner the Agency has funds to provide subsidies or low interest loans. Formerly, two grades were used in the listing of buildings. Buildings of specially great architectural or cultural-historical value were placed in Class A and the rest in Class B with different levels of control and financial assistance relating to each. Now the legislation allows for only one grade of listing so that the same conditions generally apply, irrespective of when the building was listed (Saaby, 1984). The Agency's responsibilities for scheduling buildings and responding to applications for permission to carry out work or demolition are handled by the national Historic Buildings Council which also has to prepare every five years a list of the buildings preserved under the Act, together with a report on the activities of the Council, and both these are submitted to Parliament.

D5.62   Areas rather than individual buildings can be conserved but the emphasis on individual buildings and monuments has meant that only by linking individual buildings together, which all fulfil the demands of the legislation, that it can in rare cases be applied to the protection of coherent architectural environments. Prior to the passing of the latest Act in 1986 it had been hoped that provisions would be made for the protection of wider architectural or urban environments but these continue to be conserved by the general planning legislation. From February 1977, when the Municipal Planning Act came into force, it has been possible to give protection to groups of buildings and larger areas in towns and villages. Although the Act does not give emphasis to conservation of the environment, it has been described as marking a

decisive improvement in the possibilities of practising integrated conservation (Skovgaard, 1978). Areas are preserved by means of conservation plans which municipalities adopt and which may subsequently be incorporated into local plans. With the exception of very special buildings it has only been since the passing of the Municipal Planning Act that legal protection of villages and buildings in open country has been possible; the former town planning byelaws applied only to towns, but local plans within the provisions of the Act may of course be adopted for rural zones.

### Resource Use

D5.63   The Raw Materials Act requires that the recovery of raw materials be subject to an integrated plan which takes into account other social interests. These Resource Utilisation Plans are administered at central government level by the National Forest and Nature Agency, but the counties are responsible for giving permission for quarrying and mining. The counties, again under the overall control of the Ministry of Agriculture, have to evaluate the soil as a production factor for the agricultural and horticultural industries with a view to ensuring that those areas with the most consistently high yields are used primarily for cultivation rather than any other purpose.

## Comments and Conclusions

D5.64   The current Danish planning system was established in a period of economic growth when urban expansion was active and thought likely to continue in the future. The situation changed dramatically in the late 1970's and the economic stagnation continues today. Thus, the present requirement is not to plan for the accommodation of growth but for the management of change, in both urban and rural areas. The reproduction sector in terms of urban renewal and housing improvement is taking over from the production sector involving development on new sites (Lemberg, 1982). It could be argued that the Urban and Rural Zones legislation in particular is less relevant now than when it was enacted because of the reduced pressures for development as well as because the planning system at the municipal level can now bring about effective controls. At the time it was introduced however there were rapidly increasing pressures for development and inadequate controls available for countryside protection. The countryside has thus been preserved by legal restrictions rather than by financial management. There has also been a change in emphasis from physical planning to environmental protection and management and the current supplements to the regional plans devote particular attention to this aspect especially in the countryside. Thus, the sector planning of the different agencies is taking on a relatively more significant role and as this has to be integrated with physical planning the ability of the system to cope is coming under increasing scrutiny.

D5.65   The principle of framework management and

control relates not only to the vertical integration of policies and plans produced at different administrative levels but also to the horizontal co-ordination of plans from the different sectors through the physical planning system. This has proved difficult. Sector plans are often formalised in a different way, with only about 50 per cent of them produced at municipal level, with different time scales etc. Thus co-ordination has not been easy to achieve, especially between physical and economic planning which is a particularly significant relationship. Here the different time horizons are mainly responsible (Christiansen, 1986). The problem has been succinctly expressed in terms of 'just think of a property which an intended road will cross, which has good agricultural land, which is situated in an area of great scenic beauty, and which is suitably situated for urban development' (Østergård, 1981).

D5.66   It was the issue of co-ordination and the necessity for flexibility for its achievement that was mainly responsible for the establishment in 1984 of a national committee on land use and real property regulation, the chief purpose of which is to streamline and evaluate the legislation on physical planning and its relationship to the sectoral planning acts. In 1985 a draft unpublished report was submitted to the Minister of the Environment in which the committee agreed to submit proposals for comprehensive changes in the formal requirements as to sectoral planning, and for a simplification of, and great flexibility in, regional planning. It was suggested that sector planning should be more independent and whilst regional planning should continue to be comprehensive, the current re-adjustment of regional plans relating to the different sectors should be less complicated. Some of the other points made by the committee suggest that

(i) planning systems within the Ministry of the Environment be simplified

(ii) regional planning be made flexible, but it must continue to function as binding, comprehensive planning

(iii) sectoral planning be made less constraining, and more problem and action-oriented

(iv) the results of planning be presented in a more understandable form in order to strengthen dialogue with the public, and the planning process should be more closely connected with the decision-making procedure in the counties and municipalities.

The committee is still in session but it is expected that a new bill will be presented in parliament this year which will attempt to achieve the general objective of 'harmonising' the system.

D5.67   The system of framework management and control has also proved difficult to operationalise in the 'vertical dimension' in terms of the links between the plans produced at the different administrative levels. The notion is based on the hierarchical principle whereby plans at the lower level fit in with those at the higher. This is predicated on policies and plans being firstly agreed at national level and then filtering through as a control to the municipalities. Often however local plans are stimulated by proposals which, if felt to be desirable, may necessitate changes in the plans at a higher level. This problem was eased in 1985 when the procedures were changed whereby all plans could be simultaneously modified, thus allowing faster and better co-ordination between them.

D5.68   Public participation in the planning process was intended to be another major feature of the reformed system. Plans are prepared and agreed through a dialogue with the public, and those produced by the municipalities, including the local plan which can be regarded as the most important in that it is legally binding, are adopted locally through democratic involvement. The procedures for plan preparation and adoption allow great opportunities for public comment and on this basis any inquiry on, or appeal against, their contents is not regarded as being necessary. The reality of public participation is rather different. The extent to which the public's views are adopted depends on the strength of their political influence. There were many thousands of objections to the proposed Hotel Scandanavia in Copenhagen but it was nevertheless allowed by the city council. Also there is the problem of how to handle comments which are outside what is regarded as the acceptable agenda (Gottschalk, 1984).

D5.69   There seems to be general agreement about the value of the annual report produced by the National Agency for Physical Planning in the Ministry of the Environment. It is useful as a review, but more particularly as a vehicle for presenting the attitudes and policy stance of the Ministry. It is widely distributed and helps to stimulate an exchange of views between central and local government and the public in general. It has also been used to provide guidance to the counties and municipalities in the carrying out of their planning responsibilities.

D5.70   The national planning directive instrument likewise has proved its effectiveness especially as a safeguarding and control mechanism. It can be introduced promptly, although parliamentary approval is required, and even more quickly cancelled when no longer appropriate. As it is easy to introduce and apply and is effective, other ministers would like to see it increasingly used in the future in relation to the spatial dimension of their sector planning activities and where there is a clear relevance for physical planning. Guidelines of the same status as planning directives can be introduced by other ministries, but because they lack a spatial dimension they are less favoured if co-ordination with physical planning is required. They are known as planning pre-conditions (*planlægnings forudsætringer*).

D5.71   It has been argued that regional planning has such a limited contribution that regional plans could be dispensed with (Allpass, personal communication). There is little doubt that the original plans that were approved by the Ministry between 1979 and 1982 were too detailed, with the exception of that for the metropolitan region, and therefore were too inflexible and restrictive for the municipalities to undertake their planning responsibilities in relation to their own particular circumstances. This occurred despite the innovation of the counties having to initiate the regional planning process by obtaining proposals from the municipalities, in order to achieve the decentralisation that the planning and local government reform was designed to achieve. The need to publish the various draft proposals and the assumptions on which they were based certainly opened up the debate but encouraged the detail that eventually emerged (Kerndal-Hansen, 1983). A more general and flexible approach has become apparent in the two-year statements that were submitted to the Ministry by the end of 1985, and the regional plan supplements that are now being produced and submitted continue this trend, thus making it easier for the municipalities to achieve their objectives and for the sectoral interests to be co-ordinated.

D5.72  The municipal planning operation has been the subject of delay because of the required integration with regional planning. The current position in relation to municipal structure planning is indicated in Plate 7 with over 100 structure plans having been adopted and another 150 municipal councils having published their proposals. There have been variations in content. Some have been more successful than others in producing a corporate policy document integrating the various sectoral interests, some are simply land use plans. Generally structure plans have been too detailed and inflexible and often it has been found necessary to amend the structure plan to conform to the proposed local plan to remove the inconsistencies which would prevent the local plan from being adopted.

D5.73  The local plan produced by the municipality is the most significant policy document because it is legally binding and provides the basis of control. It is also the final stage in the framework management and control operation and thus is meant to reflect the outcome of strategic national and regional goals and objectives. Over 10,000 plans have so far been adopted. They vary considerably in terms of the area they cover, with some being prepared in relation to proposals for one or two buildings, and also in terms of their content. Some are extremely detailed and therefore are capable of exercising very strict control, while others, particularly in Copenhagen, are very imprecise with such a wide variety of activities permitted that the term 'empty local plans' has been used to describe them. The advantage of this sort of plan is that they maximise flexibility but in so doing allow the politicians to make pragmatic and politically expedient decisions. Examples of both approaches were illustrated in Figures D3 and D4).

D6.1   The general provisions relating to appeals against decisions on physical planning matters are contained in the Municipal Planning Act. The Building Act contains regulations relating to appeals on building permits and the Urban and Rural Zones Act allows appeals on decisions given under its authority. Under the Municipal Planning Act, appeals to the Ministry of the Environment can generally be made only on the legal aspects of decisions, typically whether the legislation allows the council to take specific decisions or whether it has exceeded its powers. Thus, the merits of a decision cannot normally be challenged because this is regarded as a question of local politics and so exclusively the responsibility of the council.

## The Municipal Planning Act and Appeals

D6.2   Under Section 48, the decisions of a municipal council concerning matters comprised by the Act may, if legal problems are involved, be the subject of appeal to the Minister for the Environment. These matters include

(i)   a compulsory purchase order made under Section 34 which allows the municipal council to acquire real property in private ownership or private rights in respect of real property when the compulsory purchase will be of material importance for the implementation of a local plan. The appeal must be lodged within four weeks from the date when the order came to the appellant's knowledge

(ii)   an order of the municipal council to remove advertisements, light installations and similar parts of a building which cause inconvenience or have a disfiguring effect in relation to the surroundings (Section 44)

(iii)   an order of the municipal council to the owner of a site where commenced building operations are given up or postponed for an extended period and the partially constructed building has a disfiguring effect in relation to the surroundings to remove the building or to bring it into such a condition that it will no longer have a disfiguring effect. The same provision for an order exists where a building which is partially demolished or ruined remains for a long period in such a condition and has a disfiguring effect in relation to the surroundings. These provisions do not apply to farm buildings (Section 44)

(iv)   an order of the municipal council to the owner of a site used for storage, refuse disposal or similar purposes if this results in a disfiguring effect in relation to the surroundings to undertake fencing, planting or other measures which may relieve the condition (Section 44)

(v)   an order of the municipal council forbidding in an area where there are buildings for residential purposes, the construction of industrial or commercial buildings detrimental to the environment or the use of existing residential buildings or unbuilt sites in the area for industrial or commercial purposes detrimental to the environment (Section 45)

(vi)   similarly, if in an area there are industrial or commercial buildings the council may issue an order to forbid that residential buildings are taken into use for residential purposes, if residential buildings in the area for environmental considerations are deemed to be irreconcilable with the carrying on of industry and commerce (Section 45).

D6.3   These last two proceedings under Section 45 where an appeal might arise following a prohibition or an order do not apply when the area in question is covered by provisions about the building use in a local plan. In all the above matters an appeal must be lodged within four weeks of the prohibition or order being received. There is no appeal provision against the content of the municipal structure plans and local plans produced by the municipality nor the regional plans produced by the county, nor is there any form of inquiry into the proposals in plans.

D6.4   Legal proceedings concerning decisions made by the municipal council or the Minister of the Environment under the provisions of the Municipal Planning Act may be instituted before the local court under the jurisdiction of which the property is situated. However, legal proceedings concerning compulsory purchase under the Act have to be brought in the first instance to the High Court. Proceedings must be instituted within six months from the date when the plaintiff was notified of the decision or when the decision was publicly announced (Section 49).

D6.5   A civil action against a person who has committed an infringement relating to building percentages, height of buildings, advertisements, light installations and similar parts of a building or in relation to the provisions of a local plan may be pursued through the courts.

## Decisions on Building Permit Applications

D6.6 Under Section 23 of the Building Act, decisions of the municipality on applications for building permits can be appealed to the county authority or in the case of Copenhagen and Frederiksberg to the Minister of Housing provided the decision concerns a matter of interpretation of the Act or of a regulation issued in pursuance of the Act. An appeal can also be allowed to proceed if in the view of the appeal authority the decision in question is of general interest or has substantial consequences for the person making the appeal. Decisions of the county authority can be appealed to the Minister of Housing provided the decision concerns a matter of interpretation of the Act or of a regulation issued in pursuance of the Act. A ruling by the county authority on whether the conditions of appeal (in relation to whether the decision in question is of general interest or has substantial consequences for the person making the appeal) have been satisfied cannot be brought before any other administrative authority. Decisions of the municipal and county authorities cannot in other circumstances be appealed to any other administrative authority.

D6.7 In the guidance notes included in the Building Regulations for Small Buildings, the position is clarified. It is stated that objections to a decision made by a municipal authority, including the municipal authorities of Copenhagen and Frederiksberg, are permitted only in matters of interpretation. Thus, appeals cannot normally be made against decisions on building permit applications which challenge an authority's assessment of a given matter and the merits of its decision.

D6.8 The time allowed for an appeal to be lodged is four weeks from the date on which the decision was communicated to the person concerned (Section 24). A civil action to examine decisions made by the administration in accordance with the Act or regulations issued under the Act has to be instituted not later than six months after the date on which the decision was communicated to the person raising the action (Section 25).

## Decisions Based on the Urban and Rural Zones Act

D6.9 The Urban and Rural Zones Act generally requires, as has already been noted (see paragraph D3.20), permission to be given in rural zones for the

(i) parcelling-out or sub-division of land

(ii) construction of new buildings and the conversion or extension of existing buildings

(iii) taking on existing buildings and areas not built on into use for other purposes than agriculture, forestry or fishing.

D6.10 If permission is refused, the Environmental Appeal Board set up under the Protection of the Environment Act is the body responsible for making the decision on any subsequent appeal. Appeals must be lodged within four weeks from the date when the appellant concerned was notified of the decision. The Minister for the Environment is entitled to lay down rules to the effect that an appeal shall be sent to the authority which has made the decision.

D6.11 Appeals can also be made against permissions granted under the urban and rural zones legislation and a permission cannot be made use of before the expiry of the period of the appeal. If a permission has been appealed against, it cannot be made use of unless the Environmental Appeal Board rejects the appeal. The authority to which the appeal is sent must ensure that the person to whom the permission has been given is notified of the appeal. A building permit under the building law must not be given as long as the permission of the county council or the municipal council cannot be made use of. The decisions of the county and the municipal councils must contain information about the right of appeal against the decision to the Environmental Appeal Board (Section 11). Thus appeals to the Environmental Appeal Board can be brought by

(i) the applicant

(ii) anybody who must be supposed to have an essential, individual interest in the result of the appeal

(iii) the county council and the municipal council.

D6.12 The Board consists of a chairman and four other members whom the chairman selects from a group of experts according to the nature of the particular case. These experts are appointed by the Minister for the Environment for a period of up to four years on the recommendation of the following

(i) the National Association of Local Authorities

(ii) the National Association of County Councils and the Metropolitan Council jointly

(iii) the National Agency for Physical Planning and the National Agency for Forests and Nature jointly

(iv) other specially interested State authorities.

D6.13 To consider the individual appeals, the chairman of the Board nominates four appointed members, so that one from each of the four groups takes part. In connection with the consideration of appeals the Environmental Appeal Board must ensure that the decision is based on the regional and municipal planning documents and other land use planning aspects. The Minister for the Environment is entitled to lay down detailed rules about the procedure of the Environmental Appeal Board for appeals against decisions under this Act, including rules about the consideration of appeals and about fees to be paid for bringing appeals before the Board. In this connection the Minister is able to lay down rules to the effect that the chairman may decide indisputable appeals and the types of appeals that frequently occur (Section 10).

D6.14 The majority of the permissions given under the provisions of the Urban and Rural Zones Act must be published, so that it is possible for interested parties to appeal. Thus permission granted under the Act may be the subject of an appeal by a third party who feels aggrieved and so permissions cannot be exploited before the expiry of a period for appeal. Appeals against decisions made under the provisions of the legislation comprise both the legal questions and the planning policy, the discretionary part of the decisions. The decision of the Environmental Appeal Board is final as to the merits, but the legality of the decision may be brought before the courts (Christiansen, 1986).

## Environmental Legislation and Appeals

D6.15 As well as in relation to decisions on permits for new developments and activities in rural and summer house zones under the Urban and Rural Zones Act, including the use of dwellings in summer house zones for all-year habitation, the Environmental Appeal Board acts as court of appeal for several administrative decisions under the jurisdiction of the Ministry of the Environment. Amongst the most important are

(i) implementation of environmental protection legislation, particularly in so far as approvals of and orders to heavily polluting enterprises are concerned

(ii) re-use of paper and beverage-packaging and reduction of waste

(iii) extraction of raw materials (excluding oil and similar)

(iv) certain cases involving water supply

(v) certain decisions concerning the use of chemicals and chemical products, for instance pesticides

(vi) certain decisions on protection of the marine environment

(vii) certain decisions concerning the use of dwellings in summer house zones for all-year habitation.

D6.16 The Board acts on decisions made by the National Agency of Environmental Protection in cases concerning environmental protection, re-use, water supply, chemical substances and products, and protection of the marine environment. As for the extraction of raw materials, the Board acts as the court of appeal in so far as the decisions of the National Agency for Forests and Nature are concerned. In the cases concerning the legislation on urban and rural zones where the Board acts on decisions made by the county councils and the municipal councils, the National Agency for Physical Planning acts as secretariat for the Board, as it does also

in cases concerning decisions on year-round dwellings in summer house zones.

D6.17 With reference to nature conservation there is, as already indicated, a separate appeals system (see paragraph D5.56). The decisions of the Conservation Board can be challenged through an appeal to the national Superior Conservation Board, and where the issue is related to compensation a further appeal can be made to a special valuation commission.

## Urban Renewal and Appeals

D6.18 In urban renewal proposals either the owner or one quarter of the tenants may appeal against a decision to a redevelopment board. There is a redevelopment board in each county and in the municipalities of Copenhagen and Frederiksberg. These boards are composed of one judge, two building experts, one representative of the freeholders' association and one representative of the tenants' association. An appeal may, for instance, be made concerning the content of a renewal decision, including the demolition of properties, the scope of modernisation works, a housing improvement proposal etc. The same applies to the decisions of local authorities concerning the right of tenants to take over a property in co-ownership and local authority orders to owners.

D6.19 A former tenant who is not assigned a flat in a property after redevelopment may lodge an appeal alone, and decisions about redevelopment rent subsidies or an individual veto can also be the subject of an appeal by individual tenants. Otherwise, an appeal normally has to be supported by at least one quarter of the tenants. When the new, higher rent has been fixed on completion of the building accounts, there can be an appeal against it to the local redevelopment board. The owner can appeal against an expropriation order to the National Housing Agency and can take the case on to the courts. A decision on compensation handed down by a valuation commission can be appealed against by the owner to the superior valuation commission and then to the courts. An appeal to a redevelopment board must be lodged in writing not later than six weeks after the decision in question has been made public. The decision of the redevelopment board can be the subject of an appeal to a housing court within four weeks. An appeal does not normally have a suspensive effect, ie implementation of the decision appealed against is not automatically postponed by an appeal to a higher authority. However, on request, the redevelopment board can postpone, say, a demolition, if an appeal has been made against a decision on this. A complaint by an owner about an expropriation order must be submitted to the National Housing Agency within four weeks, and the Agency's decision must be appealed to the courts within six months. The decision of a valuation commission cannot be appealed to the courts before the superior valuation commission has reached its decision. This decision can then be appealed to the courts within six months.

## The Ombudsman

D6.20 Since the amended Constitution Act was passed in 1953, Parliament is able to appoint an ombudsman to supervise the administration of the State and of local government councils to the extent to which their decisions may be appealed against to state authorities. Any citizen is at liberty to bring a case before the ombudsman, but normally the ombudsman will not start to examine a complaint until the right to bring a decision before a higher administrative authority for final decision has been used. The ombudsman may, however, on his own initiative, take up matters and call the attention of Parliament or the minister concerned to disadvantageous conditions or similar circumstances. He sometimes does this in the light of information in the press. The ombudsman submits an annual report to Parliament, which includes examples of the complaints which have caused him to express criticism or make suggestions to the authorities (Christiansen, 1986).

## Comments and Conclusions

D6.21 Clearly there is a variety of appeal systems and procedures related to planning, building and associated environmental legislation. The result is that the public is generally confused. Appeals therefore fall into the general criticism of the Danish planning system that it is too complicated and difficult to operate. In the review of the entire planning system that is currently being undertaken (see paragraph D5.66) there is a proposal to rationalise and simplify the appeals procedures as part of the general aim of 'harmonising' the system. It is not envisaged, however, that the significant features of the system, such as the absence of any appeal or inquiry mechanism in relation to the content of plans and of the right to challenge decisions on building permits, will be altered.

# D.7 SUMMARY AND CONCLUSIONS

D7.1 Denmark is a constitutional monarchy with a population of 5.1 million, one third of whom live in the Greater Copenhagen area. The constitution protects private property rights but expropriation is allowed if necessary in the public interest. Compensation is payable when property is compulsorily acquired but not when restrictions are placed on its use, except in rural areas where the lawful use may be impaired by conservation policies. Government is exercised at the county and municipal levels, as well as at the national.

D7.2 Physical planning is undertaken at the three governmental levels with central government and the counties mainly responsible for strategic planning and the municipalities for more detailed planning and the control of development. The counties, along with the national ministries, also have particular responsibility for many aspects of sectoral planning with the aim of achieving co-ordination through the physical planning system. Integration between the different governmental levels is an important objective so that authority can be decentralised as appropriate.

## Planning Legislation

D7.3 The first town planning act was passed in 1925 and by the 1960's there had evolved such a proliferation of legislation, plans and controls that the term 'the planning jungle' became common parlance (Østergård, 1981). The need to simplify, achieve greater co-ordination and produce more effective controls was accentuated by the affluence and growth that were present at the time. Thus, major changes were introduced creating the physical planning system that remains in place today.

D7.4 The changes were brought into being in three stages.

(i) A more effective means of controlling the spread of development was regarded as one of the most important initial requirements and this was achieved by the passing in 1969 of the Urban and Rural Zones Act. It divided the entire country into urban zones where proposals might be allowed and rural and summer house zones where they would not be allowed without special permission unless they were related to the functioning of these areas. These restrictions were relaxed in 1987 so that redundant farm buildings may be used for business enterprises if

the maximum number of people employed is no more than five.

(ii) In 1973 the National and Regional Planning Act and an Act for the Regional Planning of the Metropolitan Region of Copenhagen were established to provide the framework for the production of strategic planning policies and control at the national and county levels.

(iii) Finally in 1975 the Municipal Planning Act and Building Act were enacted to enable local councils effectively to plan and control development in their areas.

## Local Government Structure

D7.5 The modifications to the planning system were accompanied in 1970 by a reform of local government whereby the two tier structure below the national level was retained but the 25 counties were reduced to 14, and in reality 11, because a special act was passed to create a Greater Copenhagen metropolitan authority that included three counties. More drastically the 1,388 municipalities were reduced to 275. The main intention was to rationalise the structure and establish larger units with more resources to enable them to cope more effectively with their responsibilities and so allow a much greater degree of decentralisation than had been possible previously.

## Plans and Policy

D7.6 Physical planning policy is produced at the three levels of government.

(i) There is no national plan in terms of a finite document but the National Agency for Physical Planning under the overall responsibility of the Minister of the Environment provides policy direction mainly through

(a) an annual report on national planning which combines being a review and providing guidance to the counties and municipalities with statements on future policies

(b) more informal policy statements to the local government authorities

(c) its approval of the regional plans of the counties

(d) national planning directives which in a negative way support policy through restricting activities in particular locations and safeguarding other areas for a specific purpose

(e) circulars which complement and amplify the legislation but which largely relate to procedures and legal interpretation.

(ii) The counties and the Greater Copenhagen Council have to produce a regional plan which has to be reviewed formally every two years with supplements published every four.

(iii) The municipalities are obliged to prepare a municipal structure plan covering their entire area, and local plans where and when appropriate. The structure plan is in two parts, the first laying down the general planning framework, and the second providing a more detailed context for those parts of the municipality where local plans will be required. The local plans are the most important for control purposes because they are legally binding documents. They can contain provisions relating to virtually all aspects of planning and also to administrative and legal matters. Both the structure plan and the local plan, unlike the regional plan, are adopted locally, but they have to conform to the regional plan.

D7.7 Physical planning cannot be discussed in isolation from other forms of planning. It has to be carried out in the context of economic planning and the sectoral planning of a wide range of agencies and central government ministries as well as that of the counties and municipalities themselves. Physical planning is also seen as the vehicle for the co-ordination and integration of these sectoral interests at each of the three governmental levels within the overall framework of economic policy. Thus, the various plans produced through the physical planning process are meant to operate as the corporate policy documents for the counties and municipalities.

D7.8 The principle that has been adopted to attempt to achieve this integration is known as *rammestyringssytemet* (framework management and control system) which implies that physical planning policy at one level should be in agreement with that at a higher level. It also means that there must be co-ordination not only between levels but between the various sectors, and this is especially important at the regional level where many of the sectors manifest their interests.

## The Control of Development

D7.9 There is not a single comprehensive definition of development which provides the basis for exercising planning control. The full picture of what building work, demolition, change of use and other operations which in aggregate identify the scope and content of development that is subject to control can be established only by examining the various regulations, plans and statutes that

are in force. These precisely specify what may be undertaken or what may not be undertaken unless permission is granted in particular circumstances.

D7.10 The need for planning permission in the English sense is not required, only the need, in most cases where building or demolition work or change of use is proposed, for a building permit. The building permit procedure usually involves the following

(i) a permit for the proposed work

(ii) notice given before the commencement of work

(iii) notice provided at various stages of the work, if requested in the building permit

(iv) report/announcement of the completion of the work

(v) permit to use the building.

Minor proposals may involve only stages (i) and (iv) above. The building permit may be granted without conditions, but normally conditions are attached. Only very rarely is a building permit application refused as the requirements are clearly identified and the proposal is therefore unlikely not to conform. The decision is made by the building department of the municipality.

D7.11 In deciding applications for building permits, the municipality has to ensure that the provisions of the Building Act have been followed, together with the national Building Regulations 1982 and the Building Regulations for Small Buildings 1985, with subsequent amendments, issued in pursuance of the Building Act. The presence of adopted local plans, however, partly supplants and partly supplements the general national building regulations in the area that they cover, and where they exist, building permits can be granted only if the proposal conforms to their provisions.

D7.12 Local plans must be prepared before the sub-division of large areas or before major building work, including demolition, is carried out. The definition of 'large' and 'major' is not specified and sometimes subject to debate. Thus, development proposals often stimulate the preparation and adoption of local plans. As the coverage of local plans increases, with over 10,000 now having been adopted, so they are replacing the Building Regulations as the most important instrument of planning control.

D7.13 The normal building permit procedure does not always have to be followed.

(i) Some proposals are not subject to the Building Act and the Building Regulations, eg some engineering work carried out by statutory undertakers.

(ii) Minor changes to small buildings do not require a building permit.

(iii) The construction of small extensions to garages etc and minor demolition work have only to be reported to the municipality.

(iv) More major work requires a building permit for the proposal and an announcement to the municipality of its completion.

(v) Some proposals, such as for outdoor swimming pools and agricultural buildings, are subject to only a limited number of provisions in the Building Regulations.

(vi) Applications in certain cases can be made for dispensation or exemption from some of the regulations.

The provisions in the Building Regulations relating to the above variety of procedures may be over-ruled by the content of an adopted local plan.

D7.14   Before granting a building permit, the municipality has to ensure that the proposal is in conformity with a wide range of other legislative requirements, and where additional permissions have to be obtained before the development can proceed these have to be stipulated in the conditions on which the permit is issued. These will vary depending on the location and sensitivity of the proposal. Amongst the most important permissions are those

(i) relating to developments in rural areas not associated with their traditional function. Under the Urban and Rural Zones legislation an application for zone dispensation has to be made. In summerhouse zones the more major development proposals can take place only if in conformity with an adopted local plan and this is now becoming increasingly true in rural zones also

(ii) in urban renewal and housing improvement plan areas to ensure conformity with their provisions

(iii) from the Historic Buildings Council in relation to buildings scheduled under the Preservation of Buildings Act

(iv) for quarrying and mining if consistent with Resource Utilisation Plans.

D7.15   The mechanism exists for very rigorous control over physical and more general environmental change. Any infringement against the provisions of the legislation, Building Regulations, local plans and the conditions of building permits renders the offender liable to a fine. The system can however be operated with some degree of flexibility. This is possible mainly through the local plan since it predominates over the Building Regulations when it has been adopted. Thus a proposal which infringes the Building Regulations may eventually receive a building permit when a local plan is eventually adopted. In the absence of a proposal, the municipality may achieve flexibility through preparing, as is currently frequently occurring, very general local plans which will enable building permits to be granted for a wide range of different purposes, in buildings of different architectural styles etc. There are however checks and balances in the system and the scope for municipal councils to depart

from regional and national frameworks is restricted by the requirement that local plans have to be consistent with those at a higher level, and the Minister's power to call in and if necessary veto local plans.

## Appeals

D7.16   The planning system generally may have been confusing in the 1960's but the appeals situation is certainly confusing today. In general terms appeals may be made against

(i) decisions on building permits but only in terms of legal aspects and not in relation to the substantive merits of the decision except where the decision is of general interest or has substantial consequences for the person involved. The appeal, or whether it can be allowed to proceed, is decided by the county authority. If the appeal is concerned with legal matters, it may be taken to the Ministry of the Environment if it was dismissed by the county. County decisions under the Building Act are appealed directly to the Ministry

(ii) orders to enforce compulsory purchase to prevent the disfigurement of the environment and to stop unauthorised uses, except in adopted local plan areas where fines are immediately invoked. These appeals are decided by the Ministry

(iii) the legal aspects of decisions made by municipal councils or the Minister under the Municipal Planning Act. These are heard before the local court or the High Court if compulsory purchase is involved

(iv) the granting of permissions and their refusal under the Urban and Rural Zones Act and under other environmental legislation to the Environmental Appeal Board

(v) people who have infringed regulations, legal provisions and the policies in an adopted local plan and who may be the subject of a civil action in the local courts.

It should be noted that there is no opportunity for an appeal or inquiry into the contents of a plan, even the legally binding local plan, except in the case of an urban renewal or housing improvement proposal because the public participation procedures are regarded as adequate.

## Conclusions

D7.17   The main general characteristics of the Danish planning system are usually stated as being

(i) decentralisation of decisions and control

(ii) opportunities for public participation in the planning process

(iii) increasing politicisation of decision-making

(iv) rigorous control of development outside urban zones

(v) the use of the framework management and control system to achieve integration and co-ordination between administrative levels and sectors, all within the context of economic planning, so that corporate policy documents will result.

In practice each of these features, except for the politicisation aspect, is present, but to a lesser extent than might have been expected.

D7.18 The rigidly interlocking nature of the system as required by the framework approach is proving particularly difficult to work effectively and is the main subject of deliberation on the part of the national committee on land use and real property regulation, which is currently carrying out a comprehensive review of the planning system. Other areas of concern which this committee is addressing include

(i) the need for simpler and stronger safeguards of nature and environmental protection interests in the open countryside

(ii) a general requirement to streamline the planning system

(iii) the limited flexibility available at the regional level

(iv) the complexity of the appeals system.

It is hoped to introduce the necessary changes before the end of 1987 but this will depend on the progress in evaluating the land use regulation and control system which is the current stage that the review has reached.

# GLOSSARY

**Amstrådsforeningen i Danmark**
Association of County Councils in Denmark which is mainly concerned to bring the work of the county councils to the notice of a wider public, submit proposals for changes in local government legislation, ensure a uniform standard of administration by counties throughout Denmark and to represent counties in negotiations with employees

**Amt**
County responsible for the regional plan

**Amtsborgmesteren**
County mayor, elected and then becoming the paid full-time chairman and chief executive

**Boligministeriet**
Ministry of Housing responsible for housing and building, slum clearance and urban renewal

**Borgmesteren**
Town (municipal) mayor with the same characteristics as the *amtsborgmesteren*

**Byggestyrelsen**
National Building Agency responsible within the Housing Ministry for building permit procedures

**Byggetilladelse**
The general building permit operated by the municipality enabling the development to proceed

**Bygningsreglement**
Building Regulations which lay down the requirements in relation to the building permit application and approval system

**Bygningsrelgement for Småhuse**
Building regulations for minor development

**Cirkulære**
Circular which provides a commentary on and interpretation of the legislation to assist counties and districts in fulfilling their planning responsibilities

**Folketing**
Parliament consisting of a single assembly of 179 members

**Højesteret**
The Supreme Court in Denmark

**Ibrugtagningstilladelse**
The permit to use the building after construction has been completed which is required for the more major developments

**Kommune**
Municipality (local authority) which prepares the municipal structure plan and local plans, and administers the building permit procedures

**Kommunernes Landsforening**
Association of Local Authorities representing the municipalities, and set up to handle their interests, promote relations between them and help them perform their duties

**Kommunesplan**
The municipal structure plan which is mandatory and covers the whole municipality

**Landsplandirektiv**
National planning directive issued by the Ministry of the Environment usually to restrict development in particular areas

**Landsplanredegørelse**
The annual report or statement produced by the Ministry of the Environment reviewing the previous year's planning activities and suggesting policies for selected topics in the future

**Lokalplan**
Local plan produced and adopted by the municipality

**Miljøankenaevnet**
Environmental Appeals Board which is the court of appeal for many decisions made under the overall juridiction of the Ministry of the Environment

**Miljøministeriet**
Ministry of the Environment responsible for physical planning, nature conservation and protection of buildings, natural resources, environmental protection, re-cycling, foodstuffs and forest management

**Offentlighedsperiod**
The 'hearing' period for public participation in the plan preparation process

**Ostre Landsret**
The High Court (Eastern Division)

**Planstyrelsen**
National Agency for Physical Planning, one of five agencies within the Ministry of the Environment

**Råd**
Council which is the local government decision-making body

**Regionplanskitse**
The regional plan produced by the counties

**Regionplantillaeg**
The supplement to the regional plan

**Statens Byggeforskningsinstitut**
National Building Research Institute which carries out
both building and planning research

**Underretter**
Lower Courts

**Vestre Landsret**
The High Court (Western Division)

# REFERENCES

ANON (1982), **The Human Settlements Situation and Related Trends and Policies**, Publication, No 57, National Agency for Physical Planning, Ministry of the Environment, Ministry of Housing, Copenhagen

ANON (1983), **Rent Policy in Denmark**, Economic Statistics Division, Ministry of Housing, Copenhagen

ANON (1985), **The Greater Copenhagen Region Here and Now**, The Greater Copenhagen Council, Copenhagen

ANON (1985), **Bygge og Boligpolitisk Oversigt 1983-1985** (Building and Housing Policy: an overview), Ministry of Housing, Copenhageb

ANON (1986), **Statistical Yearbook 1986**, Danmarks Statistik, Copenhagen

ANON (1987), **Government Finances in Denmark**, Ministry of Finance, Copenhagen

CHRISTIANSEN, E ed, (1986), **Statistical Ten-Year Review of the Municipality of Copenhagen**, Copenhagen: Copenhagen Statistical Office

CHRISTIANSEN, O (1985), 'Rammestyringssystemet i den sammenfattende fysiske planlaegning', **Nordic Administrative Periodical**, No 4, pp 306-324

CHRISTIANSEN, O (1986), 'Comprehensive Physical Planning in Denmark', Chapter 4 in Garner, J F and Gravells, N P eds, **Planning Law in Western Europe**, Oxford: North Holland

COMMISSION OF THE EUROPEAN COMMUNITIES (1983), **Regional Development Programme** (Second Generation) Denmark 1981-85, Brussels

CUTERA, A ed, **European Environmental Yearbook (1987)**, London: DocTer International

DANISH MINISTRY OF THE ENVIRONMENT (1983), **Danish Physical Planning Acts**, Vol 1, National and Regional Planning Act, Regional Planning Act for the Metropolitan Region, Vol 2, Municipal Planning Act, Vol 3, Urban and Rural Zones Act, Copenhagen

DANISH NATIONAL BUILDING AGENCY (1983), **Building Regulations 1982**, Publication No 66, Copenhagen

DANISH NATIONAL BUILDING AGENCY (1985), **Danish Building Regulations for Small Buildings 1985**, Publication No 84, Copenhagen

DARIN-DRABKIN, H (1977), **Land Policy and Urban Growth**, Oxford: Pergamon Press

DEMKO, G ed, (1984), **Regional Development Problems and Policies in Eastern and Western Europe**, London: Croom Helm

DIRECTORATE FOR TRANSLATION AND TERMINOLOGY SERVICES (1987), **Terminology of Town and Country Planning**, European Parliament, Luxembourg

EILSTRUP, P (1986), **Regional Self-Government**, Danish Association of County Councils, Copenhagen

GOTTSCHALK, G (1984), 'Public Participation in the Danish Planning System', **Scandinavian Housing and Planning Research**, Vol 1, pp 65-80

HANSEN, P & Larsen F A (1981), **How the Danes Live**, Ministry of Foreign Affairs, Copenhagen

HAYWOOD, I (1984), 'Denmark', Chapter 7 in Wynn, M ed, **Housing in Europe**, London: Croom Helm

ILLERIS, S (1983), 'Public Participation in Denmark: Experience with the County Regional Plans', **Town Planning Review**, October, pp 425-436

JENSEN, K (1984), **The Green Wedges of the Capital**, Greater Copenhagen Council, National Agency for Physical Planning, Ministry of the Environment, Copenhagen

KERNDAL-HANSEN, O (1983), 'Ideplan 77', Chapter 12 in Davies, R L and Champion, A G eds, **The Future for the City Centre**, London: Academic Press

KERNDAL-HANSEN, O (1984), 'Denmark', Chapter 10 in Williams, R H ed, **Urban and Regional Planning in the EEC**, London: Allen and Unwin

KNAPP, V ed, (1972), International Encyclopedia of Corporative Law vol 1, National Reports : **Mogens Doktvedgaard Denmark**, International Association of Legal Science, The Hague: J C B Mohr (Paul Siebeck)

KRISTIANSEN, K (1984), 'Denmark', Chapter 3 in Cleere, H ed, **Approaches to the Archaelogical Heritage**, Cambridge: Cambridge University Press

KRISTOFFERSEN, E et al (1982), **Energy Supply Systems and Urban Patterns**, Report of the Steering Group, National Agency for Physical Planning, Ministry of the Environment, Copenhagen

LEMBERG, K (1973), **Pedestrian Streets and Other Motor Vehicle Traffic Restraints in Central Copenhagen**, City Planning Department, Copenhagen

LEMBERG, K (1982), 'National, Regional and Local Planning', Chapter 11 in Uuisti, F ed, **Nordic Democracy**, Copenhagen: Det Danske Selskab

LESLIE, D (1985), 'The Future for Metropolitan Strategic Planning : Comparisons with Greater Copenhagen', **The Planner**, October, pp 16-17

LOHOR, A A et al (1980), **Planning in Denmark**, Student Project Report, Glasgow: Department of Urban and Regional Planning, University of Strathclyde

LYAGER, P (1973), **Copenhagen**, City Engineer's, Architect's, Planning Departments, Copenhagen

MAGNUSEN, J (1979), **Urban Policy and Change in Denmark**, Department of Geography, University of Copenhagen, Costs of Urban Growth (CURB) Project, organised by the European Co-ordination Centre for Research and Documentation in Social Sciences, Vienna

MARKLAND, J and WILKINS, C (1978), 'Professional Planning Agencies in the EEC', **The Planner**, March, pp 44-45

MATHISEN, K (1983), **Local Government in Denmark**, The National Association of Local Authorities in Denmark, Copenhagen

MATTHIESSEN, C W (1986), **Greater Copenhagen : De-Industrialization and Dynamic Growth based on New Business Activities**, Paper given at a European Congress in Rotterdam

ØESTERGÅRD, N (1981), **The Overall Danish Physical Planning System and Planning for the Sectors of Land Use and National Resources**, Ministry of the Environment, Copenhagen

PEDERSEN, T O (1980), 'The Danish System - Some Lessons for British Planning', **The Planner**, September, pp 124-125

SAABY, L (1984), **A Guide to the Preservation of Buildings in Denmark**, Ministry of the Environment, Copenhagen

SKOVGAARD, J A (1978), 'Conservation Planning in Denmark', **Town Planning Review**, Vol 49, No 4, pp 519-539

SKOVSGAARD, C J (1982), 'Danish Planning and Consensus Politics', Chapter 6 in McKay, D H ed, **Planning and Politics in Western Europe**, London: Macmillan

SMITH, B (1978), 'Impressions of Planning in Europe', **The Planner**, March, pp 35-37

SVENSSON, O (1981), **Dansk Byplan Guiden**, Ministry of the Environment and the Danish Town Planning Institute, Copenhagen

SVENSSON, O (1986), **An Introduction to Danish Urban Renewal**, The Danish Building Research Institute, Hmrsholm

TOLSTRUP, F (1975), 'Town and Country Planning Law in Denmark', Chapter 5 in Garner, J F ed, **Planning in Western Europe**, Oxford: North Holland Publishing

WILLIAMS, R H and Crawford, H W F (1984), 'Why the EEC Matters to Planners', **The Planner**, October, pp 12-14

WOOD, C and Lee, N (1978), **Physical Planning in the Member States of the European Economic Community**, Occasional Paper Number 2, Department of Town and Country Planning, University of Manchester

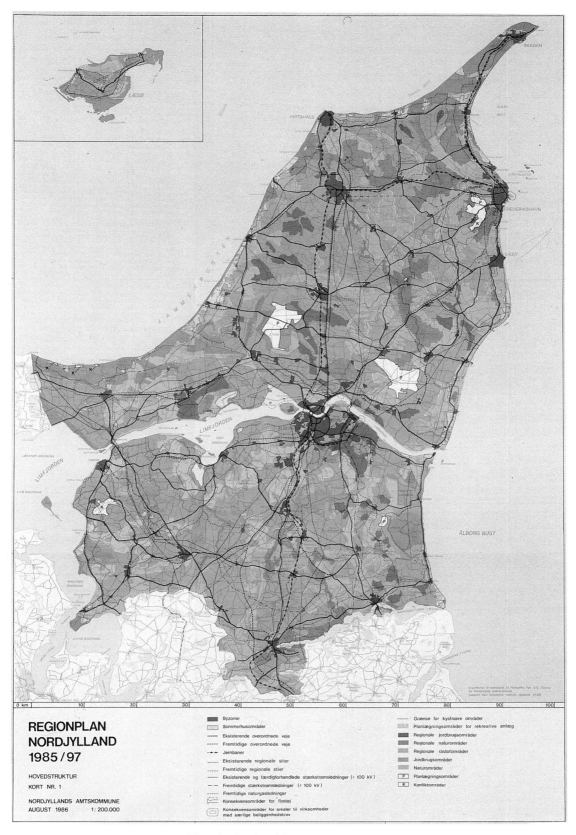

**Plate 4   Regional Plan for Nordjyllands**

Plate 4 is a reproduction of the regional plan for
Nordjyllands, the most northerly county in Jutland. The plan
identifies the extent of the urban zones and the summer house
districts and also the uses allocated to the rural areas. These
include agriculture, recreation, mineral extraction and areas
where nature conservation will be given priority. These
'natural' areas are very extensive. The plan also establishes
the routes of future roads, major footpaths, electricity
transmission lines and natural gas pipelines. The future use of
all land is established except for a few areas, denoted by K
and P on the plan, which are controversial and which still
require a planning decision.

145

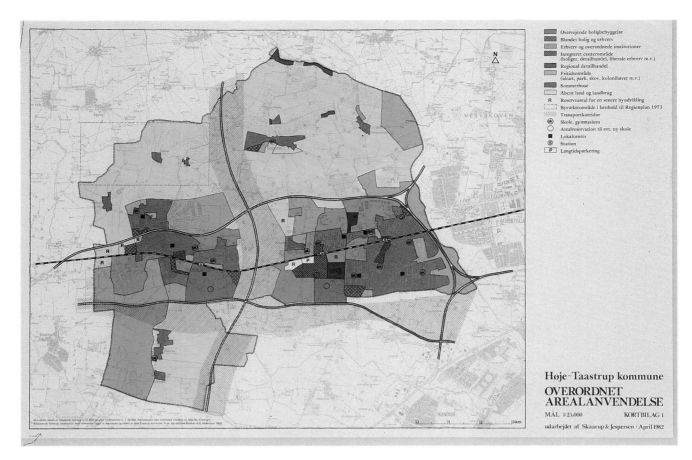

**Plate 5    Høje-Taastrup Municipal Structure Plan**

Plate 5 reproduces the structure plan for the municipality of Høje-Taastrup, which is within the Greater Copenhagen Metropolitan Region and so has to be in conformity with the metropolitan council's regional plan. Høje-Taastrup has been identified as a significant growth area and thus the plan proposes urban development for many greenfield locations. This is the part of the plan which establishes the general structure of the municipality. Sites are allocated for a variety of urban purposes although the zoning proposals are expressed in general terms such as mainly residential, mixed residential and service trades, industry and offices, central area uses, recreational areas etc. Agricultural land and summer house districts are identified and also land that is reserved for future urban growth (denoted by R on the plan). The new railway station is the focus of the plan and this explains why a long term parking area is allocated (marked by the notation P).

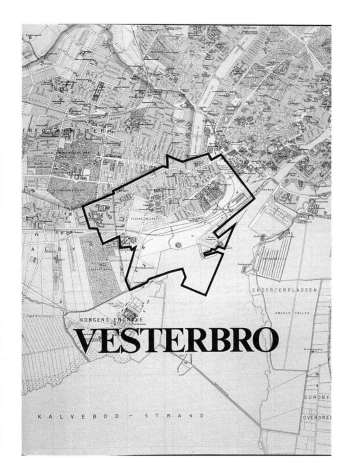

## Plate 6   Structure Plan Proposal for Vesterbro, Copenhagen

Plate 6 shows the proposals for the Vesterbro district in the
city of Copenhagen. This was a proposal emanating from the
Lord Mayor's Department and was followed by comments
from the various political parties including the publication by
the People's Socialist Party of an alternative plan in 1987. The
plan is intended to provide the framework for a legally
binding local plan and proposed areas should be zoned for
such purposes as flats and multi-storey residential units,
housing and service activity, service industry and mixed
employment, manufacturing industry etc. The boundaries of
the plan area are shown in (a) and the proposals in (b).

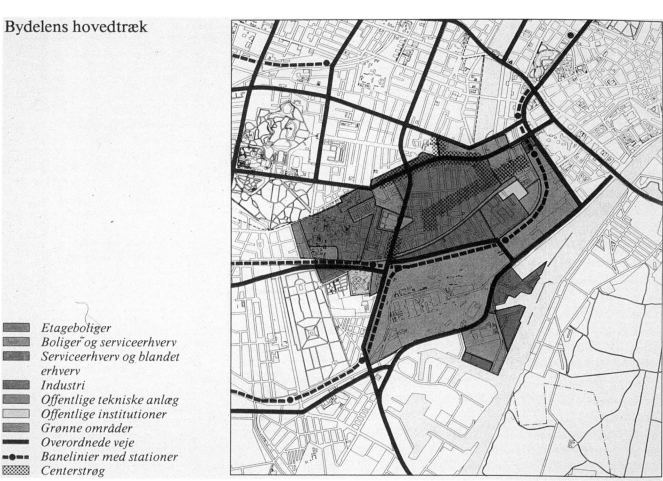

Bydelens hovedtræk

- Etageboliger
- Boliger og serviceerhverv
- Serviceerhverv og blandet erhverv
- Industri
- Offentlige tekniske anlæg
- Offentlige institutioner
- Grønne områder
- Overordnede veje
- Banelinier med stationer
- Centerstrøg

147

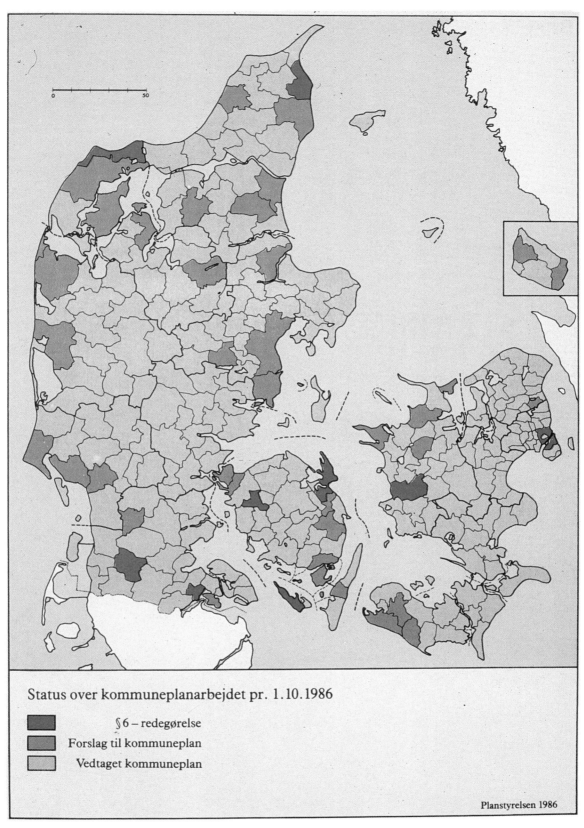

Status over kommuneplanarbejdet pr. 1.10.1986

§ 6 – redegørelse

Forslag til kommuneplan

Vedtaget kommuneplan

Planstyrelsen 1986

**Plate 7   Progress of Municipal Structure Planning**

Plate 7 is reproduced from the 1986 Annual Statement on National Planning published by the Ministry of the Environment and shows the progress that had been made by the 1st October 1986 in municipal structure planning. The notation identifies those municipalities where the plan had been adopted, those where a proposal had been made, and those where the Section 6 account, indicating the problems and possibilities, had been published.

148

# FRANCE

JV Punter

# F1 BACKGROUND

## Introduction

F1.1 This account of the French planning system follows the general structure formulated to describe the English planning system. The definition of development and the system of control are described, the hierarchy and role of plans, and the role of the courts and the ombudsman are outlined in a similar sequence. However it is a necessary preliminary to outline the context of planning in France in terms of the nature of the country being planned, its system of administrative law, and its systems of national, regional and local planning, and the roles of land and housing policy. All of these provide an essential context without which the French system can easily be misinterpreted.

## Geography, Government and Administrative Law

F1.2 France has a similar size of population to Great Britain, but it inhabits a country twice as large with a much greater geographical diversity. In many senses it is still a largely rural country with a pre-industrial settlement pattern. However, in the post-war era the agricultural component of the work force has been reduced from 35 per cent to 8 per cent, and the urban population has increased from 53 per cent to 75 per cent of the total. The scale of this rapid urban and industrial transformation in the post-war era has been greatly increased by population growth and immigration. The post-war baby boom ended seventy years of population stagnation, and the birth rate remained high into the early seventies, producing a post war increase of more than 14 million people, nearly three times that in Great Britain. Over a quarter of the population increase consisted of migrants from Southern Europe and North Africa largely concentrated in the suburbs of the major cities. The construction of 10 million housing units in 30 years to accommodate both this growth and dispersal from the dense central cities was a major focus of planning concern until the late 1970's.

F1.3 The primacy of Paris, and the Paris region with 18.5 per cent of the population and almost 50 per cent of the office space remains the most salient fact of French urbanism. Only the Lyon, Marseille and Lille agglomerations surpass one million, while a further five cities with populations of approximately half a million act as regional metropolises. A further 10 cities act as regional capitals. The twenty largest cities account for 40 per cent

of the French population. Medium sized towns between 20,000 and 200,000 persons are relatively under-represented in the French urban hierarchy, but it is these towns which are now growing fastest as expansion slows in the major urban regions. Small towns of under 20,000 population remain an important element in the urban system, accommodating one sixth of the population. Overall the distribution of population in France is remarkably uneven with sharp contrasts between the major urban regions and the sparsely populated agricultural and upland areas of the interior, contrasts heightened by the rural to urban migration of the post war era. The nature of the urban hierarchy and the pattern of population distribution are important considerations when it comes to analysing the provision of planning expertise and explaining the preoccupations of French regional and local planning.

F1.4 France has remained, at least until 1983, a strongly centralised state, partly in response to a recent rural past and rapidly urbanising present. But centralism is also a direct response to regional separatist movements, and to recent French history. The comparatively recent acquisitions of territory in the north, east and south east, and three invasions and occupations since 1870, have intensified the "French obsession with the need for a strong and centralised state authority" (Lagroye and Wright, 1979, p.4).

*Government*

F1.5 France is a unitary state with a written constitution. Since 1958 it has possessed the essential features of a parliamentary regime. However, it retains some features of its authoritarian or Bonapartist past with its reliance on a powerful and centralised bureaucracy, and with severe restrictions placed on parliamentary power to overthrow the government. Only Parliament has the power to enact statutes (*lois*), and the Prime Minister remains responsible to Parliament. However, as the powers of Parliament have been weakened those of the executive, particularly those of the President, have strengthened (this is not the case currently with a Socialist President facing a right wing majority coalition in Parliament).

F1.6 While only Parliament has the power to enact statutes and laws (*pouvoir législatif*) the executive has powers to regulate by decree (*pouvoir règlementaire*). In France one cannot make the distinction between legislation and subordinate legislation that is normally made in England, and decrees and orders carry just as much legal weight as the laws themselves. Ministers and other public authorities, including in some instances the prefect and the mayor at lower levels of government, possess regulatory powers (*pouvoir règlementaire*) to complete by regulations the framework of the legislation (Brown and Garner, 1983, pp.6-9).

F1.7 A Constitutional Council established in 1958 acts as a watchdog on Parliament to adjudicate upon the validity of elections, and to pass an opinion on the legality and constitutionality of parliamentary laws. As Brown & Garner (1983, p.14) suggest, it constitutes "a powerful and prestigious body to uphold constitutional norms, especially the fundamental rights and liberties of the individual". The Constitutional Council was placed alongside the *Conseil d'État*, which dates back to 1799, as the supreme administrative court in the land, protecting the rights of the ordinary citizen in his dealings with the state. Decisions based upon *pouvoir législatif* and *pouvoir règlementaire* are both subject to challenge and review in the system of administrative courts.

*The system of administrative law*
F1.8 The French system of administrative law is of crucial significance to an understanding of the operation of the planning system. Based upon the Napoleonic code of law, it attempts to provide clear rules of procedure and decision-making and to protect the rights of the individual through promoting the "immutable principles of liberalism" (Brown and Garner, 1983, p.15; Weil, 1965). It is a vital component of the French system of bureaucratic and technocratic control and it grants discretion not to politicians but to the administrators, emphasising the general mistrust of political control in France. The French are content to leave the details of government to the experts, and hence the truism that the French "are not governed but administered" (Brown & Garner, 1983, p.18).

F1.9 The *Conseil d'État* is the senior court, staffed by the élite of the Civil Service and enjoying an unrivalled prestige and *esprit de corps*. It provides advice on all Bills and delegated legislation as well as acting as legal adviser to government and its ministers. It also performs a judicial role, reporting on current problems of administration, and an appellate and supervisory role over twenty five *Tribunaux Administratifs*, to which any aggrieved citizen or group may take his complaint against any administrative decision. This provides a system of legal safeguards and judicial control that is accessible to all. Despite its inegalitarian origins, French administrative law (*droit administratif*) has survived to provide one of the most systematic guarantees of the liberty of the individual against the state known to the present day.

F1.10 The French administrative courts obviate the necessity found in Britain for a complicated system of administrative tribunals or inquiries to oversee the operations of the welfare state, housing provision or town planning, and reduce the need for public participation of all kinds. So in town planning, for example, it is not necessary to establish a specific appeal system because the normal remedies of *droit administratif* are available to the applicant for planning permission and the local citizen. The supervisory machinery provided by the courts allows the French safely to entrust the details of government to the experts, while in Britain the cult of the amateur in the jury system, on local councils and in various tribunals, reflects a distrust of officialdom and a need to provide some checks on their actions (Brown & Garner, 1983, p.16).

F1.11 *Droit administratif* has the weaknesses of being a relatively slow, *a posteriori* system of control largely used by civil servants, and with judgments that are sometimes difficult to execute. But it has the strengths of being accessible, cheap, and flexible and of providing simple remedies which have public confidence. It also provides a systematic and cohesive system for the intelligent policing of complex administration, all this in marked contrast to English law (Brown & Garner, 1983, pp.172-181). It is important to bear these points in mind when assessing criticisms of the lack of public participation in French local government and administration.

*The* Code de l'Urbanisme *and related codes*
F1.12 As befits the French centralised state, its constitution, its bureaucratic traditions and the legal controls on the same, planning powers are comprehensively codified into a single text, the *Code de l'Urbanisme* which is constantly updated. First executed in 1954, it was made progressively more comprehensive by decrees in 1972-3 and 1977, and it now consists of three parts. The first part (legislative) consists of actual laws passed by parliament, and defines the principles and details of the rights of different actors and procedures. The second part consists of government regulations which elaborate and detail the legal procedures involved. The third part contains *arrêtés* (ministerial orders) and more minor regulations. This last part serves essentially as a kind of development plans manual, as well as providing the details of application and declaration forms, procedures, consultative committee structures and the like. The Code is continually updated as new laws, regulations and orders are issued. The whole is cross referenced by a numbering system so that the regulations and articles applicable to the relevant laws can be established, but it is at best an inconvenient and at worst an impenetrable system (Tribillon, 1985, pp. 27-31).

F1.13 There are six books within the code dealing with the general rules of planning and development, land reserves and expropriation, land development, rules applicable to the act of construction and diverse modes of land use, the siting of services, establishments and enterprises, and administrative organisms etc (see Appendix A). Some 1500 articles provide a comprehensive framework for plan-making, development control, and plan implementation. However, it is an intimidating and complex document that requires highly skilled legalistic interpretation before any project can be undertaken. Essentially the code defines the relationship between public power, finance capital, raw land, property owners, builders and so called 'users'. Knowledge of the code, for so long largely confined to those in the Ministry, is a pre-requisite for effective use of planning powers and development rights.

F1.14 Most importantly, in contradiction to a general principle in France that "any act by the public authorities resulting in direct material and indisputable loss shall give a right to compensation" (*Conseil d'État*, 1924), Article 160.5 of the *Code de l'Urbanisme* generally excludes the possibility of compensation for loss resulting from the imposition of zoning restrictions, public works servitudes or land use regulations (Institute for Environmental Studies, 1987, p.543).

F1.15 There are other codes of major significance to planning. These include the *Code de l'Environnement, the Code de la Construction et de l'Habitation, Code Rural et Code Forestier, Code Minier* (Mining), etc. The Environmental Code, for example, includes all relevant laws and decrees on environmental protection associations (the equivalent of British amenity groups); public information; water, noise and air pollution; protection of flora, fauna, forests, coasts, national parks, national monuments, open and wooded space, listed buildings; waste disposal, and environmentally dangerous installations. These clearly overlap and extend the control of development. Similarly the Code of Construction and Housing (see Appendix B) regulates building construction, fire protection, heating and clearing of buildings as well as regulations on the construction industry, housing finance and housing improvement, the regulation of housing associations and of insalubrious buildings.

F1.16 Laws, regulations and articles provide the operational framework for planning and development control. But the all-important interpretation of the law is dependent upon decisions of the *tribunaux administratifs*, and of course ultimately upon the *Conseil d'État*. The build-up of case law defines the real limits of planning powers and the rights of those affected by planning decisions. With the flurry of reforms between 1983 and 1986, which have transformed many elements of local government and planning practice, there is only now slowly emerging relevant decisions which clarify the real nature of the available powers and define the limits of 'acceptable' planning practice.

F1.17 The *Code de l'Urbanisme* embraces legislation that normally would be described as land policy, regional planning, or public service servitudes in a British context. It is important in discussing development control in France to remember that National Plans, regional planning, land and housing policy all impinge upon the management of urban growth in ways quite different from the British system. Similarly the French planning system has evolved in stages, developing key attributes as responses to particular physical and political pressures, culminating in the complex of laws, regulations and articles in the *Code de l'Urbanisme* and *Code de*

*l'Environnement.* The second half of this background chapter describes the emergence and nature of these parallel strands of planning policy at large.

## National and Regional Planning

### National plans
F1.18 Since 1946 National Plans have been produced every five years by the *Commissariat Général du Plan (CGP)* under the direct control of the Prime Minister. These plans have prompted and overseen the modernisation of France but have been indicative in nature, intended to provide a more informed framework for government and business decisions, and a basis for regional planning designed to achieve decentralisation of development from Paris. National plans were more strongly *dirigiste* from 1946 to 1965 concentrating first upon rebuilding the basic industries and preparing French industry for foreign competition, before switching to substantial investment in housing, health, education and transport in the early 1960's.

F1.19 Since 1965 major uncertainties in the world economy have weakened the utility of the plan which has more and more assumed the character of generalised market research. Plan strategies have been abandoned in the face of social unrest, balance of payments crises, oil crises and currency problems. Attempts by the socialist government to revive national planning in the early 1980's foundered on just such an economic crisis and the progressive *deplanification* or de-emphasis upon state planning that has characterised the last two decades seems set to continue.

### Regional planning
F1.20 Regional planning began in the mid-1950's as a response to the over concentration of development in the Paris region clearly identified in 1949 by J-F Gravier's book *Paris et le désert français*. It was based upon 22 planning regions, responding both to regional political pressure and to the overcentralisation of investment in the Paris region. Controls on industrial expansion in the Paris region were introduced in 1955 along with regional action programmes and plans, and while these were not well coordinated with national programmes they did introduce a territorial dimension into economic planning.

F1.21 *The Délégation à l'Aménagement du Territoire et á l'Action Régionale* (DATAR) was set up to steer mobility for social purposes, and to co-ordinate large scale public development and investment projects. However, it never controlled the Finance Ministry's regional funds and retained only a limited budget that was often used to manipulate election votes in crucial constituencies. Consultative bodies composed of local Chambers of Commerce and Agriculture, Trade Unions and other community groups have been established to comment upon both national and regional plans, but elected regional governments were not established until 1984. The major metropolitan areas were given general

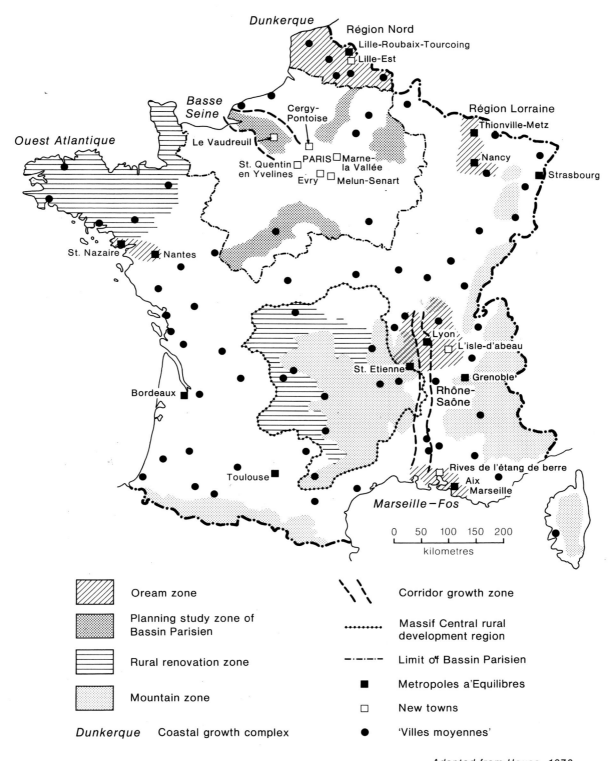

Dunkerque
Région Nord
Lille-Roubaix-Tourcoing
Lille-Est

Basse
Seine

Région Lorraine
Thionville-Metz

Ouest Atlantique

Cergy-
Pontoise

Le Vaudreuil
Nancy

Strasbourg

St. Quentin
en Yvelines
PARIS
Marne-
la Vallée

Evry
Melun-Senart

St. Nazaire
Nantes

Lyon
L'isle-d'abeau

St. Etienne
Grenoble

Bordeaux
Rhône-
Saône

Toulouse

Rives de l'étang de berre

Aix
Marseille

Marseille−Fos

0   50   100   150   200
kilometres

| | Oream zone | | Corridor growth zone |
| | Planning study zone of Bassin Parisien | | Massif Central rural development region |
| | Rural renovation zone | | Limit of Bassin Parisien |
| | Mountain zone | ■ | Metropoles a'Equilibres |
| | | □ | New towns |
| *Dunkerque* | Coastal growth complex | ● | 'Villes moyennes' |

*Adapted from House, 1978.*

**Figure F1   Spatial management policies in France in the 1970's**
**(after House, 1978, p. 318)**

This map reveals the complex regional planning programmes operating in France in the 1970's. Several of the programmes revolved around the decentralisation of development from Paris, particularly the *metropoles d'équilibre* and the *villes moyennes*, which promoted development of major and then medium sized towns as growth poles. The promotion of major industrial growth was the main purpose of the corridor and coastal zones while rural development was encouraged

through rural renovation zones and special mountain zones.

The OREAMs undertook physical planning at a regional scale. New town developments were a feature of their attempts to structure metropolitan growth, particularly in the Paris Region where the new towns were supplemented by the definition of four support zones and nine existing towns for accelerated development.

planning agencies in 1964 with the establishment of eight *Organisations d'études d'aménagement des avis metropolitaines (OREAMS)*, these being superseded by regional governments in 1984.

F1.22 The regional component of national plans became more evident in the 1960's with the promotion first of the *métropoles d'équilibres* to act as major growth poles to attract development away from Paris. These were part of a package of spatial management policies which attempted to restructure suburban growth in the Paris region through new town development (5 declared in 1964; 41 new urban regions in 1969), and to encourage industrial resurgence in the north and industrial development in the west and the Rhone and Saône Valleys. In the 1970's the focus shifted to promoting the development of medium sized provincial towns with the central government funded *ville moyenne* policy, and financial assistance for this policy has continued into the 1980's (Figure F1).

F1.23 The limited links which had been developed between national, regional and local physical planning in the 1960's were progressively weakened in the 1970's as a result of political factors, international economic crises, and the broadening of economic concerns and policies to the EEC level. Economic liberalism gave way to active *deplanification* and although the national plans included specific regional expenditure programmes they still constituted a further stage in the 'unplanning' at the national level. The revival of national and regional planning under the socialists in the early 1980's was short-lived as the government was forced to adopt an austerity programme to counteract rising inflation and unemployment. With the return of governments of the right in 1986 even the future of DATAR has been questioned, and the policies to restrict development in the Paris region have been significantly reduced to encourage national economic recovery.

F1.24 Ironically the creation of directly elected regions in 1984 has given a new initiative to regional planning by allowing the region significant funds for economic development. The region was also permitted, subject to the advice of DATAR, to establish contracts with central government (who provide the funding) for four year programmes of infrastructure investment linked with planning targets. These *Contrats de Plan* are therefore both an expression of national economic priorities and regional ambitions, and while not all regions have accepted the challenge others have eagerly embraced them as a means of building regional identity and prosperity.

F1.25 Most verdicts on the French national planning experience recognise the utility of indicative planning and the benefits that accrue to a market economy and a traditional administrative system as a result of such an activity. Administrative rivalries and ideological aversion both contributed to the demise of the National Plan, but there was a general failure to coordinate national and regional planning. Certainly by the late 1970's there were no real operational links between the CGP and DATAR and the coordination between national and regional planning was very tenuous (Estrin & Holmes, 1983, p.114; Ullmo, 1976, p.35). The reforms of 1983 have provided a mechanism for more closely linking national and regional planning, and for developing regional plans to express commitments to the development of public infrastructure, and the latter are being actively used by a number of regions. But with the persistent economic recession, and now the return of a government of the right, the general prospects of coherent regional planning are not good. However, a recent speech by the Minister, and the recent Guichard report on regional planning (*aménagement du territoire*), suggest a continued commitment to some regional goals through the revival of DATAR as an inter-ministerial body under the Prime Minister's authority.

## The Emergence of Town Planning and Land Policy

*Precedents for town planning*

F1.26 J.W. House has argued that while town planning has been better integrated with regional and national planning in France than in Britain, development control and land use planning have been "tardy, partial, and had but little success" (House, 1978, p. 324). Certainly a coherent land use planning system emerged only in stages, even in the post-war era.

F1.27 Nineteenth century precedents for town planning continue to give French law much of its flavour. Early nineteenth century provisions protected street alignments, listed historic monuments and their vistas, and encouraged uniformity in street architecture. In the late nineteenth and early twentieth centuries the listing process (statutory and supplementary lists) was extended and the protection of the setting of listed buildings increased in 1913 and 1930. However, compulsory 'extension and embellishment' plans for towns of more than 10,000 persons introduced in 1919 were very rare, though powers to control *lotissement* (allotment of plots or subdivision) were significant.

F1.28 It was not until the middle of the Second World War that the Vichy government resumed the task of creating an effective planning system. The law of 15 June 1943 essentially re-enacted the 1919 law re-establishing the principle of a development plan with street and zoning designations. However, it adopted a complicated approval machinery involving several ministries. This was the first time that the term *urbanisme* (urban planning) appeared in the law. The 1943 law introduced the major tool of development control, the *permis de construire*, which controlled building works ensuring compliance with the regulations and also ensuring that they were not contrary to any town planning provision. The means for attaching conditions to any permission was also provided. The cumbersome procedures for gaining exceptions to the regulations to obtain a *permis* were a source of much delay and criticism and were amended in 1969 and 1970. Essentially it was legislation to prevent rather than to promote development (D'Arcy, 1970, p. 37).

*The two tier* Plan d'Urbanisme *1958*
F1.29 A more comprehensive codification of town planning law was completed in 1954 (*Code de l'Urbanisme et de l'Habitation*), but actual plans remained limited in scope serving only to preserve the traditional appearance of towns and prevent the haphazard development of housing estates (Lemasurier, 1974, p.118). The purpose of plans was broadened by a decree in 1955 which emphasised their new 'prospective, operational, coordinated, and complete' character. Another decree of 31 December 1958 established a two tier system of plans that were the forerunners of the contemporary SD ('Structure') and POS ('Local') plans, the *Plan d'urbanisme directeur*, and the *Plan d'urbanisme de détail* respectively, the latter dealing with segments of the town and providing the framework for decision-making on the *permis de construire*. These reforms, and the creation of the ZUP procedure (see below) to establish and implement priority urbanisation areas, were responses to rapid urban expansion and the need for both control and implementation procedures that could be quickly prepared and applied.

*The* Zone à Urbaniser en Priorité (ZUP) *1958-1969*
F1.30 While the *plans d'urbanisme* lacked a time frame and failed to establish strong and lasting guidelines for urban land management, it was the *Zone à urbaniser en priorité (ZUP)* which became the prime means of positive planning from 1958 to 1967. It provided the instrument for the local authority to compulsorily acquire and service land for development (House, 1978, p. 325) with funds provided by the state controlled lending bank. The purpose of the ZUP was to speed up major housing development and to ensure economies of scale in the provision of infrastructure and the reduction of urban sprawl (even private housing developments of over 100 units had to be built within ZUP's). They achieved this by giving local authorities the powers of comprehensive development and the rights of first refusal on the land within the ZUP, essentially at existing use value.

F1.31 In all, some 169 ZUPs were declared, and an average of some 2500 to 3000 dwellings were erected in each. Their *gigantisme*, and the high rise, high density and socially segregated *grandes ensembles* that they created, were in part responsible for widespread social problems in, and public disenchantment with, the suburbs in recent times. The ZUP procedures were largely abandoned after 1969 even though they were not abolished until 1976. It was these conspicuous failures, and the incidences and suspicions of corruption in the application of the ZUP, that fuelled a political controversy that led to reforms of the system in 1975.

*Controls on speculation: the* Zone d'Aménagement Différé (ZAD) *1962-present*
F1.32 Attempts to tax the appreciation in land values were also made. Until 1962 expropriation values were determined by tribunal according to prevailing market prices, A new law set acquisition values at the level of the previous year to ignore any increased value added by impending compulsory purchase, although judges invariably set higher values because of valuation difficulties. To overcome the problems of land speculation, particularly just outside the boundaries of the ZUP, and to avoid the necessity for local authorities to have to purchase large tracts of land for development all at once, the instrument of a *zone d'aménagement différé (ZAD)*, a zone of deferred development, was introduced in 1962. This froze all development rights in anticipation of major public development proposals for a period of eight years (later extended to fourteen), and gave the local authority preemption rights on the land on prices based on its value a year before declaration. This land banking procedure had encompassed some half a million hectares of land by 1977, about 1 per cent of the nation, but only 600,000 hectares by 1983, notably concentrated in a few departments in mountainous and coastal France, although also along transport axes and on the periphery of urban areas.

F1.33 These reforms came too soon to anticipate the problems of very rapid, often uncoordinated, suburban expansion and the crises of the large, high density suburban public housing estates. From 1958 onwards the government progressively withdrew from the public financing of housing, and the private sector was encouraged with tax incentives to investors. The result was a property boom from 1962-65 which had dramatic effects on suburban sprawl, infrastructure provision and land prices. Reforms mooted in 1964, and implemented in 1967, were an attempt to provide a better framework for urban growth. But they eschewed the legal regulation of land use except to 'guide and coordinate' public authority programmes, leaving the private sector relatively unencumbered.

*Conservation concerns rural and urban - 1960 and 1962*
F1.34 Important conservation concerns were given legislative expression in the early 1960's. First the National Parks Act was passed in 1960 and six national parks have been created by central government since 1963, five of them in the mountainous areas on the mainland (Figure F2). In urban conservation the French have maintained their innovative responses by providing a system of grants for restoring 'group value' historic buildings in 1958 (*opérations groupées de restauration immobilière d'initiative publique*) and a system of designating positive conservation areas (*secteurs sauvegardés*) in 1962 (which were the inspiration for the British 1967 Civic Amenities Act). These have been given a greater impetus through regionally funded rehabilitation programmes since 1977.

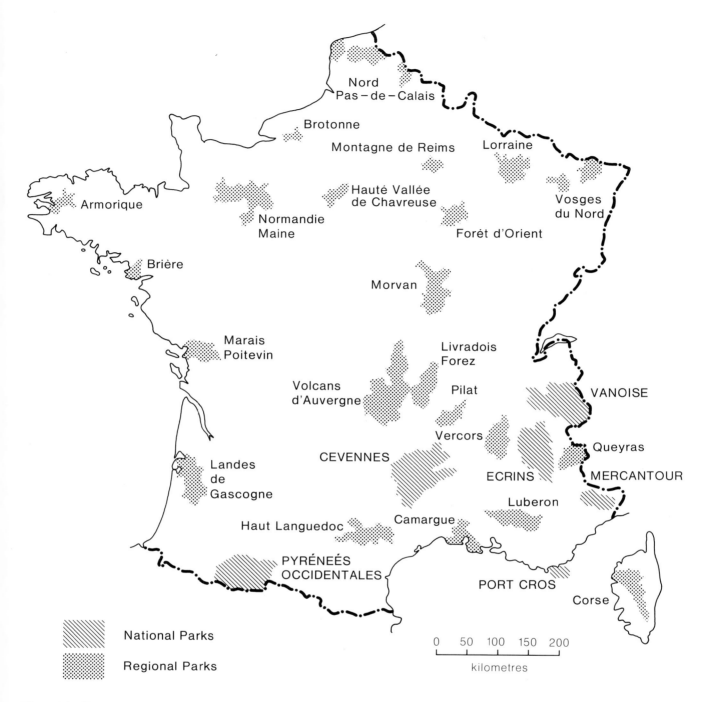

**Figure F2   The National and Regional Parks of France**

The six national parks were declared between 1963 and 1979. In theory they consist of an inner *réserve intégrale* or full reserve, protected by a central zone where little development is allowed, and surrounded by a peripheral zone where certain developments can be admitted and even financially encouraged. In practice no full reserves have been designated, and the buffer zones have been subjected to significant pressure from ski-resorts etc. Regional nature parks are defined as special land management zones designed to safeguard the natural and cultural heritage and encourage development compatible with their appreciation and conservation.

*Reforms of plans and the creation of the* Zone d'Aménagement Concerté *(ZAC)*

F1.35  The *Loi d'Orientation Foncière et Urbaine* passed in December 1967 introduced the *Schéma Directeur d'Aménagement et d'Urbanisme (SDAU)* and the *Plan d'Occupation des Sols (POS)* to replace the *plan directeur* and *plan de détail*. The SDAU is usually compared with the British structure plan with its long term projections (10-30 years), its broad guidelines for development and land use allocations, and its protective policies. The POS took the form of a detailed local land use plan precisely defining the areas for development and protection, and fixing the densities of development through the *Coefficient d'occupation des sols* (COS) or plot ratio. In an amended form these remain the basic framework for planning at the present day.

F1.36  To replace the ZUP's compulsory acquisition and control mechanisms the 1967 law introduced the concept of the *Zone d'Aménagement Concerté (ZAC)*. This was a less draconian measure designed to achieve cooperation between public and private interests in large scale urban development whether in central areas, new industrial zones, suburban housing schemes or resort communities. These took the form of contractual agreements *(convention de ZAC)* between the local authority and private developers, and provided a much more flexible planning mechanism, more reliant on private finance and, correspondingly, offering greater concessions to private developers. The size of a ZAC was set at a maximum of 2000 dwellings in the larger towns in a bid to ensure a more human scale of urbanisation.

F1.37  ZAC procedures have been heavily criticised as giving a free rein to private speculators by placing at their disposal expropriation procedures to remove the smaller land owner. Although in theory the developers benefitting from ZAC procedures were required to plough their profits back into the development by financing various public works like nurseries or schools, these investments were rarely commensurate with their profits. Furthermore the mere offer to undertake public works encouraged many local authorities, particularly the financially weak communes (Flockton, 1983, p.75), to create a ZAC, often undermining the POS completely (ZAC procedures automatically suspended the POS). Thus major developers were able to circumvent the planning system through the system of exceptions *(dérogations)* while the small developer remained constrained by the POS.

F1.38  ZAC procedures were an effective means of coordinating and concentrating urban development since they provided a framework for land acquisition, a means of providing necessary infrastructure, and a way of mobilising finance. For these reasons they were extensively used for commercial redevelopment, where the state traded higher plot ratios for additional finance and other concessions. This tended to result in overdevelopment, reductions in planned social housing provision, and the driving out of lower income groups.

*The* Zone d'Intervention Foncière (ZIF*) 1975 - present*
1.39  The shortcomings of the ZUP, ZAC and ZAD procedures in curbing land speculation, and the growing strength of the environmental movement and the anti-development lobby, led to more concerted measures to control land prices in the *Loi Foncière* (Urban Land Law) of December 1975. This introduced the concept of a *Zone d'intervention foncière (ZIF)* designed to allow local authorities to prevent over-building and to implement social housing, conservation and open space provisions in redevelopment schemes. It empowered the local authority to purchase land at the previous year's prices, and it applied automatically to all *'zones urbaines'* (see paragraph F5.27) in any published POS. Over 600 ZIFs were declared within two years, and by 1983, 1676 ZIF had been created embracing 558,000 hectares. So they were being widely used to control land speculation particularly in socialist controlled areas like Nord/Pas de Calais. With the change of government in 1986 the ZIF became no longer compulsory, but merely optional, as part of an attempt to avoid unnecessary bureaucracy.

*Legislative expression of environmental concern 1976*
F1.40  Environmental concerns about the quality of new development and urban life, and the destruction of coast and countryside, led to further reforms in 1976. The 1976 *Reforme de l'Urbanisme* brought a more qualitative emphasis to development plans and generally strengthened the power of the public sector to control development. Special provisions were introduced to protect coastal areas and the countryside from *le mitage* (urban sprawl), and environmental impact studies were introduced for major developments, making France the first European country to follow the USA's model for environmental protection. The *étude d'impact* was also extended to a wide range of hazardous developments and included in the preparation of *Schémas Directeurs* and *Plans d'Occupation des Sols*.

F1.41  The 1976 laws provided a new procedure for consultation of the public recognising the powers of environmental associations (amenity groups) to be consulted on the application of certain planning laws. These were extended in 1983 and 1985 by further democratisation of *enquêtes publiques* (public enquiries), and by obliging those promoting major planning schemes to conduct a *concertation préalable* (preliminary consultation) to prevent clandestine development and expropriation decisions (Jacquot, 1987, pp.45-7). The 1983 law also provided for better public information, longer enquiries, and more impartiality and guaranteed competence of inspectors.

*Decentralisation to deregulation? 1983 - 1987*
F1.42  The main reform implemented by the socialist government of 1981-6 was the transfer of planning controls to the commune in 1983, giving the smallest unit of local government the powers to prepare development plans, and then to issue permits for planning and demolition permission and all manner of minor authorisations. Public accessibility to the planning process

was significantly improved in plan making and development control.

F1.43   The return of a government of the right in 1986 has brought the question of deregulation of land policy and planning procedures to centre stage. Laws of preemption have been relaxed, taxes on the density of development made optional, and progressive decontrol of rents has been initiated. Each of these is an attempt to encourage private developers to produce more housing. Three Enterprise Zones have been established and more than 40 local authorities have undertaken initiatives with *technopoles* or science parks. Employment issues now clearly take priority over environmental issues, a shift of emphasis begun in the middle of the socialist term, but intensified since 1986.

## Housing Policy and the Development Industry

F1.44   One of the principal objects of land policy in France has been to ensure an adequate supply of serviceable land to meet the demand for housing for a rapidly urbanising population. Housing policies have been critical in determining the quantities, locations and type of land acquired and have strongly influenced both the demand and supply of housing in France.

### Three forms of state aid

F1.45   With rent control operating since 1914, tenants in France enjoyed both extremely low rentals and security of tenure through to the late 1940's. The 1948 Rent Act codified existing legislation and made it more effective, but its impact was confined to some six million existing buildings in the inner areas of the larger cities. A personalised subsidy system *Allocation Familiale de Logement (AFL)* was introduced to provide means tested housing assistance to low income households, and this was subsequently extended in 1971 to special categories of housing need. In 1977 such direct subsidies to tenants *Aide Personalisée au Logement* (APL) became the principal means of housing assistance.

F1.46   A second major area of state subsidy has been support for *the Habitations à Loyer Modéré* (HLM). HLM's built between a quarter and a third of all homes in most years, but increasing percentages were built for sale (26-43 per cent, 1958-1976). Originating as philanthropic housing trusts, some 1,200 HLM agencies emerged as the main instrument of social housing in post-war France providing over two and a quarter million rental dwellings, largely on a non-profit basis, and over one million dwellings for sale on a limited profit basis.

F1.47   A third source of state aided housing for sale is provided by the *Secteur Aidé* whereby the State bank *Crédit Foncier* lends the builder or prospective owner up to 70 per cent of the construction costs at fixed interest rates, as well as providing fixed grant incentives on a square metre basis. Mid 1960's schemes signalled a relaxation in state financial control of the housing

market, and the entry of new financial institutions into a fast growing mortgage market. Thus the *Secteur Aidé* quickly exceeded the HLM production to constitute about 40 per cent of total production. Private housing production was increasing all the time and in 1974 for the first time it surpassed the *Secteur Aidé* to constitute 40 per cent of production (Pearsall, 1984, p.14) (Figure F3). By the mid 1980's private single family dwellings accounted for two thirds of housing production.

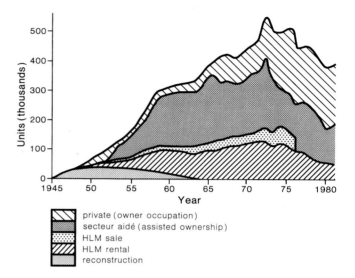

**Figure F3   Housing production in France 1945–1981**

The most striking feature of housing provision in France is the sustained high levels of production from the late 1950's through until the early 1980's with a pronounced 1973 peak immediately curtailed by the 1974 oil crisis. Even in what are perceived as production crisis years in 1985 and 1986 production remains close to 300,000 units per year. Secondly it is clear that since 1972 private sector provision has been increasing dramatically and now constitutes two thirds of all production, with a similarly dramatic increase in the number of single family dwellings.

### Post-1977 policy and the aid to owner occupation

F1.48   The Housing Act of 1977 marked a watershed in the development of French housing policy. The government, predisposed towards reducing public sector involvement, set out a residual role for the state in housing, arguing that general housing provision should be left to the market, and that householders should meet the full market costs of rents and interest rates. Much simpler forms of housing finance were introduced with a general set of loans for any subsidised rental dwellings *(Prêts Locatifs Aidé (PLA)* for new or rehabilitated property) and available to all builders not just the HLM. However, the most important reform was the introduction of a new system of direct financial subsidy to the householder rather than to construction itself. This is a much more discriminating policy aimed only at those who cannot afford market rents or purchase costs and financed by the new *Fonds National de l'Habitat (FNAH)*.

F1.49   With the public sector de-emphasised, home ownership was promoted much more vigorously, and the

*Secteur Aidé* was reformed. New sub-market rate loans were introduced to assist lower income groups to purchase their own homes, and these can be used by HLMs, mixed public-private developers, builders or purchasers themselves. More important still were loans for middle income groups subsidised by the state, for up to 80 percent of the cost with no restriction on income. Specific rehabilitation subsidies have also been developed.

F1.50 The owner-occupied sector has increased from approximately one third of all housing in the mid-fifties to approximately half today. The proportion of housing built for sale has risen to over 85 per cent, with about half of this production being state-aided for low income groups, and with all home owners having tax relief of 25 per cent of their mortgage interest. The demand for a *pavillon* (a detached house on its own plot) is strong, particularly amongst the lower middle classes, long frustrated in their housing demands by the *grands ensembles* (Scargill, 1983, p.98). Production has been encouraged by financial aid given to the multiplicity of small scale house builders in 1969 and 1975, and planning policies have been modified to facilitate this kind of development (in 1976 *lotissement* or sub-division provisions were reformulated as a more appropriate tool than the ZAC for controlling such development (Scargill, 1983, p.99)). The French pattern of housing provision, the reliance on the private sector, and the planning problems that this generates are now more similar to those in Britain than they have ever been. However there is still the fundamental difference that 80 per cent of the production of French housing is subsidised by the state in some way, over and above mortgage tax relief. The particular subsidies for those near the minimum wage to purchase new housing create significant distortions in the market, and can make resale extremely difficult.

*Landownership, finance and the role of the developer*
F1.51 Finally, in setting the background for a discussion of the French system for the control of development, it is necessary to describe some of the salient characteristics of the development industry. As in many other continental European countries, owner-occupation of commercial property has been the norm, although condominium or *co-propriete* ownership has become significant in the post war era. Speculative office or industrial space was initially introduced by British developers in the late 1960's and as a result there is now an active commercial property investment market in France. But it is still more common for users to acquire the buildings they occupy in staged payments from initial reservation of the building, making development essentially auto-financing (Jones Lang Wootton, 1987, p.6).

F1.52 The role of the developer or *promoteur* has traditionally been to organise a development up to the point where finance is sought, taking a small percentage of the equity in the development as his 'profit'. Finance has usually been organised by assembling a group of investors to put up a share of the short term finance, but

taking a capital gain when the building is sold, and generally not retaining an interest in the building once it is occupied. French institutional investment in commercial property developed in the latter half of the 1970's transferring out of residential development, although it suffered setbacks in the early 1980's owing to the arrival of a socialist government. But there is still a lack of long-term money for investment in property in France.

F1.53 The state has played an important role in the provision of finance for development within and beyond the housing sector. Semi-public development corporations have been set up to bring local authorities and financial institutions into partnership with the state, and these have developed the necessary expertise to implement development programmes, and provided flexible and financially powerful instruments. Such companies developed rapidly after 1950 and in 1963 new legislation permitted the establishment of *Sociétés d'Economie Mixte* (SEM). These were joint private-public companies with a combination of state powers like compulsory land acquisition, and private sector financial resources and entrepreneurial skills. They were widely used in housing and commercial development and, having the virtue of encouraging private investment in urban development, were particularly encouraged in the more market-oriented, 'deplanification' years of the 1970's. Mixed companies could also undertake renewal schemes with landowners as partners, and further flexibilities allowed the public authorities to lease land from private landowners to avoid laborious expropriation procedures. The supply of industrial building land is particularly dependent upon the activity of the SEM, while a special investment company was set up to invest in industrial buildings in 1967.

F1.54 The state also established two companies of its own. The *Société Centrale d'Equipement du Territoire* (SCET) and the *Société Civile Immobilière de la Caisse des Depôts* (SCIC), founded in 1953 and 1954, took on the responsibilities for the provision of social housing and the management of clearance and urban renewal respectively. The Ministry of Construction also played an important role in modernising the French construction industry, and encouraged the adoption of new techniques and management procedures, aiding research into new practices and encouraging mergers to achieve economies of scale in production.

F1.55 The state has played a role in financing urban development through direct payments for infrastructure costs (estimated at 30 per cent of the costs in 1975), and through low interest loans to local authorities from the public banks for land acquisition and development works, to prepare land for public and private development. So the finance of development is a combined effort of national and local government in partnership with private interests who actually carry out the construction. Furthermore the right of preemption has been a crucial tool in controlling land prices and assembling land for suburban development or urban redevelopment as well as protecting environmentally sensitive areas.

F1.56 Finally it is important to recognise the role of landowners in France and the part they play in the provision of development land. Individual land owners, often family farmers, tend to hold much of the land designated for development and extensive developer land banks are not common. The cadaster is complex and there is a general obsession with land holding that is encouraged by a tax régime that discourages the preparation of land for development. Furthermore a multiplicity of individual landowners may be involved in development themselves, subdividing their property or selling off parcels on existing roads. In much of ex-urban and rural France such development activity is much more significant than that of professional developers. The latter are becoming increasingly important in France, and a more powerful lobby group, but they play a much smaller role in housing provision than their British counterparts.

F1.57 For all these reasons the state has been forced to assume a more prominent role in land assembly for development, and developers have often supported the municipalisation of land and land banking procedures to obviate the problems of land assembly. Land policy has been closely related to positive planning practice in the new towns and in urban redevelopment projects. Meanwhile the question of betterment levies remains a live issue as developers are confronted with an increasing share of servicing costs by the communes. The mechanisms for a more equitable sharing of servicing costs, *Associations Foncières Urbaines* or a *taxe des riverains* (levied in parts of Alsace-Lorraine), provide only localised solutions, and the issues of land supply and servicing costs continue to loom large in any evaluation of French planning practice.


## Conclusions

F1.58 This introductory chapter has described the background to the French planning system outlining the geographical, governmental and legal factors which are prerequisites to understanding the French planning system. It has also outlined the evolution and interaction of French national, regional and local planning, and the different roles played by land and housing policies as well as the development industry in the development process. From the perspective of comparative analysis of development control systems the most important factors to emerge are as follows.

  (i) The relatively rapid post war urban growth in France, its concentration particularly in the Paris region, and the continued existence of vast tracts of small-town/rural France sparsely populated and not pressured by any development.

 (ii) The existence of national, regional and two tiers of physical plans as a basis for planning and control. Although never a clearly hierarchical or carefully integrated system of planning each of the four levels is of importance in establishing a context for development and the management of major urban growth.

(iii) The progressive emergence of a system of development control based upon a two tier system of plans, but incorporating selected elements of alignment controls and other public servitudes, architectural control and conservation, and environmental protection with increasing recognition given to public consultation.

(iv) The codification of French planning law providing a comprehensive if complex source of all legislation, regulations and procedures, and integrating the functions of legislation, ministerial directions, circulars (in part) and development plan manuals in British terms.

 (v) The importance of the administrative courts in providing legal checks and balances on this system and recourse for the ordinary citizen or other aggrieved parties to appeal against planning decisions. Above this the *Conseil d'État* as the highest court plays a crucial role in interpreting the law and establishing binding precedents.

(vi) The integral relationship between development control and land policy with land assembly instruments, land speculation controls, and taxes on development forming key instruments for the implementation of planning policies, and the planning law becoming the main means of facilitating public land acquisitions.

(vii) The important role of the state in housing finance and provision even in today's private-ownership dominated new housing market.

(viii) The important role of the state in assembling land and financing urban development through public-private partnerships, the only recent emergence of speculative development in France, and the continued importance of the small landowner in the provision of land for development.

Many of these factors will emerge as important in the discussion of the structure of local government and the administrative framework of planning which follows in section F2.

## Introduction

F2.1   The structure of French local government is complex, and that complexity has been increased by the series of reforms introduced by the socialist government since 1982, reforms which have been barely digested and have yet to be fully operationalised by the French themselves. It is not merely a matter of explaining the functions of the four levels of French government - national, regional (22 units), departmental (96), and communal (36,433) - although France is the only European country with four such levels. It is more the multifarious interactions between the various levels, the intricacies of ministerial services provided at the departmental level, and the special role of the French bureaucracy, which give the French system its complexity. The fact that French politicians often hold office at several levels, some occupying key positions at both mayoral and ministerial level, also makes the French system unique. These governmental characteristics have a very significant impact upon the land use planning system.

## National Government

F2.2   Most commentators on French government concentrate upon its centralism, long considered a political necessity to hold together France's diverse feudal territories and provinces. The Jacobins and Napoleon emphasised this and their preoccupation was the creation of the 'one and indivisible Republic,' replacing local self-government with local administration. But in recent years it is the gradual creation of a system of local government, given a very significant boost by the 1982 and 1983 decentralisation reforms, that has been the main feature of national administration. The chief debating point amongst political commentators is the extent to which this makes for another Jacobin reform, or creates a real independence for the three tiers of government.

*Twin executive and the* cumul des mandats
F2.3   At the head of national government the twin French executive of President and Prime Minister is a source of some confusion to British eyes. Since 1962 the President has been elected by universal suffrage and is able to promulgate laws, sign decrees, appoint senior officials and the Prime Minister, and dismiss the National Assembly. The scope of presidential policy-making has expanded steadily since 1958 into economic and industrial policy and social and environmental issues. The

Prime Minister is in general charge of the work of the government, but the size and shape of the government is determined jointly with the President, and it is the latter who fixes the agenda and timetables and has direct access to ministers (Ridley, 1977, pp.67-83).

F2.4   The Prime Minister's real autonomy lies in the policy area, where the President has no interest, and in supervision and coordination of the various ministers and junior ministers. The Ministers themselves can have considerable autonomy and policy initiating powers depending upon their status. Lagroye and Wright (1979, pp.33-5) emphasise the historic and continuing importance of deputies (members of parliament) acting not so much as servants of the national interest but as 'delegates' of their constituencies and protectors of their local interests. Booth (1985, p.7) emphasises the significance of the fragmentation of majorities in parliament, and the tendency for members to seek favours from central government for their constituencies. Such behaviour is of course intensified and even formalised by the *cumul des mandats* (multiple political office holding) which means that one politician may be representing several levels of government at different times. In practice members of Parliament can sometimes control both appointments at the Departmental level and local state jobs and other favours, and the accumulation of offices can increase the number of direct appeals to Ministries bypassing the Prefect (now the *Commissaire de la République*). Sometimes the most powerful deputies are themselves past, present or future ministers, and may also be the mayors of large cities to emphasise their power and influence.

*Ministerial functions: staff and structure*
F2.5   As national, regional, and urban planning have evolved throughout the post war-era, so too has the Ministry primarily responsible for controlling planning and development. Because of its role as the executant of policy at the local level, in the general absence of strong local government bureaucracies outside the large cities, some understanding of the way in which the Ministry is structured and staffed is essential to appreciate the character of planning control. The Ministry responsible for local planning has, as in Britain, undergone significant changes in title in the post-war era reflecting its changing scope and preoccupations. It began as the *Ministère de la Construction* in the immediate post-war era, and evolved to the *Ministère de L'Urbanisme* in the early 1960's. A more significant development occurred in 1966 when it was merged with the relatively weak Ministry of Housing

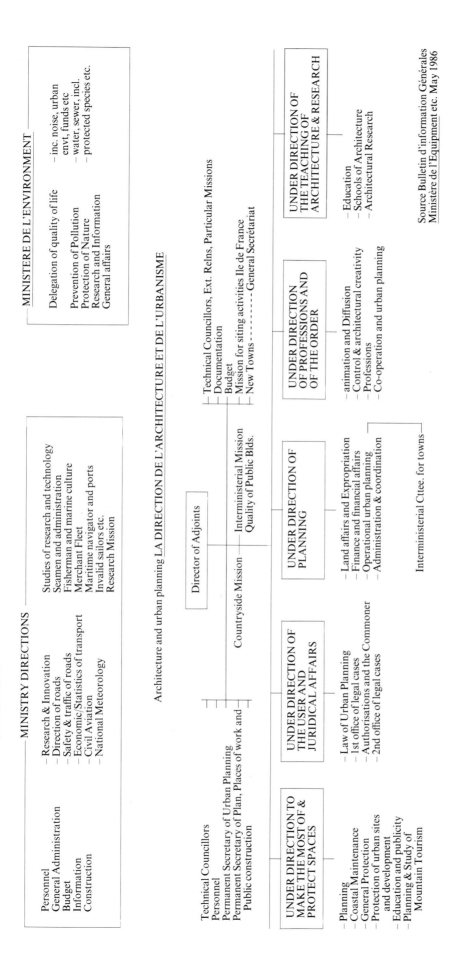

**Figure F4 The relevant planning sections of the Ministry of Infrastructure, Housing, Planning of land, Transport etc.**

Given its responsibilities, the complexity of the structure of the Ministry is understandable. The main section/*direction* of interest is the *Direction de l'Architecture et de l'Urbanisme* which has five sub-directions to embrace conservation, legal affairs, urban planning, the professions and the teaching of architecture and research. Each of these is split into several offices with specific responsibilities. Mention should also be made of relevant sections of the *Ministère de l'Environnement* which play an important role in environmental issues including urban improvement, pollution control and nature/landscape conservation. Two related boards of importance are the High Committee for the Environment, to provide advice on major projects of national importance, and the Interministerial Committee for the Quality of Life, which promotes environmental quality issues from a position close to the Prime Minister.

163

to become the *Ministère de L'Équipement*, as part of a general move to create superministries. This brought Public Works together with Transport and Construction to coordinate infrastructure provision, and to provide a global view of urban growth.

F2.6   It was also in 1966 that the field services of the Ministry were established at Department level, the *Direction Départementale de L'Equipement* (DDE) becoming responsible for plan preparation and development control in all areas which did not have their own planning agency. It was a necessary move prior to thoroughgoing reforms of the planning system implemented in 1967 which placed renewed emphasis upon strategic and local plan preparation; it was also designed to strengthen State control countrywide.

F2.7   In 1978 the Ministry's title was widened to *Ministère de L'Environnement et Cadre de Vie,* a move designed to emphasise quality of life considerations and wider environmental planning concerns, but in 1981 the *Ministère de l'Environnement* was separated off (with its special interests in coasts, mountains, rural areas, water pollution and hazardous installations). In 1984 a further split occurred separating the Ministry of Transport (except roads) from the *Ministère de l'Equipement, du Logement, de l'Aménagement, du Territoire, et des Transports.* In 1986, with the return of a government of the right, the *Ministère de l'Environnement* found itself back with the *Ministère de l'Equipement* etc. But despite all the changes in the period from 1981 to 1986 the basic environmental policies have changed little.

F2.8   The *Ministère de l'Équipement* etc was divided into some 16 directions and delegations in 1985, but planning control is brought under the *Direction de l'Architecture et de l'Urbanisme* (DAU), though there are several other directives with interests in environmental planning at large (Figure F4). The DAU possesses five subsections embracing historic buildings and landscape protection; legal affairs dealing with the *Code de l'Urbanisme* and its use and disputations; development tax and finance including land and expropriation; the professions, both architectural and planning; and the director of architectural education and research (Figure F4). The second subsection is also responsible for controlling the activities of the DDE who in turn elaborate and supervise the application of the regulations of land use.

F2.9   The Ministry as a whole is very much in a transitional stage and further changes can be anticipated. Legal affairs are being given much more prominence and decisions of the *Conseil d'État* on the application of the new planning procedures are being given careful scrutiny as a prelude to possible changes in the *Code de l'Urbanisme.* A private audit is now under way to analyse efficiency, and there are ideas about developing a less hierarchical more pedagogical system.

F2.10   Although more emphasis is now being placed on legal expertise in the Ministry the traditional structure,

and its clear expression in the field services of the DDE, has had an important impact upon both the practice of planning and the development of the planning profession.

*The* Directions Départementales de l'Equipement (DDE)
F2.11   The origins of the DDE lie in the creation of the *Corps d'Ingenieurs des Ponts et Chaussées* by Louis XV, which became one of the most influential of the *Grand Corps* (the élite civil servants). The historic engineering mentality and the *grand école* traditions were combined in the senior civil servants at the DDE, and they assumed leadership and control over the emerging planning system, linking it to the administrative apparatus at large. The actual development of planning practice and expertise has been placed in the hands of contractual staff, educated at less prestigious institutions, who seldom progress into management and who do not belong to the civil service. D'Arcy and Jobert (1976, p.300)

> "suggest the hypothesis of a division between the management level and the specialists, whose work and terms of reference are defined by the former. The imbalance sometimes appears enormous between a directorate that remains deeply embedded in the State, and the specialists with a precarious status on the periphery of the traditional administration"

DDE functions embrace plan making, development control, transport planning, public housing, public utilities and public building (Figure F6). Within the DDE the plan making functions of the *Groupe d'études et programmation (GEP)* in particular are performed by multi disciplinary contractual staff who do not belong to the civil service, while the other functions remain dominated by the engineers (Wilson, 1982, p.161)

*The planning profession*
F2.12   So while civil service procedures and engineering traditions generally dominate the mentality of the DDE's operations, the actual planning staff occupy very much a secondary, inferior position, and this has in turn impaired the development of a strong planning profession, reflected in the fact that only 25 per cent of all planners in France belong to the professional organisation the *Société Francaise des Urbanistes* (Markland & Wilkins, 1978, p.45). In fact relatively few planners are trained as such. A 1982 survey of personnel in communes revealed the multi-disciplinary nature of French planning. Among the more senior urbanists 25 per cent were engineers, 21 per cent were architects, 11 per cent legal experts, 10 per cent urbanists (urban planners) and 6 per cent each geographers, geomorphologists and economists. Some 16 per cent had subsequently followed some urban planning training (Pesce, 1983, p.37).

*Contrasts with Great Britain*
F2.13   There remain two fundamental contrasts between the role of the Ministries in French and British planning respectively. Firstly the French Ministry maintains direct links with planning at the local level through its field services at the departmental level, who

perform the plan-making functions and advise on development control in all but the largest cities. With the advent of decentralisation, and the grant of planning powers to the local authorities who have completed a *Plan d'Occupation des Sols*, the field services work for the local mayors rather than for the state. However, it is the state which determines their appointment and which ultimately commands their loyalty. So through the field services the Ministry remains in touch with and to a large extent in control of planning and development control at the local level, especially in medium and small sized towns and rural areas. One Ministry official suggested that between 1978 and 1982 about 90 per cent of the day to day planning work of the Ministry itself was dealing with problems brought to it by the field services.

F2.14  On the other hand, central control is in some senses weaker than in Great Britain because the Minister does not perform an appellate function for aggrieved developers, the appellate function operating through the administrative courts, and being oriented more towards third parties than unsuccessful applicants for a *permis de construire*. So the minister's power to enforce policy changes through issuing circulars is much more circumscribed. The Ministry does issue circulars which provide key interpretations of legislation, or which advise on new policy directions, and which can be implemented directly through DDE practice, but significant changes have to be translated into laws or decrees to ensure their effectiveness. This in part accounts for the frequent amendments made to the *Code de l'Urbanisme*.

## The Regions

F2.15  The regions originated in 1955 as 22 economic planning regions based upon groups of departments (Figure F5). Reforms in 1972 emphasised the region's role as 'an appropriate tier for the co-ordination of the country's economic development' and its role was seen as principally one of investment. One of the first socialist government reforms was to grant much fuller powers to the region and to give it an equal status with the commune and the department. But it was its planning function which remained paramount. In the words of the Minister of Planning and Regional Development:

> "Without planning, regionalisation would degenerate into petty interests; conversely, without regionalisation, planning would tend to become a uniform and centralising straitjacket" (Rocard, 1981, pp.137-8 quoted in Kofman 1985, p.17).

F2.16  The law of March 1982 transformed the same regions into directly elected, self governing bodies free from the control (*tutelle*) of the regional Prefect (now *Commissaire de la Région*) and under the control of the President of the Regional Council who acts as its chief executive. The region was given the power to levy taxes of up to FF150 per inhabitant, the power to finance its own regional projects, and the ability to initiate others in

conjunction with the communes and departments, as well as the power to negotiate planning contracts funded by the national planning agency. The most important fiscal power given to it was the ability to allocate significant sums for direct economic intervention.

F2.17  The regional government now participates in elaborating the National Plan and the guidelines for regional development (*aménagement du territoire*). In addition the region now develops and approves a *contrat de plan* that runs concurrently with the four-year national plan, specifying medium and long term objectives and programmes of implementation in conjunction primarily with public, but also with private, investors (Keating, 1983, pp. 244-5). But these reforms have generally not gone nearly as far as most autonomist and separatist movements would have wished, and only Corsica was granted a *Statut Particulier*. Otherwise no region was able to undertake any programme that undermined the unity of the Republic.

### The Contrat de Plan

F2.18  The content of the new regional plan, the *Contrat de Plan* is specified by various laws and decrees passed in 1982 and 1983 and it is essentially a contract between the state and region defining strategic choices, objectives and major developments which conform to the National Plan and which defines the means of execution of economic, social and cultural projects. As an example the *Contrat de Plan* of the region Ile de France sets out 39 articles defining action in six fields - economic development, housing, environment, new towns, water, transport and circulation. These articles are a mixture of policies, firm objectives, planned developments, public expenditure commitments and research programmes. Where appropriate they specify planned national and regional expenditures, and the plan also includes more detailed contracts laying out the joint participation in the major programmes. In sum the plan outlines a five year programme of public investment that can link economic planning and development with infrastructure provision and environmental enhancement, a programme totalling over FF1,600m (£16m) shared almost equally by state and region (*Prefecture de la Région d'Ile de France*, 1986). It is an expression of both the significant commitment to public investment and the desire to provide a planned regional infrastructure to accommodate major growth. However, for a variety of reasons, few regions have yet produced such a contract and supporting documentation.

## The *Département*

### The *Prefect and* tutelle.

F2.19  The third level of French government is provided by 96 Departments. These territorial units were established by Napoleon to unify the disparate regions of France, and each Department's area was broadly defined as within a one day's horseback journey of the principal town after which the Department was named (Figure F5). A Prefect was appointed by the state to control each

**Figure F5   The 22 regions and 96 departments of France**

The 96 Napoleonic departments were less organs of local government than convenient administrative units for centralised power channelled through the Prefect. Despite democratic reforms and decentralisation they retain much of this flavour, particularly through possessing the departmental field services of the various ministries including urban planning, now under the control of the *Commissaire de la République*. The departments are now grouped into 22 regions established in 1955 as economic planning regions. Since the 1982 reforms the regions have become fully elected self-governing bodies with significant fiscal powers.

Department and he became the key figure in this system of "rational, hierarchical and bureaucratic control" (Lagroye and Wright, 1979, p.29), a personal representative of Government. Nowadays the Prefect is appointed by the Prime Minister, but he is also an official of the Ministry of the Interior, and his role is "essentially that of a hinge between the different levels of government" (Garrish, 1986, p.7), a mediator between grass roots pressure and Parisian imperatives. In a highly centralised state the Prefect's role was crucial because he administered all the state services at the Department level, as well as exercising the power of *tutelle* (tutelage) over the sub-prefects (for *arrondissements*) and mayors and councils of the communes, with powers to dissolve councils, veto budgets and control law and order.

F2.20 Even before 1982 these powers were progressively limited (by *a priori* budget control), and the Prefect's lack of security of tenure and frequent moves to new appointments often meant he was unable to exert much influence over local notables who could often deal with Paris over his head. But it is his technical and financial *tutelle* which remains of importance, although these powers have been significantly eroded by the 1982 decentralisation.

F2.21 The first of the 1982 decentralisation reforms ended the necessity for Departments and communes to obtain the Prefect's prior approval for all their administrative and financial transactions, and substituted *a posteriori* control of legality only, removing the power of annulment to the *Tribunal Administratif* (the administrative court). The *Chambres Régionales des Comptes* was established to provide a new regional audit system to enforce statutory budgeting rules and the Prefect's influence has been further reduced by giving his executive powers to the President of the *Conseil Général*, the elected Departmental Council, with one representative drawn from each canton. However, the Prefect's powers have not been entirely eroded and he has been given the position of *Commissaire de la République* within the Departmental administration, re-establishing his authority over the field services which he now directs rather than coordinates, and over the mayors who do not follow national regulations or procedures. Most observers see the Prefect continuing to exert very significant influence especially over the majority of small communes and their mayors, and especially in planning matters.

*The field services of the* Directions Départementales de l'Équipement (DDE)
F2.22 The existence and importance of the field services in finance, labour, agriculture and infrastructure is another manifestation of the centralism and hierarchical nature of French government. In planning it is the field services of the *Directions Départementales de L'Équipement (DDE's)* which provide the technical expertise in housing, town planning and transport at the local level. They remain the primary source of planning expertise for most smaller communes and Departments, leaving central government as the executant authority and not

merely the policy maker (Booth, 1985, p.15). Another crucial factor blurring the technical-political and advisory-executive dichotomies so important to British planning practice is that senior civil servants may also stand for local office. As a result they are well-represented in the Departmental and communal councils throughout the country (over 400 are mayors). Similarly the *Grand Corps* (the élite civil servants) colonise senior positions in the nationalised industries, utilities and state finance agencies. Collectively the importance of the field services' role is emphasised by the general weakness of local government, the centralisation of power and finance, the legalistic system of administration, and the discretionary power which is vested in the administrator and not the politician.

F2.23 Table F2.1 gives some appreciation of the staffing of the DDE plan making and control sections in 1983. Its very structure emphasises the importance of the civil servant - contractual staff split outlined above, and the importance of professional status. So while most DDE plan-making departments have perhaps six senior planners (half civil servant, half contractual) these will be supported by perhaps a further 16 staff. For control perhaps five senior planners will be available, mostly civil servants but with about 40 other employees. Each DDE

**Table F2.1  The planning staff within the DDE 1983**

| Groupe d'Etudes et programmation (GEP) | | | | Urbanisme Operationale et Construction | | | |
|---|---|---|---|---|---|---|---|
| Civil Servants | | Contractual | | Civil Servants | | Contractual | |
| Prof. | Others | Prof. | Others | Prof. | Others | Prof. | Others |
| 312 | 989 | 253 | 509 | 444 | 2390 | 62 | 840 |
| 3.3 | 10.4 | 2.7 | 5.4 | 4.7 | 34.6 | 0.7 | 8.8 |

*Source:* RISU 1985, pp.56–58

**Table F2.2  Use of credits for studies for urban planning purposes 1983** in FFmillions

| | | | |
|---|---|---|---|
| General Studies | | | |
| Methodology | 0.14 | | |
| Inventory of National | | Land Policy | 0.82 |
| resources | 0.40 | Existing neighbourhoods | 1.64 |
| Studies for Planning | 3.16 | New neighbourhoods | 1.23 |
| *Total* | *3.70* | Executive studies for | |
| | | Planning | 1.16 |
| Documents of Urban Planning | | *Total for implementation* | |
| SD | 0.21 | *Studies* | *4.85* |
| POS (+ZEP) | 9.12 | | |
| Cartes Communales | 1.18 | Technical Missions | 8.12 |
| *Total* | *10.51* | Topographic works | 1.87 |
| | | Coastal Servitudes | 0.14 |
| Material Expenses tied to Studies | | | |
| Documents of Urban | | | |
| Planning | 12.10 | | |
| Other Studies | 2.37 | | |

This table outlines the main expenditure of central government funds by the DDE and the other local planning agencies. The large share devoted to the production of POS and other urban planning documents is obvious, and these 1983 figures pre-date the very marked increase in POS production.
*Source:* RISU 1985, pp.34–49

167

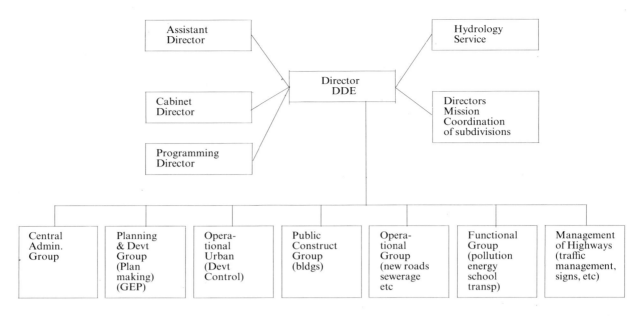

14 REGIONAL SUBDIVISIONS
(each with a team including a planner)

**Figure F6    Structure and Staffing within the DDE: the example of the Department of Loiret**

Loiret has 900 employees divided between the nine services listed above, with some 27 employees for assisting local authorities with preparation of *Plans d'Occupation des Sols*, and some 40 employees overseeing the issuing of *permis de construire* and related development control decisions. The Loiret DDE is further subdivided and regionalised into 14 territorial divisions each with an average of 24 communes, but the plan-making support consists of only a dozen planners split between five broad areas. Seven of the 14 territorial divisions have left complete responsibility for development control with the DDE, and the DDE staff continue to offer advice and expertise in all the others.

The City of Orleans, the main city of Loiret and the regional capital, and a further 17 communes have the benefit of the advice of an *Agence d'Urbanisme* (see paragraph F2.39). The urban agglomeration covered by this accounts for some 200,000 of the Department's 550,000 population, but the DDE planning staff of 67 (including administration etc.) remains the only source of planning expertise for a population of approximately 350,000. In addition their advisory functions in control embrace the agglomeration itself since this is not a main function of the *Agence d'Urbanisme* whose main role is plan preparation and planning studies of various kinds.

deals with an average of 400 communes and some 7000 permit applications a year. There are of course quite significant variations from Department to Department. A more revealing insight into the structure of the DDE can be gained from from figures which explain the situation in the Department of Loiret centred on the city of Orleans in central France (Figure F6).

F2.24    The number of DDE employees is declining with decentralisation but at an almost imperceptible rate (1 per cent per annum was suggested by Ministry officials). Given the unlikelihood of small local authorities preparing their own plan-making functions or of employing sufficient skilled advice to administer development control, the DDE will continue to function as the main source of local planning expertise outside the large cities of France. The state will also continue to finance most planning studies through its system of credits for planning studies. Some idea of the nature and extent of expenditure can be gained from Table F2.2, although one would have expected expenditure on POS production to have at least trebled since 1983.

F2.25    The Department remains something of an anachronism with the creation of fully fledged regions, and its main *raison d'être* would seem to be the existence of the field services at this level. But this in turn is ambiguous since these field services are staffed by civil servants representing the state, while serving the local mayors and communes. This ambiguity is in practice slightly less of a problem.

## The *Commune*

F2.26    The basic unit of local government remains the commune of which there are over 36,000 (Figure F7). Unlike most European countries the administrative boundaries of French villages did not change when they were swallowed by towns, and many have retained their status as communes and independent political units. This has left the central commune of most towns or cities with essentially pre-industrial boundaries, while the surrounding villages have in turn become completely developed suburbs with no change in political units. The State assumed the role of the provider of most key public services, leaving the local elected representatives to act principally as intermediaries between the largely uncoordinated activities of the various state departments and the consumers of these services, the local electors.

F2.27 The number of communes has stubbornly resisted efforts at consolidation or abolition, and despite the 1982 decentralisation of powers their potential for independent action remains slight. Lacking the financial and technical resources for real autonomy they will continue to depend upon the civil service and the Department level of local government for advice and expertise. Most communes have only a budget for one full time employee at the minimum wage, and a small sum (FF80,000 or £8,000 in 1976) for public works (Flockton, 1983, p.71). A glance at the table in Figure F7 reveals that five out of six communes have less than 1500 residents and two out of three less than 500.

F2.28 Attempts to bring together communes of the same agglomeration have always been vigorously resisted by the local elected representatives for fear of losing office and because of local pride. Furthermore such reforms have never been strongly supported by the central state, since the latter retains important advantages with the continued fragmentation of political power at the local level. With the recent fine balances in national electoral support most governments have concluded that they could only lose politically from attempting to reduce the number of local mayors and elected representatives (of whom there are over half a million), the traditional basis of local power (Ashford, 1983, p.31). However, several mechanisms for intercommunal cooperation have been established.

*Intercommunal organisations and planning agencies*
F2.29 From 1884 onwards syndicates could be established across communes to administer single public services, but these were inadequate. In 1959 legislation created the multi-purpose commune *Syndicat intercommunal vocation multiple (SIVOM)*, and in urban areas the district. By 1975 some 1021 single purpose syndicates had been created. More significantly, by 1985, 1,980 multi-purpose syndicates had been established (each syndicate being managed by a committee made up of two delegates for every member commune and acting as a kind of local federation of mayors). These now embrace over half of all *communes* (1983) and serve a population of over 20 million, while 147 districts have brought together 1,284 communes and a population of 4.5 million. As of 1983, however, nearly half the French population remain governed by single communes, and the district remains a rather limited instrument of intercommunal cooperation with an ill-defined legal basis and a weak authority in the larger agglomerations (Lagroye & Wright, 1979, p.102). A 1971 law permitting mergers only produced a reduction of 2000 in the total number of communes.

F2.30 In 1966 a further mechanism for intercommunal cooperation was introduced by the Gaullists with the *Communauté Urbaine*. This was not a directly elected territorial authority but rather a public administrative institution with wide powers for metropolitan planning, and public works. Communes which entered into a *Communauté Urbaine* were unable to withdraw at a later date. The new powers accorded to the *Communauté Urbaine* allowed them significant planning independence and the ability to appoint their own planning staff thereby freeing them from dependence upon Ministerial planning staff. Four cities initially established the *Communauté Urbaine*: Lille, Lyon, Strasbourg and Bordeaux (as an example, in the case of Lille it provided an administrative body linking 88 communes). By virtue of proportional representation the larger central communes were given an influence commensurate with their size. However, only five additional cities of much smaller size subsequently united voluntarily (Figure F8). In general it can be argued that the State has been reluctant to impose collaboration on the communes, and even the power to create districts compulsorily was withdrawn in 1970 (Garrish, 1985, p. 11).

F2.31 *Communautés Urbaines* do not necessarily reflect the extent of the built-up area, nor do they embrace potential expansion or primary catchment areas. Their boundaries are fixed only by the extent to which individual communes wish to participate. While preparing plans is a major part of their functions they can also have responsibility for housing production and renovation; servicing land and land banking; public transport, road maintenance and car parks; sewerage and water supply; waste treatment; fire fighting; cemeteries, abbattoirs and important markets. Other functions can be added.

*Agences d'Urbanisme*
F2.32 Finally, to provide a measure of inter-communal planning in the larger cities and urban regions most major cities and the nine new towns possess their own planning expertise (Figure F8). This is usually known as an *Agence d'Urbanisme* which is the closest approximation to a British planning department that exists in France. However, each is independent of any political control and can discuss planning problems with any group or individual or with a *Communauté Urbaine*. They provide a classic example of a French approach to governmental reform since an additional new structure was created rather than directly tackling the problems of the existing one. Although it was a centrally inspired reform, and the State contributed 50 per cent of the annual costs of each agency (now less than 20 per cent in most established agencies), the majority have subsequently been established by SIVOMs. Even where established by a SIVOM the agency performs a largely plan making function, and plays only an advisory role in the control of development when requested to do so by the communes. Each agency provides the technical expertise for the revision of *Schémas Directeurs* (structure plans) but spends most of its time helping the constituent communes prepare, modify, or revise their *Plans d'Occupation des Sols* (local

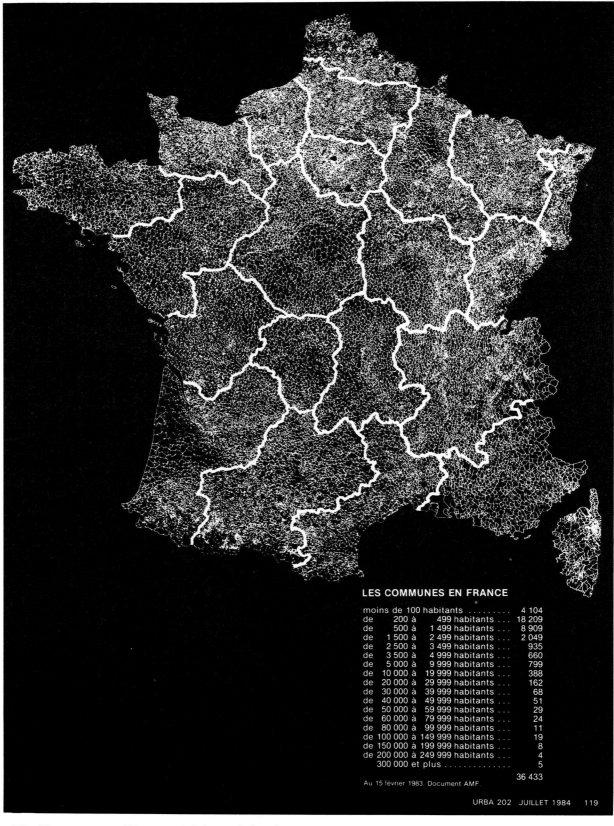

LES COMMUNES EN FRANCE

| moins de 100 habitants | | 4 104 |
| de | 200 à | 499 habitants | 18 209 |
| de | 500 à | 1 499 habitants | 8 909 |
| de | 1 500 à | 2 499 habitants | 2 049 |
| de | 2 500 à | 3 499 habitants | 935 |
| de | 3 500 à | 4 999 habitants | 660 |
| de | 5 000 à | 9 999 habitants | 799 |
| de | 10 000 à | 19 999 habitants | 388 |
| de | 20 000 à | 29 999 habitants | 162 |
| de | 30 000 à | 39 999 habitants | 68 |
| de | 40 000 à | 49 999 habitants | 51 |
| de | 50 000 à | 59 999 habitants | 29 |
| de | 60 000 à | 79 999 habitants | 24 |
| de | 80 000 à | 99 999 habitants | 11 |
| de | 100 000 à | 149 999 habitants | 19 |
| de | 150 000 à | 199 999 habitants | 8 |
| de | 200 000 à | 249 999 habitants | 4 |
| | 300 000 et plus | | 5 |

36 433

Au 15 février 1983. Document AMF.

**Figure F7   The communes of France and their population size (1983)**

Commentators on local government in France often refer to 'the map' as a shorthand for the complex of issues that embrace the system of communal government in France. What 'the map' graphically demonstrates is a 'parish' scale of local government in France where the advantages of local representation, sentiment and responsiveness have to be set against the disadvantages of extreme fragmentation and inadequate resources.

Figure F8 *Agences d'urbanisme* and *Communautés urbaines* in France

Although portrayed together, these two institutions should not be confused, nor their role equated with municipal planning departments. The *Communauté Urbaine* is a public administrative institution with wide powers for metropolitan planning and public works that operates across the constituent communes, but is not a metropolitan government. In five of the *Communautés Urbaines* the actual plan-making functions (including improvement or redevelopment plans, urban design projects etc.) are carried out by an *Agence d'Urbanisme*. The *agences d'urbanisme* are independent of political control although they are usually established by syndicates of communes and part-funded by central government. Neither agency/institution plays a significant role in development control.

171

plans). Agencies will also undertake planning studies for specific conservation, rehabilitation, development and redevelopment projects and conduct preliminary or impact studies for ZADs, ZACs etc. But again it is important to recognise that it is the communes who will perform the development control function once a POS is approved.

## The role of the Mayor
F2.33   If the persistence of the commune is a key feature of the French system then the importance of its political leader, the mayor, only further emphasises the uniqueness of the French local government system. Only very rarely does a local political leader in Britain achieve similar authority, prestige or power to a French mayor. Although the elected leader of the municipal council, the mayor is really constrained only by the annual budget meeting of the council, and his financial control and his privileged relationship with the council and prefect of the Department give him immense local influence. He is at once both leader of the Council and Chief Executive in British terms, emphasising the overlap of political and administrative functions which is alien to Britain. He is both elected leader and agent of central government answerable to the Prefect and civil servants, and authority is vested in him as an individual rather than in the council.

## The cumul des mandats
F2.34   The influence of the mayor is intensified by the political tradition of *cumul des mandats*, whereby politicians can accumulate offices (and their stipends). The mayors of larger cities are often key figures in national politics, occupying the posts of Deputies, Ministers and even the post of Prime Minister itself. They are frequently members of the *Conseil Général* of the Department, but only comparatively rarely do they belong to political parties relying on 'personality' more than politics (Wilson, 1983, p. 159). Many Mayors remain in office for decades. While the Socialist reforms of 1982 did reduce the attractiveness of multiple office holding by reducing the combination of salaries to only 1.5 times that of a deputy, and while it improved salaries, pensions, time release and training for all local elected officials, it failed to limit the *cumul des mandats*, thus retaining one of the great anomalies of French politics (Kesselman, 1985, p.174).

F2.35   So the mayor often continues to exercise political power at many levels. He also continues to exercise full control over the local council. French local authority meetings are often very short and produce very little discussion. There is little or no delegation of powers and the council acts as little more than a rubber stamp. In part this was due to the election procedures which meant that councillors were elected by lists (drawn up by parties or coalitions) and voters could cast a vote only for one or other list, that which obtained an absolute majority being elected en bloc for six years. So there was little possibility of political debate and the council acted like the mayor's cabinet or a board of directors (Pugsley, 1982, p.147). However, the voting system was changed for the 1983

elections combining the old system with a form of proportional representation. A slim electoral majority can still mean a large majority of seats but now there is at last an opposition on town councils to ensure a measure of democratic control, even if such oppositions are often denied vital papers and facilities (Hanley et al, 1985, p.146). However, the role of the mayor remains undiminished.

## Arrondissements *in large cities*
F2.36   Machin (1985, p.136) has argued that traditional commune structures are a positive asset in large urban agglomerations since they can provide a focus for local political participation and community life. The 1982 reforms provided for such foci at the *arrondissement* level in Paris (20), Lyon (16) and Marseille (9) with elected councils and a neighbourhood mayor. The *arrondissement* is responsible for provision of a range of minor services (day care, cultural and social services, public paths and small parks and one third of all public housing allocations), but must also be formally consulted on any major project of urban development within its boundaries. This is a clear response to the need for local participation and accountability, albeit only in the three largest cities. It was also a move to create socialist strongholds at the local level in the traditionally conservative cities of Paris and Lyon, a gambit that failed utterly, to judge by the 1983 elections.

## Newly decentralised powers for the communes
F2.37   One of the main results of the 1982 legislation has been to devolve more power to the communal level, and the devolved powers are particularly relevant to planning and development. The commune will be responsible for local infrastructure, local planning, and development control, but these powers are sharply curtailed if the commune chooses not to prepare a *Plan d'Occupation des Sols*, and there are safeguards for the planning interests of higher levels of government. These powers will be the focus of sections F4 and F5 of this report.

## The problem of communal finances
F2.38   Hanley et. al. have succinctly noted that even with the decentralisation of powers to the local level "autonomy ... is proportional to resources" (Hanley et.al. 1985, p.144), leaving the vast majority of communes with very little scope for independent action and reliant upon the field services of the various ministries for expertise. Local authority finances are in something of a state of crisis as the costs of local services have risen 27 per cent above the rise of inflation between 1979 and 1984, and the opposition to increases in local taxes has grown. With the deepening economic crisis in 1983, and a congested legislative timetable, the Government was forced to postpone further reforms to the system which would have provoked bitter opposition. However, the Government did finance some decentralisation through a transfer of tax resources and a direct grant. It retained the block grant *(dotation globale de fonctionnement)* which provides just over one third of all revenue costs, and established a new *dotation globale d'Équipement* which provides for just

over one quarter of the capital (servicing) costs. But almost half of the revenue expenditure is met from local taxes (mainly on property but also on business activity which now accounts for half of all direct income), which are shared with the Department two thirds/one third respectively. The importance of the payroll tax (*tax professionnelle*) in providing half the local taxes is particularly significant in terms of the relative acceptability to the communes of commercial/industrial as opposed to residential development (Table F2.3).

F2.39 The public bank, the *Caisse des Depôts* has always provided below market interest loans for almost one third of the commune's capital costs (Hanley et al, 1985, p.142) but their rates are now less than one per cent below market rates with the result that interest rates absorb 20 per cent of reserves, and reduce new investment to below 1976 levels in real terms. Recent research suggests that although decentralisation has greatly contributed to the removal of intercommunal inequalities, these still persist and undermine the efforts of many communes to govern themselves (*Metropolis*, 1986, pp.59-90). Reform of the payroll tax which is complicated by state tax relief to local businesses produces particular spatial inequalities and differential attraction of firms and is considered by many to require reform. The picture is further complicated by the state giving partial tax relief from the payroll tax to some companies, compensating the communes through grants which are not necessarily equivalent.

Table F2.3   Direct local taxes 1983 (product of general rolls in FF millions)

| Land Tax on built properties | Tax on non built properties | Product of the tax of dwelling | Product of 'taxe pro-fessionelle' (payroll tax) | Global product of principal taxes and annexes |
|---|---|---|---|---|
| 19,092 | 5,603 | 25,788 | 50,524 | 119,799 |

*Source:* RISU 1985, pp.246–7

F2.40   Thus, few of the smaller communes have been able to develop their own planning expertise even if they have assumed development control powers by preparing a *Plan d'Occupation des Sols (POS)*. While a 1982 survey postulated that all towns over 80,000 population had their own planning services, for communes with populations between 10,000 and 20,000 only one in four had qualified personnel to control urban development. A few progressive communes have developed an *Atelier Public* to assist their own planning activities, and to interact with the local public, but the majority do not have the resources. The function of processing applications for the *permis de construire*, and all associated development control functions at the communal level, is usually performed by staff within the Technical Service departments of the commune, in close cooperation with the mayor. These departments will seek advice from the DDE and perhaps the *Agence d'Urbanisme*, if one exists, as necessary but development control is largely a matter of

ensuring that the *Plan d'Occupation des Sols* (POS) is adhered to.

F2.41   In conclusion, it is conventional for external observers to emphasise the negative aspects of 'the map' of communes, and these are obvious. However, it is important to recognise that the system is "neither ineffective nor profoundly undemocratic" since it provides a local voice and very localised representation and ensures a mutually cooperative and interdependent relationship between the centre and the periphery (Machin, 1983, p.135). However, while the vertical relationships between centre and periphery are strong and diverse, horizontal relationships are "almost non existent or .. very difficult" (Meny, 1983, p.18) especially where town versus country issues enter into the scene. This factor makes regional planning or co-ordination of local planning difficult if not impossible, especially with recent reforms to the procedure for preparing the equivalent of 'structure' plans.

## The Key Elements in French Local Government and the Impact of Decentralisation

F2.42   Any verdict upon the post 1982 decentralisation is necessarily premature. However, it is important firstly to emphasise that the reforms are, in certain key aspects, merely an extension of previous attempts to decentralise power from Paris, most notably the right wing plans of the Barre Government in 1978. Secondly the reforms were intended to be irreversible. There may have been no clear conception of what would transpire, even amongst key Ministers responsible for the reforms, but there was a determination to initiate significant change. Thirdly the reforms are widely recognised as a long overdue response to a highly centralised and top heavy French bureaucracy where the State was attempting to regulate all aspects of French life. Fourthly although the reforms are often couched in terms of weakening the State, they are in effect an attempt to increase the efficacy of state intervention, to allow it to organise key industries and plan the economy while divesting itself of social regulation. This is intended to achieve more local participation, more local identity and a more pluralist society.

F2.43   It is important to examine the key features of French local government, as identified by numerous writers, against the changes wrought or not wrought by decentralisation. The first of these features is what Lagroye and Wright encapsulate as the "interpenetration of national and local elites" in France. This is of immense significance to the relationship between the different levels of government enforcing centralism, while at the same time sensitising both government and bureaucracy to the needs of the provinces (Lagroye and Wright, 1985, pp.7-8). But with the general weakness of political parties at the local level there are few restraints on the power of local political personalities, while the elected council still plays a subordinate role to the mayor.

173

F2.44 It is unlikely that these complex interactions and relationships between civil servants and notables, between prefects and mayors, between regional commissioners and ministers will change dramatically as a result of decentralisation, and the system will still be dependent upon a 'honeycomb structure' which demands conciliation, compromise and concessions between the different political actors (Meny, 1983, p.20). As Wright argues they are

> "condemned to live together in a chaos of surreptitious bargaining, illicit agreements, hidden collusion, unspoken complicity, simulated tension and often genuine conflict". (Wright, quoted in Hanley et al (1985) p.150).

F2.45 The second key observation that is frequently made is that in France historically "central government has only reluctantly conceded power to the local authorities", at least until 1982. So whereas British local government has a specific sphere of operation defined as "residual domain" the French have only "conceded domain" which does not provide a clear cut division of competences (Booth, 1985, p.4). It was not until 1983 that the transfer of powers was set out clearly in two laws to be implemented over the period to 1986. These powers are listed in Table F2.4. As far as possible the new laws sought to transfer complete responsibility for a given function to a single level of government to promote specialisation, and to prohibit one level of sub-national government establishing supervisory powers over anoth-

**Table F2.4    The division of competences in French government and planning 1986**

| FIELDS OF INTERVENTION AND DATES OF TRANSFER | COMMUNE | DEPARTMENT | REGION | STATE |
|---|---|---|---|---|
| REGIONAL PLANNING DISPOSITION OF LAND ECONOMIC – DEVELOPMENT (15–5–1983) | ELABORATION AND APPROVAL OF TAXES ON THE LAYING OUT OF LAND AND ON DEVELOPMENT | | participation in the elaboration and implementation of the National Plan | ELABORATION OF THE NATIONAL PLAN |
| | | Advice PROGRAMME OF AID FOR RURAL INFRASTRUCTURE | ELABORATION OF THE REGIONAL PLAN | |
| | Agreement | Agreement | REGIONAL NATURE PLAN | |
| URBAN PLANNING CONSERVATION (1–10–1983 & 1–4–1984) | INTERCOMMUNAL ELABORATION OF SD | | | Fixation of the perimeter of the SD |
| | ESTABLISHMENT OF THE POS | | | control of the prescriptions and demands for revising or modifying the POS |
| | Advice | Advice | Advice | ELABORATION OF *SCHEMAS* FOR USE OF THE SEA |
| | DELIVERY OF THE *PERMIS DE CONSTRUIRE* ETC. IF THE POS IS ALREADY APPROVED | | | |
| | Agreement | | Advice of the college of the patrimony and of sites | ZONE OF PROTECTION OF THE URBAN AND ARCHITECTURAL PATRIMONY |
| ENVIRONMENT AND CULTURAL ACTION | CULTURAL SERVICES, LIBRARIES, MUSEUMS | | | technical control of the State |
| ENVIRONMENT 1986 | Advice | DEFINITION OF THE DEPT. PLAN OF WALKING AND RAMBLING ITINERARIES | | |

Adapted and translated from *Les Cahiers Français* No. 220 March-April 1985.
Lower case indicates an advisory/supervisory rather than an executory role.

er thus preventing the development of hierarchical relationships. The latter reform was also interpreted as avoiding the issue as to the priority of the region and department, a particularly vexatious matter to many *notables*. Certainly the reforms now provide a clear cut division of competences, the lack of which has for so long bedevilled French local government. But for various reasons central control is omnipresent, especially in the planning sphere with the continued influence of the state services at the local level.

F2.46   The third key observation is related to both the structure of local government and the system of administrative law. There is no clear division between administrative and political power in the French system as there is in Britain. This is evident in the role of the mayor, and in the discretion granted to the administrator, but not to the politician, and in the ability of civil servants to hold political office.

F2.47   All three factors make the French system quite different from that operating in Britain, and they undermine any apparent similarities between procedures and practices in the two countries that can be established. However, it is clear that many more varied relationships between levels of local government, and greater variation in competences, scope, and styles of government, will emerge at each level, further eroding the Jacobin ideal of uniformity, even if the power of the state is still omnipresent. The role of elected politicians has been enhanced making the system more democratic, but the failure to tackle the 'map' of communes, the inability to tackle the problems of multiple office holding, and the failure to reform local finances remain key weaknesses in the programme.

F2.48   An effective critical summary of the impact of decentralisation is provided by Moulinard (Institute for Environmental Studies, 1987, p.546)

> "The unclear wording and the difficulties of interpretation, the lack of suitable personnel and the inadequate financing of studies and of the preparation of plans have led to harsh criticism of the law of 7 January 1983 (decentralisation). Will the difficulties which municipalities face in dealing with technical problems lead to the restoration to the state of legislative powers which have been transferred? ... is it not likely that control by the state will be replaced by *de facto* control by the *Départements*, something which it was the law's intention to rule out altogether?
> In a country like France where centralisation is rooted in political and government tradition, there can be no doubt that a reform which is strongly decentralising will be implemented with great hesitancy and some resistance".

## Conclusions

F2.49   This chapter has described the governmental structure of France and the administrative framework for planning. It has explained some of the complications of central-local relations, and of multiple office holding that persists in spite of a thoroughgoing programme of decentralisation which has dramatically increased the responsibilities of the communes. From the perspective of comparative analysis of development control systems the most important factors to emerge are as follows,

(i)   the existence of four levels of government in France, more than any other European country, and the curious non hierarchical relationship between the levels,

(ii)   the interpenetration of local and national élites made possible by multiple office holding and the 'honeycomb' structure that this produces, yielding very complex political bargaining at all levels but with the benefits of strong linkages between all levels of government.

(iii)   the absence of a clear division between administrative and political power in local government and the planning process, evident in the role of the mayor at the local level, in the ability of civil servants to hold political office, and in the discretion granted to the administrator

(iv)   the very minor role played by the municipal council in decision making, including planning, which contrasts sharply with Great Britain

(v)   the advent of decentralisation which has greatly strengthened the regional level, and more importantly the commune, providing the commune and the mayor with full development control powers once a *Plan d'Occupation des Sols* has been approved.

F2.50   These major points about the structure of local government can be supplemented by a number of important observations about each level of government.

*At central government level*

(vi)   the inability of the Ministry to enforce policy through the appeal process as in Britain, and hence their recourse to frequent amending of the *Code de l'Urbanisme* through laws, decrees and *arrêtés*

(vii)   the compensatory power of the Ministry through its field services to control planning at the local level, but the significant erosion of this power by the development of *agences d'urbanisme* in the large cities, leaving the field services in control of rural and small town France

(viii)   the particular standing of the planning profession which is of secondary status within the civil service (at national and departmental level) but within which there is a much broader representation of disciplines than in Great Britain.

*At the regional level*

  (ix) the role of the region in economic planning, and of the *Contrat de Plan* in prescribing five year plans of infrastructure development in partnership with the State

*At the Departmental level*

  (x) the ambiguous role of the DDE as civil servants and representatives of the central state, guardians and providers of planning expertise for the smaller communes combining ministerial, legal, and local advisory roles

  (xi) the role of the *Commissaire de la République* as the representative of the central state, the person responsible for supra communal interests, and the controller of the legality of local planning decision making

*At the communal level*

  (xii) the persistence of the commune as the fundamental unit of local government with boundaries that make a nonsense of contemporary realities but which retain significance as expressions of community in the minds of the population and elected officials

(xiii) the lack of resources of the majority of communes which make them incapable of meaningful action and unable to retain adequate personnel for the execution of services including planning

(xiv) the role of the *agence d'urbanisme* as a plan making and advisory body in major cities but its minor role (except in Paris and a few other cities) in development control, and its separation from political control

  (xv) the role of the mayor at the local level as the most powerful political figure in plan making, and when a POS is approved, the signatory of development control decisions.

So there is no real equivalent in France (except perhaps in Paris) of the British local authority planning department staffed by professionals and performing both plan-making and development control functions in an integrated manner, under the control of a planning committee with delegated powers from an elected municipal council. In addition there remain doubts about the ability of most municipalities to assume and effectively exercise planning powers.

# F3 THE DEFINITION OF DEVELOPMENT

## Introduction

F3.1 The control of development in France is tight and comprehensive and based essentially on obtaining a permit for virtually all construction works, even for minor buildings. These controls are prescribed in the *Code de l'Urbanisme*, though there are additional procedures enshrined in the *Code de l'Environnement*, and the building regulations are laid out in the *Code de la Construction* etc. Minor exemptions from obtaining a permit do exist, and are an increasing preoccupation of French planners and developers, but control remains all-embracing. These controls are reinforced by widespread demolition controls and controls on the subdivision of land. A lighter scheme of control is applicable to a range of minor developments, often of a public service nature but now including minor residential extensions, while a range of largely rural activities, enclosures, caravans, excavation, tree felling and land clearance, are subject to special authorisations. In addition there are specific controls on hazardous installations, major shopping developments and industrial/commercial development in the Paris region.

## The *Permis de Construire* - the Control of Building

F3.2 The French control building rather than land use and a material change of use becomes significant only when it results in a change to the fabric of the building. The crucial law (*Code de l'Urbanisme* L421-1) states

"Anyone who wishes to construct or have constructed a building, to be used as a dwelling or not, whether with foundations or not, must in the first instance obtain a *permis de construire* (building permit) ... this same permit is compulsory for any work carried out on an existing building when this work would change the usage, where the exterior appearance or volume would be modified, or extra storeys would be created".

F3.3 This law applies to public services (and their concessionaries), to the state, region, department and commune as well as to private individuals.

F3.4 The building permit is the means by which the administration can ensure that the proposed building conforms to the National Rules of Urban Planning (the RNU, see paras F4.97-102), or the *Plan d'Occupation des Sols (POS)*, or local plan, to certain Building Regulations, and to rules for safety, hygiene, fire and public safety. However, it neither confirms nor verifies legal rights that the applicant may have over the ground, such as rights of way or of view, which is why the permit is delivered subject to the rights of third parties (*sous réserve du droit des tiers*). The permit is applicable to the land in question and not only to the applicant for the permit or title holder of the land.

F3.5 The definition of development is in fact very comprehensive and embraces chicken runs, garden shelters, and wooden chalets, as well as any new openings in a façade. It was significantly tightened in 1976 but in January and March 1986 the extent of exemptions was considerably widened. The complete list of exemptions (R421-1) is given below.

1. Underground operations for storing gas, pipelines or cables
2. Any work pertaining to infrastructure of railways, rivers, roads, footpaths; public or private works in ports or airports
3. Temporary installations on a site directly necessary for carrying out works or linked to the commercialisation of a structure being built
4. Model buildings for use during an exhibition
5. Street furniture in a public area
6. Statues, monuments or works of art equal to or less than 12 metres above ground and less than 40 cubic metres in volume
7. Terraces less than 0.6 metres high
8. Posts, pylons, streetlamps, or windmills of a height equal to or less than 12 metres, as well as aerials of less than 4 metres diameter
9. Walls of a height of less than 2 metres (without prejudice to the *régime* referring specifically to enclosures/fences)
10. Any work not included above whose ground surface is less than 2 square metres and whose height is no more than 1.5 metres from the ground

F3.6 Longer standing exemptions have been afforded to buildings covered by the Official Secrets Act (R422-1), largely munitions works, defence establishments and military communication lines.

## Development Controlled by a *Déclaration Préalable*: a Lighter Form of Control

F3.7 A second set of 'exemptions' are placed in a rather different category requiring a *déclaration préalable* or preliminary declaration, which must be made to the responsible authority one month before the start of work explaining the nature of works to be undertaken. These exemptions from the general regime of the *permis de construire* (see below) are still subject to an approval process, but if the mayor has not opposed the declaration within one month the work can proceed, and the declaration serves the same purpose as the *permis de construire*. Those added to this list in January and March 1986 include works producing and distributing energy, school buildings, post and telecommunication buildings, and non technical buildings within ports, stations and aerodromes.

The list of those developments exempt from *permis de construires* but subject to *déclarations préalables* (R422-2) are as follows

1. Cleaning/maintenance of façades (*ravalements*) (under French planning law this is required every ten years)
2. Reconstructions or works on listed buildings under the historic monuments legislation
3. Machinery necessary to the functioning of public services within ports, aerodromes, railways
4. Technical works necessary to maintain the security of navigation of ships, railways, aircraft etc
5. Technical works for telecommunications of less than 100 square metres and posts and pylons of more than 12 metres
6. Electrical distribution of less than 63kV and less than 1 kilometre in length, and transformers of less than 20 square metres and 3 metres in height
7. Portable classrooms for schools and educational establishments up to 150 square metres, but up to a total maximum of 500 square metres on the same land
8. Light leisure dwellings of less than 35 square metres replacing similar dwellings
9. Uncovered swimming pools
10. Greenhouses and frames of between 1.5 metres and 4 metres and less than 2,000 square metres on the same land
11. Constructions or works not covered in 1 above, but not changing the future use of an existing building, not creating a new storey, or not creating additional floor space of more than 20 square metres gross

None of the above is exempted from a *permis de construire* if the building concerned is included on the supplementary list of historic monuments.

## Controls on 'Development' other than Building

F3.8 The *permis de construire* controls only building and alterations to buildings. Taking the broader British view of development there is a wide variety of other actions that are controlled. These include

(i) demolition

(ii) subdivision (*lotissement*)

(iii) enclosures and fences

(iv) raising and undermining the soil

(v) camping

(vi) cutting and felling of trees

(vii) clearing of land

These diverse uses and activities are subject to national regulations in which planning looms large, and authorisations or *déclarations préalables* are required. These authorisations are subject to the same rules of decentralisation as the *permis de construire*.

### Demolition controls

F3.9 Demolition controls originated in 1913 and 1930 legislation protecting historic buildings, and the 1945 legislation protecting sites. In 1976 these precedents were unified into a single *permis de demolir* as a counterpart to, but never a substitute for, the *permis de construire* and their use extended beyond the protection of historic sites and buildings. Like the latter they apply to all public services and levels of government, but by way of contrast not to all national territory. They apply to three types of area: firstly to all communes within 50 kilometres of Paris, all communes with more than 10,000 inhabitants, and all those listed by decree; secondly to zones delimited for the protection of historic sites and monuments and listed buildings (actual demolition of listed buildings themselves requires a separate *déclaration préalable*); thirdly they apply to conservation areas and other sensitive areas designated in a POS.

F3.10 The definition of demolition can include rendering the building uninhabitable, or unhealthy, or dangerous in any way (L430-2), but there are also exemptions for demolitions which conform to conservation plans, the implementation of a POS, for road alignment, or service easement. Demolition can be refused in order to protect housing in urbanised communes, or to ensure occupants have been rehoused, or to maintain the character of protected zones.

### Lotissement *or subdivision controls*

F3.11 *Lotissement* (allotment or the creation of building plots) may be conceived as the subdivision of raw land into building plots. Originally a private operation, it was widely abused immediately after the first world war and brought under some control in 1919 and 1924. It was the main means of controlling, if that is not too strong a word, suburban development in the inter-war period. In the post-war era it was largely replaced by larger scale urban operations, but by the 1970's it was regaining its role as single family housing increased in importance. The law and regulations were reformed in 1976 and 1977 confirming the singular character of *lotissement* as a private operation of urban planning under public powers, but making the procedures more environmentally conscious.

F3.12 The definition of *lotissement* is complex but precise and relates to five characteristics

(i) the property 'island' allowing parcels separated by public roads or third parties to be treated separately

(ii) the terms of reference of the division, creating full property rights

(iii) the period of division - calculated over 10 years

(iv) the goal of division - the construction of buildings

(v) the number of divisions - more than 2 or 4 lots depending on the future use of the original plot retained by the owner, and whether or not the property is being divided among successors in title.

F3.13 These latter provisions protect the patrimony and reassure families that their division of properties amongst heirs remains unaffected. Activities such as *remembrement* (the reallocation of agricultural land holdings) are also excluded, along with operations like ZACs or urban renewal/rehabilitation procedures and cooperative building ventures which are controlled by the *permis de construire*. Again the approval process is decentralised and will be discussed in the next section.

*Enclosures and fences*
F3.14 Recognising the damage to the urban or natural landscape that can be caused by unaesthetic walls or fences the old *Code de l'Urbanisme* required a *permis de construire* for the erection of enclosures. However, as of 1976 a specific *autorisation de clôture* (enclosure authorisation) was required, only to be replaced by a system of *contrôle allégé* (light control) as part of the 1986 deregulation. Control is applied only in areas with a POS, designated environmentally sensitive areas, and in communes listed by the *Commissaire de la République* in consultation with the mayors. In these areas all forms of enclosure are controlled - walls, palings, metal fences etc, but hedges and ditches are excluded as are enclosures habitually necessary for agricultural activity, forestry and building. Since 1986 a *déclaration préalable* or preliminary declaration is required with standard application forms and procedures.

*Diverse works (including amusement parks and sports areas, raising and undermining soil, vehicle depots etc)*
F3.15 Since 1962 these diverse works have required a *déclaration préalable*, modified in 1977 and 1984, but only in communes with a POS and in those communes listed by the *Commissaire de la République*. They include parking areas with more than ten car spaces, and caravan garages, and soil works of greater than 100 square metres and/or 2 metres in height or depth, except in works on public land. These rules may well be covered by other procedures if they involve waste disposal or tipping, or are part of other operations like caravan sites.

*Camping, caravanning, and leisure parks*
F3.16 Originating in 1959 and 1968 these controls have been progressively refined and apply to all national territory except fairs, markets, streets and public places. The stationing of six caravans for up to three months in unmanaged areas is not controlled, but above this it is necessary to obtain an authorisation, and certain prohibitions exist to conserve the coast, protected sites, wooded spaces etc. In addition, zones where the stationing of caravans is forbidden can be delineated for purposes of conservation in the broadest sense. Similar rules apply to camping, for more than six tents or 20 campers, and specific authorisations must be obtained.

F3.17 Light buildings for leisure are more tightly controlled and confined to caravan parks, holiday villages or special residential parks the opening of which requires special authorisation.

*Wooded spaces and the cutting and felling of trees*
F3.18 First considered in 1943, but more especially in 1958 and 1976, local plans (*POS*) and other local planning documents were able to protect woodland, forests, parks and street trees and encourage their protection and creation. *Autorisations préalables* are required for cutting and felling trees and may be required alongside a *permis de construire* etc, and failure to achieve an authorisation would make a *permis de construire* inadmissible.

F3.19 The authorisation is principally required in territories covered by a POS (even those not rendered public or approved), or in those areas designated as environmentally sensitive, or forests woods or parks designated by the *Commissaire de la République*. It is not required for dangerous trees or those covered by a forestry authorisation under another code (such as state forests which conform to *arrêtés* of the Ministry of Agriculture). Again there are set forms, consultations, publicity and approval procedures.

*Clearing of land*
F3.20 This kind of clearance (which refers to the ending of forestry use) has been subjected to control for a long time and this was refined in 1969. It requires an *autorisation préalable*. For those spaces outside classified woodland, clearing is ordered by the Forestry Code of 1985 (L311-1) under several sections. However the reasons for refusal are limited to the impact on soil erosion, water courses, dune protection etc. In the classified woodland in principle any clearance is forbidden, and any rejection is verified by the Commissaire de la République.

F3.21 The classified woodland designation prevents not only clearance but any use which compromises conservation or afforestation. However in order to facilitate the public acquisition of such woodland, authorisation of construction or stationing of caravans on one tenth of the surface can be permitted if the remainder is transferred to the local authority, if this is compatible with the *Schéma Directeur*. The application for such development has not been decentralised and must be made to the *Commissaire de la République*. It must include an impact study or an explanation of how environmental considerations have been taken into account.

## Other Controls on Major Development

F3.22 Over and above the *permis de construire* and associated permissions on demolition, subdivision, tree felling etc there are additional authorisation procedures required for various major developments. These have arisen from rather different sources. Environmental concern has led to the introduction of special controls on a variety of hazardous or environmentally noxious enterprises. The concern of local shopkeepers has produced a special control procedure on major shopping developments. Finally regional planning concerns have led to the establishment of special authorisations for industry and offices in the Paris region.

*Installations classées - a control for environmentally hazardous installations*
F3.23 A separate control procedure exists for all development proposals which may threaten danger or nuisance, under the title of *installations classées*. These procedures are set out in the *Code de l'Environnement*, but most of the actual control lies with Ministry of Industry officials. The 1976 law defined *installations classées* as any factory (including some factory farms), workshop, depot, etc which can present danger or inconvenience to the neighbourhood, both its security and public health, be it land reserved for agriculture, the protection of nature or the conservation of sites and of monuments.

F3.24 The register of *installations classées* defines all such activities and land use, and the register is kept up-to-date by the *Conseil d'État* and a council for *installations classées* through Ministry of Environment reports and the issuing of decrees. Over 400 headings are currently on the register and quantitative thresholds are established on the size, capacity or volume of stocks of a plant or installation. This determines whether such installations can be issued with a declaration granting them effective permission, or whether the scheme is a sufficiently significant threat to the environment to require an authorisation procedure which involves an *enquête publique* (public enquiry). The public inquiry procedure is used only for such instances and is not a feature of the general system of control and the issuing of *permis de construire*.

*Special controls on major shopping developments*
F3.25 Special authorisations are required for shopping developments with a sales area of more than 3000 square metres, or for single shops (essentially hypermarkets) of 1500 square metres. These figures become 2000 and 1000 square metres respectively in communes with a population of less than 40,000. Major extensions or conversions of commercial establishments realising a single shop of more than 200 square metres also require authorisation, and transformation of existing retail buildings reaching these limits is also controlled. Warehouses selling by mail order or wholesale or service stations are not included.

F3.26 The decision is taken by a departmental commission of urbanism composed of 20 members, nine elected officers from the commune, nine representatives of commercial and artisan activities (including six from independent business) and two consumer representatives. The *Commissaire de la République* presides over the commission but does not vote, but the commission must abide by the principles of commercial and artisan law which are an interesting mixture of value for money, protection for the independent retailer, and development appropriate to urban growth (Jacquot 1987, p.416).

*Special controls on industry, offices, research establishments in Ile de France*
F3.27 Finally there is a set of controls on enterprises and establishments proposed in the region Ile de France (Paris region) that date back to 1955 decentralisation and regional planning policies. A decentralisation committee was set up initially to examine public developments, but it was extended to the private sector in 1958. This created parallel committees of *agrement* (consent) approval and for *permis de construire* and the two were fused in 1967. The new committee had to give approval to public installations, but offered advice to the Minister only on private enterprises. Its field of application grew beyond industrial premises to embrace tertiary activities and even research. However in 1985 these restrictions were lessened, exempting offices and industrial premises, in order not to harm the economy of the region in a period of economic recession, and in order to ensure that such controls did not lead to foreign companies avoiding Paris in favour of some other European metropolis.

F3.28 Development in new towns, office groupings, shops of less than 2000 square metres; industrial, scientific or industrial units of less than 3000 square metres, specific offices of less than 2000 square metres, warehouses of less than 5000 square metres were all excluded from *agrement* control.

F3.29 The committee of 20 members has eight ministerial representatives, one for the Region and one from the Department of l'Oise, and the other 'competent' personalities. They are advised by DATAR. Conditions can be affixed to the use and size of premises, siting and abandonment of buildings. A *permis de construire* is still required.

## Conclusions

F3.30 This section has described the definitions of development which are controlled by different legal procedures in the French planning system. While the controls on the act of building constitute the main thrust of French control, there is a range of other authorisations that provide a rather lighter control on development acts which are not strictly building. In addition there are strategic and regional controls on certain kinds of development, as well as controls on hazardous installations, to broaden out the actual practice of development control. From the perspective of comparative analysis of development control systems the key factors to emerge are that

(i) the control of the act of building avoids the legal problems of defining development and provides a much more comprehensive control over the modification of built form, townscape and landscape

(ii) the range of exemptions offered are in fact minimal and the most significant (eg the 20 square metre extension) are very recent (1986), and while there is much interest in further relaxation the current system provides the potential of tight control

(iii) the widespread application of demolition control, in the medium to large size urban areas and in all areas with a POS, provides another indication of the high level of control exercised, in contrast with Britain where it is available only in conservation areas and on listed buildings

(iv) both construction and demolition controls apply to all levels of government and all public services, contrasting somewhat with British practice. The only clear exemptions are for developments which may be considered to be state secrets, and for infrastructure for transport including pipelines and minor electricity transmission, although these have also been extended in 1986

(v) the control of hazardous installations offers a control transcending environmental and industrial inspectorate procedures, and incorporating the requirement of an impact study and a limited form of public enquiry

(vi) agricultural landscape control remains a non-issue, and agricultural activities remain clearly outside control except for clearing wooded land. Local plans frequently exempt agricultural buildings from control

(vii) the controls on major shopping developments seem anachronistic until one considers Ministerial involvement in the approval of major out-of-town and city centre schemes in Britain. It provides a clear manifestation of the economic and political issues in control, and it is significant that it is only in the arena of the retail trade that such issues have surfaced

(viii) the controls on offices and industrial establishments in *Ile de France* have many similarities with the old Office Development Permit and Industrial Development Certificate practice in Britain, and have drawn a similar response in the proliferation of units designed marginally smaller than the lower limit. In a period of economic recession the days of these controls are numbered.

## Introduction

F4.1    The control of development in France is a process clearly prescribed in considerable detail by the *Code de l'Urbanisme*, using nationally standardised forms and following nationally prescribed rules. These substantive rules are also written into the *Règles Nationales Urbaines* (RNU), or national planning regulation procedures, and they apply to all urbanised areas without a *Plan d'Occupation des Sols* (POS). In non urbanised areas without a POS, development is now largely forbidden. The POS itself incorporates these national rules of urban planning and generally weaves them into a much more comprehensive and detailed document, conformity to which gives a legal right to development. To enforce these procedures authorisations for development have to follow a bureaucratic but clearly prescribed application-examination-approval-delivery-enforcement process with set time periods. Numerous other approval procedures and tax levies complicate this essentially clear-cut process, and these will be discussed in turn.

F4.2    Thus this section begins by outlining the changes introduced by decentralisation. It then describes the procedures for obtaining a *permis de construire*, both the types of permit that can be granted and their effects. Some analysis of the speed and efficiency of decision-making is attempted. Other development control authorisations associated with the *permis de construire* are discussed and their incidence briefly analysed. Enforcement procedures are explained as are the provisions for obtaining a kind of outline permission. Finally the control of land sub-division (*lotissement*) and its importance are described. Having outlined the basic procedures of control, attention then shifts to the planning considerations enshrined in the *Code de l'Urbanisme*, and to the special controls for urban and rural conservation, for hazardous installations and major shopping developments, and the procedures for ensuring that supra-communal and national interests are taken into account. Finally, since the delivery of a *permis de construire* results in the levying of a series of associated taxes that crucially affect the costs of development, these fiscal devices are briefly discussed.

## Decentralisation and the *permis de construire*

F4.3    Until the law of January 1983 the *permis de construire* was always delivered in the name of the state. Now the state delivers the permit only if the commune has not approved a *Plan d'Occupation des Sols* (POS). If the

commune has approved a POS then the mayor takes the decision and signs the permit, but the *Commissaire de la République* retains important checks on decisions and the power to overrule decisions.

F4.4    The key recent innovation in French planning has been the decentralisation of the power to grant a *permis de construire* to the mayor of the commune, if a POS has been approved for six months, and it is sufficient if the POS covers only the urbanised and 'urbanisable' area of the commune. The transfer of powers is definitive and irreversible. The mayor may however delegate the powers (along with all other authorisations) to an *établissement public de coopération intercommunale* (EPCI), which would be appropriate where a syndicate of communes or a *communauté urbaine* exists, but this 'delegation' has to be renewed each time the municipal council changes and a new president of the EPCI is then elected. Approval for development realised by the State, region or Department or their concessionaires, works relating to energy production or transmission, and works in the national interest remain delivered in the name of the State (L421-2-1), and the mayor or president of the EPCI only offer advice in such cases.

## Obtaining the *Permis de Construire*

*The request/application for a* permis de construire
F4.5    The application for a permit must be made by one or other of the owner of the land, a person mandated by him/her, or person with a title that allows him/her to build, or a person able to benefit from the expropriation of the said land for public utility (R421-1). A standard form is provided for the application (with different forms for individual houses, works not creating new floor space, or electrical transmission lines) and four copies must be deposited. This form (Plate 9) requires specification of the identity of the applicant and the builder/developer and the architect if necessary, the address of the property and its current use (details of buildings etc), and its legal status with regard to *lotissement* (see paragraphs F4.84-F4.89).

F4.6    Details of the project must be given including the nature of works, future use of premises, density of construction and floorspace created and height of construction. Full details are also required of the nature and colour of façade materials, all external woodwork etc, roofs, and walls/fences, and full details of car parking. Full details of the residential component of any development must also be supplied including composition, occupation, financing, numbers of rooms and whether of individual or collective construction.

F4.7 The forms must be accompanied by a site plan, sections to convey all three dimensions, and elevations. The application for a *permis de construire* must be accompanied by applications for any other necessary authorisations such as tree felling, clearing land, occupation of public land, demolition or hazardous installations.

### The requirement for architectural advice

F4.8 A request for a *permis de construire* cannot be investigated unless a qualified architect has "established the architectural project", which is defined as preparing "the plans and written documents, the siting of buildings, their composition, their organisation and the expression of their volume as well as the choice of materials and their colours" (L421-2). So the architect must design the buildings but he does not have to have control of the whole project.

F4.9 Specific exemptions to this rule include most importantly all those people who want to build for themselves, or modify a construction "of little importance". This exemption applies to any building of less than 170 square metres. Thus most individual houses are excluded from this control except those frequently repeated designs or standard plan dwellings which, since 1981, must be established by an architect at the outset. Certain agricultural buildings (less than 800 square metres) and greenhouses (less than 2000 square metres) are also exempted. So too are shopfronts, the developing or equipping of internal spaces, or any modifications which cannot be seen from the outside.

F4.10 In general, architectural quality is given greater priority throughout French development. This is evidenced by the greater emphasis upon owner-occupied/custom built development, a system of architectural competitions for almost all public projects, the wide use of design guidance, the promotion of architecture by the Ministry of Culture, and a mechanism for public education about architecture (see paragraph F4.58). In the control process it would be normal for an architect within the DDE or control authority to deal with the application to facilitate informed negotiations on design matters. While the requirement for architectural advice occasionally produces architects who are prepared merely to sign drawings, it clearly does promote the wider use of architectural skills. Furthermore, as Booth has argued, the very nature of the French control of building, rather than land use, and the emphasis on the greater appreciation of urban design, building form and urban design in the POS, tend to encourage a greater emphasis upon architectural quality and a better quality of urban design generally (Booth, 1987, pp. 16-17).

### The deposit and transmission of the application

F4.11 The applications (four copies) are deposited with the mayor of the commune. The mayor opens a file (*dossier*), gives it a registration number and processes it in conformity with the rules laid out in the *Code de L'Urbanisme*. The procedures are dependent upon whether the mayor, the president of the EPCI or the state in the form of the *Commissaire de la République* have the power to authorise the permit, but the latter requires a copy under all circumstances to control the legality of the decision, receiving three copies if he retains the power to grant any permit.

F4.12 Since 1983 (although the government advised it in a 1976 circular) the application for a permit must be advertised in the town hall within 15 days of its acceptance (previously publicity was forthcoming to neighbours only after the grant of a permit), and must record the name of the applicant and the address, net floorspace, height and future use of the proposed development. So a short term planning register is established in poster form at the town hall. However there is no requirement for the application to be advertised on site, and this is regretted by many (Jacquot, 1987, p.339). Nor can the *dossier* be consulted before the decision is reached.

### Examination of the application

F4.13 The examination of the application is carried out by the field services of the state (essentially the relevant section of the DDE), or the technical staff of the commune or group of communes. As of May 1984 over 95 per cent of all communes with a POS had opted to use the DDE (Jacquot, 1987, p.340) but their access to the DDE is now (as of August 1986) slightly restricted to technical studies of those applications which appear to the DDE to justify their technical assistance. The decision itself remains the responsibility of the mayor or commune. In those communes without an approved POS the DDE provides the technical support and the state remains responsible for the decision.

F4.14 The *recevabilité* (whether or not it can be officially received) of the application is first examined to establish whether or not the file is complete, and the applicant must be notified (within 15 days) by registered mail of the date by which a decision must be notified to him. If the file is incomplete the applicant is advised in a similar way and the time period for decisions does not start to elapse until the file is complete (Figure F9).

### Time period for consideration of the application

F4.15 A key aspect of the control process is that a specified time period for decision is written into the *Code de l'Urbanisme* and, since 1983, if a decision has not been forthcoming in that period the applicant is entitled to a *permis tacite* or tacit permission. However the applicant who has not received any notification from the relevant authority within two weeks of transmitting his application must seek a request for examination, and in a further week, if there has been no reply, he/she will be entitled to a *permis tacite* within two months of his request (the letter notifying the receipt of his application then constituting permission).

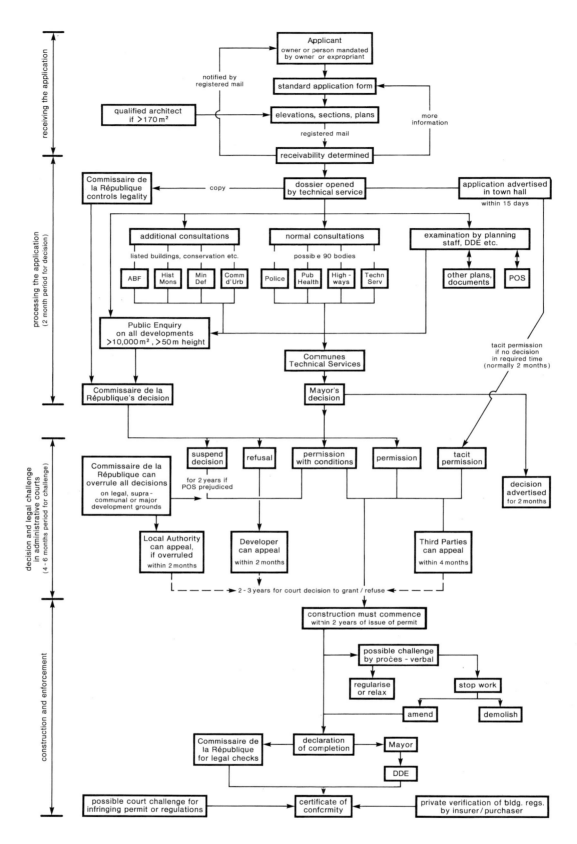

**Figure F9   The processing of an application for a *permis de construire***

In portraying the process of planning control it is important to emphasise the four stages: determining the 'receivability' of an application: processing it through the consultation and technical review procedures to a decision; having the decision checked for its legality and displayed to allow legal challenges; and finally a stage of enforcement through the construction phase, checking the permit and the building regulations against the construction as it is completed.

F4.16  Two months from the date of deposit of a complete application is the normal period for a decision, but three months is allowed for developments of more than 200 housing units, or for premises of over 2000 square metres, or to allow consultation of regional or departmental commissions, ministerial or public persons (three months also for a scheme which deviates from or modifies a mining restriction).

F4.17  A six month period is provided for developments requiring consultation of a national commission or commercial developments requiring consultation with a *Commission Départementale d'Urbanisme* (CDU). This is extended to ten months if the application is a recourse against a previous decision of this commission. Three month periods are also provided for applications involving historic monuments, architectural heritage protection areas (ZPPAU) or nature reserves, and this becomes five months if any additional consultations are required.

*Consultations*
F4.18  The 1981 manual of the *permis de construire* lists 90 possible consultations on any permit. As an example normal practice in Paris is enshrined in Plate 8 and constitutes some 25 bodies. A range of specialist commissions exists to protect certain interests, and the role of these is discussed in paragraphs F4.56-F4.59 and F4.108. The object of the consultations is to take into account the collective interest other than that of those delivering the permit, so that the mayor is consulted if the *Commissaire de la République* takes the decision and *vice versa*. In addition the consultations are designed to embrace legislation other than urban planning and public servitudes which fix the regulations of health and security. The advice may be obligatory (*avis confirmé*) or not (*avis simple*). The consultations have to be completed in one month, or two months for a national commission, and if not the advice is presumed to be favourable towards granting the permit.

F4.19  Until April 1985 these consultations were purely internal to the administration, but the law of July 1985 requires that buildings of more than 50 metres height and commercial buildings of more than 10,000 square metres be subjected to an *enquête publique* or public enquiry. The procedures for a public enquiry are outlined in paragraphs F4.48-F4.55 below.

F4.20  Having received the various responses, the planning agency charged with examining the application ensures that it conforms to the relevant rules - the POS or other planning documents, the rules of *lotissement*, the national rules of urban planning, or the servitudes of public utilities. Minor adaptations or derogations do not require a specific authorisation. The control service recommends a decision to the mayor or *Commissaire de la République*.

*The decision*
F4.21  The decision is taken and signed by the mayor, the President of the EPCI, or the *Commissaire de la République* according to the conditions outlined in paragraphs F4.2 and F4.3 above. If the mayor takes the decision in the name of the State, because no POS is approved, then he must follow the advice of the DDE, and if he does not agree then the *Commissaire de la République* takes the decision. If it is a government or public service development it is the *Commissaire de la République* who takes the decision, and he is also responsible for approving other major developments (commercial buildings of over 1000 square metres, high buildings, buildings related to energy production, changes of use) and developments in certain conservation zones or near airports or military installations. He can also delegate his authority to sign approvals to the director of the DDE. The Minister also has a right to intervene in any case.

*The form of decision*
F4.22  A formal decision (*la décision expresse*) is presented in the form of an *arrêté* and follows a ministerial model, and a copy must be sent to the *Commissaire de la République* and to his deputy in the *arrondissement* or (since 1982) the *arrêté* does not become executory. If the *arrêté* rejects the application or places conditions upon it, or suspends a decision, the *arrêté* must be justified and the reasons explained to the applicant. The same applies if a permit is only forthcoming following a derogation or minor adaptation of a plan, when it is necessary to explain the decision to third parties.

F4.23  Since 1983 the decision or its letter of notification must mention the means of appeal against the decision and the periods for such action. If the decision is positive a poster must be displayed in the town hall and on the site within one week of the delivery of the permit, and it must be displayed for two months in premises accessible to the public. Interested persons can also consult the actual file. In the event of a failure to display notification in the town hall or on the site, the two months cannot begin to elapse until both notices are displayed.

*A tacit permission (*permis tacite)
F4.24  In the general rules of French administrative law the lack of response of an administration to the administered indicates a rejection, but with a *permis de construire* it is the opposite. This is not a traditional solution but was the object of a specific reform in 1970 removing the need for the applicants to ask for a reply before tacit permission can be assumed. Now a *permis tacite* is automatic, but there are certain dangers with regard to protecting the rights of third parties that have led to limitations on its use. Reforms in 1977 and 1983 afforded particular exemption from tacit permissions to historic monuments, nature reserves, and protected historic architecture zones to prevent them being adversely affected by official negligence. Also any development requiring an *enquête publique* cannot be authorised by a *permis tacite*.

F4.25 Apart from the above-mentioned exceptions a *permis tacite* is automatic if no decision is forthcoming in the allotted period for the decision, and the letter notifying the receipt of the application constitutes the permit. Since 1983 it has also been possible, in the absence of even a letter notifying the date of decision, to use a letter requesting examination of the file as a *permis tacite*, if no reply has been forthcoming within two months of its postmark. These letters must be affixed to the works to notify third parties for the same period as the *décision expresse* (two months).

F4.26 The applicant has the possibility of asking the authority to attest that no negative decision will be forthcoming, or indicate the prescriptions required, in order to ensure that the tacit permission is not illegal, and this can be done in the delay allowed for *recours contentieux*, appeals to the administrative courts.

*The refusal of permission*
F4.27 The terms of refusal are laid out in the *Code de l'Urbanisme*:

"... the *permis de construire* cannot be given unless the proposed buildings are in conformity with the legislative and regulative dispositions concerning the siting of buildings, their use, nature, architecture, dimensions and the development of their surroundings (L241-3) ..."

and it adds that for high rise and public buildings conformity to the rules of public safety are required. A refusal may also be forthcoming if the public authority cannot say when a development will be serviced. In these cases the authority is bound to issue a refusal.

F4.28 In other cases the competence of the administration is only 'regulatory', that is to say it can refuse authorisation but it is not obliged to do so. If the rules in the plan or the Code are permissive this allows the administrative authority room to judge for itself, but its ability to refuse a permit is limited by the administrative judge who applies 'normal' legal control on refusals.

*The grant of permission*
F4.29 The significance of this latter point is not obvious until one considers the legal control on the grant of a permit. In this case administrative judges apply 'minimal' control, giving a dissymmetry to the law and a privileged position to the applicant, perhaps because traditionally

the *permis de construire* has been considered a right (Jacquot, 1987, pp.354-5).

F4.30 There are a variety of additional permits granted which are not straightforward. These include

*permis conditionnel*, conditional upon certain financial contributions to the cost of services (as of July 1985) or the observation of appearance, health, safety conditions laid out in the POS or national rules of urban planning (RNU)
*permis dérogatoires*, which contradict the POS or RNU. These have been severely limited by the December 1976 laws, but are still granted although under very strict control by the administrative courts
*permis de construire*, authorising division of property not controlled by *lotissement*
*permis à titre précaire*, (in precarious title) can permit provisional construction on land reserved for development by a plan where the ownership will change
*permis de régularisation*, which regularises unauthorised work.

F4.31 Besides refusing a permit, the administrative authority can issue a *sursis à statuer*, a suspension of a decision for two years, in order to prevent a development compromising either the implementation of a plan under preparation or any projected public works. Beyond two years the administrative authority must issue a definitive decision when requested. Another suspension can be implemented but only for a different reason, and the total delay cannot exceed three years.

*The implementation of a* permis de construire
F4.32 Third parties have six months, and other parties four months, from the issue of the permit in which to attack it through the administrative courts (there is the two months required for advertisement of the application, and a further two or four months for an appeal to the courts to be launched). The administrative authority can withdraw the permit during the period prior to the appeal. This appeal procedure provides an important legal check on the control process, and an opportunity for the public to challenge the substantive and procedural legality of the decision. On the debit side it creates very significant uncertainties for the developer, effectively extending the length of time required to obtain an unequivocal permission.

Table F4.1 Applications for a *permis de construire* 1983

| | Applications | Refusals | % Refusals | Suspensions | Permissions delivered | Tacit permissions | % Tacit |
|---|---|---|---|---|---|---|---|
| Number creating new floorspace | 408,864 | 19,095 | 4.7 | 824 | 371,610 | 1,757 | 0.5 |
| Number with no additional floorspace | 160,297 | 7,231 | 4.5 | 236 | 143,097 | 655 | 0.5 |
| Total Applications | 640,183 | 31,111 | 4.9 | 1,149 | 584,826* | 2,709 | 0.5 |

* discharged
*Source:* RISU 85, p.117

F4.33 Because of the frequent changes made to the content of plans, both their regulations and land use provisions, it is necessary to limit the validity of the permit, and a period of two years is allowed within which to start the work. If work is interrupted for more than a year the permit is withdrawn. A three year interruption is allowed for a small, phased development if a second phase is larger than the first, and the first does not exceed 100 square metres. The permit can be extended for one year if planning circumstances have not changed in the interim.

*Modification and transfer of a permit*
F4.34 Modifications can be made to a permit, providing the proposals conform to the prevailing rules at the time of the modification, under the same kind of procedure as the original. A permit can also be transferred to another individual by an *arrêté modificatif* but this does not alter its validity in terms of expiry dates. Otherwise a *permis de construire* does not run with the land although a form of outline planning permission does (the *certificat d'urbanisme*, paragraphs F4.77-F4.83).

*The* permis de construire *and development control performance*
F4.35 The 1983 statistics on the applications for a *permis de construire* (Table F4.1) reveal some 640,000 applications of which perhaps only half are implemented, if the *certificat de conformité* is a reliable measure of uptake, and if there were no major variations in application rates 1981-1984. Of the applications one quarter involve no additional floorspace created, but merely the modification of existing premises. Refusal rates run at just below five per cent of all applications, so the system would seem to produce applications which can easily conform to the rules and regulations. The proportion of tacit permissions runs at less than half a per cent, indicating that local authority or DDE negligence is comparatively rare (and may not be negligence at all). The number of derogations or minor adaptations to the plan or RNU numbered just under three per cent of all applications.

F4.36 Statistics on the granting of permits reveal that about 85 per cent are delivered within three months, but these figures are substantially less in most highly urbanised and high application rate regions, and the regional variations are quite considerable. Ninety five per cent of all applications are dealt with within a period of five months to indicate that the system is a reasonably efficient one, and that the statutory periods for issue of permits are taken seriously, However, these statistics only tell part of the story (see paragraph F4.145).

## Other Control Procedures

F4.37 Having outlined the basic process for application, examination and decision related to a *permis de construire* there are several related procedures and control mechanisms that merit brief discussion. These include developments now subject to a preliminary declaration

rather than a *permis de construire*, and a special category of controls on hazardous installations, which demand both an environmental impact statement and public enquiry before a *permis de construire* can be issued. In addition there is a variety of circumstances under which specially created bodies are consulted on applications for a permit - major shopping developments, situations where conflicts arise between consultees, or particular architectural projects. Each of these is briefly described.

*The* déclaration préalable *procedure*
F4.38 In a bid to deregulate controls, the *déclaration préalable* or preliminary declaration procedure was introduced as a lighter method of control than the *permis de construire* on a variety of minor developments from cleaning facades to minor extensions, from commercial swimming pools to small electricity pylons (see paragraph F3.7). A preliminary declaration must be made one month before work commences and if no objections are forthcoming work can proceed. However, if there are objections, notably from the *Architectes des Bâtiments de France* (see paragraph F4.104), prescriptions may be attached to the works. A *déclaration préalable* is not exempt from the appropriate taxes (see paragraphs F4.127-F4.138).

F4.39 The *déclaration préalable* has a one-page application form but must include a site plan, section and elevations showing the proposed modification, and full details of the applicant, landowner, lot size, situation, nature and future use of the building and density. The declaration must be sent recorded delivery to the mayor or inter-communal president who sends it on to the DDE if the commune does not possess an approved POS. The DDE is also consulted if the mayor opposes or wishes to modify the declaration. If the DDE oppose it or modify it and the mayor disagrees, the application is referred to the *Commissaire de la République*.

F4.40 But a *déclaration préalable* can be opposed only on points of law related to the Code or to a POS based upon the code, limiting the discretionary element of control. Limited consultations also take place. If the declaration is approved it has a validity for two years. In sum the *déclaration préalable*, while not an exemption from control, creates a category of developments where rapid approval is much more likely to be forthcoming. However, far from simplifying control procedures it has tended to make them more complex.

*Other development authorisations and procedures*
F4.41 The procedures for the delivery of a *permis de démolir* or demolition permit are largely similar to those for a *permis de construire* with a set form (and the requirement of stating the reasons for demolition), plans, sections and photographs of the building. Some authorisations like enclosures and diverse works require a *déclaration préalable*, others like caravan and camping installations, have specific authorisations. These do not merit detailed discussion here, but the broad statistics of such applications and the relative refusal rates are of some

interest (Table F4.2). Demolition permits for example are almost always granted and it is only caravans, camping and rubbish dumps which have high refusal rates and are apparently strictly regulated.

**Table F4.2   Other development control authorisations, applications and refusal rates (1983)**

| AUTHORISATION | APPLICATIONS | REFUSAL RATES | MISCELLANEOUS |
|---|---|---|---|
| Demolition Permits | 13,953 | 1% | |
| Diverse Works | 326 | 13% | |
| Enclosures | 24,000 | 2 | 11 suspensions |
| Individual Caravans | 1,535 | 36% | 15 suspensions |
| Caravan Sites | 44 | 13% | 3 suspensions |
| Campgrounds | 782 | 25 | 7% suspensions |
| *Installation classées* | 5,536 | ? | |
| Clearing Land | 3,516ha | | FF 7.6 millions recovered |
| Tree cutting etc | | | |
| *Lotissement* | 404 | 7 | 669ha involved |
| demands | 8,184 | 4 | 38 suspensions |
| units | 76,545 | 4 | |

*Source:* RISU 1986, pp.139–167, 173–183

## *Installations Classées*: Hazardous Installations and the Use of Environmental Impact and Public Enquiry Procedures

F4.42   The control of *installations classées* or hazardous developments is enshrined in the *Code de l'Environnement* rather than the *Code de l'Urbanisme* and so it is not strictly speaking a development control procedure. Nonetheless for a variety of reasons it is important to discuss it in the context of development control, since it is an authorisation that is required in addition to a *permis de construire*. Furthermore, in common with major planning applications and some other planning prescriptions, the procedure involves both an environmental impact study (*étude d'impact*) and public enquiry procedure (*enquête publique*).

F4.43   In cases of minor development involving an *installations classées* an application (*déclaration*) containing details of the site, plans, nature of the works and activities, and the measures for environmental protection must be made to the *Commissaire de la République*. The regulations can be modified and relaxed for the benefit of the applicant or tightened at the insistence of third parties of the Health Council of the Department. The applicant can make representations to the latter, and the final *arrêté* and details of the regulations can be scrutinised by the public.

### Étude d' impact - *environmental impact studies*
F4.44   If the proposal to develop and operate such an installation is defined as being a significant threat to the environment then a more elaborate application for authorisation must be submitted to the *Commissaire de la République*. The application must detail the site, nature and extent of activities and the manufacturing processes and provide plans of the works. A *permis de construire* has

also to be applied for and this must be accompanied by an *étude d'impact* or impact study introduced as part of the 1976 Environment Protection Law. The municipal council and the relevant field services of the Ministries are also consulted, particularly the various emergency services, and sometimes even the *Conseil Général*. Two types of impact analysis are required, a full *étude d'impact* for major developments such as those over 3000 square metres where no POS exists, or for all major *installations classées*, and a simplified procedure called *notice d'impact* for more minor schemes.

F4.45   A full *étude d'impact* has four components including an environmental appraisal of the site at the present, an analysis of the impact on the environment (natural environment, biological equilibrium, flora and fauna, monuments, landscape, amenity and public health are specifically mentioned), the reasons why the development is being promoted, and the measures proposed to reduce its environmental impact and the costs involved. The *notice d'impact* is a mini-impact study and it applies to a variety of rural land transformations including tree-felling, land reclamation in mountain areas, dune stabilisation and avalanche control, ski-lifts and smaller camping sites etc.

F4.46   The *étude d'impact* can be consulted during an *enquête publique* if one is held, but if not then the study is only published after the application for a *permis de construire* has been decided. This emphasises again the low value placed on public participation in the French planning process.

F4.47   It is important to note that an *étude d'impact* is also required for major projects of 3,000 square metres floorspace, or six million French francs or more, for land reorganisation, high voltage transmission lines, large camping and caravan sites etc. However, there have been criticisms of some of the exempted categories, particularly the development and exploitation of forestry which is not subjected to these controls.

### Enquête Publique *or public enquiry procedures*
F4.48   An *enquête publique* has been used to enquire into public works that might cause a nuisance since 1791, and they are used whenever land is compulsorily purchased for a project, whenever there is a proposal to develop a major *installation classée*, and for the purpose of publicly examining a POS. The current regulations on procedures for an *enquête publique* were laid down by decree in 1976, and are the same for all types of enquiry. However, according to Macrory and Lafontaine (1980, pp.21-5) they are far from being but into practice as a whole.

F4.49   When the application for an *installation classée* is complete the Prefect issues an *arrêté* (which is published in the local newspapers eight days prior to the enquiry) which announces the date, time and place of the enquiry, and where and when the public may inspect the application. A poster notifying the public of the enquiry must be displayed outside the town hall but the Prefect is strongly recommended to display them more widely (but rarely does so). All the papers relating to the enquiry including the *étude d'impact* are available for public consultation, but any information considered to prejudice trade secrets will be withdrawn from public view. The materials for public consultations may include models and exhibitions.

F4.50 The name of the *Commissaire* who will conduct the enquiry will also be announced. The *Commissaires* are drawn from an annual Department list and are usually retired magistrates, ministry officers, *auxiliaires de justice*, civil servants or even members of chambers of commerce, but can include specialists in ecology or architecture.

F4.51 The *enquête publique* generally lasts from two to eight weeks. The enquiry for an *installation classée* must last for at least a month with the *Commissaire* in attendance at least three hours per week. A register is provided in which any person or group can write their observations at specified times, and for all but the smallest enquiries, during the last two or three days, or at other specified times, the *Commissaire* will listen to oral comments (a provision established in 1957).

F4.52 For *installations classées* at the end of the enquiry the *Commissaire* has one week to communicate all the observations (which have been made in writing or in person) to the applicant who in turn has 22 days to send a written reply. Within a further eight days the *Commissaire* must send all the papers together with his reasoned conclusions to the Prefect, who will in turn issue the draft *arrêté*.

F4.53 The inspectorate for *Installations Classées* are drawn from the *Direction Interdépartemental de l'Industrie* (DII). They have to consider the results of these consultations, the enquiry papers and the *Commissaire's* report and submit a report to the *Commissaire de la République*. The Health Council of the Department also consider the report and the applicant is entitled to appear before them. Both the reports of the DII and the Health Council are considered by the *Commissaire de la République* who issues a draft *arrêté*, and the applicant has 15 days to respond to the Prefect. In some cases it is the *Ministère de l'Environnement* or the *Ministère de Défense* who issues the final report and *arrêté*.

F4.54 The *arrêté* can be inspected by the public and is advertised. While there is no legal requirement for further consultation, the public can make comments upon it. The *arrêté* will include any standards that the *installation* is required to observe, including emission standards for pollutants, but these can be modified by subsequent *arrêtés*. The inspection service for *Installation Classées* is continually updating and revising its standards, but the applicant has the opportunity to make representations to the Health Council or to the *Commissaire de la République* on the draft *arrêté* and in the event of a change in this *installation* may be required to submit a fresh application

for authorisation which would involve an *enquête publique*.

F4.55 For other forms of compulsory purchase enquiries the *Commissaire* produces a reasoned set of recommendations which are usually ready within one month, and which can be examined at the *mairie*. The *déclaration utilité publique* is pronounced by an *arrêté* of the *Commissaire de la République* or Ministry. If the *Commissaire de la République* disagrees with the *Commissaire's* report for the public inquiry, or with the local authority on the advisability of the scheme, then the Ministry arbitrates and the Minister may consult the *Conseil d'État* before issuing an *arrêté*.

## Consultative Bodies in Development Control

F4.56 Having outlined the multiplicity of procedures in development control it is important to note a variety of other consultative bodies that operate in the development control arena. Three of these were not re-established when the decentralisation procedures were introduced in May of 1982 and hence do not exist as of June 1984, though the procedures for consultation still remain in the *Code de l'Urbanisme* until the Code is revised again. The three commissions concerned were

(i) the *Commission départementale d'urbanisme* (CDU), which acted as an advisory body to the prefect of the Department on controversial or difficult decisions

(ii) the *Conférence permanente du permis de construire* (CPPC), which provided a departmental advisory body for minor derogations engendered by *permis de construire* applications

(iii) the *comité d'aménagement de la région d'Ile de France* (CARIF) which was consulted on the *schéma directeurs* for Paris and Paris region, and on the determinations of *Zones d'Aménagement Concerté* (ZAC) and natural protection areas.

F4.57 To replace these in part *Commissions de conciliation en matière d'urbanisme* (CCU) were established in September 1983 to research solutions to the conflicts that arise between the various levels of governments, particularly in the elaboration of the SD and the POS (see paragraph F5.13).

F4.58 The CCU deal with conflicts within the plan making process. More relevant from a control perspective are the *Conseils d'architecture, d'urbanisme et d'environnement* (CAUE) which were instituted in 1977 to promote the quality of architecture, urban planning and the environment. They consist of a general assembly, an administrative council and a multi-disciplinary technical team on a Departmental basis. They are funded by a Departmental tax and hence their services are free. They have two assignments. The first is to sensitise the public and to educate them about architectural and urban quality by teaching, putting on exhibitions and organising local groups. The second is to offer advice to the construction industry and to the public collectives. From January 1982 they were to have been consulted on all projects of construction which had not been established by an architect, but lack of resources prevented this. However, they can still play an important role if they are consulted on difficult cases.

F4.59   The *Architecte Conseil*, instituted in 1980, is the sole real survivor of the original consultative organisms. These are single architects appointed to assist the DDE in providing judgement on difficult and important projects. They are appointed by the minister and are independent, but they have had to adopt new relationships with local authorities since decentralisation. They have also had to adapt to the new preoccupations of national policies in terms of the standards of public construction, housing design and places of work.

### The control of major shopping developments

F4.60   Perhaps the most important specially constituted consultative bodies that remain are the *Commissions Départmentales d'Urbanisme Commercial* which, since the 1973 *Loi Royer*, have had to authorise all major shopping developments (major is currently defined as greater than 3,000 square metres gross floorspace or 2,000 square metres in smaller communes). The criteria for decision making are laid down in the Law of Commerce and Artisans as follows: competition clear and fair, satisfaction of the needs of consumers as far as price, level of service and products offered; expansion of all forms of enterprises (independent, grouped or integrated); avoiding disordered growth of new forms of distribution and the crushing of small enterprise and the wastage of commercial services; adaption to the ordered development of agglomerations and the requirements of planning.

F4.61   The Commission is composed of twenty representatives, nine locally elected councillors including the mayor of the commune in question, nine representatives of commercial and artisan activities (six independent businessmen, two representatives of major and multiple stores, and one artisan representative), and two consumer representatives. The Commission is presided over by the *Commissaire de la République*. A quorum of two thirds is required to deliberate and a majority required to reject any demand for an authorisation.

F4.62   A vote of one third of the commission can lead to an appeal to the Minister of Commerce, and he will seek the advice of the national commission on commercial urban planning. The Minister does not have to give reasons for a positive decision, only for a negative one. Any authorisation given is personal and not transferable and is valid for two years or until a *permis de construire* is obtained.

F4.63   These procedures are of interest since they represent a kind of corporatist decentralisation predating political decentralisation by a decade (Keeler, 1983). The balance of powers seems to lie with the small shopkeepers who have held an average of 39 per cent of the seats nation wide, giving them a powerful position from which to negate new large scale developments. However, the government specified that a majority present had to vote against the scheme for a project to be rejected, significantly shifting the balance of power away from the traditional shopkeepers.

F4.64   Evidence of the impact of the commissions on major retail applications reveal that less than half (48 per cent of projects: 41 per cent by area) the supermarket proposals and less than one quarter (22 per cent of projects and areas) of the hypermarket proposals were accepted between 1974 and 1980 with authorisation rates markedly lower in the more rural departments.

F4.65   However, these restrictive policies have been significantly moderated by the Ministry of Commerce overturning 24 per cent of all local commission rejections, and 35 per cent of all rejections appealed. Analysis reveals that Ministers took a very restrictive line in sensitive pre-electoral periods, but at other times allowed major retail development to proceed apace (in 1979 more than half the rejections appealed were overturned) (Keeler, 1983, pp. 279-283). However, the impact of commission decisions has significantly slowed the development of larger scale retailing, and produced a spate of developments designed to evade such controls. Nevertheless, France is still second only to West Germany within the EEC in terms of super-market/hypermarket density. The socialist government initially adopted an extremely restrictive attitude to large scale retail development, but their intention to introduce much tighter limits particularly in rural zones (perhaps as low as 400 square metres) was abandoned in the face of pressing needs to counter inflation and improve the business climate. Consequently they took a very permissive attitude from April 1982 onwards and supermarket chains have become major contributors to local political parties to ensure approvals. Even so, figures for 1983 reveal the major constraining influence of the process, while the Minister found with the commission in two thirds of cases and allowed only an additional one million square feet of floorspace (Table F4.3). The importance of these provisions to the nature of French urbanism, and to the cost of French goods, will be appreciated by every British tourist.

F4.66   The *Loi Royer* marks a rather anomalous decentralisation of decision-making power not to traditional administrative or elected structures but to particular interest groups and professional representatives. Much the same happened in agriculture as a centre-right government sought to respond to the political difficulties of over centralisation without devolving power to local governments, which were more often than not controlled by the left. This corporatist decentralisation helped to defuse social protest and to produce a *rapprochement* between the state and the forces

**Table F4.3   Authorisation of major shopping developments 1983**

| | Decisions of *Commission Département d'Urbanisme et Commerce* | | Appeals | | Minister of Commerce Decisions | | | Definitive Decisions | |
| --- | --- | --- | --- | --- | --- | --- | --- | --- | --- |
| | Authorised | Refusals | on Authorisations | on Refusals | Confirming | Annulling and Refusing | Annulling and Approving | Authorised | Refused |
| Number | 167 | 264 | 32 | 134 | 85 | 7 | 42 | 198 | 204 |
| 000 sq.metres | 442 | 1,042 | 148 | 579 | 404 | 28 | 125 | 523 | 820 |

*Source:* RISU 1985, p.239

of traditional commerce, but it has also fragmented planning at large and created marked anomalies in retail provision over time and space.

## The Control of Works and Enforcement

F4.67   Having discussed the various procedures for obtaining *permis de construire* and other development authorisations, and the special consultation bodies that are involved, it is now possible to examine enforcement provisions and the control of works which assume rather more significance in the French planning system than they do in the British. The French system for achieving a form of outline planning permission is also explained at this point, prior to a discussion of subdivision (*lotissement*) controls which assume great importance in France.

F4.68   A declaration that work has started has to be made to the mayor, the *Commissaire de la République* and, if different, the authority who granted the permit. The *Code de l'Urbanisme* makes provision for the mayor or *Commissaire de la République* or their agents to visit the construction at any time to verify that the plans and regulations are being/have been followed, and this right extends to two years after completion, and applies to those authorities charged with protecting road alignments and surveying (L460-1). If infractions are detected work can be stopped immediately by instructions to a judge or the administrative authority. These infractions are discussed in paragraph F6.27 and Table F6.3 but they constituted only three per cent of all completed developments in 1983.

*The* certificat de conformité - *a built-in enforcement system*
F4.69   A declaration is also required that the work is complete and if the development meets the terms of the permit a *certificat de conformité* is issued. The *déclaration d'achèvement* (completion declaration) must be made within 30 days of completing the work on a standard form in triplicate, and signed by the applicant and the architect (if there was one) that the works conform to the permit. The declaration is circulated to the mayor, President of the EPCI, *Commissaire de la République* or DDE as necessary.

F4.70   An inspector of the Departmental field services must satisfy himself that the building conforms to the permit in terms of its siting, use, nature, external appearance, dimensions, and development of the surroundings, but the control does not embrace the building regulations enshrined within the *Code de la Construction* (see Appendix F2). The inspection can include verification of works but this is only optional, except in three instances

(i)   in the case of historic monuments, when it is verified in liaison with the ministerial representatives charged with historic monuments

(ii)   in the cases of high buildings and public buildings, when it is verified in liaison with departmental fire and emergency services

(iii)   in the case of works within a national park or nature reserve.

F4.71   If the works conform satisfactorily a *certificat de conformité* is issued, but if the work is not in conformity the builder must be advised as to how it fails to conform. If the authorities fail to reply within three months of receipt of a *déclaration d'achèvement* a tacit certificate is assured, but the applicant must seek it from the appropriate administrative authority, who must reply in one month or it is automatically granted.

F4.72   A certificate of conformity can be attacked at any time, unless it is a tacit certificate. Most importantly, except for the cases outlined above where consideration of the *Code de la Construction* is obligatory, the obtaining of a certificate of conformity constitutes an obligation on the builder. He has attested that the building has conformed to the permit which authorised it. This shelters the building from penal sanctions, and puts whoever acquires it under a favourable fiscal regime. A certificate of conformity is necessary before a building can be sold.

F4.73   Some 308,000 certificates of conformity were issued in 1983 and 78 per cent of these were favourable, 20 per cent provisionally refused, and 2 per cent definitively refused. These figures suggest the importance of the system to ensure that buildings conform to the law, although it suggests most errors are minor.

*Enforcement of the building regulations*
F4.74   The building regulations (*Code de la Construc-*

*tion*) consist of a thousand page tome divided into six books (see Appendix B). The first book embraces general dispositions of construction, fire protection, heating and cleaning of façades, aid to the building industry, control and penalties. The law gives the state representative, mayor or ministerial representatives the right to visit the constructions at any time from inception up to two years after completion and prescribes the penalties for non observation (L151, L152). The second book sets out rules for construction companies and the sale of buildings, while the third specifies all the housing support mechanisms including rehabilitation and improvement procedures. The fourth book specifies the organisation of low income housing, and the fifth the procedures for dealing with unhealthy or dangerous buildings. The final book sets out measures available to remedy the 'exceptional difficulties' of housing.

F4.75   The building regulations are enforced by visits from technical inspectors of the DDE who make a very thorough and precise inspection of the property. In reality however, only a sample of completed buildings are inspected. The really effective mechanism of control lies in the fact that it is the developers' and architects' responsibility to ensure that the code is adhered to, and action can be taken against them under civil law should any deviations be detected. In fact technical inspections by engineers working for major insurance companies ensure adherence to the regulations and while minor deviations might be allowed, any significant anomalies would render the building unsaleable and uninsurable. So effectively there is a privatised building inspection service.

F4.76   There are powers to refuse to connect services to any building that does not satisfy the building regulations, but there does not appear to be any checking mechanism, and the *Electricité de France* do not actually observe this. The system relies more upon the public acting as watchdogs and a system of neighbourhood denunciations, but in fact abuses of the control do not appear to be a significant issue.

## The *Certificate d'Urbanisme* - an Outline Planning Permission

F4.77   To complement the *permis de construire* control system a second set of procedures provides a means of determining in advance the development rights on a site and the possibilities of developing the applicant's proposals. So the *certificat d'urbanisme* is a form of outline planning permission. Unlike the *permis de construire* it can be obtained by the future acquirer of land in order to determine the *constructibilité* of the land or the feasibility of a particular project. These are distinguished as certificates 'A' normal or 'B' detailed, the only difference being that in case A the local authority merely states what is acceptable while in case B the applicant specifies in broad terms the project he has in mind and the local authority responds to this. Such a project may include subdivision as well as building, camping etc. The

resultant certificate provides protection for both the landowner and potential developer by clearly establishing the development rights and this is its first object. It has important fiscal advantages and tends to be requested with each sale of the land, and plays a key role in defining its actual price. From being a notarial practice encouraged since 1950 it became part of planning law in 1971.

F4.78   The second object of the *certificat d'urbanisme* is to define the possibility of realising a certain development and the form demands specific information concerning the operation that the applicant proposes to realise. (It may not be concerned exclusively with building *per se* and a third and fourth role of the certificate is to control land division (*lotissement*) and subsequent construction). It allows the developer to question the administration on the feasibility of the project before commencing the architectural and technical studies.

F4.79   The application is relatively simple, containing details of site, proprietor, cadaster, existing divisions of the site and their use, as well as details of the applicant and what the object of his request is (whether merely testing *constructibilité* or specifying the kind of development anticipated - broadly the type of use of buildings or division of land).

F4.80   A positive certificate spells out what is realisable, what agreements are necessary, any rights of preemption or public servitudes, the nature of planning restrictions, densities, public services and taxes applicable, with any additional observations or particular prescriptions. In bald terms it merely copies out the parameters defined by the plan or the RNU, the administrative limits on the property and the state of public infrastructure.

F4.81   A positive certificate constitutes a promise of permission to develop that is valid for at least six months (for A certificates indicating the limits of constructibility) or 18 months (for B certificates authorising a specific project) with the possibility for an extension of another year if the rules have not changed. A *permis de construire* subsequently issued cannot contradict what is stated in the *certificat d'urbanisme*. A clarification of this latter rule in 1983 states that while the planning regulations could not be changed administrative regulations and public servitudes (including for example the extension of a zone of protection of an historic monument) might well change. This somewhat undermines the utility of the certificate.

F4.82   So a *certificat d'urbanisme* is essentially a notarial device that has become a means of establishing initial contacts between developers and controllers, ensuring that land can be acquired with some security, and providing basic guidance for the preparation of an application for a *permis de construire*. It also serves as a declaration of intention on the part of the developer and the local authority. It serves some of the same purposes as the British outline planning application although it has quite different origins. While the British outline permission sets out the reserved matters of siting, layout and design etc the *certificat d'urbanisme* merely lists the relevant land servitudes, planning documents, servicing requirements and taxes. However, the net effect is rather similar.

F4.83 Some one quarter of a million *certificats d'urbanisme* were issued in 1983 (Table F4.4), emphasising their role in land transactions rather than development intentions *per se*. Only a sixth of these were for pre-specified projects. The rates of refusal in all categories for certificate are high compared with the *permis de construire*, rising from 16 per cent for division of a plot from a property, to 31 per cent for general tests of '*constructibilité*'. Fifty eight per cent of all certificates were delivered within two months although regional variations are extreme, some departments achieving only seven per cent, others 95 per cent in this time period.

## Lotissement

F4.84 Finally in the armoury of development controls there are the *lotissement* procedures. *Lotissement* is essentially the process of subdividing and servicing land in advance of building on it, so it is necessarily prior to the application for a *permis de construire* on most undeveloped land. It is an increasingly significant element of housing provision as demand increases for single family *pavillon* housing. In 1983 some 8,184 authorisations yielded 71,000 building lots, only four per cent of applications being refused.

F4.85 The control of *lotissement* follows essentially the same process as the *permis de construire*, except that the use of an architect is not required. However, an architect-planner or highway engineer, or at least the services of an expert surveyor, may be required if the project is large enough. The *lotisseur* can do all the work on his own but a preliminary meeting with the authority is recommended. A standard form requires full details of the applicant(s), the terrain and its ownership, the project, the phasing, sale procedures as well as the signatures of applicant, proprietor and an attestation of mandate to the applicant. The application must include a reasoned justification for the scheme, including its respect for the

site and its architectural quality, a plan of the scheme and a plan of the land in its present state, details of the organisation and treatment of collective space etc. But since 1977 the plan does not have to show the division of parcels in order to allow those who acquire the lots to create their own environment (the actual building is of course controlled by the *permis de construire*).

F4.86 The application may need to be supported by its own set of regulations to control future development. It may need a plan of collective works, an agreement with the commune over services, an *étude d'impact* if it exceeds 5,000 square metres floorspace in certain communes, and even a dossier for the *cahier des charges* (or land registry) fixing the relations between *lotisseur* and those purchasing the lots. The competent planning authority must notify the receipt of the application within 15 days and publicise the same. It has three months to process a *lotissement* of less than five lots and five months for larger divisions, with additional time for more complex consultations or an *étude d'impact*.

F4.87 Permission may be refused on the basis that the scheme does not conform to the POS or is contrary to the RNU, or if it is considered to compromise the *equilibrium* of development of the commune (R315-28). If permission is granted it is subject to conditions and regulations, financial agreements, technical matters, financial contributions and lists of works. The authorisation reveals the plan, composition and the total amount of floorspace, or merely the road layout and a maximum number of lots and lists the conditions etc. Any tacit permission requires confirmation from the competent authority.

F4.88 The *lotisseur* then has 18 months to begin the work and 3 years to finish, although phasing can be written into the agreement. The same enforcement procedures apply via the *certificat de conformité*. Minor derogations are possible to allow some works to be completed after sales to reduce costs and technical difficulties (damage from construction etc). Finally there are controls on the advertising, sale and pre-payments before authorisation, and any advertisements must conform completely to the authorisation. There are also arrangements for syndicates of proprietors to control the *lotissement* after sale, and after ten years the *lotissement* rules are superseded by those of the POS.

**Table F4.4** *Certificats d'Urbanisme* **issued in 1983**

| | Applications to test 'constructibility' | Applications for a determined operation (incl. buildings) | Applications involving plots detached from main ownership | Applications involving divided plots but not *lotissement* | Cases not aired |
|---|---|---|---|---|---|
| Positive | 73,462 | 44,686 | 55,540 | 50,728 | 35,428 |
| Negative | 33,342 | 17,127 | 11,990 | 16,304 | 14,116 |
| Total | 108,698 | 61,989 | 75,056 | 69,952 | 57,875 |
| % refusal | 31% | 28% | 16% | 23% | 24% |

*Source:* RISU 85, p.108

F4.89 There is probably more negotiation in *lotissement* than in any other sphere of development control in France and it is very much a question of deciding whether the community or the *lotisseur* will pay for certain services, make contributions towards certain works or agree to adopt or maintain certain properties. The authorisation formalises this negotiation (Tribillon, 1985, pp. 67-72), but the negotiations themselves can be very prolonged and the cause of much delay.

## Planning Considerations in Issuing a *Permis de Construire* etc

F4.90 Having outlined the process and procedures of subdivision and development control it is now appropriate to examine the criteria used to decide whether or not to grant the various kinds of authorisations or permits. These embrace a wide variety of planning considerations, enshrined in the national rules of urban planning (RNU) or the *Plan d'Occupation des Sols* (POS) (see section F5), as well as a range of conservation considerations both urban and rural, and control of hazardous installations. The latter introduces the environmental impact study and public enquiry procedures which apply to all major developments. Finally under planning considerations it is important to discuss both public utility servitudes and national or supra-communal interests which can prevent or facilitate the issuing of an authorisation for development.

F4.91 The reasons for refusing a *permis de construire* are specified in paragraph F4.27 and they emphasise the need for conformity with 'legislative and regulatory provisions'. There are two main sources for these provisions, the POS and the RNU, while the rule of *constructibilité limitée* provides very strict control outside built-up areas in the absence of a POS.

F4.92 If a POS has been published this will provide details of all the legislative and regulatory provisions. Conformity to the prescriptions of the POS guarantees the issuing of a *permis de construire* in normal circumstances, so the POS itself elaborates all the rules of location, servicing, size, siting, volume and external appearance of development. It also contains details of all public servitudes for roads, sewers, public utilities etc as well as all the details of special conservation restrictions that might apply. The POS, the typical content of which will be outlined in the next section F5, provides very clear rules which dramatically reduce the scope for discretion and almost remove the notion of dealing with a proposal 'on its merits'. There remains significant room for negotiation and discretion, however.

*The rule of* constructibilité limitée
F4.93 As prescribed in the *Code de l'Urbanisme*, in the absence of a POS no development may take place outside built-up areas. This is the rule of *constructibilité limitée* (limited construction) which has a double function. Firstly it protects the countryside and undeveloped areas

in the absence of a local plan. Secondly it provides a major incentive for local communes to prepare a POS to avoid this restriction on development. In the latter context the rule of *constructibilité limitée* was a key innovation in the 1983 decentralisation. But it was a controversial provision that aroused strong opposition in the Senate because it was considered to remove the 'right' of the proprietor to develop his land. The new government did make some concessions in August 1986 by widening the exceptions to this constraint, and by retarding its introduction.

F4.94 The area of limited construction is defined as all those parts of a commune not covered by a POS or any other urban planning document and not partly urbanised, ie non built-up areas, or areas with very dispersed development. Hamlets and villages cannot be included in an area of limited construction but ribbon development is nonetheless expressly prevented. The identification of the partly urbanised areas to be excluded from this restriction is left to the DDE in liaison with the local communes (Jacquot, 1987, p.121).

F4.95 Exemptions to this tight control include the adaptation, repair or extension of existing buildings, constructions necessary for collective servicing, agriculture or exploiting natural resources and constructions incompatible with inhabited areas. A second set of exemptions can be justified in the interest of the commune and are much less precise. They were introduced by the National Assembly in 1986 to give more flexibility particularly in the smallest, least populated communes, opening the possibility of isolated or even small groups of buildings. The municipal council must give reasons and the development must not threaten the natural landscape, public health, security or the local economy.

F4.96 The rule of *constructibilité limitée* was supposed to be implemented throughout the country in October 1984, but two sets of exemptions have been applied postponing this until October 1987 in those communes which had declared a POS in preparation by October 1984 and in those communes where the *Modalités d'application du RNU* (MARNU) are in operation (see paragraph F4.101).

*The* règles nationales d'urbanisme (RNU) *or national rules of urban planning*
F4.97 In built up areas where no POS has been published the *règles nationales d'urbanisme* (RNU) or national rules of urban planning apply, and these are enshrined in the *Code de l'Urbanisme* itself (Figure F10). First prescribed in 1919 and encoded in 1955, they were spelt out by decree from the *Conseil d'État* and revised in 1966 and 1977. They are not precise rules since they have to apply to all territory, but they are grouped under three rubrics - the location and servicing of buildings, the rules establishing the siting and volume of constructions, and general rules for external appearance.

F4.98 Under location and servicing a *permis de*

*Régles Nationales d'Urbanisme* (RNU) R111-1-R111-27
National Rules of Urban Planning

*Location and servicing of buildings*

These rules permit the refusal of a *permis de construire* on a locational basis if it would:
  * – endanger local security or public health
  * – lead to development in areas subject to flood, erosion or avalanche
  – cause a grave nuisance such as noise etc.
  – be within 50m (40m for housing) of motorway or 35m (25m) of arterial road outside built-up areas
  – threaten natural spaces in the broadest sense incl. agriculture
  – threaten harm to natural landscapes flora and fauna
  – threaten conservation of an historic or architectural site
  – threaten sprawl into unauthorised areas or compromising agriculture, forestry etc. or the National Directives of Planning

These rules permit the refusal of a *permis de construire* on a servicing basis if:

  * – the site is not open to a public highway, or access is difficult or dangerous (can lead to parking requirements)
  – the site is required for green space or childrens play (can lead to these being required)
  – the building is not served by drinking water, sewerage, industrial drainage etc.
      (some exceptions for unserviced development)
  * – if constructions damage the environment or contradict national directives
  * – the services required are beyong current resouces (can lead to demands for service provisions etc)

*The siting and volume of constructions*

These rules permit the refusal of a *permis de construire* on a siting or volume basis if they infringe the rules of:

  – minimum distance between buildings
  – minimum angles of view of the sky
  – minimum sunshine for buildings
  – maximum height to width ratios with the street
  – minimum set backs

unless the proposals improve or have no effect on the conformity with buildings in the area.

*The General rules relating to external appearances*

These rules permit the refusal of a *permis de construire* on an aesthetic basis if:

  – 'the situation, architecture, dimensions or natural appearance would damage the character or interests of surrounding areas, to natural or urban landscapes'
  – the buildings do not conform to particular heights in partially developed areas
  – if the surrounding walls/fences are not built in harmonious materials
  – if industrial or temporary buildings are not set back and screened

Those items asterisked remain in general force, even in areas with an approved POS, and enforce national directives on the environment and the public order provisions of health and safety.

**Figure F10    The National Rules of Urban Planning**

The National Rules of Urban Planning (RNU) apply in the absence of a POS, and they provide discretionary, permissive controls over the location, servicing, siting, volume and external appearance of development. Some of the regulations, notably those on servicing, siting with regard to a highway, and volume, are however imperative.

*construire* can be refused if it constitutes a danger to public health and security, if it threatens natural spaces or conservation areas, or if it constitutes urban sprawl (*le mitage*). Under the servicing restrictions, highway, water and sewerage, and play space restrictions apply, and if the servicing required to permit the development is beyond the current resources of the community a permit can be refused. This latter provision allows the authority to demand various contributions from the developer to supplement such resources.

F4.99 Under the rules establishing the siting and volume of constructions are restrictions on daylighting and sunlighting and angles of view, the spacing, set back and position on the plot of the building. If buildings do not conform to these rules then a permit will be forthcoming only if the proposals improve or have no effect on the conformity of siting or building envelopes in the area.

F4.100 The general rules relating to external appearance have essentially aesthetic objectives and ensure that the development accords with the surrounding environment. Permits can be refused if the "situation, architecture, dimensions or outward appearance ... would damage the character or interests of surrounding areas ... or natural or urban landscapes". There are particular rules on the conformity of height in partly developed areas, and requirements that surrounding walls be built in materials harmonious with the façades. There is also insistence on set backs and screening for temporary or industrial buildings.

F4.101 While the RNU only apply in the absence of a POS they may have a supplementary quality when other local planning documents (which predate the 1976 reforms) are in force. The rules for their application are prescribed in the *Modalités d'Application du RNU* (MARNU) and certain key provisions (asterisked in Figure F10), such as those dealing with national directives on the environment or with public health or security, still then apply. These are essentially transitional arrangements.

F4.102 Most of the regulations have a permissive character, leaving to the competent authority a measure of discretion in their imposition, or the addition of conditions (there is no such thing as a standard list of conditions) but never compelling them to do so. Other regulations are imperative such as those which relate to servicing, siting and volume, and distance from major roads, although the *Commissaire de la République*, with advice from the DDE, can alter the siting and volume restrictions. Under this provision the commune may substitute a guide to the application of the RNU defining its applicability to different parts of the territory by means of a map. This is an important step towards a POS, and suspends the rule of *constructibilité limitée* for two years, but it is not binding on third parties, only on public authorities (see Plate 14).

## Urban Conservation Controls and the Processing of Applications

F4.103 Modifications to the process of examining the application for a *permis de construire* ensue if the proposals affect historic buildings, urban conservation areas, or protected natural areas. These longstanding controls, the evolution of which was briefly outlined in section F1, enforce a different set and timetable of consultations, and require consideration of additional planning factors.

*Immeuble classé or inscrit - listed buildings*
F4.104 The legislation of 1913 provided two kinds of protection for historic buildings and these are still in operation. The first, *classement*, provides a full level of protection by ensuring that all works affecting any building so listed have to be approved in advance by the Minister of Historic Monuments. The second control, *inscription*, is applied to buildings of less interest, and on these it is only obligatory to declare in advance all the modifications proposed, and these can be rejected or modified as necessary. These controls also provide protection within a 500 metre radius of all designated buildings, as well as certain surrounding buildings, A rather more easy to operate procedure introduced in 1943 allowed control of 'any transformation or modification of a nature that affects the aspect' within the 'field of visibility' of a listed building. This can be controlled by a preliminary authorisation. These two rules provide very strict controls on the environs of listed buildings, and bring in the *Architecte des Bâtiments de France* (ABF), the department concerned with protecting historic buildings, as the key consultee in the control process in conservation issues. These new rules also extend the consultation period on applications by one month.

F4.105 The reach of these controls is in part accounted for by the fact that there are only 31,000 protected historic buildings in France, one twentieth of the number in Britain. Nearly 12,000 of these are the fully protected classified buildings, with 19,000 on the supplementary list. Of these fully protected historic monuments nearly half are churches, chapels or monasteries, a quarter are antiquities or castles, and only 15 per cent are dwelling houses. As with conservation areas (see below) French designations on historic buildings are much more selective and the controls correspondingly tighter than in Britain.

*Listed sites*
F4.106 A now (since 1983) defunct process applied the same kinds of protection to listed sites with much wider zonal protection but only some 50 were designated because of the cumbersome procedure. The controls however still apply.

## Secteurs Sauvegardés (SS) and Zones de Protection du Patrimoine Architectural et Urbain (ZPPAU) - Conservation Areas

F4.107   While *Secteurs Sauvegardés* were introduced in 1962 and constituted not merely a preventative mechanism to protect the character of an historic area, but also a positive commitment to restoration and improvement, only some 80 have been designated, and many have yet to embark upon the full programmes and procedures proposed (Figure F11). In 1983 new conservation zones were introduced to protect the urban and architectural patrimony, the *zones de protection du patrimoine architectural et Urbain* (ZPPAU). These are easier to implement and more locally responsive. They are less reliant on the interventions of the ABF and allow the commune to play a more positive role. They have been criticised as potentially weakening the pre-existing controls exercised over the environs within 500 metres of listed buildings by providing more flexible criteria and less discretionary power to the ABF.

F4.108   The *Secteurs Sauvegardés* must possess

> "a character historic, aesthetic, or of a nature justifying conservation and making the most of all or part of an ensemble of buildings" (*Code de l'Urbanisme*, L313-1)

and they are established by the deliberations of communes, DDE and the *Commission Nationale des Secteurs Sauvegardés* (with 8 ministerial representatives and 14 experts). The basic effects of designation are to set in motion studies for a plan, to suspend the issuing of *permis de construire* until the plan is published, and to introduce the detailed surveillance of the *Architectes des Bâtiments de France* (ABF). It is the latter who have the important powers to preserve aesthetic character and conserve historic buildings, through advice on permits for building or demolition, through conditions, or through the suspension of decisions (*sursis à statuer*). While the *permis de construire* is issued in the usual way, any decision must conform to the advice given by the ABF. The *Plan permanent de sauvegarde et de mise en valeur* (PSMV) provides more detailed guidance than even the POS in this respect, often descending to the level of the parcel and defining the exact nature of desirable improvements.

F4.109   The ZPPAU also has a plan base but is prepared by the commune in conjunction with the *Collège regional du patrimoine et des sites* put together by the *Commissaire de la République*. Here again the ABF plays the most important consultative role, and there is no appeal against their advice, although the *Commissaire de la République* may intervene if the commune and ABF disagree. The ZPPAU has the added benefit of sweeping away any public utility servitudes. All of these procedures and consultations increase the decision period for a *permis de construire* by one to three months.

## Rural Conservation Controls and the Processing of Applications

F4.110   Rural conservation controls in France are complex, overlapping and difficult to disentangle. They embrace national prescriptions for land use and planning which affect the coasts and mountain zones, and which control specific environmentally sensitive areas that are designated at a Departmental level, and which include acquisition powers and a tax base for the same. There is special protection for all woodlands, especially in any POS, and the latter have now embraced a range of environmental protection zones (ZEP). Beyond these there are National Parks, Regional Nature Parks and nature reserves and a whole complex of *schémas* for the coasts. The latter will be discussed in section F5 since they can be considered as plans even though, strictly speaking, some of them are not.

Prescriptions d'aménagement et d'urbanisme - *prescriptions for land and urban planning*
F4.111   The diverse national laws and prescriptions for land use and urban planning originated in the *directive d'aménagement et d'urbanisme* being converted into law in 1983.   They supplement the RNU with national prescriptions to protect agricultural land, preserve forests and landscapes etc, but also to provide sufficient land for economic activity and housing needs.  Of more interest, there are the specific controls on the airport noise zones, mountains and coasts, which first emerged in 1977, 1979 and 1980 respectively.  There is the prospect of a fourth directive on rural areas.  These prescriptions have been progressively defined by the judges who have demanded that they clearly emanate from the government, have more precise dispositions and more publicity to allow them to be opposed by third parties. From 1987 all POS have to be made compatible with their prescriptions.

F4.112   These national prescriptions on the planning and protection of sensitive spaces protect the mountains and coasts by applying a nationally defined set of planning rules.  These protect the traditional economy and agricultural/fishing practices of these areas, their cultural landscapes, their urbanisation, and their tourist development.    They embrace special controls on mountain roads, the prevention of development within specified distances of lakes and coasts, and development outside built-up areas. Special planning operations are applied to new tourist developments like ski-resorts or coastal villages. In the same family of controls there are special provisions for the land around aerodromes where different forms of development are forbidden within three different zones, according to their tolerance of, and susceptibility to, noise.

F4.113   Most procedures to protect environmentally sensitive natural areas date from the reforms of 1976, including the *Zones d'Environnement Protégé* (ZEP) which operated like simplified POS.  These were superseded by the prescription of a POS itself from 1983 onwards, as part of the creation of a decentralised and

**Figure F11  Conservation Areas in France and the progress of conservation plans 1984**

As of October 1984 only 61 *Secteurs Sauvegardés* had been designated, a mere 15 per cent of those identified as potential designations in the Malraux Act of 1962. All 61 had a prescribed plan (1) but the majority had only proceeded to the second stage, having it accepted by the local collectives (2). Less than half had proceeded to the publication stage (3) and only one quarter actually had approved plans (4). Only

Montpellier had already revised an approved plan (5). However, as with the POS, the mere prescription of the *Plan de sauvegardé et de mise en valeur* carries significant legal effects, and this may explain why the majority of plans, which were accepted by local collectives in the late 1970's, have not been further progressed.

comprehensive planning system. The rest have been brought together under general controls for the protection of naturally sensitive areas introduced in 1985 (which must be interposed before July 1987) and by new controls of *déclaration préalable* or preliminary declaration introduced in April 1986.

### Périmètres sensibles *or sensitive areas, woodlands and leisure plots*

F4.114 These were introduced to protect the Côte d'Azur in 1959. Inside these designated areas exceptional measures could be taken both to safeguard natural areas and especially woodland, and if necessary to acquire threatened spaces (using the TDEV tax outlined in paragraph F4.132). These developed into a complete instrument for implementing an active policy of green space provision at the Departmental level, and were progressively extended in 1976-7 when the procedures were decentralised and accelerated. The requisite policy must protect, manage and open to the public such areas. As of July 1987 a new regime will be introduced (July 1985 law) but it will apply only in 44 Departments covering the coasts and mountains (Figure F12).

F4.115 The new controls provide strict controls on land uses and an optional consultation with the *Commission Départementale d'Urbanisme* (CDU) (which is being reformed, see paragraph F4.56). Lower thresholds apply on camping and caravan authorisations and all enclosures require declaration. Special controls are placed on wooded areas and in certain areas demolition permits apply and in others no building at all is allowed. Before 1983 zoning was always prescribed, sometimes supported by regulations, but the procedure for adoption was kept simple. However the new regimes of protection cannot be implemented in the absence of a POS and are not decentralised leaving the *Commissaire de la République* in control.

F4.116 Acquisition powers are bestowed on the Department (financed by the TDEV: see paragraph F4.132) which has a priority option in any sale, and if they do not choose to use it the national *Conservatoire du Littoral* or the communes can do so, the land becoming part of the public domain and inalienable. These acquisitions have to be compatible with the *Schéma Directeur* (SD) or *Charte intercommunale* plans (intercommunal rural development charters; see paragraph F5.74), and they cannot be exercised outside of delineated areas as previously.

F4.117 *Espaces Boisés Classés* or classified woodlands date back to 1958 as a directive subsequently translated into a law in 1973 and 1976. The designation subjects the lopping and felling of trees to a preliminary authorisation procedure, thus giving them significant protection if outside a POS, and practically definitive protection within. Provisions are now included to allow forest owners to give the land to the communes in return for building land of one tenth of the size (see also paragraph F5.53).

F4.118 Additional *déclaration préalable* procedures were introduced in July 1985 against *lotissement sauvage*, the division of small rural parcels to take advantage of the rights to station caravans or mobile homes for a limited period. The proposals were significantly weakened by the National Assembly and only apply to preemption zones. However, the Senate enlarged its application to all natural zones necessitating protection "because of their quality of site or landscape". These additional zones of control conform to the NC and ND areas of the POS, if there is one (paragraph F5.27), and apply to all divisions of land by sale whether for building or not.

### National Parks, Regional Nature Parks and Nature Reserves

F4.119 In the six National parks (2.3 per cent of the nation, which are designated, funded, and administered nationally) there are in theory three zones with a full reserve (*réserve intégrale*) within a central zone, within a peripheral or buffer zone. The former were designed to offer extremely tight protection for the most ecologically interesting areas, but as yet no such zone has been established. Meanwhile the peripheral zones have had to absorb more development as a result of the tight development constraints imposed on the central zones, particularly against winter sports development. Since 1977 these peripheral zones have had to be covered by a *Plan d'Occupation des Sols*. In general there has been a conspicuous failure to co-ordinate financial incentives, infrastructure provision and control practice, and therefore to provide a transition between the core of the parks and the surrounding countryside. Furthermore, French statutory undertakers like the Forestry Commission and the Electricity Board are largely able to circumvent or even ignore park regulations and rules. The latter govern the protection of flora and fauna, the maintenance of traditional agriculture, the restriction of recreation to walking and hiking, and the banning of most industry and commerce. A *Conseil d'Administration* composed of ministry and local government officers with conservation groups represented is the key decision-making body and policy maker within the core zone. This body has no jurisdiction within the buffer zone which is supervised by an inter-ministerial committee (Aitchison, 1984; Institute for Environmental Studies, 1987, pp. 362-7).

F4.120 The 22 regional nature parks (4.5 per cent of the nation) are something of a misnomer being precisely defined as *zones d'aménagement fin du territoire* or special land management zones. They have a dual aim of safeguarding the heritage and preserving the traditional economy, supporting it with tourism, craft industries and other cultural/educational activities. The park bodies are important consultees in the plan-making and environmental impact analysis processes as a result. The administration usually works as a *syndicat mixte* under a specific charter, with local representation. They are funded by three levels of government to undertake various management and improvement programmes and

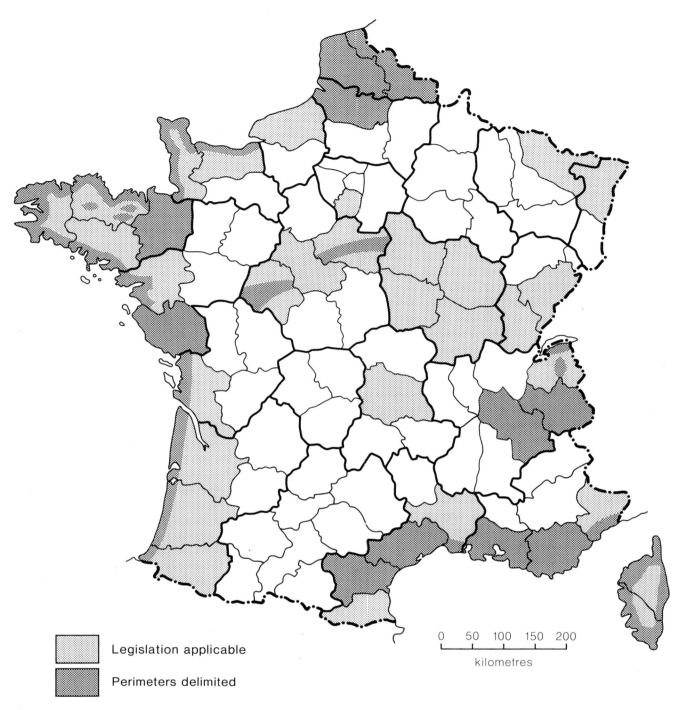

**Figure F12    Sensitive Perimeters designated in France
(December 1983)**    (Source: RISU 1985, p. 90)

Sensitive perimeters (*pèrimètres sensibles*) provide the means of designating areas within which exceptional measures can be taken to conserve the natural environment, including land acquisition. Most of coastal France and the high Alps are now protected in this way but little designation had taken place in eligible areas in Burgundy or north-eastern France by the end of 1983.

Legislation applicable

Perimeters delimited

to provide advice on building design, agricultural practice etc.

F4.121 Regional nature reserves provide specific site protection for rare or interesting ecosystems with similar protection to that accorded to national parks. But they are more flexible instruments varying in size, ownership and management. Since 1983 they can be protected by buffer zones or *périmètres de protection*.

## Public Utility Servitudes and National and Supra-communal Pre-emptive Planning

F4.122 A penultimate set of special controls which influence the delivery of a *permis de construire*, and which must be clearly enshrined in every POS, are the *servitudes d'utilité publique* which delineate areas reserved for all manner of public services. These carry restrictions on the exercise of the law of property in the general interest. They are established for the benefit of neighbouring property but they are more often conceived as simple administrative limitations on the law of property established in the general interest.

F4.123 They are established in the framework of legislation which pursues the goal of planning, and they are servitudes which affect the use of land. Sixty *servitudes* are listed in R126-1 revised in August 1986 and they are grouped into four types

(i) *servitudes* relating to the conservation of the patrimony - including the forests, coasts, woodlands, water, nature reserves and natural parks, historic monuments and sites as well as sports grounds

(ii) *servitudes* relating to the use of certain resources and equipment - including mines, energy and fuel and its transmission, hydro electricity, pipelines and canals, tele-communications, and transport of all kinds

(iii) *servitudes* relating to national defence including military establishments etc

(iv) *servitudes* relating to public health and security including cemeteries, oyster beds and navigation.

Once created by the various public utilities, *Commissaire de la République*, and government ministries, they are imposed on the local authorities elaborating the POS who must annex the list of all those which are applicable to both the plan and the delivery of any permit.

*Projets d'intérêt général* (PIG) and *opérations d'intérêt national* (OIN)
F4.124 Finally the delivery of a *permis de construire* and the initiation of major development projects may be prevented or facilitated by the intervention of special procedures permitting the initiation of projects of general or national interest. The former was introduced in January 1983, the latter legalised then, to provide supra-communal provisions and to allow the state, region, or departments to impose their schemes on the

communes. the *Projets d'intérêt général* (PIG) have a very comprehensive definition and embrace projects of infra-structure provision, planning operations (like *remembrement* and rehabilitation), conservation and protection. Eligible projects must be destined to function as a public service for the use of the disadvantaged population, must seek to protect the patrimony, seek to prevent risks or to develop natural resources. Finally the project must be sufficiently elaborated, in the sense of belonging to a public body or performing a public service, and be written into one of the planning documents at national or regional level. It cannot be an initiative of the local executive.

F4.125 The *Schéma Directeur* (SD) and the POS must accommodate and prepare the necessary provisions for such projects, and the *Commissaire de la République* must ensure that they are taken into account in the preparation of an SD and POS, or ensure that the latter are modified to accommodate them, or ensure that a *schéma* is prepared to accommodate the PIG. However the *Commissaire* still cannot compel a local authority to prepare a POS.

F4.126 The *opérations d'intérêt national* (OIN) are the important operations decided by national policies of planning (including New Towns, La Défense, the ports at Dunkerque etc). Their realisation is put under a regime that suspends normal legal practice, and the *permis de construire* within their defined perimeters are delivered in the name of the state. They are also possible to initiate in zones of *constructibilité limitée*.

## Taxes Associated with the Delivery of the *Permis de Construire*

F4.127 The granting of a *permis de construire* renders the holder of the permit liable to a wide range of taxes and contributions towards the costs of servicing and providing amenities for the proposed development. With mounting costs of service provision, and increasing pressures on local authority budgets from central government and local electors, these taxes are assuming even greater importance to development feasibility and local authority policy making. Embracing as they do questions of planning gain, community benefit and social equity the levying and level of these taxes have become central issues in planning and development in France, particularly since decentralisation. They are an integral part of the development control process.

F4.128 The financing of infrastructure remains one of the key elements of public finance and in France has been supported by *Dotations Globales d'Équipements* (DAG) from the state, and low interest loans from state banks. However, these subventions have been expected to contribute to an ever wider sphere of activity as quality of life concerns have been transmitted into various environmental and social programmes. These expenditure pressures have come at the same time as the state is attempting to disengage itself at the local level and is contributing less. The communes are therefore looking more and more to developers and builders to pay the costs of infrastructure in a bid to prevent the fiscal load falling on the local taxpayer.

## Taxe Locale d'Equipements *(TLE) or local service tax*
F4.129   In 1958 the law defined a range of taxation procedures which were linked to authorisations for *lotissement* (subdivision) and the *permis de construire* and allowed communes to negotiate on a case by case basis to cover the costs of infrastructure/services. However, this slowed development so much that it was reformed in 1967 and replaced in part by the *Taxe Locale d'Equipement* (TLE). The TLE was linked to the issue of a *permis de construire* and designed to meet some of the costs of servicing the development in its broadest sense. Set at one per cent of assessed value it failed to meet a significant proportion of the costs. Meanwhile the pre-existing taxes remained in place, and continued to be discretionary. It was completely impractical for developers to appeal against the levies placed on them since if they were successful the actual authorisation to build or subdivide to which the levies were attached became automatically invalid.

F4.130   The latter was one of the key reforms instituted in July 1985, separating the validity of the taxes and 'contributions' from the validity of the permit, and removing much of the discretionary element by specifically listing the taxes and contributions that would be forthcoming.

F4.131   Since the 1985 Reforms the communes have essentially three mutually exclusive choices. Firstly they can level a general TLE across the whole commune, although this tax has traditionally covered only 20 per cent to 30 per cent of infrastructure costs. Secondly they can create a specific area (*secteur de participation*) where taxes will be collected as contributions to specific infrastructure laid down in the *Code de L'Urbanisme* (L332-1 to 16). This includes "all or part of the costs of providing the public infrastructure costs corresponding to the needs of the actual or future residents of the sector concerned and rendered necessary by the development". It can include schools and crèches as well as highways, sewerage and other services. In practice this has been little used because it obliges the municipality to realise the necessary infrastructure to which the developer is contributing. Or, thirdly, the commune can neither levy the TLE, nor establish a *secteur de participation* but content itself with the three other groups of taxes (Figure F13).

### Other tax provisions
F4.132   These include a general group of taxes and fiscal fees. Two are set by the department. They include the *Taxe Départementale d'Espaces Vert* (TDEV) or green-space tax to pay for open space, and the *Taxe Départementale pour le financement des Conseils d'Architecture, d'Urbanisme et de l'Environnement* (TDCAUE)

which pays for architectural and planning education (public) and advice from the Council. Both these taxes are payable on a square metre of net floorspace basis, as is the TLE.

F4.133   In this same group of levies there are additional taxes payable for exceeding the plot ratio (*Coefficient d'Occupation du Sol* (COS)) in any plan. This is now imposed at the discretion of the *commune*, and is worked out on the basis of the cost of the land that would be required legally to support the level of development. Finally in the city of Paris there is the *redevance*, a tax on the creation of new office space, currently set at about £44 per square metre but likely to be doubled if not trebled in the near future.

F4.134   More importantly as Figure F13 shows, there is the *Plafond Légal de Densité* (PLD) introduced on a national basis in 1975 as a simple plot ratio (1.0 throughout France except in Paris where it was set at 1.5) the exceeding of which resulted in a floor area tax for new development. It was intended to discourage overbuilding and encourage rehabilitation, to counteract speculation, and to ensure resources to the local collectives. However it yielded less than a quarter of the estimated tax revenue and resulted in some lowering of densities in central areas. The levying of the tax was decentralised to the communes in 1982 allowing them to fix their own level. Paris for example set it at 3.0 to reduce the disincentive to development but, as part of the December 1986 deregulation provisions of the new government, the PLD has become optional for the communes. Paris, for example, is about to abandon it altogether, while other communes will only use it to tax certain kinds of development.

### The provision of infrastructure
F4.135   Two other payments due with delivery of a *permis de construire* relate entirely to infrastructure, services or provisions within the development (L332-6). The first includes the 'payment of contributions towards the expenses of public infrastructure' and these are now clearly defined. They include the costs of connecting up the sewerage, the requirement of parking (which can become a commuted fee of FF50,000 (£5,000) per parking space, special infrastructure for commercial or agricultural activities, or free grants of land to local authorities or public activities. The second includes private infrastructure and relates to the needs of all the beneficiaries of the authorisation such as service networks, waste disposal, lighting, parking, open or play spaces and landscaping. These provisions were formulated in 1985 in a bid to clarify the whole question of what contributions could be reasonably expected from developers towards the costs and quality of infrastructure and communal services at large. They followed a refusal by Parliament to implement a reform that would have placed a charge on each property according to its interest in the realisation of appropriate infrastructure.

F4.136   These reforms by no means resolve the problem,

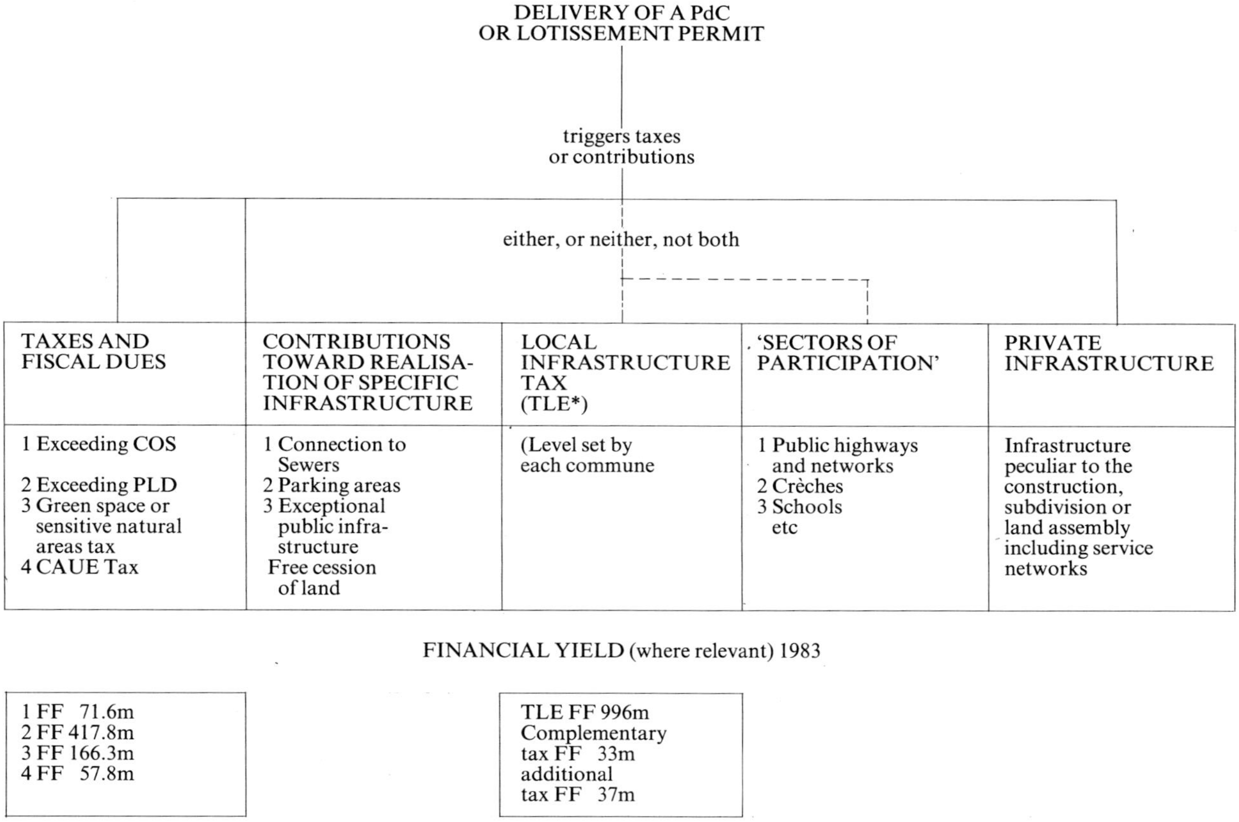

DELIVERY OF A PdC
OR LOTISSEMENT PERMIT

triggers taxes
or contributions

either, or neither, not both

| TAXES AND FISCAL DUES | CONTRIBUTIONS TOWARD REALISA-TION OF SPECIFIC INFRASTRUCTURE | LOCAL INFRASTRUCTURE TAX (TLE*) | 'SECTORS OF PARTICIPATION' | PRIVATE INFRASTRUCTURE |
|---|---|---|---|---|
| 1 Exceeding COS<br><br>2 Exceeding PLD<br>3 Green space or sensitive natural areas tax<br>4 CAUE Tax | 1 Connection to Sewers<br>2 Parking areas<br>3 Exceptional public infra-structure<br>Free cession of land | (Level set by each commune | 1 Public highways and networks<br>2 Crèches<br>3 Schools etc | Infrastructure peculiar to the construction, subdivision or land assembly including service networks |

FINANCIAL YIELD (where relevant) 1983

| | |
|---|---|
| 1 FF   71.6m<br>2 FF 417.8m<br>3 FF 166.3m<br>4 FF   57.8m | TLE FF 996m<br>Complementary tax FF  33m<br>additional tax FF  37m |

(adapted from Jacqot, 1987, pp.555–7 and RISU, 1985, pp.217–231).

\* Taxe locale d'équipment

**Figure F13   Taxes associated with delivery of a *permis de construire* or *lotissement* authorisation**

Since 1985 these development 'taxes' have been reformed and communes can either levy a TLE to recover a proportion of infrastructure costs, or create a specific *secteur de participation* to collect contributions towards specific infrastructure (which must be provided), or do neither. Other taxes remain the same but major taxes like those for exceeding the COS or the PLD are now optional for the communes. All these changes increase the variations in development levies from commune to commune.

though they do limit the kind of 'planning gains' that can be imposed on developers. What they have achieved is the creation of a patchwork of different levels and types of tax/payment across France, so that one commune can have a quite different regime from another. It has made development prohibitively expensive or particularly onerous in some communes, and dramatically cheaper and easier in others.

F4.137    But one factor is abundantly clear. These taxes and payments constitute one of the most significant issues in development control today and developers in general are concerned at the delays their negotiations produce, their impact on development costs, and increasingly at their deliberate manipulation to discourage development. While the 1985 reforms have addressed the issue of the imposition on developers of costs and requirements not clearly related to the development being undertaken, a major bone of contention, they have of course done nothing to resolve the essential problem of the local fiscal crisis. As this intensifies so will the demands on the developers. Nowhere is this clearer than in the negotiation over suburban subdivisions or *lotissement*.

F4.138    Finally to complicate matters further, for fiscal reasons residential development at large is subject to a further set of constraints. This is partly because industrial and commercial developments yield more in land taxes (assessment) and cost less in services than residential development. But it is also because in France the *taxe professionelle* (pay-roll tax) levied on businesses and industry constitutes over 40 per cent of the direct taxes levied by the local authority (see Table F2.3). These fiscal disincentives to residential development can be, and often are, reinforced by local political factors which discourage new residents because of opposition to development, or because new voters might threaten traditional voting patterns. Given the small size of many communes such factors can be very significant over the short term, and can significantly influence local planning practice and housing policy.

## Discretion and Accountability in Development Control

F4.139    The French system of development control is very closely prescribed by the laws and regulations in the *Code de l'Urbanisme*, supplemented by the *Code de la Construction* and the *Code de l'Environnement* where necessary. The procedures of application, scrutiny, consultation, decision and implementation are standardised, clear cut, and highly regulated and bureaucratic, and they have a precise timetable. As Booth emphasises, the process of applying for a *permis de construire*;

> "... lays no stress on consultation before approval, but puts a premium on the visibility of administrative and legal propriety afterwards ... the procedure stresses statutory consultation with other ministries and the legal and administrative correctness of

decision-making at the expense of policy debate and public participation" (Booth, 1985 p.17).

F4.140    From a British perspective this is an observation that can be applied across the full range of development control and captures the essence of the contrast between the two systems. However, it by no means exhausts the important points of contrast and comparison. There is most notably the question of development control staff for this sophisticated technical and legal advisory process. The level and complexity of control emphasises the importance of skilled staff and therefore the continued reliance of the small communes on the staff of the DDE. It emphasises the necessity for mayors and their assistants to master the *Code de l'Urbanisme*.

F4.141    The complexity of control is reduced to some extent by the potential comprehensiveness and precision of the POS in prescribing the rules for development. This will become evident in the next chapter. But while conformity to the POS provides a legal right to a *permis de construire*, there remains considerable scope to interpret the rules. Discretion and interpretation are still important parts of the process and detailed case studies reveal how important this discretion can be (Booth, 1987b, pp. 7-8).

F4.142    So despite the theory that the system of administrative law and the highly codified planning law should dramatically reduce discretion it remains an essential part of control practice. There is the discretion to decide whether an article or regulation applies or not, and there is often discretion within the article itself (eg RNU or POS regulations). The French controller of development is still required to interpret the complex rules and regulations and this provides significant discretionary powers. It is not the wide discretion provided in Britain by the ability to take into account 'any other material considerations', for both the POS and the RNU do clearly prescribe the range of planning considerations. However, all administrative systems require discretion in order to operate and the French planning system is no exception.

F4.143    As important as the question of discretion is the question of who exercises this discretionary power. As has been noted (paragraph F1.8) French administrative law grants discretion to administrators not to politicians. Thus despite decentralisation the planners of the DDE retain their powers and the mayors have not been free to manipulate control decisions, though they might more successfully change the regulations themselves through modifying the POS. Furthermore there remains the control of the legality of decision making which is exercised by the *Commissaire de la République*. This severely constrains the action of the mayors, preventing total anarchy in the view of some, retaining the effective *tutelle* of the Prefect in the view of others. This legal control itself is a discretionary power executed selectively and thereby creating policy that is beyond local control (see Booth, 1987a, pp. 19-23).

F4.144    Public 'participation' in the control of development in France can only take place after a decision has been taken. Notwithstanding the advantages of the system of administrative courts in providing a system of appeal freely and relatively easily available to all (see section F6), this is no substitute for actual participation in the decision-making process. Decentralisation, despite being predicated upon increasing local accountability, has not extended the public's ability to make itself heard, nor has it increased the transparency of the decision-making process. Public enquiry procedures have been made more democratic and there are obligations to hold preliminary discussions (*concertations préalables*) on major development projects, but these reforms have not dramatically altered the situation. Meanwhile the appeal system encourages recourses against *permis de construire* to protect the financial interests of neighbouring property owners, rather than the wider civic interests of amenity. This increases suspicion about the motives of litigants, intensifies complaints about selfish or irresponsible challenges, and generally further undermines the democratic process (see Booth, 1987c, pp. 21-23).

F4.145   Finally the average speed of control decision-making is slightly slower than that in Britain, but the real time taken to get development under way may be significantly longer. Prior negotiations can be very protracted and many applications drag on interminably as potential legal obstacles are ironed out (Booth 1987b and personal correspondence). Even after a permit has been issued there is the four month period for challenge of a permit, six months for third parties with a direct interest, and if a court case ensues it will take two to three years for a decision to be reached.

## Conclusions

F4.146   From a comparative perspective the following points emerge as of major significance in the French control system

(i)   the reliance on the POS clearly to define development rights, ensuring that conformity of a proposal to the plan guarantees a *permis de construire*

(ii)   in the absence of a POS the imposition of a 'no building' restriction on all rural areas as both an anti-urban-sprawl mechanism, and as an incentive to the commune to prepare a POS

(iii)   in the absence of a POS the reliance, in built-up areas, on the national rules of urban planning (RNU) governing location and servicing, siting and volume and external appearance of development, though these rules are discretionary

(iv)   the reliance on the DDE technical staff to advise on the development control function in all but the largest towns, even though the mayor and his technical services teams are responsible for the decisions once a POS is in place. This provides for the continuing influence of the state, a useful flow of information for the Ministry, and careful adherence to the complexities of the *Code de l'Urbanisme* which are nationally prescribed. All of these reduce the room for manoeuvre at the local level and limit the real nature of decentralisation

(v)   the role of discretion in the control process is greatly reduced by both the Code and the RNU, but the latter contain no fewer than 13 discretionary 'can be' clauses out of a total of 33. So there is considerable scope for the exercise of technical expertise and political will, but the latter can be influential only if the system is properly monitored and mastered.

(vi)   the role of negotiation in the control process would seem at first sight to be greatly reduced compared with Britain. In fact it can be of equal importance since advice is often sought prior to making an application, and the *recevabilité* (receivability) of an application is often denied until certain information, studies and refinements are forthcoming. Negotiations also revolve around questions of external appearance (see x) and taxes (see xvi)

(vii)   questions of negotiations and discretion are however significantly restrained by the fixed time periods for deciding on an application (usually two months), and by the ability to obtain a *permis tacite* if no decision is forthcoming within the time period. This ensures a measure of administrative efficiency although 'legal' reasons can be found when necessary to delay or suspend a decision. Equally tacit permissions are rare, and 85 per cent of all applications are decided within two months

(viii)   delay has not become an issue in development control in France in the way it has in Britain, despite the fact that on average applications take slightly longer, and despite the significant problems outlined in (vi), (x) and (xvi). What is of concern is the fact that once approved a permit can be challenged over a period of four months and, more importantly, that if a recourse to the court ensues it will be two or three years before a decision will be reached.

(ix)   architectural expertise is required for all applications for development of over 170 square metres, providing important exclusions for the single house but encompassing all medium to large scale developments

(x)   external appearance is given clear recognition as an important element in control, reinforced by allowing detailed consideration of the materials and colours of proposed buildings, and the requirement of architectural expertise, by demolition controls, and by architectural councils to promote both visual education and development quality, and by extensive conservation controls and design guidance

(xi)   urban conservation controls provide comprehensive protection for listed buildings and their wide environs, and while conservation areas number only 80 they are the focus of major programmes of rehabilitation and improvement. Demolition controls reinforce general urban conservation principles

(xii) rural conservation procedures still concentrate upon the built environment and are a rather late development, but national prescriptions introducing special controls on the coasts and in the mountains potentially offer much tighter regimes in the most sensitive areas. These are linked to important land acquisition powers and funds, while rural *remembrement* offers some prospect for sharing development rights equitably in the interests of conservation

(xiii) an outline application procedure is available for use, but is of primary importance to land valuation and sales

(xiv) the *lotissement* procedure provides important controls on the subdivision of land and the preparation and servicing of plots for development, but with each individual building subsequently requiring a *permis de construire*

(xv) an effective, built-in enforcement system exists in the *certificat de conformité*, while building regulation enforcement is the responsibility of the architect and builder and policed by sale and insurance provisions. Some 20 per cent of developments do not initially conform but only two per cent are definitively refused

(xvi) the issue of taxes on delivery of a *permis de construire* looms large with communes able to use various taxes as a means of discouraging development. The tax issue has become much more important with the decentralisation of planning and the growing fiscal crises at the local level. In the matter of *lotissement,* where contributions to services are much more negotiable, it is becoming particularly crucial. It is the issue of planning gain in another guise

(xvii) the issue of public participation in control has only been briefly discussed. There is the publicity of applications (but no right to consult the file) and decisions. Even the *enquête publique* procedures for *installations classées* and major developments allow only public comment rather than open debate. The crux of the matter is that the citizen can appeal against the grant of a permit only on the grounds of its not adhering to the rules set out in the plans or the RNU, or not following the proper procedures. Participation is *a posteriori*, a matter for the courts, and therefore the subject of section F6.

# F5 DEVELOPMENT PLANS AND THEIR ROLE IN THE FRENCH PLANNING SYSTEM

## Introduction

F5.1   In theory there are four levels of plans in France beginning with the National Plan, and its component regional plans, the latter now supplemented by the *Contrat de Plan*. At the city-region scale there is the *Schéma Directeur* (SD), or the strategic long-term development plan, while at the local level of the commune there is the *Plan d'Occupation des Sols* (POS), or land use zoning plan. The relationship between these plans seems at first sight to reveal a clear hierarchical logic, but in practice the relationships are not hierarchical, the plans are not synchronised, and in the majority of instances the linkages seem non-existent. This is in part a reflection of the post 1983 emergence of regional plans and the *contrat de plan*, and in part a reflection of the fact that most *Schémas Directeurs* are seriously out of date. But it is also a reflection of the curious division of powers and relationships between the four levels of government in France, and the fact that decentralisation has not yet produced a clear-cut and coherent new structure.

### The National Plan
F5.2   The national plan has progressively abandoned notions of comprehensive economic planning or normative projections of economic growth or business activity. Through the 1970's the plans were progressively emptied of any operational substance and replaced by a small number of specific programmes targeted at selected locations and issues. By the 1980's some 25 action programmes were complemented by some 80 regional priority action initiatives spelling out joint state-region expenditure programmes. The programming of public investment projects seemed to reaffirm the importance of planning, albeit in a budgetised form, but the reality was rather different (Green, 1981, pp. 115-120). Under the Socialists in the 1980's the National Plan was given much greater emphasis but rising inflation and unemployment forced the government to reverse its avowed policy, particularly its attempts to save traditional industries in Northern France or to discourage growth in the Paris region.

### The Regional Plan and the Contrat de Plan
F5.3   As part of the national planning process each region already produced its own regional plan consisting of a statement of objectives and priorities agreed between the state and region, embracing both economic and social priorities and programmes. As part of their attempt to revive both national and regional planning, and to provide for the better coordination of both, the socialists introduced the instrument of the *Contrat de Plan*. This takes the form of a contract between state and region based upon the regional plan but setting out a programme of works or investments and detailing the contribution of state and region.

F5.4   The new round of national-regional planning did not work as intended in the early 1980's and became a much abbreviated process. As a result the *Contrat de Plan* often preceded the regional plan with the state's allocation of funds playing a determining role in shaping regional investment. Many regions failed both to develop adequate expertise or to display the political will to shape their own strategies, and few have developed a coherent industrial policy. However, as Garrish argues, the regional planning process is only just beginning and the regional plan and *Contrat de Plan* are in their infancy (Garrish, 1986, pp.48-49, 66-71). The potential role that can be played by the *Contrat de Plan* in defining programmes of economic development, housing, open space provision, water/sewerage, highways and transport may be seen by reference to the *Contrat de Plan* for Ile de France (see paragraph F2.18), the region best able to take early advantage of the new procedures. What the *Contrat de Plan*'s exact relationship with development plans will be remains to be seen, and to a large extent depends upon the extent to which the *Schéma Directeur*, or structure plan, is revitalised as a planning tool.

### The two-tier development plan
F5.5   The *Loi d'Orientation Foncière* in 1967 established a two tier system of plans to provide the basis for French town and country planning. This set up the *schéma directeur d'aménagement et d'urbanisme* (SDAU) at the city-region scale to set out the broad parameters of future growth restrictions and land use, and to articulate a comprehensive system of infrastructure provision. At the local level the *plan d'occupation des sols* (POS) was established to be the regulating document at the scale of the street and the individual property, the essential reference point for development control decisions. One of the key objectives of the reform was to separate the forecasts of future land use and infrastructure provision from the regulations affecting the law of the land and the nature of construction, which had been combined in the old *plans d'urbanisme directeurs*. In bald terms these two plans corresponded to the structure and local plans of British town and country planning but their preparation, content, and legal effect were fundamentally different.

While a two tier system of plans continues to operate, the law of January 1983 has decentralised their establishment and refined and widened their content.

## Schéma Directeurs (SD)

F5.6   In 1983 the SDAU was retitled the *Schéma Directeur* (SD) and became no longer compulsory for all urban areas of more than 10,000 population, but purely a matter for the communes to decide upon between them. The old SDAU remained in effect according to their various stages in the approval process, but the content and function of the new SD changed very little. They are still required to fix the fundamental limitations of planning over a 30 year time frame, and to ensure an equilibrium between urban extension, agricultural activity and other economic activities and the protection of natural sites (L122-1). They are intended to harmonise the policies of central and local governments, to specify a future pattern of land use, and to designate a pattern of infrastructure provision. However, the content can be varied within certain broad requirements, and they can be complemented by *schémas secteurs* (SS) which can spell out proposals in greater detail.

### Content
F5.7   The SD has two components - a report and a set of graphic documents. The report contains an analysis of the current planning situation including demographic, economic and social change analysis. It is required to justify the delimitation of the area of the SD and to consider its relationship with surrounding areas. It must analyse the existing state of the environment and explain how its qualities are to be protected, and indicate the major phases of development foreseen and how these are to be serviced. Finally it must illustrate the compatibility between its dispositions and the implementation of projects of general interest (*projets d'intérêt général* (PIG).

F5.8   The graphic documents must include three maps depicting the current situation, that likely to prevail in 30 years time and that anticipated after ten years of the operation of the plan (Plate 10) Depending upon whether they cover urban or rural areas the maps must also show urban extensions, redevelopment or renovation zones, zones protected for agricultural or forestry uses, protected sites, proposed public infrastructure, the location of major activities, the general organisation of circulation and transport and the essential elements of the water and sewerage network. The maps are normally at a scale of 1:50,000 and the system of symbols and colours is standardised by the *Code de l'Urbanisme*.

F5.9   The report and graphic documents are complementary and they establish the basic priorities in terms of infrastructure provision and the allocation of land to different uses. But they do not fix the precise dates of implementation or land release though they do indicate the order of priority of schemes, and the broad strategies of development to be adopted.

### Elaboration of the SD
F5.10   The law of January 1983 decentralised the elaboration of the SD to groups of communes which could come together within an *établissement public de coopération intercommunale* (EPCI), removing the state from the process almost altogether. However, the state, along with the region, must still be consulted. An alternative approach could be to set up a special syndicate, either a *syndicat intercommunal d'études et de programmation* specifically to prepare the plan, or a *syndicat mixte*. The municipal councils of the various communes formulate a proposition for SD preparation to the *Commissaire de la République*, and if more than two thirds of the communes with more than half the population (or half and two thirds respectively) agree then the *Commissaire de la République* fixes the perimeter of the scheme, after consulting with the regional and departmental councils if more than 100,000 persons live in the area of the SD.

F5.11   The organisation set up to elaborate the plan must include representatives of the other levels of government, state, region, Department; other intercommunal planning establishments; and chambers of commerce, trade and agriculture etc, if they wish to be included. The law leaves a great deal of scope to the communes to decide the composition of this planning commission but requests to participate must be notified within two months and the composition of the commission publicised. The commission is led by the appointed president of the EPCI who can seek advice from any competent body which has jurisdiction over building, planning or town planning including HLMs, environmental groups or mixed development companies (SEMs). The plan itself is prepared by the DDE or an *agence d'urbanisme*. It must then be declared by *arrêté* and all participants consulted. In the event of conflict a commission of conciliation is brought in to resolve the issues (see paragraph F5.13). The *Commissaire de la République* ensures that the scheme does not compromise the *projets d'intérêt général*, and the plan is subject to public consultations for one month, a provision added in 1983. However, there is no public enquiry.

F5.12   Modifications may ensue and the plan will then be adopted, but it only becomes executory 60 days after approval to protect the supra communal and communal interests. The former interests are largely entrusted to the *Commissaire de la République* who ensures respect of planning law and the interests of state, region and department. The communes themselves can withdraw in this 60 day period if they feel that their essential interests are being compromised, or they can take their case to a commission of conciliation. Under exceptional circumstances the state can use its old prerogatives to establish or modify an SD, but the communes have two years to respond to a 'request' to do so from the state, before a joint elaboration (*elaboration conjointe*) process is initiated.

F5.13 Each department has established a *commission de conciliation* or conciliation commission to resolve conflicts that may arise in the content of the SD or the POS. These may intervene to iron out differences between any local authority or interest group, but their role is limited because access to them is limited. Thus they perform the role of heading off any potential conflicts that might result in either the interventions of the *Commissaire de la République*, or a resort to the administrative court (Besson-Guillaumot, 1987, pp.160-1). The commissions help to compensate for the lack of an hierarchical relationship between the various levels of government. Six elected representatives, and six qualified planners or environmental experts independent of any group, constitute the commission with a president and vice-president elected from amongst the commune's elected representatives.

F5.14 The *Commissaire de la République* represents the interests of the state in the preparation and content of both the SD and the POS, and is empowered to apply sanctions to either or both. He/she may require either plan to be amended to allow the implementation of 'schemes of public benefit', or require an SD to be prepared or amended to meet the 'application of planning requirements'. The authorities have two years in which to accept the advice after which it can be imposed by decree. Furthermore the *Commissaire de la République* can revise a POS after a public enquiry to ensure that it takes account of new planning requirements (Institute for Environmental Studies, 1987, p.545).

*Modification of an SD or SDAU*
F5.15 The modification of existing SDAUs is for a variety of reasons (see paragraphs F5.20–23) a much more important issue than the preparation of new SDs, and new laws in July 1985 facilitated this by modifying the *elaboration conjointe* procedure. The *Commissaire de la République* initiates the procedure, usually because of a request from a commune or EPCI preparing a POS which is going to be incompatible with the SDAU, but sometimes because the SDAU threatens a project of general interest. For the modification a *commission locale d'aménagement et d'urbanisme* (CLAU) is established, with similar composition to the commission for preparing a new SD, and they direct the work. But the *Commissaire de la République* leads the study team be it the DDE, an *agence d'urbanisme* or a private planning consultancy (the commune's planning staff can be involved). Similar consultations are undertaken, but when the plan is completed there is a three month period for opposition to be voiced. The crucial difference since 1985 is that while opposition is very much taken into account in the final refining of the SDAU it can now be passed over more easily (in the modification process). The approval rests with the state authorities and is not entrusted to the communes, though if a quarter of the communes (or those

with a quarter of the population) object, the *Conseil d'État* will have to approve the final arrêté.

*The judicial effect of the SD*
F5.16 The most important fact about the judicial effect of the SD is that it is enforceable upon the administration, and public or private corporations which administer public services, but not upon third parties. Since 1967 it is the POS which contains those elements which are enforceable upon third parties. This clear division is however complicated by the fact that the SD and POS are in no sense synchronised, and thus many areas covered by an SD are not covered by a POS. This renders the SD's prescriptions for the conservation of farmland or environmentally sensitive areas unenforceable on third parties, and the possibility of direct enforcement was specifically removed by decree in July 1977.

F5.17 However, the enforceability of the SD on the public collective and public agencies has been reinforced in January 1983 and July 1985 by demands that their programmes and decisions, including the POS itself, must be compatible with the provisions of an SD and/or SS. The issue of compatibility has been thoroughly tested in the courts, and in a key decision in February 1974 the *Conseil d'État* distinguished 'compatability' from 'conformity' giving a measure of flexibility to the former by accepting that minor deviations (eg road lines) did not threaten the fundamental options of the SD. But there is still the opportunity for private persons to take cases of violation of the SD by public authorities to the courts.

F5.18 Despite attempts to ensure that the issuing of *permis de construire* and *lotissement* authorisations conform to the SD, the *Conseil d'État* has maintained (notably in March 1977) that the SD remains unenforceable on third parties. However, in New Towns and in the matter of the development of wooded spaces (*espace boisé classé*) the SD can be enforced.

F5.19 Finally the modifications to the law made in 1983 and 1985 have made provision for an SD to take effect before it has been fully approved, and provision has been made to suspend a POS in the event of an unconformity, or to ensure that public actions conform to the new SD rather than the old SDAU. By contrast a provision to ensure that public actions specifically conform to the POS has been expressly excluded by recent legislative changes.

*The coverage of SD*
F5.20 The reality of the situation is that very few new SD's are being prepared and reliance is placed upon the 187 SDAU which cover some 5,000 communes, some two-fifths of the population but less than one seventh of the land area of European France. While more than twice as many SDs have been delimited covering nearly four fifths of the population, no attempt is being made to approve most of these (Figure F14). So coverage is intermittent and while most significant urban areas have a delimited SD there are major omissions in terms of approval. Hence Bouches du Rhône, which includes Marseille, Aix, Arles and the growth areas of the Étang de

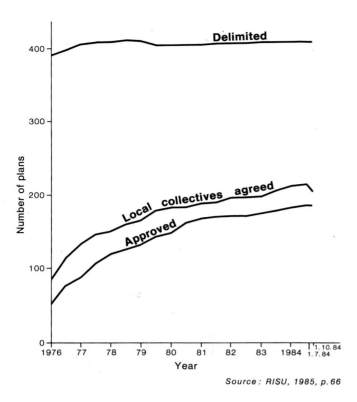

Source: RISU, 1985, p.66

**Figure F14   The progress of preparation of the SDAU/SD 1976–1984**

This graph shows that the production of *Schémas Directeurs* has ceased since 1978. The number delineated (having a perimeter fixed by the *Commissaire de la République*) has remained virtually unchanged, and less than half of these had been approved as of mid 1984, with only a handful being approved each year. The relatively high approval rates of the late 1970's have disappeared as it has become increasingly obvious that the plans are outdated, and as impending decentralisation has made their role more ambiguous. In fact the number of local collectives actively engaged in production shows an actual downturn in the second half of 1984.

Berre has twelve SDs delimited covering more than two thirds of the department, but only one approved covering 1.4 per cent of the territory (Figure F15).

F5.21   The failures to complete the approval process pale into insignificance when one examines the dates of the preparation of most plans. More than half of the plans were completed before 1975 and based on 1960's forecasts, while two thirds were approved before 1979. It is often suggested that the SDAU's of Brittany contained enough land allocated in different uses to accommodate the growth of all of France to the end of the century. Most SDAU's are thus based upon notions of population and economic growth that are entirely inappropriate to the recession-dominated 1980's. The plans are technically redundant and yet they continue to produce juridicial effects on public agencies and upon the POS. It is this fact which emphasises the importance of modifying the existing SDAU.

F5.22   The reason for the failures to prepare new SD are related to several factors including the tedious process of preparation and approval, and the ability of communes to withdraw from the venture very late in the process. Above all it is unlikely that supra communal interests, which are the very point of the SD, will be given priority when the essence of decentralisation consists in giving enhanced planning powers to the communes. As a result the process of POS production has become the almost total preoccupation of communes and planning agencies, and their numbers have increased accordingly. Meanwhile the production of SD's has completely stagnated and they have become, if anything, a hindrance to constructive planning even if they have not entirely lost their supra-communal coordinating role.

F5.23   Some groups of communes have begun to look afresh at the SD to give it a stronger policy dimension, better promotion of public-private partnership, good urban design and environmental protection provisions, better functional planning and more encouragement for economic investment. But inevitably decentralisation has made consultation and approval more problematic, and viable boundaries and areas have become an important issue. As an example in Orléans, whereas 40 communes participated in the old SDAU, only 21 are going to be involved in the preparation of the new one, and of course each now has the right of withdrawal at a later stage. Thus the future of the SD is still very much in the balance.

## Plans d'Occupation des Sols (POS)

F5.24   The POS spell out the orientations defined by the SD, if they exist, but they have a regulatory character, defining the rules and servitudes which govern land use and construction, and are enforceable on third parties. Since October 1983 they are no longer obligatory on all communes with more than 10,000 population and, as a key part of decentralisation, it is entirely up to the commune whether or not they prepare a POS. They are of course strongly encouraged to do so by application of the rule of *constructibilité limitée* in the absence of a POS, which greatly restricts their powers to allow development, and by the non transfer of powers to deliver a *permis de construire*. POS procedures have been simplified to encourage many more small communes to prepare them, and made more flexible to allow them to replace pre-existing *documents d'urbanisme*. Meanwhile the state has provided more funds to finance the preparation of more POS.

### Content of the POS
F5.25   Previous laws permitted a partial POS although they gradually became more uniform over time. The January 1983 reforms defined an obligatory or minimal content for all POS, which includes a zoning map and regulations and rules governing the siting of buildings, and an optional content which might embrace issues such as the regulation of the external appearance of development, reserved sites for development, or plot ratios. While a commune can choose whether or not to produce a simplified or a more complete POS, this choice

Paris & Ile de France

Val-d'Oise ④
Seine St. Denis
Paris ① *1*
Val de Marne ① *8*
Hauts-de-Seine *6*

*9*
*15*
Yvelines ④
⑦ *7*
③
*17*
Essone
①

*8* Pas-de-Calais ⑦   *6* Nord ④
*2* Somme ②   *7* Aisne ①   *3* Ardennes
*5* Seine-Maritime ④   *9* Oise ③
*7* Calvados ⑤   *4* Eure ③   *20* Seine-et-Marne ⑱
Manche ②
*2* Orne   *4* Eure-et-Loir ①
Finistère ③   *10*   Côtes-du-Nord ④   *5* Ille-et-Vilaine ③   *1* Mayenne   *3* Sarthe ③
*6* Morbihan ①   *2* Loire-Atlantique ①
Maine-et-Loire ①   *2* Indre-et-Loire ①
*5* Loir-et-Cher ②   *3* Loiret ②   *2* Yonne ②
Cher ①   *2* Nièvre ①
*4* Marne ④   *3* Meuse ②   *4* Meurthe-et-Moselle ①   *6* Moselle ③   Bas-Rhin *9* ⑥
*3* Aube ①   *3* Haute-Marne ①   *3* Haute-Saône ①   *5* Vosges ④   Haut-Rhin *5* ⑤   *1*
*3* Côte d'Or ①   *2* Doubs ②   Tre-de-Belfort
*5* Vendée   Deux-Sèvres *3*   *1* Vienne ①   *2* Indre ②   Creuse *1* ①
*7* Saône-et-Loire ③   Jura *3* ②
*6* Charente-Maritime ③   *2* Charente ③   Haute-Vienne *2* ①   *2* Puy-de-Dôme ②   Allier *3* ③
*3* Loire ②   Rhône *3* ①   *4*
Ain ①   *2* Haute-Savoie *2*
*5* Gironde ①   *3* Lot-et-Garonne ④   Dordogne *5* ①   *2* Corrèze ②   Cantal *2* ①   *1* Haute-Loire ①
*2* Lot ②   Aveyron *2* ①   Lozère *2*   Ardèche ②   *3* Drôme ①   *3* Isère ③   Savoie *3*   *2* Hautes-Alps
*4* Landes ①   *3* Tarn-et-Garonne ③   *2*   *1* Gers   Tarn *2* ②   Hérault ②   *3*   *4* Gard ①   *4* Haute-Provence   Alpes-de- *4*   Alpes Maritimes *3* ③
Vaucluse ②   *6* Var ①   Corsica
Pyrénées Atlantiques ①   *8* Hautes-Pyrénées ①   *5* Ariège ③   ①   *2* Aude ①   Pyrénées Orientales *6*
Bouches-du-Rhône *12*
Haute-Corse *2*
Corse du Sud *4*

0   50   100   150   200
kilometres

*4*   Schéma directeur delimited
③   Schéma directeur approved

*Source; RISU, 1985, p 67.*

**Figure F15   The coverage of *Schémas Directeurs* by Departments 1984**

This map showing the numbers of SD delineated and approved by Departments reveals several factors. First many departments only have one or two SD, usually covering their major towns and immediate hinterland. Secondly there are wide differences between departments in numbers of both delineated and approved plans. For example in the whole of the Languedoc region there is not an approved SD. By contrast, in Ile de France all departments have a high number of delineated plans, as might be expected, but Seine et Marne has approved 90 per cent of its SDs while Essonne or

Yvelines have only approved 6 and 20 per cent respectively.

(Ile de France is also the only region with a *Schéma Directeur* that covers its entire territory. This is prepared by *L'Institut d'Aménagement et d'Urbanisme de la Région d'Ile de France*, which is an *Agence d'Urbanisme*. First produced in 1965 (a forerunner was introduced in 1960) it was reformulated in 1976, and a more strategic, less specific revision is currently under consideration. The inset map illustrates the existence of conventional SD in the Ile de France).

211

is also governed by the judge according to the commune's importance. Basically a POS always consists of a dossier with four components, a report, graphic documents, a set of regulations, and some annexes/ appendices covering a variety of *servitudes* and restrictions.

F5.26 The report must set out the demographic, economic, and social development of the area, analyse the environment and prescribe measures to protect it (including the agronomic value of the land), determine the perspectives on urban development and the means of implementation, and define different zones for development. It must justify the dispositions of the POS ensuring its compatability with planning law and projects of general interest. However, the report has interpretive value only in relation to the maps and regulations which are the crucial elements of the POS.

F5.27 The graphic documents include one or more maps at a scale between 1:2,000 and 1:10,000 which must show the delimitations of the zones and *servitudes* included in the plan as well as property boundaries. These zones embrace a variety of different land uses and combinations of land uses as well as zones falling under different types of environmental protection, protected lines for road widenings, or new highways, public utilities, services or sites for the same, and protected sites of all kinds (Plates 12 and 13). Again the cartographic representation is defined by the *Code de l'Urbanisme*, although many POS have developed their own specialised maps for particular phenomena. Many have detailed plans for particular areas, or massing plans for areas of comprehensive redevelopment; others define the protection of historic vistas and views, or co-ordinated green space planning. The principal categories of zoning in the POS are listed below.

*(a) Urban Zones:*

ZU — Zones in which the capacity of existing services or those in the course of realisation permit immediate construction (ie can be raw land where services are in the process of being provided, and this has always been flexibly applied) so a *permis de construire* cannot be refused on these grounds. Cultivated land, classified woodland and zones can be protected.

Sub-categories of ZU are not prescribed by the *Code de l'Urbanisme* and depend upon the sophistication of the POS. They can relate to specific categories of land use, to particular mixes of land use, to specific locations with a particular character, to a particular built-up character, or to special operational zones. The Paris POS has 15 different ZU zones each with their own set of regulations (Plate 11), that of Lyon has 11 although these represent the ultimate in sophistication and responsiveness to the complexity of built up areas. Many POS in rural areas will operate with a single ZU category.

*(b) Natural Zones:*

ZN — Natural zones which must stay temporarily or permanently urbanised. These are divided into four major groups as listed below.

NA — Zones of future urbanisation, not serviced but destined to be equipped in the longer term. These could be urbanised immediately if the builder/developer/ subdivider is prepared to assume the full costs of infrastructure. Priority can be accorded to each zone (short, medium, long term) to protect agriculture for as long as possible.

NB — Ordinary natural zones not destined to be developed further where some supplementary development will be permitted, but minimum amounts, and where *lotissement* is forbidden (largely low density unserviced development may well be accepted).

NC — Zones of natural riches which support actual or potential productive activities, usually agriculture or forestry. Where economically productive these areas must be protected from urbanisation and only buildings required for active production are permitted.

ND — Zones of environmental sensitivity, risk or special nuisance including noise zones, flood zones etc, but more importantly areas of landscape or ecological interest, or areas where construction would damage the environment. Some transfer of development rights may be possible

F5.28 The graphic documents are supported by a set of regulations which prescribe the rules and *servitudes* enforceable on third parties. They spell out the uses that are acceptable and those which are forbidden in any zones, and also the prescriptions relative to the siting and construction of highways and other works. These are obligatory even in simplified POS, but the regulations can go much further. They can provide edicts and prescriptions relative to the access to services; expropriation; height and external appearance of construction; and the parking, green space, and play space associated with them. They can prescribe areas of demolition control and fix plot ratios for different land uses, and the rules for exceeding or transferring them.

F5.29 The regulations take a form clearly prescribed by the *Code de l'Urbanisme* (A123-2). They fall under three titles. The first consists of a set of general dispositions which prescribe the area of the POS, the reach of its dispositions, the different zonings adopted and the minor adaptations which are possible. The second and third set embrace the rules applicable to the urban and rural/ natural areas respectively (paragraph F5.33).

*(a) Structure of regulations of the POS*

Title 1:      General dispositions

    article 1:    Territorial application of the plan
    article 2:    Range of regulations and other legislation
               relating to the occupation of the land
    article 3:    Division of territory into zones
    article 4:    Minor adaptation of certain rules

Title 2       Dispositions applicable to urban zones

    chapter 1:   Regulations applicable to Zone UA
    chapter 2:   Regulations applicable to Zone UB and so
               on UC, VE etc

Title 3:      Dispositions applicable to natural zones

    chapter 1:   Regulations applicable to Zone NA
    chapter 2:   Regulations applicable to Zone NB and so
               on NC, ND etc

*(b) Structure of regulations of each zone*

Section 1:   Nature of the occupation and use of the
             land

    article 1:    Types of occupation or use of the land
               admitted with or without condition
    article 2:    Types of occupation or use of the land
               forbidden

Section 2:   Conditions of the occupation of the land

    article 3:    Access and highways
    article 4:    Service networks (water, sewerage,
               electricity)
    article 5:    Characteristics of land (form, surface etc)
    article 6:    Siting with regard to highways and public
               easements
    article 7:    Siting with regard to neighbouring
               buildings
    article 8:    Siting of buildings with regard to other
               buildings on the same property
    article 9:    Coverage of the land
    article 10: Height of construction
    article 11: External appearance (form, materials etc)
    article 12: Parking
    article 13: Open spaces and planting

Section 3     Maximum possibilities of land occupation

    article 14: Coefficient of land use (COS)
    article 15: Exceeding the COS

So each POS contains a set of general dispositions, and then each zone UA, UB, NA, NB etc contains specific regulations under the fifteen headings. A better grasp of the nature of the regulations or articles 5 - 15 can be obtained by referring to Plate 11 and Table F5.1 which provide illustrations from the Paris POS.

F5.30   The annexes or appendices to the POS include six categories of restrictions including lines and sites reserved for highways and public works, a list of public utility operations, details of water and sewerage provision, a table of public utility *servitudes* which can number up to 60 different kinds, the national land use prescriptions which apply, and the list of *lotissements* where the rules of urban planning have been maintained. These annexes are of the greatest importance, and give the POS a certain uniqueness, for it is often argued that the POS is not a plan, in the forward looking sense of the term, but a regulatory document enshrining property rights and restrictions on the same. Thus it must enshrine all the *servitudes* affecting property rights, and if any *servitude* is not included in the plan at its approval (or added to it within one year) it is not enforceable on third parties.

F5.31   Superimposed on the zoning are special statutes relating to particular spaces. These include reserved sites (*emplacements réservés*) for all manner of infrastructure (everything from pedestrian routes to hospitals, from green space to administration buildings) and classified woodland (*espaces boisés classés*) protected for ecological or recreational reasons.

F5.32   The actual zoning pattern is subject to certain 'superior' rules which were reduced in 1983 as part of decentralisation, but which have multiplied since. These can be very general in terms of prescribing disposition of land in an economical fashion, protecting the landscape and natural environment, and promoting equilibrium between urban and rural populations and between environmental protection and development. But they also contain some more precise prescriptions to enforce the recognition of the SD's, *chartes intercommunales* (see paragraph F5.74), public utility servitudes and the like. These obligations vary through conformity to compatability to simply being taken into account. They are subject only to the minimal control of the administrative courts (since 1979) thus giving the communes a significant discretionary competence, but as Jacquot suggests it is possible that the measures of decentralisation will lead to stricter legal control by the courts (Jacquot, 1987, p.181).

*The rules of land use*
F5.33   Each zone classified in the POS is subjected to some 15 regulations which relate to the nature of the land uses accepted or forbidden, the conditions of development including servicing, siting, massing, aspect and landscaping, and the plot ratios and densities applicable (paragraph F5.29 and Plate 11). The first two regulations determine the different occupations and use of the land which can be authorised, conditionally authorised, or forbidden, and practically the only excepted uses are of a purely agricultural nature. The actual uses permitted depend upon the category of the zone. The second set of regulations, rules three to thirteen, have a morphological and servicing function. Rules three to five prescribe access, water, sewerage and energy provision. Rules six to eleven determine the conditions of siting in terms of relationship to roads, neighbouring buildings, siting, height, and external aspect, while rules twelve and thirteen define the parking, open space and landscaping requirements. Some understanding of what these rules can mean for different kinds of zone can be obtained from Plate 11. Their relationship with *the règles nationales urbaines* (RNU) are obvious (see Figure F10), and their

**Table F5.1**  The *Coéfficient d'Occupation des Sols* for Central Paris

| | HOUSING BUSINESS PUBLIC BUILDINGS | OFFICES | ACTIVITIES | | HOUSING | OFFICES | ACTIVITIES BUSINESS PUBLIC BUILDINGS |
|---|---|---|---|---|---|---|---|
| | | | | ZONE UM | | | |
| ZONE UA | 3.00 | 1.50 | 2.00 | SECTEUR UMa | 2.70 | 1.30 | 3.00 |
| ZONE UC | | | | SECTEUR UMb | 2.50 | 1.30 | 3.50 |
| SECTEUR UCa | 2.70 | 1.00 | 2.70 | SECTEUR UMc | 2.00 | 1.00 | 3.50 |
| SECTEUR UCb | 2.70 | 1.00 | 2.70 | SECTEUR UMd | 1.50 | 1.00 | 3.50 |
| SECTEUR UCc | 1.80 | – | 1.80 | ZONE UN | | | |
| ZONE UF | 3.50 | 3.50 | 3.50 | ZONE UR | | | |
| ZONE UH | 2.70 | 1.00 | 2.00 | SECTEUR URa | 2.70 | – | 2.70 |
| ZONE UI | – | – | 3.50 | SECTEUR URb | 2.70 | – | 2.70 |
| ZONE UL | – | – | – | ZONE UO | – | – | – |

Since March 1976 it has been possible for the COS to be differentially applied to different land uses within the same zone. In the Paris POS full opportunity is taken of this generally to discourage office development in central Paris (except in UF zones), and to encourage the retention of housing, retail and similar businesses. Higher plot ratios are also retained for public infrastructure like schools or hospitals. As a result, in the UC zone which covers much of central, historic Paris the plot ratio for housing, retail public infrastructure is 2.7 whereas that for offices is 1.0. Thus the COS is a very effective instrument controlling redevelopment and the invasion of office uses.

application as the basic principles for development control can be easily appreciated.

Coéfficient d'Occupation du Sol (COS) *or plot ratios*
5.34   The COS is essentially a plot ratio defining the area of floorspace that can be constructed in relation to the surface of the building plot. It is exactly defined by the *Code de l'Urbanisme* as new floor-space (but excluding parking, animal shelter, storage, attics, balconies etc) as a ratio of the surface of the parcel (which can include public utility or highway easements if the land is donated to the relevant authority).

5.35   The COS is fixed for each zone as a function of the capacity of the collective infrastructure that is existing or in the course of construction, and a function of the nature of the buildings that are to be built. However, the COS does not overrule the other 13 rules applied to each zone. The setting of the COS is of course determined by non-technical factors such as the landowners' desire to raise land values and the communes desire to raise taxes (for exceeding the COS). In general there is only one COS per zone or part of a zone, but since March 1976 it is possible to apply different COS to different approved land uses within the same zone to encourage more flexibility in the urban fabric and to encourage, for example, the provision of housing in mixed commercial areas. Thus the POS for Paris prescribes different COS for housing, offices and commercial activities and collective infrastructure in each zone, generally discouraging the conversion of properties to offices and encouraging the maintenance of retail and residential functions (Table F5.1).

F5.36   It is possible to exceed the COS by paying a tax (which is set at the land value of the site hypothetically required legally to construct the additional floorspace) or by acquiring other development rights through the transfer of a COS. But it can be approved only in order to ensure an architecturally satisfying or conforming building better to protect natural zones, or because the

services/public facilities are reinforced in some way. The transfer of development rights is a very controversial area, since it violates the general principle of non-indemnity for imposition of planning *servitudes*. However it is of great interest particularly in rural areas where it has been used to allocate farmland (*remembrement*) more rationally and to share out development rights, to provide a restrictive but fair allocation of development land. It has been similarly used to protect areas of quality landscapes, and may yet be used to protect more ordinary landscapes in rural communes at large.

*Establishment of the POS*
F5.37   Since 1983 the establishment of the POS has been at the initiative of the commune or an EPCI for a group of communes. However although these decentralisation reforms are significant their effect must not be overestimated because the state continues to be associated with the project, ultimately controls its legality through the law, and more often than not provides the technical assistance through the DDE. It is also possible that private consultants can be used (estimates suggest about ten per cent of occasions, although legal redrafting by the DDE is often necessary) or an *agence d'urbanisme*.

F5.38   If the role of the actors has in principle changed, no fundamental modification has been made to the structure of the procedure. This falls into three parts, that of elaboration/publication, that of approval, and that of revision (major change) or modification (minor change) (Figure F16). These procedures are irreversible since once prepared a commune cannot abrogate the POS, and even if its approval or modification is annulled the published plan still applies with significant legal force.

F5.39   The initiation of the scheme is prescribed by the mayor or the President of the EPCI. The funds for the preparation of the plan have to be found by the commune who often have very limited financial means, but more often from the *dotation générale de décentralisation*, a

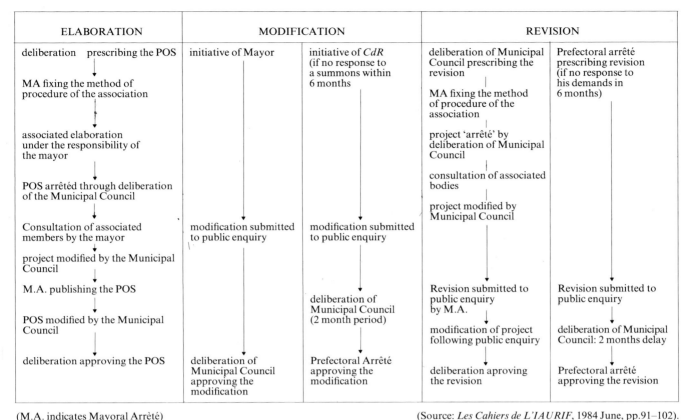

| ELABORATION | MODIFICATION | | REVISION | |
|---|---|---|---|---|
| deliberation prescribing the POS ↓ | initiative of Mayor | initiative of *CdR* (if no response to a summons within 6 months | deliberation of Municipal Council prescribing the revision ↓ | Prefectoral arrêté prescribing revision (if no response to his demands in 6 months) |
| MA fixing the method of procedure of the association ↓ | | | MA fixing the method of procedure of the association ↓ | |
| associated elaboration under the responsibility of the mayor ↓ | | | project 'arrêté' by deliberation of Municipal Council ↓ | |
| POS arrêtéd through deliberation of the Municipal Council ↓ | | | consultation of associated bodies ↓ | |
| Consultation of associated members by the mayor ↓ | modification submitted to public enquiry | modification submitted to public enquiry | project modified by Municipal Council | |
| project modified by the Municipal Council ↓ | | | | |
| M.A. publishing the POS ↓ | | deliberation of Municipal Council (2 month period) | Revision submitted to public enquiry by M.A. ↓ | Revision submitted to public enquiry |
| POS modified by the Municipal Council ↓ | | | modification of project following public enquiry ↓ | deliberation of Municipal Council: 2 months delay |
| deliberation approving the POS | deliberation of Municipal Council approving the modification | Prefectoral Arrêté approving the modification | deliberation aproving the revision | Prefectoral arrêté approving the revision |

(M.A. indicates Mayoral Arrêté)

(Source: *Les Cahiers de L'IAURIF*, 1984 June, pp.91–102).

**Figure F16    The process of POS elaboration, modification and revision**

A comparison of the three processes of initial elaboration, and subsequent modification or revision of the POS illustrate the short cuts provided by the modification process. The chart also illustrates the possible interventions of the *Commissaire de la République* to update the plan.

central fund under control of the *Commissaire de la République* who decides allocations on a priority basis. The scheme must be publicised in two local newspapers and in the relevant town halls, and the state, *Commissaire de la République*, region, Department, chambers of commerce, trades, agriculture, industry and environmental groups etc can all be associated if they desire. A good deal of scope is left to local discretion, but in practice the plan is most often developed by a working group (*groupe de travail*) which brings together the various representatives of local groups and levels of government with the technicians of the DDE, under the leadership of the mayor or the president of the EPCI. This is legally constituted by an *arrêté* which must be the subject of publicity.

F5.40   The mayor can hear the views of all represented and any qualified persons, while the *Commissaire de la République* is responsible for transmitting all the information on supra-communal constraints such as the national prescriptions for planning and a project of general interest.

F5.41   Once the plan is completed it is declared by a municipal *arrêté* and made the subject of a number of consultations that must include the associated public persons, over a three month period. Conflicting advice has to be resolved by a *commission de conciliation*. Other persons consulted but without resort to the commission are the various presidents of the agreed associations and the mayors of neighbouring communes. Once completed, the plan is decreed by the President of the EPCI, approved by the municipal council(s) concerned, and then rendered public. The plan is then placed in the town hall for two months, and advertised in local newspapers, after which it becomes executory providing there are no challenges to its legality. In this respect the scrutiny of the *Commissaire de la République* is critical.

*Public Enquiry (* Enquête Publique*) on the POS*
F5.42   Once published, the plan must be submitted to a public enquiry to allow all property owners to discuss the restrictions placed upon them. The July 1983 law significantly decentralised and democratised public enquiries, and the appointment of the commissioner of the enquiry is now controlled by the president of the *Tribunal Administratif*. The enquiry is prescribed by a mayoral *arrêté* which states its object, opening, location and duration (at least one month), hours of opening, hours of public access to the complete file and register of comments, and hours when the commissioner will receive oral comments from the public. Since the 1983 reforms the commissioner must play a more positive role in helping the public to be well informed and to articulate

their comments, and he can visit places, hold public meetings and listen to any person he judges useful to this end.

F5.43  All observations are annexed to the register and the commissioner provides a resumé of all comments within his report which is transmitted to the mayor (and to the president of the *Tribunal Administratif* and the *Commissaire de la République*) within one month of the end of the enquiry. The authorities elaborating the POS are not bound by the commissioner's conclusions but the courts are likely to enforce them if appeals result from the failure to take his/her advice. This represents another strengthening of the rights of citizens to be heard.

F5.44  If major modifications to the plan ensue it may be necessary to consult the various persons associated with the project again, and if major changes result a new public enquiry may be necessary. Assuming neither of these, the plan is approved by a deliberation of the municipal council or the president of the EPCI. It is subject to the same publicity as before and becomes executory after two months. After six months have elapsed the mayor of the commune is able to take on responsibility for granting the *permis de construire* controlled by the plan.

*Process of adaptation of an approved POS*
F5.45  The law of December 1976 reformed the procedures for adaptation of the POS by substituting for the single process of modification two distinct procedures of revision and modification. The reforms of 1983 have retained these two procedures but decentralised them, while allowing the possibility of the *Commissaire de la République* initiating the procedure in the event of negligence by the local authorities, or a failure to acknowledge supra-communal constraints. In addition there is a procedure for updating the plan, which allows the alteration of public utility *servitudes* created outside the planning framework.

F5.46  Revision is the most formal and thorough-going procedure and follows the procedures of approval except that the phase of publication has been omitted. It is a slow process but it is not necessary rapidly to develop a document that can be enforced on third parties since the old plans perform that function. Modification is a more simple procedure for more partial and limited changes that do not affect the 'general economy of the plan' including changes of zoning and regulations, or changing the nature of protection of rural areas, or the classification of wooded spaces. These procedures are outlined in Figure F16.

F5.47  The reality of the situation is that in most growing urban areas the POS is the process of perpetual modification or revision, particularly in an era of economic recession when more flexible responses to development initiatives are necessary. Between October 1983 and July 1985 some 2,300 modifications were approved, a number almost comparable with the number of new prescriptions. There were many fewer revisions,

some 375 in the same period, and their number showed signs of decrease, but the continual process of modification and revision remains a key aspect of keeping the POS relevant and flexible.

*The effects of the POS*
*(a) with prescription of a POS*
F5.48  The POS is effectively a regulatory administrative act, but its effects vary according to the stage of its elaboration and it has judicial effects long before it is definitively approved. For example once the project of the POS is prescribed, the administration is able to take certain conservation measures (particularly with regard to woodland) and to safeguard future dispositions of the plan by using the *sursis à statuer* provisions (see paragraph F4.31), though this can now no longer be used with plans being modified. But the prescription of a POS allows the commune to escape from the restrictions of *constructibilité limitée* until November 1987 if the MARNU are in operation.

F5.49  The *sursis à statuer* allows suspension of authorisations of a *permis de construire* during the period between prescription and publication of a plan, and during the period of revision between prescription and approval. It can be applied to any request for an authorisation likely to "compromise or render more onerous the execution of the future plan" (L132-5), and it can even use studies for the plan as a basis for suspension if the plan is insufficiently specified. The suspension can last for only two years, and a further suspension can only be for a different reason and only for an additional year.

F5.50  On the other hand it is again now possible, since the modifications of December 1986, to obtain a *dérogation anticipation* allowing an application to be authorised if it conforms to the revisions in a plan not yet approved, even though it does not accord with the old plan which is legally still in control. There are even limited possibilities for such anticipations where the plan is being elaborated for the first time.

F5.51  Minor adaptations to the plan have long been the source of abuse and judicial reaction, and major reforms were initiated in 1976 allowing only minor adaptations rendered necessary by the nature of the land, the configuration of the parcel, or the character of neighbouring buildings (L123-1). There is some possibility of obtaining derogations by using alternative rules through the differential application of the regulations of the zone, but this is also prone to abuse. More important are the possibilities of using a *déclaration d'utilité publique* to vary the content and application of a plan if the proposals fit this category of development. But such procedures have to be submitted to a public enquiry and consultation with all those bodies associated with the elaboration of a POS.

*(b) with publication of a POS*
F5.52  A published POS is enforceable for three years and its enforcement is very general embracing public

bodies and private persons, and ensuring a general compatability of operations. But conformity is required where public utility *servitudes* are concerned (the principle of non-indemnification of such *servitudes* has operated since June 1943 and admits only two minor exceptions - where it harms a previously acquired right, or directly harms a place).

F5.53 In addition to these general effects of the publication of a POS, there is a range of particular effects relevant to certain specified areas. One of these allows the owner of a site especially reserved for a particular use (*emplacement reservé*) to force acquisition of his/her property within one year, resorting to the judge of expropriation if this is not achieved. Further protection is afforded to classified wooded spaces (*espaces boisés classés*) forbidding the changing of the attributes of the land and any use likely to compromise its character. It also offers the possibility of the commune trading building land for ownership of the woodland, or for authorising construction on one tenth of the woodland if the rest is conveyed free to the commune. This can only be done with woodland which has been owned by the vendor for five years, and if the land offered in compensation is equal to, or less than, the land ceded to the collective (these restrictions controlling speculative acquisitions and abuses of the system), and if it is compatible with the SD. Even then it requires a decree from the *Commissaire de la République* who seeks the advice of three ministries.

F5.54 Finally the publication of a plan gives the local collective a right to preempt sales (*un droit de préemption urbain*) which can be exercised in ZU and NA areas. This procedure replaces the old *Zone d'Intervention Foncière* (ZIF) procedure by which all land sales were notified to the local collective and they had the right to preempt them. This preemption was rarely used in most communes and as part of deregulation it is now necessary for communes to delimit such zones of preemption if they wish to retain this right. As important as the ability to preempt is the information which knowledge of land transactions can bring to the planning function, and it could usefully be retained for this reason alone.

### (c) with approval of a POS
F5.55 If the plan is not approved within three years of its prescription its provisions are suspended, but if it is approved then the effects engendered by publication are reinforced. Additional powers include the sweeping away of the necessity to check the classification and declassification of roads with their consequent restrictions. Most importantly it is the approval of the plan and not its publication which gives the commune(s) the power to deliver the *permis de construire*, *lotissement* and other land use authorisations. This transfer is definitive and irreversible and brings new competences in the matter of operational urban planning, allowing the creation of *zones d'aménagement concertés* (ZACs), restoration areas, and associations of urban land-owners for public-private enterprises of implementation.

### Appeals against the POS
F5.56 Finally the POS can be the subject of an appeal to the administrative courts but the *Conseil d'État* ensures only that there are no manifest errors of assessment in the provisions of the POS, exercising a minimal control but being empowered to annul parts of the document. As an example, an important recent case before the *tribunal administratif* in Poitiers annulled a POS because it provided an inadequate analysis of the initial state of the environment.

### The coverage of the POS
F5.57 In considering the extent of coverage of the POS in France as a whole (Figure F17) it can be seen that as of 1 October 1984 over much of rural France this was less than 15 per cent. In the rapid urban growth departments like Ile de France, Bouches du Rhône and Loire-Atlantique coverage rose to between 60 per cent and 90 per cent indicating a necessary plan-making response to development pressure. However, much of rural France is under minimal development pressure and could be considered to hardly to merit the preparation of even a simplified POS, so that the apparently poor coverage creates a rather misleading picture.

F5.58 More revealing are the statistics on coverage by size of population (Table F5.2) and here the October 1984 statistics reveal that three quarters of urban agglomerations of over 10,000 persons had approved POS, while only a small percentage, decreasing from 14 per cent to 6 per cent as the agglomerations increase in size, are not in the process of preparing them. So these figures portray a much more satisfactory state of affairs.

F5.59 The steady increase in POS production through the 1970's showed signs of levelling off in the early 1980's in terms of the number of POS prescribed. However, production was given a very significant boost by the decentralisation reforms of July 1983. The immediate response was a sharp upturn in the number of POS prescribed (Figure F18) and these trends have continued. As of January 1 1986 98 per cent of urban communes, and 85 per cent of those with a population between 1,000 and 2,000 were engaged in producing a POS. Since 1984 a large number of small communes have also been engaged in the process, 71 per cent of the communes between 700 and 1,000 population, and even 27 per cent of those with populations of less than 700 persons.

F5.60 The actual rate of POS prescription in the last three years is five times that of the previous three. More significantly a much higher proportion of the POS is being fully approved, allowing the communes to exercise the development control functions if they so desire. As of January 1986 80 per cent of the communes of over 2,000 persons had an approved POS as had 52 per cent of those with populations between 1,000 and 2,000 and 36 per cent of those between 100 and 1,000.

F5.61 While these figures are useful measures of a successful decentralisation there is some disappointment

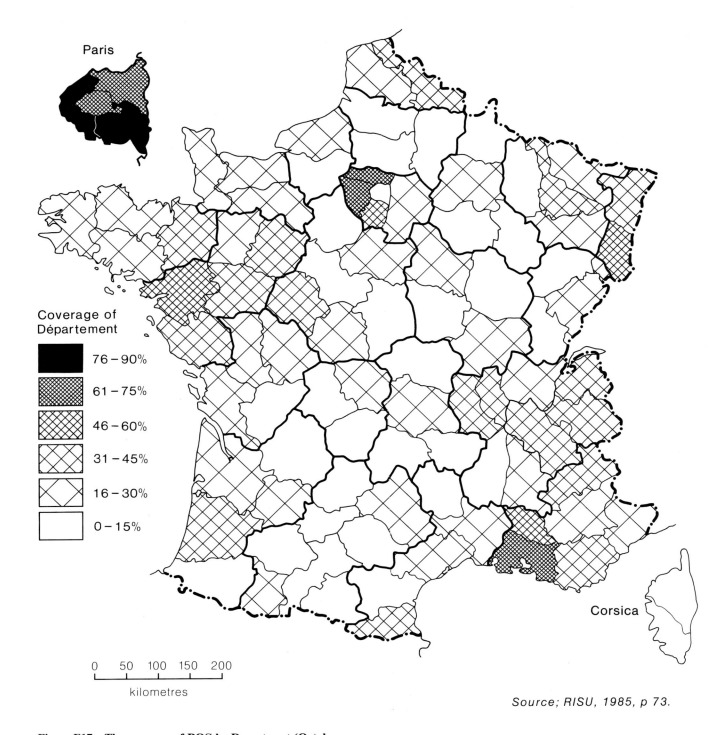

Coverage of
Département

| | |
|---|---|
| ■ | 76 – 90% |
| ▨ | 61 – 75% |
| ▨ | 46 – 60% |
| ▨ | 31 – 45% |
| ▨ | 16 – 30% |
| □ | 0 – 15% |

Paris

Corsica

0   50   100   150   200

kilometres

*Source; RISU, 1985, p 73.*

**Figure F17    The coverage of POS by Department (October 1984)**

This map reveals the differential coverage of the POS throughout France, though it does not record the major increase in coverage that has taken place in the last three years. In much of rural France less than 15 per cent of the territory is covered by a POS although in the west of the Ile de France or in Bouches du Rhône coverage exceeds 60 per cent. In both the Paris region and Provence/Côte d'Azur/Languedoc/Roussilion there are notable variations in approval rates. Some other regional patterns can be discerned with higher rates of approval in the Alps, Lyon and Lower Loire valley areas.

**Table F5.2   The progress with POS production by size of commune 1984**

| SIZE OF POPULATION | NUMBER OF COMMUNES | POS rendered public | | POS approved | |
|---|---|---|---|---|---|
| | | No | % | No | % |
| More than 100,000 habitants | | | | | |
| 50,000–100,000 | 36 | 34 | 94 | 26 | 72 |
| 40,000–  50,000 | 68 | 64 | 94 | 50 | 74 |
| 30,000–  40,000 | 69 | 62 | 90 | 52 | 75 |
| 20,000–  30,000 | 162 | 142 | 88 | 109 | 67 |
| 10,000–  20,000 | 412 | 356 | 86 | 302 | 73 |

185 approved POS (out of 582) have been put into revision.
*Source:* RISU 1985 p.77

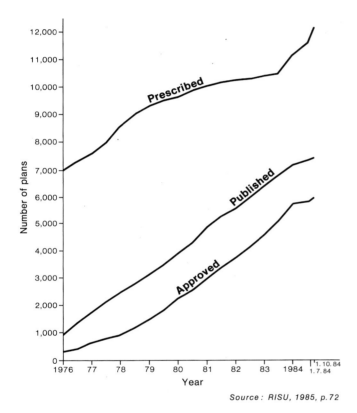

Source : RISU, 1985, p.72

**Figure F18   The production of POS 1975–1984**

Although this data predates the major increase in production of the POS, the effects of decentralisation can be detected in a marked increase in the number of POS prescribed from the second half of 1983 onwards. The number of published POS equals the number prescribed by 1977, while the number approved by October 1984 was the equivalent of those published by the beginning of 1982. This gives some indication of the time periods involved in the full approval process before decentralisation, but it must be remembered that with the POS having legal effect from the moment of prescription many will not proceed to full approval. However, the reforms of 1983 now provide significant incentives for much more rapid production and for complete approval, and there is evidence of both in the fact that by the beginning of 1986 there were some 12,000 approved POS.

at the plan-making reluctance of the smaller rural communes, and a recognition that inadequate financial means and inadequate technical resources have slowed down the process. But in spite of the dramatic increase in the number of POS nearly three quarters of the communes are still managed under the rule of *constructibilité limitée* with the *Commissaire de la République* delivering the various authorisations. The approved POS do not provide comprehensive coverage, but rather reveal a patchwork pattern even in urban fringe locations (Plate 14). However, decentralisation has not resulted in an explosion of *permis de construire* authorisations, and there has been a large number of refusals of *certificats d'urbanisme* although each of the regions has fared rather differently. Some flexibility in the rule of *constructibilité limitée* is available if the interests of the commune are genuinely at stake, or if the general principles of urban planning are not threatened. The strict interpretation of these two clauses (L111-1-2, 111-1-3) has prevented an explosion of one-off decisions (Bouzely, 1986).

*Criticisms of the POS*
F5.62   The most common criticism of the POS procedure was always the long period of gestation before a plan was approved. Wilson's 1981 survey of over 100 POS approvals in three departments revealed an average of five and a half to six and a half years from initiation to approval (Wilson, 1981, pp. 155–173). The fact that there was no prescribed timetable, and the obsession with a strict adherence to administrative procedures were two contributory factors. However, the shortage of DDE staff, particularly in the GEP sections responsible for basic survey and information, have always been a major constraint. In 1978 only 751 staff (of whom only 340 were contracted staff) were employed in the Departmental sections of the GEP, and the production of 1,424 POS in 1978 looked impressive set against these scant resources. By 1983 these numbers had trebled but so had the number of POS prescribed. However, there is every indication that the three year time limit between prescription and approval has obviated this problem, along with the obvious incentives of avoiding the regime of *constructibilité limitée* and assuming development control functions.

F5.63   In many ways the POS preparation procedures epitomise the local political process and central-local government relationships in France. These have been

219

characterised by Wright as "a chaos of surreptitious bargaining, illicit agreements, hidden collusion, unspoken complicity, simulated tension and often genuine conflict" (Wright, quoted in Hanley et al, 1985, p.150). Certainly the crucial importance of the senior officials of the DDE, who tend to dominate discussions, and their intricate relationships with the local mayor (who may himself be an important political figure at a higher level), are evident in POS procedures, as are the emphases upon administrative procedures rather than the substantive content of the plan. Informal meetings have an important influence on plan content and local property interests are usually well catered for in the personages of the mayor and the local *notables*. Land zonings will usually accommodate their development aspirations.

F5.64  Decentralisation has obviated another major criticism of the POS, that it allowed the mayor to hide behind the plan, blaming the DDE officials for unpopular development control decisions. The mayor's roles and responsibilities are now clear. The DDE officials still have a vested interest in proposing and implementing public works in the communes because they are paid honoraria (which can often contribute 50 per cent of their salary) for supervising the same. Thus there is a mutual interest and advantage in promoting the finance for public facilities which will ensure good commissions and electoral popularity.

F5.65  Despite the changes wrought by decentralisation, to British eyes there remains no clear division between the technical and political aspects of plan-making, or between the administrative, professional and political roles of the participants. Those with the strongest power base are clearly able to influence the outcome in their favour. There is no genuine public debate and real power is concentrated in the hands of the mayor, particularly if he holds other political offices through the '*cumul des mandats*' and is able to master the complexities of planning and exploit the internal divisions within the DDE. Booth (1987a, p.20) identifies two particular powers of the mayor with regard to the POS, the first being the new power to actually initiate or modify the POS, the second being the power to negotiate the prescription and development of NA zones, the zones of future urbanisation. However, ultimately the power of the mayor remains very dependent upon the finance and administrative control of the centre, especially the legal control of the *Commissaire de la République*, and in this way the old political system continues to exert its influence on the new.

F5.66  Obviously the POS provides a very detailed and clear basis for the practice of development control biased towards considerations of traditional form, prescribing a wide range of rules about land use, building massing, aspect, size etc, spacing and landscaping, and density and allowable floorspace that, if followed, guarantee the authorisation of a *permis de construire*. In many senses the POS is not a plan but a legal document prescribing development rights, an equivalent of the Highway Code

as one experienced planner described it. Its existence does not remove all discretion and negotiation from the control process but clearly it dramatically reduces it. Whether these benefits are eroded by the inflexibility and rigidity of the POS prescriptions is a matter for debate, but certainly the POS of Lyon and Paris have shown how it is possible to develop plans that are much more responsive to the particularities of townscape, the needs of conservation, and the desirability of retaining employment, vitality and housing in urban areas. The use of modification procedures, and the option of minor derogations, also provide flexibility without in any way returning to the abuses of the derogation procedure evident in the early 1970's.

F5.67  Essentially the POS is a negative document, and it may easily prevent appropriate development taking place because of its emphasis upon uniformity and conformity. It makes site-by-site prescription difficult and in today's development climate many developers feel more responsive rules and opportunities should be created. Some POS have already adopted a much finer grained or more flexible approach and embraced economic regeneration and employment retention goals, but density and design responsiveness are also sought by developers. There is a widespread view that the heavy 'police' measures of the POS, and of control regulations at large, were important in periods of rapid large scale development, but that more sophisticated vehicles are necessary today. Perhaps an even more pressing problem with the advent of decentralisation is how to prevent the POS from becoming an instrument for 'snob zoning', a charter for no growth. That said, in general the role of the POS is well established and well respected, and very considerable effort is being directed at both increasing its coverage in the small communes, and refining its content to make it a more sophisticated tool in the larger communes.

## Other Forms of Plan Relevant to Development Control

F5.68  Before concluding the discussion of the role of development plans in the French planning system, it is important to mention the role of urban conservation plans and the changing nature of plans for rural areas.

*Urban conservation plans*
F5.69  There are two important plans that relate to urban conservation. The *Plan de sauvegarde et de mise en valeur* (PSMV), of which some 81 are in preparation in France (although only 15 had been approved by October 1984, see Figure F11), is similar to a POS in its elaboration and approval process. The decision to create a PSMV is a ministerial one advised by the *commission nationale des secteurs sauvegardés* (CNSS). The local communes are consulted though their ability to oppose it is limited. The effects of the creation of a *secteur sauvegardé* are to set in motion the studies for the plan, to suspend the issuing of any *permis de construire* until the

plan is published, and to bring all authorisations under the control of *l'Architecte Bâtiments de France* (ABF). The latter ensures that all proposals, including repairs and demolition, respect the aesthetic character of the area and the historical continuity of the buildings. The plan is elaborated by an appointed architect liaising with the ABF, the local CNSS, DDE and local elected representatives, and is rendered public and subjected to an *enquête publique* in much the same way as the POS. Its content is similar to a POS, but it is more precise and detailed, frequently defining site specific proposals, and indicating where demolition, modification or alteration of buildings will or will not be allowed. These controls can extend to the paving of streets or the imposition of archaeological conditions. Since 1976, publication of the plan has brought its provisions into effect, and from then on all its provisions are enforced by the supervision of the ABF, although minor derogations may be allowed.

F5.70 Since 1977 these plans have been backed by central government funds from the wider *opérations programmées d'amelioration de l'habitat* (OPAH) which have given them a stronger emphasis upon improving the living conditions of the existing population, and created a tripartite structure of participation with the commune, the state and *l'agence nationale pour l'amélioration d'habitat* (ANAH), the national agency for housing improvement.

F5.71 In addition to the PSMV there is now the possibility of creating conservation-oriented plans under the *zones de protection du patrimoine architectural et urbain* (ZPPAU) which can embrace non-built up areas. These essentially replace the old regime for protecting the surroundings of historic buildings and listed sites. These plans have a similar structure to the POS, and a similar preparation process to the PSMV, except that they cannot be approved without the consent of the local council. The effect of the plan is to sweep away public utility *servitudes* but not those of urban planning, and to subject all applications to approval of the ABF. The progress of these plans is not yet documented.

*Rural conservation plans*
F5.72 With regard to the conservation of natural spaces, the pre-existing *zones d'environnement protegée* (ZEP) (of which there were some 857 covering some 37,000 sq.km in 1984), which were of the nature of simplified POS, ceased to exist as of October 1986 and must be subsumed with the POS themselves. The remainder are brought within the control of *périmètres sensibles* (see paragraphs F4.114-118). The buffer zones of national parks are also required to be covered by a POS to ensure their sensitive development and conservation.

F5.73 The French coastline has been subject to a special protection since 1971, being formally declared an area of national interest in 1977. Legislation remains complex and there is a variety of types of planning document that can form a framework for control. These include regional strategies, *schémas d'aménagement* for industrial or

tourist development which date back to 1963 in Languedoc-Roussillon; and *Schémas regionaux du Littoral*, which have no legal status but are drawn up by the three higher levels of government to provide guidelines for coastal conservation in the drafting of the SD or POS. Two other plans of relevance include the 1975 *Schéma Directeur National* which is a national master plan protecting oyster cultivation and aquaculture, and some eight *Schémas de mise en valeur de la Mer (SMVMs)*, elaborated jointly by the region and the state. The latter are in various stages of development. They attempt to afford special protection to the development of key coastal zones and management of the coastal public domain, allocating land for industrial and port development while attempting to reduce the conflicts of multiple use. They carry a stronger element of national planning prescriptions, and any SD or POS must be compatible with their provisions.

Charte intercommunale
F5.74 Finally, in a rural context it is important to mention the intercommunal charters (*chartes intercommunales*) which have been reformed and tied to the prescription of a POS. A life of only two years given in 1983 has been extended to four years in the reforms of August 1986, and this may well encourage their more frequent use. As of January 1986 some 1,600 were being developed. These documents, introduced in January 1983, fuse planning and economic development and were introduced to "define the medium term perspectives of economic, social and cultural development.. [and].. corresponding programmes of action stating their conditions of organisation and the functioning of infrastructure and public services" (Jacquot, 1987, p.239). They replaced the old *plans d'aménagement rural* (PAR), but they widened the latter's sphere of operation to include urban areas. They are not regulatory documents, but they do possess some juridical effect as reference documents for planning or economic decision making, and for conventions between the various levels of government, to fund different programmes and infrastructure development. The existence of a *charte intercommunale* must be considered in defining the area to be covered by an SD, in any reorganisation of rural land holdings, in the creation and development of regional natural parks, and in the creation of sensitive natural spaces.

## Conclusions

F5.75 If the context of national and regional plans is ignored, and they do not exert much influence upon the content of the lower tier plans, the French system of plans can be simply described as a two tier division of strategic and local plans, with the latter having to be compatible with the former. The strategic plans are enforceable only on the public authorities of all kinds, while the local plans are enforceable on third parties, and take the form of precise documents prescribing all development rights. The 1983 decentralisation placed all the emphasis upon the approval of the local plan, the POS, as a prerequisite for local authorities to avoid a very restrictive régime of preventing development outside rural areas (*constructibilité limitée*), and a prerequisite for assuming development control powers. This has resulted in a very marked increase in the preparation and approval of the POS, so that they now offer a nearly complete coverage of urban and urbanising areas, even if three quarters of all communes as yet remain under the *constructibilité limitée* régime. With the emphasis upon the POS the strategic plans have received very little attention, and their delimitation and approval has completely stagnated. If anything they have become constraints on the production of the POS because their prescriptions, largely based upon an early 1970's view of the economy and demography, are largely inappropriate today. The future of the SD remains in doubt, while that of the POS has become more than ever the keystone of French planning practice.

F5.76    From the perspective of the comparative analysis of development control systems the key points to emerge are as follows.

*The* Schéma Directeur (SD)

(i)    There is considerable utility in having a long term strategic plan, the SD, with 10 and 30 year prescriptions, to prescribe the medium and long term future of urban regions, but these have been rendered largely meaningless by a failure to update and modify them.

(ii)    The enforceability of the SD on the public sector gives an important indication of necessary infrastructure provision, even if commitment to develop it may be postponed indefinitely.

(iii)    The completion of the majority of SD's in the mid 1970's renders them significant obstacles to the development of meaningful POS, because the scale of development has so dramatically changed. They can directly impair the prospect of appropriate development.

(iv)    The problems with approval and modification of the SD lie in a tedious approval process and, more particularly since 1983, in the ability of the communes to withdraw from participation if they do not think the prescriptions are in their interest.

(v)    Far from acting as a framework for the POS, ensuring their conformity, any new SD is likely to be developed as an aggregation and rationalisation of existing POS, a bottom up rather than a top down procedure.

(vi)    There is negligible public participation in the SD preparation and no public enquiry but chambers of commerce, trades, agriculture etc do have an important role to play.

*The POS*

(vii)    The POS is less a forward looking plan than an assemblage of development rights, and it must include all public *servitudes* which affect property rights. As such it provides a very clear framework for development control and conformity to the plan's regulations will ensure the authorisation of a permit to build. But it is also a primarily negative, policing document.

(viii)    In their most complete forms the POS are extremely sophisticated documents that can respond well to the intricacies of differing urban landscapes and creatively to subtly different functional areas within a town. In their simplified forms they are very basic zoning maps which principally define the areas for immediate and future urbanisation. Each plan has a standardised format and content, but within that framework any number of zones can be defined and the rules of use and construction modified accordingly.

(ix)    The POS greatly reduces the role of discretion and negotiation in the planning process, but it must be appreciated that the regulations that they contain require considerable interpretation, that minor modifications are possible, and that on major developments where servicing is required, the POS does not prescribe the rules of financing such infrastructure.

(x)    The flexibility of the POS remains an important question and the large number of modifications emphasises the necessity of continually updating them. In this sense the POS procedure is one of continuous plan making. Some flexibility is provided by minor derogations and the abuses of the derogation procedure so widely exploited by developers and mayors in the early 1970's have been ended.

(xi)    The POS has a complex process of prescription, elaboration and approval which is highly prescribed and legally controlled at all stages. There have been criticisms that the procedure assumes more importance than the content of the plan. But it is important to recognise that each stage of the process provides increasing planning powers.

(xii)    The local plan or POS, once approved, gives the commune the powers to control development of all kinds and lifts the restrictions against development outside built-up areas (*constructibilité limitée*) which are otherwise universal.

(xiii)    Although the procedure is decentralised in most instances the state retains important technical control through the DDE and important legal control through the *Commissaire de la République* and the courts.

(xiv)    The actual process of elaborating the plan has been characterised as a process of negotiation between major land owners and the mayor, so property owners' development interests are effectively built into the plan. The technical and political roles of plan-making are in no way separated, and while there are controls on the legality of procedure, and an adherence to national planning directives, the main protection for the public lies through the courts.

(xv) Since 1983 and 1985 a stronger measure of public participation has been introduced into the public enquiry procedures, but the process remains one of consultation rather than participation. As with the SD, the chambers of commerce, agriculture, trade etc are given a direct role in the working groups, and professionally qualified groups or persons are given a stronger voice along with environmental groups.

(xvi) Since 1985 there is an obligation for preliminary discussion with the public for any important planning project, a recognition of the inadequacies of involving the public only at a late stage in project formulation.

(xvii) The complaints about a long and tortuous approval process, so common before 1983, appear to have been overcome by the new incentives in the system.

(xviii) The coverage of POS has shown a very dramatic improvement as a result of the 1983 decentralisation and now embraces most significant urban settlements, although it will be some years before all those smaller communes who desire to prepare such a plan will be able to do so.

## Introduction

F6.1    The French system of administrative law was introduced in section F1 and its basic significance to the planning system explained there. The *tribunaux administratifs* provide a first recourse for any aggrieved applicant for planning permission, for local authorities, for recognised environmental groups, and for affected third parties like neighbours. These courts, along with the higher court of the *Conseil d'État* to which parties may subsequently appeal, provide a system of legal safeguards and judicial control that is accessible to all. They provide a system of appeals for aggrieved applicants, procedures to contest the granting of various authorisations by third parties, as well as controls on the legality of decisions, the legality of the buildings and land uses themselves, and a system of enforcement. Thus the courts provide a complete system of legal control over development control decisions accessible to all actors in the control process.

F6.2    Legal control is necessary to finalise administrative control and to provide checks and balances on its application. But control is ensured through three systems of courts: firstly by the administrative jurisdictions which rule on the various authorisations themselves; secondly by the penal judge, or the repressive courts or bodies who are competent to deliver sanctions against those who break the rules of urban planning; thirdly by the civil courts who can also be brought in to force offenders to repair the damages caused to third parties.

## The Administrative Judgements

F6.3    The administrative courts have as their object the control of the actions of the public authority be it state, commune or *établissement publique de coopération intercommunale* (EPCI). They ensure that the the rules of urban planning are respected. They are there to restrain authorities from exceeding their power, and also to ensure the validity of their interpretation of the law. They are also able to indemnify individuals whose rights have been infringed by administrative acts.

F6.4    The control of the legality of decisions is the fundamental task of the courts and is an essential role performed by the *Commissaire de la République*. All decisions are systematically transmitted to him, and he can seek the opinion of the legal control sections of the *Directions Départementales de l'Équipement* (DDE) (the

establishment of which was encouraged by a circular of August 1984). But those administered also have recourse to the administrative court, in particular the aggrieved applicant for any authorisation. These grievances can embrace refusals and withdrawals of authorisations, but not the actual nature of a *permis de construire*, except its financial provisions (the latter is important for reasons outlined in paragraphs 4.129-138). In addition third parties have rights of appeal through the courts against the authority issuing the authorisation, but they must have a direct interest in the annulment of the authorisation, for example be neighbours, not necessarily narrowly defined, or environmental groups. Finally the commune can appeal 'in the public interest' against decisions delivered by the *Commissaire de la République*.

F6.5    Various time periods are prescribed for launching an appeal, generally within two months of the expiry of the period for notification of the authorisation, but third parties are allowed a further four months to attack the decision on a *permis de construire*.

*The extent and reach of control*
F6.6    The administrative judges ensure that the decisions taken by the competent authority conform to the rules of urban planning. They will rule on questions of fact, and on the imperative rules of planning. But they also rule on the discretionary decisions of the administrative authority, although here they will only exercise minimum control, covering only cases of manifest error. The judges also assess the derogations of the rules allowed by local authorities ensuring that no harm has been done to the general interest.

F6.7    Although decisions may be annulled various works originally authorised may in fact be in the course of execution, or may even be completed. Various alternatives are open to the courts to deal with such eventualities. It is relatively simple to regularise decisions which have been annulled only because they failed to follow the proper procedures (*légalité externe* or external legality). But if the annulment is because the rules of urban planning have been violated (*légalité interne* or internal legality), the possibilities of regularisation are much more limited. Derogations cannot be issued *a posteriori*, and the only possibility of regularisation lies in changing the regulations. This is fraught with difficulties, and it is generally necessary to change the development itself to accord with the law.

F6.8    In order to ensure that works are put into

conformity with the authorisation, recourse is necessary to the penal courts or *le juge répressif* who is competent to order such works (or even demolitions) if an administrative judge has condemned an infraction of the rules of urban planning. Neighbours and environmental groups can also take an action through the civil courts to obtain reparations, perhaps even to ensure that the property concerned is returned to its original state. However, they themselves cannot be indemnified for the inconveniences suffered. If the works have not been completed an interruption of works can be issued by the mayor or *Commissaire de la République*.

F6.9　Applicants for various permissions can also take action against the public authority who has issued the authorisation, because of illegal acts, because of delays in taking decisions, or because of errors in assessing the rules of urban planning. But while the administrative judges are quite liberal in ruling on the mistakes of the local authorities, they are much more parsimonious in the amount of indemnity that they will allow. Only direct and certain prejudice is allowable, and key court cases, while accepting that costs have been incurred by developers or builders, tend to minimise the amount of indemnity payable. As a response to decentralisation a circular in October 1984 required the establishment of insurance schemes to cover local authorities for these risks, and to allow judges to levy appropriate amounts of indemnity without causing bankruptcy to any commune.

F6.10　Third parties also have recourse to the courts to obtain indemnity against the local authorities or the state (since 1974) for issuing an illegal certificate that causes them prejudice. Since the reform of 1976 this has lost much of its relevance since the local authority is now under an obligation to inform the public prosecutor's office of any infractions of which they have knowledge.

*Immediate measures of legal control - the suspension of execution*
F6.11　One of the major constraints on the efficiency of the administrative courts is that a court decision does not prevent the beneficiary of an authorisation from initiating, continuing or completing the authorised works, often doing irreparable damage to a site and presenting the administration with a *fait accompli*. To obviate this problem the administrative judge has the possibility of using a *sursis à execution* or suspension of execution. This general instrument was not fully available for use against a *permis de construire* until the 1976 reforms, the provisions being further adjusted in 1983 with the advent of decentralisation.

F6.12　It is only possible to seek a *sursis à execution* against an act which has a possibility of being executed, and thus it cannot be used to oppose a *certificat d'urbanisme* (outline permission). It can be applied for by anyone entitled to seek the annulling of the administrative act, and the actions can be executed singly or together. A *sursis à execution* will be granted only if the decision attacked will produce an 'irreparable or difficult to repair

prejudice', or if there is a serious breach of the law likely to lead to an annulment of the authorisation itself. These two conditions are necessary but not sufficient for the suspension to be ordered, and the judge has considerable discretion to assess the nature of the interests involved, though he will usually proceed to order a suspension if the defects of the authorisation cannot easily be amended.

F6.13　Even with a *sursis à execution* a six month period could easily elapse before the judge reaches a decision, much more if an appeal against it is lodged, and this can defeat the whole object of the exercise. Reforms in January 1983 required that henceforth the *Tribunal Administratif* must decree a suspension within one month on any *permis de construire* brought before it (by a person other than a commune, EPCI, or the state). The reforms subjected other authorisations of land use issued under an approved *Plan d'Occupation des Sols* (POS) to the same system. As Jacquot has noted, this is not entirely logical since it exempts such decisions taken by the state in areas without a POS (Jacquot, 1987, pp.439-440).

F6.14　The effect of a *sursis à execution* is to suspend the original authorisation, and it is notified to the relevant authorities, and all those involved in the works and their execution, that any further work would constitute an infraction. In the event of non-compliance a common law action can be taken by the judge or the administrative authority, and the minister for the superior court is informed of all such suspensions to facilitate this. This will then permit the execution of works at the expense of the developer/builder to ensure compliance.

F6.15　Two new procedures were introduced in 1983. These are the 'facilitated' suspension which does not require that any works be damaging or doing irreparable harm, only that a serious breach of the law has taken place; and the 'accelerated' suspension, which is ordered within 48 hours, for illegal acts which compromise the exercise of a public or individual liberty. Both these are applicable to decisions taken in the name of the state, and open to the mayor or president of the EPCI, not just to the *Commissaire de la République*, since there are common law provisions concerning public and private liberties.

*The intervention of repressive jurisdictions*
F6.16　The control of the penal judges completes the administrative jurisdictions by imposing penal sanctions on those who commit infractions of the rules of urban planning. These sanctions are prescribed within the *Code de l'Urbanisme* (L480-1 to L480-13). A set of general charges applies to all failures to respect the régimes of authorisation, and a set of specific charges relates largely to publicity and site visit infractions. Most of the infractions have a purely material character and are not concerned with motives or fraudulent intentions.

*The procedure*
F6.17　Most of the infractions listed by the *Code de l'Urbanisme* constitute offences and it is the *Tribunal Correctionnel* which is competent to decide these

(occasionally the *Tribunal de Police*). Most of the infractions are established by verbal process (*procès verbal*) which can be set up by officers and agents of the judicial police, as well as the mayors and their adjoints, officers of the gendarmerie or commissioners of police. They can also be established by civil servants and agents of the state or communes who have responsibility for checking the legality of developments, including those overseeing historic monuments. These establish the date, place, and nature of the infraction and interrupt the work until they are disproved.

F6.18   These public actions can be pursued against all those involved in the execution of the works, including the architects, builders and entrepreneurs whose failures to observe the rules are less excusable. Since 1976, once the authority knows of any infraction, it is obliged to prepare a *procès-verbal*, and it must be transmitted immediately to the minister who decides on the proceedings to be taken. But other parties can initiate actions, notably environmental groups officially recognised as being of public utility, and they can join the minister or take direct action through the Judges of Instruction, though 'a bill of insufficient evidence' can prevent further pursuit. Since an *arrêté* of the Court of Cassation (the highest Court of Appeal concerned only with revising legal procedure) of January 1984, individuals have been able to take action as civil parties if the infraction has caused them direct and personal prejudice. Previously it had been argued that the rules of urban planning were designed to protect only the general interest, not that of individuals. The result has been that neighbours and other affected persons have more systematically challenged infractions than the environmental associations were able to do. Since 1985 the public collectives have also had a clear right to pursue infractions through this process.

*The sanctions*
F6.19   Once an infraction has been established various sanctions can be applied. The judiciary authority hears from the beneficiary of the works or decision within 48 hours, and if it considers an offence has been committed it can order an immediate interruption of all works, and the mayor or *Commissaire de la République* can implement this after 24 hours. They can take any measures necessary to ensure the interruption, even seizing materials on site or sealing the building etc. Continuation of work can be published by a fine of FF200,000 - FF500,000 (£20,000 - £50,000) and/or four weeks to six months imprisonment.

F6.20   Since 1976 infractions can also be subjected to straight fines of similar amounts, with additional imprisonment of from one to six months. Fines can also be levied according to the floor area of up to FF10,000 (£1,000) per square metre. Furthermore minor fines are imposed for obstructing site visits, failing to advertise works etc.

F6.21   Over and above these penal sanctions the judge can order the restitution of works to ensure conformity of the development to the rules of urban planning, or to return the land concerned to its original state. They are conceived as both a punishment and a civil reparation, and can be applied even after the death of the accused. Failure to conform to these provisions can result in fines of FF50-FF500 (up to £50) per day, and more if the public minister agrees, though these provisions can be modified if work is being undertaken. There is also a 'forced' execution procedure whereby the administrative authority can undertake the necessary work themselves, although damage to third party rights and evictions would have to be approved by the High Court.

## Intervention of the Civil Judge

F6.22   It is only since 1959 that the Court of Cassation has accepted the rights of private individuals to bring actions to repair prejudices caused by breaking the rules of urban planning. But the violation must be of a fundamental regulation, such as a contravention of a *servitude* or land use prescription of an imperative, not of a permissive, character. Only an approved POS can be so violated. With regard to a *permis de construire* a civil action is only possible if the authorisation has been previously annulled. Furthermore the judiciary judge must assure himself that the infraction was serious, and that it caused a direct prejudice to the applicant (for a development conforming to a *permis de construire* there is a five year period to take action after the work is completed). Other violations of the prescriptions of land use are subject to the same general provisions though here abuse of procedure will not result in reparations.

F6.23   The proof of a direct personal prejudice from the violation is required to establish a right of reparation, and while personal prejudice can be easily established, for example a spoiled view, loss of privacy or an 'overshadowed' garden, the Court of Cassation has since 1974 been very exacting in establishing such prejudice. The violated regulation must cause damage related to its original purpose, not to some other consideration. This is a severe limitation which has been confirmed by subsequent decisions. In addition, since 1972, third parties have been unable to seek possession of property as reparations.

F6.24   Finally the judge can order reparations 'in kind', or 'through equivalence' of a financial nature. Until a decision in 1979 there was a tendency to convert reparations into simple financial remedies leading to criticisms that the law was failing to deal with the *'fait accompli'* situations. However since then the right to insist upon demolition as part of reparations 'in kind' has been reasserted, restoring to the 'creditor' the right to have any contravention 'destroyed'.

## The Incidence of Court Cases

F6.25   The latest general statistics for the incidence of

**Table F6.1  Contentious appeals to the courts on urban planning issues 1983**

| | Plans | PdC | *Lotissement* | Expropriation | ZAC | Sites | Other | Totals | |
|---|---|---|---|---|---|---|---|---|---|
| *Tribunal Administratif* | 318 | 2,084 | 135 | 48 | 28 | 1 | 629 | 3,243 | |
| *Conseil d'Etat* | 50 | 286 | 19 | 22 | 6 | 15 | 41 | 438 | Recourses |
| Totals | 368 | 2,370 | 154 | 70 | 34 | 16 | 670 | 3,681 | |
| *Tribunal Administratif* | 189 | 1,767 | 94 | 67 | 27 | 3 | 349 | 2,496 | |
| *Conseil d'Etat* | 31 | 231 | 12 | 18 | 5 | 4 | 37 | 338 | Decisions |
| Totals | 220 | 1,998 | 106 | 85 | 32 | 7 | 386 | 2,836 | |

*Source:* RISU 1985, pp.184–187

court cases relating to urban planning are for 1983 and hence do not reflect the post decentralisation situation. Nonetheless they provide some valuable insights into the pattern of litigation and the incidences of abuse of the system. Table F6.1 shows the number of recourses to and decisions taken by the *tribunaux administratifs* and the *Conseil d'Etat* challenging the decisions of local authorities under various heads. It can be seen that 2,496 decisions were taken, while 3,243 recourses were made to the *tribunaux administratifs* against the administration. About one eighth of this number were being taken higher to the *Conseil d'Etat*. There is some indication of a significant increase in the number of recourses and this is the opinion of most observers. The vast majority of cases concerned the *permis de construire*, but these cases constitute only one per cent of all permits being implemented annually. Similarly the number of *lotissement* cases constitute only two per cent of annual authorisations. However, the number of plan-related cases is rather more significant with 300 recourses in 1983, though it is difficult to know whether to measure this against the 900 or so newly approved plans, and a similar number modified, or to include a larger number of plans. Generally these figures indicate a very low level of abuse of the system.

F6.26  Of the court decisions (Table F6.2) only one fifth of those taken by the *Tribunaux Administratifs*, and only one quarter of those taken by the *Conseil d'Etat*, were decided in favour of the petitioners while over half and two thirds respectively were decided in favour of the administration. But in a significant number of cases the

court decided either that it had no authority, or that there was no case to answer. These figures emphasise that local authorities were operating the system within the law in an overwhelming number of instances.

**Table F6.2  Direction of decisions of the Courts 1983**

| | In favour of admin | In favour of petitioner | No suit | No Authority | Total |
|---|---|---|---|---|---|
| *Tribunal Administratif* | 1,307 | 539 | 361 | 301 | 2,508 |
| *Conseil d'Etat* | 226 | 80 | 19 | 14 | 339 |

*Source:* RISU 1985, pp.188–190

F6.27  Analysing infractions of the legislation through the *procès-verbal* (Table F6.3) it can be seen that there were over 10,000 cases, some 3.3 per cent of the number of *certificats de conformité* issued in 1983. Over one quarter of these were regularised, but almost ten per cent resulted in demolitions, and almost five per cent in an interruption of works, after the *procès verbal*. Only one fifth were therefore the subject of judgments, and of these 2000 or so cases, nearly half were constructions to be demolished or the land to be put back into its original state. In the remaining cases the rules were mainly relaxed or merely amended. Again these figures reveal a relatively small number of infractions, predominantly those of the *permis de construire* process, but there are sufficient demolitions to emphasise the strong sanctions which are available as a deterrent.

F6.28  More recent data on the recourses to the courts is-

**Table F6.3  Infractions of the legislation of urban planning 1983**

| Process-verbal Infractions | Interruptions of work ordered | Regularisation after P.V. | Demolition to ensure conformity | Judgments | | | | | | Execution | | |
|---|---|---|---|---|---|---|---|---|---|---|---|---|
| | | | | Relaxation | Amendment | Removed or demolished | Put in conformity | TOTAL | Demolition | Put in Conformity | of office |
| 10,142 | 474 | 2,838 | 998 | 437 | 491 | 925 | 175 | 2,028 | 362 | 139 | 6 |

| *Nature of Infractions* | |
|---|---|
| *Permis de Construire* | 8,212 |
| *Permis de Demolir* | 45 |
| *Lotissement* | 59 |
| Enclosures | 586 |
| Caravans | 695 |
| Sites | 81 |
| Diverse | 464 |

*Source:* RISU 1985, p.137

unfortunately not available so it is not possible to measure the effects of decentralisation. Four basic trends are generally conceded.

(i) the number of challenges to *permis de construire* by neighbours and environmental groups are increasing markedly

(ii) many of these challenges are irresponsible, but are very effective delaying tactics and bargaining tools

(iii) there is a significant increase in the number of challenges to the POS, particularly by groups and individuals seeking to prevent urban growth

(iv) there is a rapidly increasing backlog of cases in front of the *tribunaux administratifs*.

F6.29   With reference to the latter point one can expect in the next few years to see the courts making landmark decisions on the key issues relating to the decentralisation of planning powers. These take three or four years to work through the system, and the major reforms were only initiated in January and July 1983, July 1985 and January 1986. Seasoned commentators have noted the uncertainties of the current situation since the crucial interpretation of the rules and the limits of the regulations remains to be tested. Once decisions filter through, a clearer picture will emerge. This only emphasises the crucial role played by the courts in controlling the legality of the French planning system and defining its scope, application and impact.

## The Strengths of Administrative Law

F6.30   As was argued in section F1, the French administrative courts obviate the necessity in Britain for a complicated system of administrative tribunals or enquiries to oversee the operations of the welfare state, housing activities or town planning, and reduce the need for public participation of all kinds. The supervisory machinery provided by the courts allows the French safely to entrust the details of government to the experts. None of this has changed with the advent of decentralisation.

F6.31   The accessibility to all citizens of the administrative courts is emphasised by the fact that legal representation is not required (except in the *Conseil d'Etat*), that legal aid is available for poor litigants, and that there are no fees. Furthermore the courts are inquisitorial rather than accusatorial and dominated by written dossiers with an almost complete absence of oral arguments by counsel. There is no defence of Crown privilege to protect government departments and a 'failure to reply' on the part of a decision-taking body will leave the court free to draw its own conclusions. The substantive law seeks a delicate balance between individual rights and administrative necessities and while it rests upon a body of case law it does not constitute a rigid set of rules. This is clearly evident in the balance struck between protecting the rights of third parties,

developers, and the administration on the issue of *permis de construire* infractions.

F6.32   Despite these prodigious virtues the system of *droit administratif* has its failings. Most importantly it is an *a posteriori* means of control, and this form of judicial control cannot successfully balance the absence of safeguards for citizens in the day to day functioning of the administration (Weil, 1965, p.256). In addition justice is slow and plaintiffs may have to wait three to four years for a judgement and this can cause personal hardship and create a climate of uncertainty. Furthermore the judgement of the courts may be difficult to enforce, although cases of non execution are rare (Weil, 1965, pp.252-7).

F6.33   Critics had long suggested the need for the administration to offer more safeguards to the public during the process of decision making. In 1978 the citizen's right to be notified of decisions affecting him was reaffirmed, and access was granted to administrative documents affecting him individually or generally. In 1979 a reform was instituted stating that written, precise reasons must accompany any administrative decision which directly affects the interests of an individual in a prejudicial manner. In 1980 financial penalties were introduced in cases where there was a failure on the part of the state to pay damages, and the *Conseil d'Etat* established a daily financial penalty for unexecuted judgments. In planning, important innovations have been made in 1976, 1983 and 1985 to increase the role of third parties, affected neighbours and environmental groups in contesting administrative decisions, allowing access to planning files, publishing lists of applications etc. These go some way towards obviating the criticisms of a lack of public participation, but it is still an *a posteriori* control.

## The Role of the *Médiateur* (Ombudsman)

F6.34   For many years it was assumed that the system of constitutional, judicial and administrative organisation in France left no role for an Ombudsman. However, a growing sense of public disillusionment with the operations of the *tribunaux administratifs* led to the introduction of the *Loi Médiateur* in January 1973, to provide a simple, free and readily accessible system of judicial control, even of policy. This established a national *Médiateur* with a broad remit to investigate complaints about administrative malfunctioning in central and local government and other public service agencies. Clark (1982, p.46) notes that the style of the *Médiateur* is more political than his British equivalent, and he acts as a champion of the administered, not merely upholding the principles of their administration, but advocating reforms of laws and administration procedures. Although he reviews upwards of 5000 cases per annum much of his effort is directed at proposing reforms of the regulations or laws that provoked these cases.

F6.35 The city of Paris was the first municipality to establish its own *Médiateur* in May 1977, accountable to the Mayor of Paris. Although not enjoying either the statutory protection or the guarantees of independence that are the basis of the British Ombudsman system, and having very considerable freedom to define the scope of his activity, the local *Médiateur* with a staff of eight is able to receive over 1000 complaints per year from individuals or groups. The complaints he can receive are not defined merely as 'sustained in consequence of maladministration' as in Britain, but embrace all forms of administrative malfunctioning including the fairness of compensation offers for compulsory purchased land. The only real areas of exclusion are those related to clinical matters, the administration of hospitals or homes, and the exercise of public powers. So the Paris *Médiateur* has become a very convenient point of access for Parisians to government at all levels, not merely to the Commune itself.

F6.36 The vast majority of complaints (42 per cent) are in fact complaints about the activities of central government departments and agencies, particularly the public utilities and statutory undertakers. Housing, principally allocation procedures, accounts for a further 27 per cent with social services representing another 14 per cent. Planning cases accounted for less than 7 per cent of all cases dealt with by the *Médiateur* (1977-1979) and these related to the discretionary decisions relating to *permis de construires*, and to the assessment of compensation in compulsory purchase actions. The application of the *plafond légal de densité* (PLD) 'penalties' on extensions to flats has also been a particular issue. Otherwise the complaints relate to the failure to enforce planning regulations or to implement plans, or to delays incurred in the processing of *permis de construire*. Such complaints are familiar to, but by no means as frequent as in, Britain.

## Conclusions

F6.37 As was argued in section F1 *droit administratif* has the weaknesses of being a relatively slow a posteriori system of control largely used by civil servants, and with judgments that are sometimes difficult to execute. But it has the strengths of being accessible, cheap and flexible and of providing simple remedies which have public confidence. It also provides a systematic and cohesive system for the intelligent policing of complex administration. It performs three essential functions in providing an appeal system for aggrieved applicants against the local authorities, a means for local authorities to take action against those who break the rules of urban planning, with strong enforcement procedures to back these actions up, and a means for third parties and environmental groups to appeal against the 'illegal' grant of authorisations or prejudicial developments. The appeal system and the role of the courts, so clearly differentiated in British planning, are one and the same thing in France.

F6.38 From the perspective of comparative analysis of development control systems the following points emerge as of significance

(i) the comprehensive role played by the courts as outlined in the previous paragraph

(ii) the advantages of an accessible, cheap, relatively simple and flexible system

(iii) these advantages are counterbalanced by the disadvantages of a slow, *a posteriori* system of control that minimises public participation in, or knowledge of, the actual decision making process

(iv) the evidence (admittedly lobby-group derived) that recent increases in the participation element of litigation are producing irresponsible challenges, and that environmental protection is being used as a mask for no-growth politics and social discrimination

(v) the development of the suspension of execution as a counterweight to the slowness of court decisions, but also a further source of uncertainty for the developer

(vi) the long gestation period for the courts to begin interpreting new legislation to determine the validity of current practice, and the creation of a climate of deep uncertainty

(vii) the widespread evidence (1983) of the respect for the law and the very limited abuse of the system by either developers or local authorities

(viii) the continuity and coherence of enforcement provisions and effective penalties that minimise any potential infractions by the developer.

# F7 CONCLUSION: THE SALIENT CHARACTERISTICS OF DEVELOPMENT CONTROL IN FRANCE

## Introduction

F7.1    France has a similar size of population to Great Britain but it inhabits a country twice the size. It is a unitary state with a written constitution under a Parliamentary régime, but with a Presidential executive providing a second source of regulatory power. There are four levels of government: national; regional (22 units); Departmental (96); and commune (36,433). It is not a hierarchical system. Each level has different but interactive responsibilities, while the state retains important supervisory powers over the communes through the *Commissaire de la République* (formerly Prefect) and its state services (DDE) at the Departmental level.

F7.2    Since 1983 the planning powers and control of development have been decentralised to each commune, providing they have an approved local plan (*Plan d'Occupation des Sols* (POS)). So in an apparent complete reversal of the French centralist tradition, planning is now a largely local responsibility. Most urban communes have accepted these responsibilities, and the number of local plans has quintupled in the last four years, but more than two-thirds of the communes have not yet approved plans. Given the fact that half the communes have a population of less than 500 persons there can be little prospect of comprehensive coverage. In the communes without a POS, development in rural areas is theoretically prevented, although some particular exceptions are now admitted, while within urban areas development is controlled by national regulations of urban planning and the state's technical services.

F7.3    Most communes do not have planning departments as such, and if they assume control of development they rely upon a very small technical service to issue the permits and authorisations. Most plan-making functions and development control advice is provided by the state services at the Departmental level, thus retaining strong central influence on both plans and control. Thirty or so major cities have agencies of urban planning, strong interdisciplinary teams who undertake plan-making and planning studies, but most of these are independent of political control. While syndicates have been established by the communes to produce plans their activities do not extend to development control which, if an approved plan exists, remains under the control of the mayor. Plan-making and control are thus entirely separate activities, though the state services oversee and thus respond to both. For this reason and many others there is

some doubt as to the extent to which decentralisation really will mean local control of planning.

F7.4    Planning legislation is encoded in a single legal text (since 1954) and it has been continually amended to provide a comprehensive guide to development control, plan making, land policy and servicing, even taking on the form of a development plans manual.

(i)   The first urban planning legislation was enshrined in 1919. There were important precedents for control in Paris, and in nineteenth century public health and building control, and significant listed building controls.

(ii)   In 1943 interwar legislation was codified and the *permis de construire* introduced as the main tool of development control ensuring conformity to the building regulations and development plan or general regulations. The plan system was refined to a two tier system in 1958 with important land policy provisions to produce land quickly for urbanisation, and conservation area and national parks legislation was passed in the early 1960's.

(iii)  The most important change was in 1967 when the *Loi d'Orientation Foncière et Urbanisme* established the current system of development plans, and a less draconian measure of land policy to ensure public-private partnerships in major development and redevelopment projects.

(iv)  In 1976 the control of the POS was tightened, environmental groups and citizens were given a right to participate in plans and to take legal action against infractions, and a greater emphasis placed upon environmental considerations especially in rural areas.

(v)   In 1983 development control was devolved to the communes with an approved POS, further refinements being added in 1985 to strengthen the role of the public.

(vi)  In 1986 the first attempts to deregulate planning were made, introducing some important exceptions to development control. This may well be the theme for the next few years.

F7.5    While the *Code de l'Urbanisme* provides the basic framework for plans and control, and contains very important expropriation, land policy and development tax provisions, there are important Codes of the Environment, Construction, Mining and Administrative

Law which prescribe related areas of control and the procedures of appeal for all parties.

## The Control of Development

F7.6 The basic principle of control derives from a system of legal rights to development if the applicant conforms to the provision of the POS. Since 1983, in the absence of an approved POS, it is almost impossible to develop outside built-up areas, while within the latter national rules of urban planning prescribe the location and servicing, siting and volume, and external appearance of development, providing clear guidance to an applicant.

F7.7 The POS itself is less a plan in the forward-looking sense of the term and more an enshrining of development rights. It must include all public restrictions on development - public utilities, highways, noise zones, safety zones etc - as well as all the planning restrictions and conservation zones. It is essentially a zoning map varying in sophistication from a mere six zones to perhaps twenty, each with its own set of fifteen regulations. These embrace land use (two); access and services (two); siting, height, and aspect (six); external space and parking (two); and plot ratios (two). The regulations can be of varying complexity and sophistication; the plot ratio for example can discriminate between different uses offering major incentives to certain uses and effective disincentives to others.

F7.8 The authorisation for most development is the *permis de construire* and the procedure for its considera-tion, and a rigid timetable, are enshrined in the urban planning Code.

(i) A standard form is provided, requiring full details of the various actors involved, along with plans, sections, and elevations. Almost all development over 170 square metres must be architect designed.

(ii) These forms are sent to the mayor (if the POS is approved) or *Commissaire de la République* who opens a file, advertises the application in the town hall/prefecture, and acknowledges its acceptance.

(iii) The file is examined by the state services (DDE) or the commune's technical services, and a range of consultations undertaken with some 90 bodies eligible, but normally embracing only the relevant municipal service departments.

(iv) A time period for decision is prescribed for each type of development, usually two months, with an extra month for larger developments or historic monuments, and up to six months for major commercial developments. The time period is extended for any additionally necessary consulta-tions.

(v) A failure to deliver a permit within the specified period means an automatic authorisation, but this can be challenged in the courts by third parties.

(vi) The authorisation can be conditional upon certain financial or planning considerations, a derogatory permit contradicting a particular regulation (now rare), a regularising permit, one authorising division of property or one of precarious title allowing provisional construction. But there is also the option to suspend a decision for two years on any development compromising a POS or public works in preparation.

(vii) The grant of a permit must be advertised, can be attacked through the courts in the following six months, and has a two year validity.

(viii) A certificate of conformity is required before the building can be occupied providing a built-in inspection and enforcement system with prescribed procedures and penalties. Conformity to the building regulations is checked separately and is largely reliant upon control through private insurance procedures.

*The definition of development and the scope of control*
F7.9 Development is defined nationally as any act of construction, even those structures without foundations. It is a tight, all-embracing definition though a range of minor developments (as well as schools, energy produc-tion and distribution, post and telecommunications, buildings etc) were exempted or subjected to preliminary declaration controls in 1986. It applies to public services and their concessionaires as well as to all levels of government.

F7.10 The *permis de construire* provides the means of ensuring that all buildings conform to the plan or planning regulations, public *servitudes*, building regula-tions and the rules for fire, hygiene, and public safety, but it says nothing about private rights over the land.

F7.11 An outline planning permission procedure is available with a similar process of approval, although this has its origin in, and is still widely used for, establishing development rights and consequently land prices.

F7.12 A lighter form of control is available through the procedure of a preliminary declaration which can be opposed by the authorities within one month or is deemed to have permission. This embraces much of the minutiae of control (cleaning façades, posts and pylons, swimming pools, greenhouses etc).

F7.13 Demolition controls are applied to all areas within 50 kilometres of Paris, to all communes over 10,000 population and to all zones protecting historic monuments and sites, with a *permis de démolir* process similar to that of the *permis de construire*.

F7.14 *Lotissement* or subdivision controls are of increasing importance covering the division of land into plots for development, and controlling the all-important servicing of the land and the costs of the same.

F7.15 Enclosures and fences, leisure parks, soil excavations, camping and caravanning, cutting and felling trees, and clearing land are all subject to control, leaving only most purely agricultural activities exempt. Hazardous installations of all kinds are protected under special procedures shared by the Ministries of Environment and Industry, but involving a public enquiry.

F7.16 Finally there are special controls on major shopping developments (with a unique consultation procedure) and on industry, offices and research establishments in the Paris region.

*Flexibility in control*
F7.17 Although the POS and the national regulations on urban planning provide very precise rules for development there is still considerable flexibility in the way in which they are interpreted, giving considerable scope to those exercising control. Similarly many of the national regulations have key conditional verbs in their wording which provides significant discretion for the controller.

F7.18 However, the POS itself is not really a flexible instrument, especially since the possibility of derogations has been significantly reduced since 1976 due to widespread abuse of the procedure. Now modifications to the POS must go through a shortened version of the process of approval which renders the plan in a permanent state of change and adaptation. From a development perspective the POS often prescribe far too rigid rules that negate the very prospect of development, and bear no reality to the current situation. However, there are signs of the emergence of a much more sophisticated set of plans that attempt to respond to local economic and social situations, and to provide greater flexibility and incentive for development, while still protecting environmental quality.

*Permission and refusal rates*
F7.19 The number of applications for planning permission is broadly similar to that in Great Britain. Ninety five per cent of all *permis de construire* applications are granted, and 85 per cent of these are delivered within three months, though there are wide variations in rates of approval, highly urbanised areas being notably slower. Outline planning applications to test the 'constructibility' etc. of schemes show 58 per cent delivered in two months, but with refusal rates hovering around the 25 per cent mark. In terms of development only half the permits seem to be taken up, but of those that are some 78 per cent conform immediately to the regulations, with only 2 percent being definitely refused. Refusal rates are negligible for demolition permits, *lotissement*, enclosures etc., being significant only for camping and caravan authorisations.

# Plan Preparation and Application

F7.20 In theory the French National Plan provides an overall framework for physical planning, but its influence has progressively waned. However, a stronger regional planning dimension was introduced in the 1980's and the regions are now able to elaborate a *Contrat de Plan* which spells out the full range of regionally funded programmes for housing, open space provision, transport and water/sewerage infrastructure and economic development initiatives. These are firm commitments to investment, but few regions have yet participated in the preparation of such documents.

F7.21 The *Schéma Directeur* (SD) is the French equivalent of the structure plan essentially covering urban regions, with ten and thirty year time horizons and guidance maps. Most of these plans were prepared in the high growth periods of the early 1970's, rendering their allocations of land largely meaningless in the 1980's. In recent years very few have been prepared, although there are some signs of new initiatives and directions in a number of urban regions. The significance of these plans is twofold: the POS must be compatible with the SD, and the SD provisions are binding on the public authorities but not on private individuals.

F7.22 The SD no longer provide the guidance and framework for the POS as the legislation intended, and any new SD approved will be essentially a compilation of the various POS. This 'bottom up' procedure has been given much greater impetus by decentralisation which has also allowed the communes to pull out of the voluntary SD preparation before approval, if they feel it prejudices their interests.

F7.23 The POS approval process is clearly defined by the *Code de l'Urbanisme*. Its initiative is the sole responsibility of the commune or group of communes and the plan is elaborated by a working group, composed of the representatives of a variety of local commerce, agriculture, trade, environment bodies, headed by the mayor or president of the communal group, in conjunction with a planning agency, usually the state services, the DDE.

F7.24 Once prepared, the plan is published and the views of any qualified or respected persons must be heard, and wide consultation undertaken. Conflicting advice must be resolved by a special Commission of Conciliation, but once agreed the plan is rendered public and submitted to a public enquiry which is essentially an opportunity for the public to comment not cross examine. The commissioner of the enquiry makes a report which must be taken into consideration and then the plan can be approved.

F7.25 The adaptation of the plan can take place through a process of modification (nearly 90 per cent of adaptations follow this procedure) which significantly simplifies the above procedure, or a full scale revision, which avoids only the publication stage in the full approval process outlined previously.

F7.26 The time taken to prepare a POS used to vary from five to seven years but with the advent of decentralisation, and a five-fold increase in production statistics, two years is now the average with perhaps one and a half years for a modification.

F7.27 The POS is essentially a regulatory administrative act, and its powers increase as it goes through the various stages of prescription, being rendered public, or approved, allowing the suspension of prejudicial authorisations during the period of preparation. Once approved, it becomes the essential reference point for development control with conformity to the POS guaranteeing the issuing of a building permit, but its provisions can be challenged through the courts.

F7.28 Other relevant forms of plan are the much more detailed plans for conservation areas or architectural protection areas, which are legally binding, and the *charte intercommunale*, a socio-economic development charter for rural areas which must be 'taken into account by any plan'. Special rural environmental protection plans have since 1983 been subsumed within the POS.

## The Appeal Process and the Courts

F7.29 French administrative law provides a comprehensive system of legal control protecting the rights of the individual from the consequences of any administrative act. In the context of planning it provides

(i) an appeal system for aggrieved applicants

(ii) a means for local authorities to take action against those who break the rules of urban planning, and general enforcement measures

(iii) a means for third parties and environmental groups to appeal against what they see as infractions of planning law.

F7.30 Administrative law defines the infraction, the repressive jurisdictions define and impose the penalties, while civil law provides the means of forcing offenders to repair the damages caused to third parties.

F7.31 The system has the advantages of being accessible to all, cheap to use and flexible, providing simple remedies which have public confidence. But it has the disadvantages of being slow (two to three years for a decision) and requiring a significant amount of expertise normally only available to civil servants (and fully exploited by them).

F7.32 The introduction of a suspension of execution of decisions procedure has provided a much more effective short-term measure to prevent harmful developments, and other measures have widened the possibility of private or environmental group action against development 'infractions'.

F7.33 One consequence of this is a growing number of what the development industry perceives as irresponsible challenges, which effectively freeze the prospect of development, and create a climate of great uncertainty. This uncertainty is essentially fostered by the fact that it takes three years for the courts to interpret the application of new legislation and regulations, and hence decentralised planning is awaiting just such interpretation at present. To add to these difficulties there is a large backlog of cases before the courts.

F7.34 The introduction of ombudsman procedures in 1973 has provided a more simple, free, and readily accessible system of judicial control, even of policy, but planning cases do not seem to be a large part of their workload (seven per cent).

F7.35 Despite the virtues of the system of administrative courts it provides legal/democratic checks only after a decision has been taken. It does not allow the public to observe the process of decision-making or to participate in it. Decentralisation has failed to increase the accountability of decision-making in this important respect.

*Court cases and infractions of planning law*
F7.36 Figures for 1983 on administrative court cases revealed over 3000 recourses to the courts to challenge administrative decisions, the vast majority relating to the *permis de construire*, but constituting only one per cent of all authorisations being implemented. Only a quarter of those cases, 539 in 1983, were won by the petitioners, while over half were judged in favour of the administration. This provides some indication of the adherence to the law by planning authorities.

F7.37 In terms of infractions in the development control process some 10,000 cases were recorded, about three per cent of all development. A quarter of these were subsequently regularised and a tenth demolished, leaving only 2,000 cases to go to the judgment stage. Of these a further half were demolished or the site returned to its original state, with the rest being relaxed or amended. These figures reveal the widespread respect for planning laws though they also show that the deterrents are there to be used when necessary.

## Conclusions

F7.38 In conclusion the basic characteristics of the system of control in France are

(i) the local plan (POS) enshrines development rights for developers and adherence to its provision ensures an authorisation

(ii) an approved POS is a prerequisite for the devolution of planning to the local level, and in its absence all development outside built up areas is effectively prohibited, leaving the remainder under control of national regulations

(iii) the definition of development is precisely and all-embracingly drawn, notwithstanding very recent relaxations, and covers all acts of construction, subdivision, enclosure etc, and controls the external appearance of development. Lighter forms of control prevail for some minor forms of development.

(iv) development control has a built-in enforcement system to ensure conformity to planning provisions

(v) notwithstanding decentralisation, the state keeps significant control on planning practice through the law, which prescribes the format of the POS and the procedures of decision making, and continues to exercise technical control through the state services (DDE) in most smaller communes

(vi) the existence of a comprehensive system of legal control and recourse for the applicants, local authorities, affected individuals and environmental groups provides an appeal system and legal recourse for all actors in the process.

F7.39   The main criticisms depend upon the interest group consulted but the following are widely expressed concerns about the operation of the planning system

(i) the very weight, complexity and impenetrability of French planning law to all but the specialist

(ii) the desirability of some significant de-regulation to combat the frequent inefficiencies of control

(iii) the questionable competence of the communes, especially the majority of smaller ones, in the absence of adequate funds for competent staff, to exercise control successfully over development and adequately protect the environment

(iv) the long delays in court actions, the uncertainty created for developers by some irresponsible challenges in the courts, and the time taken to interpret new legislation

(v) the lack of public participation in, the control process and its general lack of 'transparency' to the layman

(vi) the negative, 'policing' nature of the POS which tends to enforce a conformity that may run counter to economic realities and socially desirable trends, and may not be appropriate in an era of smaller scale development

(vii) the emergence of overly restrictive POS and of 'snob zoning' which reduces development options, particularly residential development, because it constitutes a significant financial burden on local authorities from a local tax perspective, and because it is electorally unpopular for the mayor and municipal council

(viii) related to (vii) above, the local fiscal crisis results in significant development taxes and demands for contributions to services, and may delay and complicate negotiations over a planning permission, as well as adding very significantly to the costs of development.

F7.40   Most criticisms of contemporary planning practice embrace the related questions of adequate land supply for developers, the opposition to growth of local residents, the local fiscal crisis and the need to ensure that development contributes positively to the tax base and/or that developers, or more important by the vendors of the land, pay the full costs of infrastructure provisions.

F7.41   While it is pointed out that the POS provide some 500,000 hectares of development land (Zoned NA), sufficient for 120 million inhabitants, there is a shortage of development land because of a complex cadaster, a general obsession with landholding, a tax régime that discourages preparing land for development, and now falling land prices that inhibit sales to the developers. The French planning system has been integrally related to land policy but while the legal aspects of land supply have been treated, the economic aspects have not. These latter issues lie beyond the scope of this study but they remain central to the debate about the purposes, effectiveness, and equity of the French planning system.

# GLOSSARY

**agence d'urbanisme**
agency of urban planning (plan-making, design studies etc) in some 32 cities, offers consultancy service to constituent communes but is independent of political control

**AFL association foncière urbaine**
associations of urban land owners who come together to develop their land and to share the costs of servicing

**APL aide personalisée du logement**
principal means of housing assistance direct to tenant (since 1977)

**alignement**
literally alignment of roads and reservations for road widening, a nineteenth century control innovation that was a forerunner of planning controls

**arrêté**
decree issued by any level of government; a legal act or resolution introducing a new provision or putting into effect a plan, working group etc

**arrondissement**
division of a Department with a sub-prefecture, given certain neighbourhood powers in Paris, Lyon, Marseille in 1982

**certificat de conformité**
certificate of conformity, which must be obtained before any newly built structure is occupied, and which ensures its conformity to the planning regulations

**certificat d'urbanisme**
certificate of urban planning which provides the equivalent of an outline planning permission, and states development rights on a property

**charte intercommunale**
intercommunal charter, a socio-economic development document

**clôture**
enclosure, refers basically to walls, fences, banks etc around buildings, not those for agricultural use

**Code de l'Urbanisme**
the code of urban planning law, kept up to date, the comprehensive reference source for planning law, regulations, and regulatory arrêtés prescribed by national government

**coefficient d'occupation du sol**
the plot ratio used to control density in the POS, and a basis for certain taxation of development

**Commissaire de la République**
prior to 1982 the Prefect, the state-appointed head of the Department, responsible to the state for legal control of all manner of administration, controller of the DDE and signatory of all planning decisions not taken by the mayors or presidents of EPG

**CGP Commissariat Général du Plan**
quasi-autonomous body under control of Prime Minister, charged with elaborating the national plan

**commission de conciliation**
a departmental commission composed half of elected officials and half of professionally qualified persons to resolve disputes resulting from consultations on plan preparations (established 1983)

**communauté urbaine**
intercommunal agency embracing a wide range of local government functions, including planning, in nine urban regions

**CAUE Conseil d'Architecture d'Urbanisme et d'Environnement**
architectural and planning council established to sensitise the public and offer advice to developers on environmental-developmental issues (established 1977)

**Conseil d'État**
the highest administrative court, point of appeal above *Tribunal Administratif*, and legal adviser to the government

**Conseil Général**
General Council now established at regional and Department levels and directly elected

**constructibilité limitée**
rule which virtually forbids any development in zones outside built up areas, in the absence of a POS

**contrat de plan**
the new five-year regional plan prescribing public expenditure, largely in infrastructure provision, housing, environment and economic development

**cumul des mandats**
multiple office holding at different levels of government

**déclaration préalable**
literally a preliminary declaration, a 'lighter' method of development control whereby the local authority is notified of usually minor development, and can control it if necessary

**déclaration d'utilité publique**
legal declaration by *Commissaire de la République* or Ministry for compulsory purchase provisions

**défrichement**
clearance of land, usually from forest or scrub

**DATAR   Délégation a l'Aménagement du Territoire et à l'Action Regional**
the inter-ministerial regional planning agency with direct access to the Prime Minister

**déplanification**
literally unplanning, usually a reference to national planning and the reduction in state control, largely in the sixties and seventies

**dérégulation**
literally deregulation, the current preoccupation with reducing the number of planning regulations and control procedures

**dérogation**
a derogation or legal contradiction of a planning document or control regulation

**DDE   Direction Départementale d'Equipement**
the field services of the Ministry located at the Departmental level which embraces highways and planning, and provides the planning service for the communes

**enquête publique**
public enquiry, required on hazardous installations or POS, but essentially a public consultation exercise

**espace boisé classé**
a classification in the POS which protects woodlands for 'biological equilibrium' and leisure reasons, and prevents their destruction

**équipement**
literally equipment, but embraces infrastructure in the widest sense from sewers to schools, from play space to roads

**EPCI   établissement publique de cooperation intercommunale**
an intercommunal organisation that brings in representatives of the other levels of government and various chambers of commerce etc to elaborate a plan

**étude d'impact**
impact study with a broad environmental remit required with major commercial developments or hazardous installations

**FAU   Fonds d'Aménagement Urbain**
fund established in 1977 to coordinate the use of credits for all levels of government for the planning and development of commercial centres and neighbourhoods

**grands ensembles**
literally large concentrations of high-rise housing in suburban areas, product of the 1960s public housing programme

**GEP   group d'études et programmation**
the plan-making and supervisory section of the DDE

**HLM   habitations à loyer modère**
originally philanthropic housing agencies, now a form of housing association under state legal and financial control: some 1200 in existence

**immeuble inscrit**
listed historic building

**installation classée**
a classified installation with some hazardous aspects by virtue of the materials/processes it uses or produces, and subject to special development controls by the Ministries of Industry and Environment

**lotissement**
subdivision (and servicing) of land for development purposes, controlled under a separate authorisation procedure

**MARNU   modalités d'application du RNU**
the instructions for applying the RNU (see below) including the transitional arrangements following decentralisation

**métropole d'équilibre**
growth pole policy of late 1960's and early 1970's designating major cities to attract developments away from Paris

**le mitage**
literally moth-eaten, a phrase encapsulating the effects on the landscape of discontinuous development or urban sprawl

**OPAH   opération programmée d'amelioration de l'habitat**
programmes for the rehabilitation of buildings and public spaces overseen by the commune but involving the FAU and ANAH - used especially in *secteurs sauvegardes*

**OREAM   organisations d'études d'aménagement des avis metropolitaines**
general planning agencies established in eight metropolitan areas in 1964, superseded by regional governments in 1984

**périmètre sensible**
sensitive perimeter delineated at departmental level to protect certain natural areas essentially on the coasts and in the mountains, and allowing much tighter control over development

**permis de construire**
a permission to build, the authorisation required for all acts of construction, the French version of planning permission

**permis de démolir**
demolition permit, required in all medium sized and large towns, conservation areas, the vicinity of listed buildings and within 50 km of Paris

**plafond légal de densité**
a plot ratio set for the basis of taxation, originally 1 (1.5 in Paris) now decentralised and optional. It provided a tax on every square metre of additional floorspace built over that allowed in the ratio

**plan de détail**
not to be confused with the old *Plan d'Urbanisme de détail*; the forerunner of the POS, these are especially detailed zoning maps within the POS regulating key areas

**POS    plan d'occupation des sols**
the local plan that enshrines development rights in a zoning map with regulations at the scale of the property or street

**PSMV    plan de sauvegarde et de mise en valeur**
as above but much more detailed, dealing with building alterations, street treatments etc for each conservation area

**PIG    projet d'intérêt général**
projects of general interest - public utilities, transport, *remembrement* - for the benefit of conservation, disadvantaged, etc; a supra-communal interest that must be discussed with the public, and taken into account in all plans

**promoteur**
'promoter' or developer

**recours contentieux**
'contentious appeals' or recourses to the administrative courts to challenge administrative decisions

**RNU    Règlement National d'Urbanisme**
national rules of urban planning established to provide a framework for control where there is no POS (see Figure F12)

**SD    schéma directeur**
the large scale plan for the 30 year development of urban regions etc, prescribing major land uses and infrastructure - the structure plan equivalent

**SDAU    schéma directeur d'aménagement et d'urbanisme**
the predecessor to the SD (1967-1983)

**SS    secteur sauvegardé**
(urban) conservation area

**SEM    société d'économie mixte**
mixed public-private development companies which execute works 'in the public interest'

**servitude**
literally a charge or condition placed upon property, usually referring to public utility easements, road widening lines etc included in the POS

**sursis à exècution**
legal suspension of the implementation of a planning authorisation

**sursis à statuer**
legal suspension of a decision on an application for a planning authorisation

**tribunal administratif**
the administrative tribunal - local courts providing a recourse for those aggrieved by administrative decisions

**tutelle**
the old system of Prefectoral control

**TLE    taxe locale d'équipement**
tax imposed on planning permissions, universal from 1968 onwards but since 1985 optional and set by the commune

**TDEV    taxe départementale des éspaces verts**
Departmental tax on planning permissions to provide a fund for purchase and management of public open space

**ZAC    zone d'aménagement concerté**
(1967-) contractual agreement between local authority (landowner) and private developers (largely funders) with expropriation powers

**ZAD    zone d'aménagement différé**
(1962-) zone of deferred development where development rights were frozen, and the local authority gives preemption powers

**ZEP    zone d'environnement protegé**
(1976-1983) a simplified POS to protect rural areas now subsumed by the POS itself

**ZIF    zone d'intervention foncière**
(1975-) zone to allow local authority to purchase at previous year's prices, worked in conjunction with the PLD's disincentives to private development

**ZUP    zone à urbaniser en priorité**
(1958-1969) instrument for local authority land acquisition for housing on a large scale

**ZPPAU    zone de protection du patrimoine architectural et urbain**
zone replacing the old 500 metres visibility protection for listed buildings with positive guidance in the form of a plan

**ZU, ZNA, ZNB, ZNC, ZND etc**
zoning designations in a POS (see paragraph F5.27)

# REFERENCES

Recent changes in the French planning system, and the complexity of French planning law mean that any up to date and reasonably complete account of the control of development must rely upon very recent French language sources. The following were indispensable to this study:

BOUYSSOU, F & HUGOT J (1986), **Code de l'Urbanisme (commenté et annoté)**, Paris: Litec

JACQUOT, H (1987), **Droit de l'Urbanisme**, Paris: Dalloz

TRIBILLON, J F (1985), **La vocabulaire critique du droit de l'urbanisme**, Paris: Les Editions de la Villette

*For statistics:*

MINISTERE DE L'URBANISME etc (1985), **Recueil d'informations statistiques sur l'urbanisme**, (RISU) Paris: Min l'Urbanisme, du logement et des transports

*As a best practice example of a Plan d'Occupation des Sols,*
COMMUNAUTE URBAINE DE LYON; SECTEUR CENTRE (1984), **Plan D'Occupation des Sols de Lyon**, Lyon: Agence d'urbanisme de COURLY

*Articles from French Planning journals*

————————, (1984), "POS et décentralisation", **Les Cahiers de IAURIF**, No 72, June, pp 81-101

BOUZELY, S C (1986), "Le POS vont bon train", **Le Moniteur**, 5 December

DELFANTE, C et al (1984), "La décentralisation", **Urbanisme**, No 202, July pp 46-90, 108-119

FREBAULT, J et al (1985), "Innover dans les POS : l'exemple du Lyon", **Urbanisme**, No 206, March, pp 42-45

FREVILLE, Y et al (1986), "Les Finances Locales Françaises", **Metropolis**, No 72, pp 59-90

LENA, H (1985-86), "La réforme de l'aménagement", **Urbanisme**, No.210, November pp 133-137, No 211, January pp 145-9, No.213, May 1986 pp 135-137

*A classified list of English language sources that were extensively used in early drafts of this paper follows. A key to the classification will be found at the end of the bibliography.*

ARDAGH, J (1982), **France in the 1980's**, Harmondsworth, Penguin                                                                  (C)

AITCHISON, J W (1984), "The National and Regional Parks of France", **Landscape Research**, Vol 9, Part 1, pp 2-9    (RuP)

ASHFORD, D E (1983), "The Socialist reorganisation of French local government - another Jacobin reform?", **Environment and Planning C : Government and Policy**, Vol 1, pp 29-44                                                                  (C)

BEAUJEU-GARNIER, J (1979), "Retail Planning in France", in R L Davies ed, **Retail Planning in Europe**, London: Saxon House, pp 99-113                                                         (ReP)

*BESSON-GUILLAUMOT, M (1986), "Town & Country Planning in France" in J F Garner & N P Gravells eds, **Planning Law in Western Europe**, North Holland: Elsevier, pp 152-187                                                                 (L)

BOOTH, P A (1984), "The problems of comparative research : the case of France", **Environment and Planning B : Planning and Design**, Vol 11, No 2, pp 131-252           (DC)

* BOOTH, P A (1985), **Decision making and decentralisation : development control in France,** Sheffield: University of Sheffield, Department of Town and Regional Planning, Occasional paper No 60                                                              (DC)

*BOOTH, P A (1987a), **The theory and practice of French development control: case studies in the Urban Community of Lyon**, Sheffield: University of Sheffield, Department of Town and Regional Planning (mimeo)                   (DC)

*BOOTH, P A (1987b), **The determination of planning applications in the Urban Community of Lyon**, Sheffield: University of Sheffield, Department of Town and Regional Planning, Occasional Paper No 73                          (DC)

BROWN, L N & GARNER, J F (1983), **French Administrative Law**, (3rd Edn) London: Butterworths           (L)

CARASSUS, J (1978), "The Budget and the Plan in France" in J Hayward & O A Narkiewicz, **Planning in Europe**, Croom Helm, pp 53-66                                                       (NP)

CASTELLS, M (1978), **City, Class and Power**, London: Macmillan                                                             (C)

CERNY, P G (1983), **Social Movements and Protest in France**, London: Frances Pinter, esp chaps 1 & 9        (C)

CERNY, P G & SCHAIN, M A eds, (1985), **Socialism, the State and Public Policy**, London: Frances Pinter        (C)

CERNY, P G & SCHAIN M A eds, (1981), **French politics and public policy**, London: Methuen                         (C)

CHALINE, C H (1981), "Urbanisation and Urban Policy in France", **Built Environment**, Vol 7, Nos 3/4, pp 233-242    (G)

CHALINE, C H (1984), "Contemporary trends and policies in French City planning : a chronicle of successive urban policies, 1950-1983," **Urban Geography**, Vol 5, No 4, pp 326-336                                                              (H)

CHALINE, C H (1986), "France" in N P Patricios ed, **International Handbook on Land Use Planning**, New York: Greenwood Press, pp 283-293                              (G)

CHAPUY, P M B (1984), "France" in R H Williams, **Planning in Europe**, London: Allen & Unwin, pp 37-48

CLARK, D (1982), "The City of Paris Médiateur - An Ombudsman a la Française", **Local Government Studies**, Vol 8, No 5, pp 45-65                                                       (L)

D'ARCY, F (1973), "France", in D C Rowat, ed, **International Handbook on Local Government Reorganisation : Contemporary developments**, London: Aldwych Press, pp 5-71 (LG)

D'ARCY, F & JOBERT, B (1976), "Urban Planning in France", in J Hayward & M Watson, eds, **Planning, politics and public policy**, London: Cambridge University Press, pp 295-315 (G)

DRABKIN, H DARIN (1978), **Land Policy and Urban Growth**, London: Pergamon Press (LP)

DUCLAUD-WILLIAMS, R H (1978), **The Politics of Housing in Britain and France**, London: Heinemann (H)

DUFFY, E (1983), "Nature Conservation in National Parks in Western Europe" in A Warren, & F B Goldsmith, **Conservation in Perspective**, Chichester: John Wiley, pp 447-463 (RP)

ESTRIN, S & HOLMES, P (1983), **French Planning in Theory and Practice**, London: Allen & Unwin (RP)

FLOCKTON, C H (1984), "France : Ambitious Gaullist designs and constrained Socialist Plans", **Built Environment**, Vol 10, No 2, pp 132-144 (RP)

FLOCKTON, C H (1983), "French Local Government Reform and Urban Planning", **Local Government Studies**, September/October, pp 65-77

GARRISH, S (1986), **Centralisation and Decentralisation in England and France**, Bristol: University of Bristol, School for Advanced Urban Studies, Occasional Paper No 27 (LG)

GEORGEL, J (1973), "Land Use Planning and Control in France", in A W Davidson & J E Leonard, eds, **Urban Development in France and Germany** Reading: College of Estate Management, CALUS pp 15-33 (H)

GREEN, D (1981), "The Budget and the Plan", in P G Cerny, & M A Schain, eds, **op cit**, pp 101-124 (G)

GREMION, P & WORMS, J P (1976), "The French Regional Planning experiments" in J Hayward, & M Watson, eds, **Planning in Europe**, London: Croom Helm, pp 217-236 (GP)

HAIGH, N (1983), "The EEC Directive in Environmental Assessment of Development Projects", **Journal of Planning and Environment Law**, October, pp 585-595 (GP)

HANLEY, D L, KERR, A P & WAITES, N H (1985), **Contemporary France : Politics and Society since 1945**, London: Routledge and Kegan Paul, (2nd Edition) (C)

HAYWARD, J & WATSON, M (1976), **Planning Politics and Public Policy**, Cambridge: Cambridge University Press (RP)

HAYWARD, J & NARKIEWICZ, O A (1978), **Planning in Europe**, London: Croom Helm (NRP)

HOUSE, J W (1978), **France : An Applied Geography**, London: Methuen, pp 305-436 (C)

*INSTITUTE FOR ENVIRONMENTAL STUDIES (MILAN) (1987), **European Environmental Yearbook 1987**, London: DocTer International

JAMMET, H & MADELMONT, C (1982), "French regulations on Environmental Impact Assessment of Nuclear Installations", **Environmental Impact Assessment Review**, Vol 3, Part 2-3, pp 259-270 (GP)

KAIN, R (1982), "Europe's model and exemplar still? The French approach to urban conservation 1962-1981", **Town Planning Review**, Vol 53, pp 403-422 (C)

KAIN, R J P (1975), "Urban Conservation in France", **Town and Country Planning**, pp 428-432

KEATING, M (1983), "Decentralisation in Mitterrand's France", **Public Administration**, Vol 61, Autumn, pp 237-251 (LG)

KEELER, J T S (1983), "Corporatist decentralisation and commercial modernisation in France : the Royer Law's impact on shopkeepers, supermarkets and the State", in Cerny, P G & Schain, M A eds, **op cit**, pp 265-291 (ReP)

*KESSELMAN, M (1985), "The Tranquil Revolution at Clochemerle: Socialist Decentralisation in France", in P G Cerny, & M A Schain, **op cit** pp.165-185

KOFMAN, E (1985), "Regional autonomy and the one and indivisible French Republic", **Environment and Planning C. Government and Policy**, Vol 3, pp 11-25

*LAGROYE, J & WRIGHT, V eds, (1979), **Local Government in Britain and France**, London: Allen & Unwin, Esp. chaps 1,3,5,7,11, on Local government, Chap 9 on Planning and Chap 13 on regionalisation (LG)

*LEMASURIER, J (1975), "Town and Country Planning in France", in J F Garner, ed, **Planning Law in Western Europe**, Oxford: North Holland - Elsevier, pp 117-154 (L)

MACHIN, H. (1981), "Centre and periphery in the policy process" in P G Cerny, & M A Schain, eds, **op cit**, pp 125-141

*MACRORY, R & LAFONTAINE, M (1982), **Public Inquiry and Enquête Publique : forms of public participation in England and France**, London: Institute for European Environmental Policy (DC)

MCKAY, D H ed, (1982), **Planning and Politics in Western Europe**, London: Macmillan, pp 1-12 (G)

*MENY, Y (1983), "Permanence and Change in the relations between central government and local authorities in France", **Environment and Planning C : Government and Policy**, Vol 1, pp 17-28 (LG)

MONBAILLIU, X (1984), "EIA procedures in France", in B D Clark, et al eds, **Perspectives on Environmental Impact Assessment**, Dordrecht: D. Reidel, pp 51-55 (GP)

*PEARSALL, J (1984), "France" in M Wynn, ed, **Housing in Europe**, London: Croom Helm, pp 9-54 (HP)

RACINE, E & CREUTZ, Y (1975), "Planning and Housing : France : A developer's view", **Planner**, Vol 61, No 3, pp 83-85 (HP)

RIDLEY, F F ed, (1977), "France" in **Government and Administration in Western Europe**, London: Martin Robertson, pp 67-106 (LG)

*SCARGILL, I (1983), **Urban France**, London: Croom Helm, esp. pp 34-133 (C)

SCHABERT, T (1985), "The Decentralisation and the new Urban Policy in France", **Urban Law and Policy**, Vol 7, No 1, pp 57-74 (G)

STUNGO, A (1972), "The Malraux Act 1962-72", **Journal of the Royal Town Planning Institute**, Vol 58, No 8, pp 357-362 (C)

SORLIN, M F (1968), "The French system for conservation and revitalisation in historic centres", in P Ward, ed, **Conservation and Development in Historic Towns and Cities**, Newcastle-upon-Tyne: Oriel Press, pp 221-234 (C)

*STEPHENSON, C P (1986), "Recent developments in French planning law", **Journal of Planning and Environment Law**, October, pp 720-726 (L)

SUTCLIFFE, A (1981), **Towards the Planned City; Germany, Britain, the United States and France 1780-1914**, Oxford: Basil Blackwell, pp 126-162 (H)

ULLMO, Y (1976), "France" in J Hayward, & M Watson, **Planning Politics and Public Policy : The British, French and Italian Experience**, Cambridge: Cambridge University Press, pp 22-51                                                    (NRP)

WEIL, P (1965), "The strength and weakness of French administrative law", **Cambridge Law Journal**, Vol 24, pp 242-259                                                                        (L)

*WILSON, I B (1983), "The Preparation of Local Plans in France", **Town Planning Review**, Vol 54, No 2, pp 155-173                                                                    (LP)

WILSON, I B (1985), "Decentralising or Recentralising the State? Urban Planning and Centre-Periphery Relations", in P G Cerny, and M A Schain, **op.cit**, pp 186-201.          (LG)

WOOD, C & LEE, N (1978), **Physical Planning in the Member States of the European Economic Community**, Manchester: University of Manchester, Department of Town and Country Planning, Occasional Paper No 2, pp 30-36              (G)

*Key to classification of reference*

| | | | |
|---|---|---|---|
| C | Context | RP | Regional Planning |
| DC | development control | SP | Strategic Planning |
| | | LP | Local Planning |
| H | History of Planning | | |
| L | Law | GP | Green Planning |
| | | RuP | Rural Planning |
| | | ReP | Retail Planning |
| G | General reviews of planning | | |
| LG | local government | | |

\*     indicates particularly useful sources

# APPENDIX A   CONTENTS OF THE *CODE DE L'URBANISME*

## 1.   GENERAL RULES OF PLAN-MAKING AND URBAN PLANNING

I  General rule of Land Use
   General rules of urban planning;  PLD, transition arrangements

II  Forecasts and rules of urban planning
   General communal dispositions for the SD+POS, SD, POS, Transitional,
   Arrangements Diverse Dispositions, servitudes of public utilities

III  Wooded Spaces

IV  Dispositions particular to certain parts of the territory
   (i)    Paris and Ile de France, SD + POS
   (ii)   Natural sensitive spaces of Depts.
   (iii)  Protection of certain communes
   (iv)   Corsica
   (v)    Dispositions particular to mountain zones - principles of planning, new
          tourist units
   (vi)   Dispositions particular to the coast
   (vii)  Dispositions particular to noise zones of aerodromes

V  Application to Departments overseas

VI  Sanctions and servitudes

## 2.   EXPROPRIATION AND LAND RESERVES (COMPULSORY PURCHASE)

I  Law of Preemption

   (i)    urban
   (ii)   ZAD
   (iii)  common dispositions to urban expropriation and ZAD
   (iv)   -
   (v)    ZUP
   (vi)   domestic gardens

II  Land reserves

   (i)    Land reserves
   (ii)   -

III  Departments Overseas

## 3.   LAND PLANNING (INCLUDING RENEWAL)

I  Operations of plan making

   (i)    ZAC
   (ii)   -
   (iii)  Building restriction and *secteurs sauvegardés*
          SS, building restriction, common dispositions
   (iv)   Protection of occupants
   (v)    *Lotissement* (subdivision)
   (vi)   Penal sanctions relative to *lotissement*

(vii) Improvement of certain *lotissements*, subventions of the State, Loans for Department banks, other participations, particular rules of the functioning of syndicate associations

(viii) dispositions relative to certain operations bringing down and transfer of property, application to businesses and artisans, particular dispositions

II Systems of Execution

    (i)    Public establishments of planning
    (ii)   Urban land associations (AFU)
    (iii)  Chambers of commerce and industry and trade councils

III Financial dispositions

    (i)    FNAFU
    (ii)   participation of builders and subdividers, participation in case of exceeding COS, participation in realisation of public servicing, exactable on the occasion of granting authorisation to builder or beneficiaries of authorisations to occupy or use land
    (iii)  payment resulting from exceeding PLD
    (iv)  diverse dispositions

IV Dispositions applicable to overseas Departments

## 4. RULES APPLICABLE TO THE ACT OF BUILDING AND TO DIVERSE MODES OF LAND USE

I     *Certificat d'Urbanisme* (outline planning permission)
II    *Permis de construire* (planning permission) general régime, exceptions, permis de construire with precarious title
III   Demolition permit
IV   Dispositions relative to particular modes of land use enclosures, installations and diverse works, camping and caravan parking, ski lifts etc
V    Diverse dispositions
Disposition for certain uses of built surfaces, communal courts, industrial constructions outside POS creation and construction of large shops (hypermarkets)
Dispositions relative to the Paris Region
VI   Control
VII  Departments overseas
VIII Infractions

## 5. SITING OF SERVICES, ESTABLISHMENTS AND ENTERPRISES

I     General administrative dispositions
II    Financial dispositions concerning the Paris Region
III   Siting outside of the Paris region of certain activities
IV   Construction or planning of buildings for industrial use in view of their resale
V    Sanctions

## 6. ADMINISTRATIVE SYSTEMS AND DIVERSE DISPOSITIONS

I     Consultative organisms
Departmental commissions of urban planning, permanent conference of *permis de construire*, committee of planning in Paris region
II    -
III   Final dispositions

# APPENDIX B    *CODE DE LA CONSTRUCTION ET DE L'HABITATION*

(Building and Housing Regulations)

This is a translation of the table of contents of the Code. However, only the sub-sections of Book One Title One are included since the other books refer to issues of more marginal interest to development control. Elsewhere only the titles in each book are included.

## BOOK 1    GENERAL DISPOSITIONS

Title I      Construction of buildings
1.      general rules
(i) dispositions applicable to all buildings (ii) dispositions applicable to housing (iii) handicapped persons, (iv) thermal characteristics (v) sound insulation (vi) responsibility of constructors/or workmanship (vii) technical control (viii) insurance of building works (ix) common dispositions

2.      special dispositions
(i) construction bordering the highway (ii) boring and underground works (iii) private property servitudes (iv) view servitudes (v) relay antennae (vi) building near munition works (vii) building near forests (viii) nuisances due to certain activities

Title II      Security and protection against fire
1.      protection against fire, classification of materials
2.      dispositions of safety relative to high buildings
3.      protection against risks of fire and panic in public buildings
4.      adaptation of buildings during time of war

Title III      Heating and decay of buildings
1.      heating of buildings
2.      cleaning of buildings

Title IV      Dispositions relating to the building industry
1.      aid to productivity: coordination of infrastructure programmes
2.      studies and technical research interesting the building industry

Title V      Control and penal sanctions
1.      measures of control applicable to all buildings
2.      penal sanctions

## BOOK 2    LAW OF BUILDERS

Title 1      Statute of building societies
Title 2      Promotion of building
Title 3      Contract for construction of individual house
Title 4      Diverse common regulations
Title 5      Construction lease
Title 6      Sale of buildings to be constructed

**BOOK 3   DIVERSE AID TO HOUSE-BUILDING, IMPROVEMENT OF HOUSING AND PERSONAL AID FOR HOUSING**

Housing aid policy

Title 1    Measures tending to favour the construction
           of dwellings
Title 2    Improvement of the habitat
Title 3    Loans for construction, acquisition and
           improvement of dwellings, replacing
           personal aid to housing
Title 4    Repayment of state aid
Title 5    Personal aid to housing
Title 6    Consultative organisms


**BOOK 4   HOUSING OF MODERATE RENT
           (HLM's)**

Title 1    General dispositions
Title 2    HLM organisations
Title 3    Financial dispositions
Title 4    Reports/accounts of HLM organisations
           and beneficiaries
Title 5    Control
Title 6    Consultative organisations
Title 8    Dispositions particular to SEM's


**BOOK 5   INSALUBRIOUS BUILDINGS OR
           BUILDINGS FACING RUIN**

Title 1    Buildings facing ruin
Title 2    Insalubrious buildings


**BOOK 6   MEASURES TENDING TO REMEDY
           THE EXCEPTIONAL DIFFICULTIES OF
           HOUSING**

Title 1    General dispositions
Title 2    Dispositions tending to facilitate and orient
           the allocation of existing housing etc
Title 3    Dispositions tending to maintain or
           augment the number of housing units
Title 4    Housing functions (requisitioning)
Title 5    Sanctions and diverse dispositions

**VILLE DE PARIS**
DIRECTION DE LA CONSTRUCTION ET DU LOGEMENT
Sous-direction du Permis de construire

# DEMANDE DE PERMIS DE CONSTRUIRE

P.C. n°

............e Arrond.          Adresse des travaux :

*SPECIMEN*

Date de recevabilité

**INSTRUCTION**

| | Délai | Date limite |
|---|---|---|
| | ............ mois ............ | |
| MODIFICATIFS | ............ mois ............ | |
| | ............ mois ............ | |
| | ............ mois ............ | |
| | ............ mois ............ | |

T. L. E. n°                    REDEVANCE n°

P. L. D. n°                    DEMOLITION n°   D

| SERVICES CONSULTES | ENVOI LE | RETOUR LE | AVIS (1) |
|---|---|---|---|
| **SERVICES MUNICIPAUX** | | | |
| **Direction de l'Aménagement urbain** | | | |
| Secrétariat ..................... | | | |
| **Direction de la Voirie** | | | |
| Service technique des Etudes.......... | | | |
| Service technique de la Voie publique........... | | | |
| **Direction de la Propreté** | | | |
| Service technique du Nettoiement............. | | | |
| **Direction de la Construction et du logement** | | | |
| Bureau des transformations de locaux...... | | | |
| Section technique de l'Habitat........ | | | |
| Service de la Politique foncière.............. | | | |
| Sous-direction de la Gestion du domaine privé.... | | | |
| **Direction des Services industriels et commerciaux** | | | |
| Inspection générale des Carrières.............. | | | |
| Service technique des Eaux............ | | | |
| Service technique de l'Assainissement.......... | | | |
| **Direction des Parcs, jardins et espaces verts** | | | |
| Sous-direction des Parcs et jardins............ | | | |
| **Direction des Finances et des affaires économiques** | | | |
| Sous-direction des Affaires économiques......... | | | |
| **Direction des Affaires scolaires** | | | |
| Affaires scolaires .............. | | | |
| **Direction de la Jeunesse et des sports** | | | |
| Jeunesse et sports ............... | | | |
| **Direction de l'Action sociale, de l'hygiène et de la santé** | | | |
| Equipements de P.M.I. ............. | | | |
| **Direction des Affaires culturelles** | | | |
| Secrétariat .............. | | | |
| **CONSULTATIONS DIVERSES** | | | |
| Section technique d'urbanisme opérationnel....... | | | |
| ...................... | | | |

| SERVICES CONSULTES | ENVOI LE | RETOUR LE | AVIS (1) |
|---|---|---|---|
| **SERVICES EXTERIEURS A LA VILLE DE PARIS** | | | |
| **Préfecture de Paris** | | | |
| D.U.E. (Aide à la Construction)................ | | | |
| **Direction des Affaires sanitaires et sociales de Paris** | | | |
| Bureau de l'Equipement.................... | | | |
| **Préfecture de police** | | | |
| Direction de la Prévention et de la protection civile. | | | |
| **Ministère du Travail** | | | |
| Inspection du Travail et de l'emploi............ | | | |
| **Ministère des Transports** | | | |
| Inspection du Travail et de la main-d'œuvre des transports ........... | | | |
| **Ministère de l'Urbanisme et du logement** | | | |
| Architecte des Bâtiments de France........... | | | |
| **R A T P** | | | |
| Direction des Travaux neufs................. | | | |
| **S N C F** | | | |
| Service du Domaine ............. | | | |
| **Ministère de l'Economie et des finances** | | | |
| **Direction générale des impôts** | | | |
| Services fiscaux ............... | | | |
| **CONSULTATIONS DIVERSES** | | | |

C.E.P. ................ du ...........
C.P.P.C. ............... du ...........
COMMISSION DES SITES. du ...........
☐ COMMISSION D'URBANISME COMMERCIAL
☐ COMMISSION SUPERIEURE D'ARCHITECTURE
☐ ...........

PERMIS DELIVRE LE...........

RECONDUCTION : ...........
TRANSFERT : ...........
REFUS DU ...........
SURSIS A STATUER DU ...........

DROITS DE VOIRIE   ☐ Oui   ☐ Non

(1) AVIS : (F) Favorable (et avec réserves).   (D) Défavorable   (C) Complexe.

**Plate 8    The dossier cover for a *permis de construire* in Paris**

Plate 8 is a standard dossier cover into which all material relating to the application is placed. It provides a convenient vehicle for describing the range of possible consultations, and it can be seen from the bottom of the form that the advice is summarised as favourable, unfavourable or complex. For each consultation the date of sending and return is recorded. The top section of the form records the all important date of 'receivability' and the prescribed period for decision-making, and assigns reference numbers for assessing the relevant taxes. Although some 90 or so consultations are possible it is mainly those with the municipal services which are significant, though in Paris the majority of applications require special consultation of the *Commission des Sites* etc. because of conservation controls.

# DEMANDE DE PERMIS DE CONSTRUIRE

## 1 DEMANDEUR (le demandeur est le bénéficiaire de la future autorisation)

NOM, PRÉNOMS OU DÉNOMINATION :

TÉLÉPHONE :

| Office public d'HLM | Société d'HLM de crédit immobilier coopérative de production | Société d'économie mixte | SCI autres sociétés vouées à la construction | Entreprise ou établissement à caractère industriel ou commercial | Collectivité locale | État ou Administration | Autre personne morale | Particulier |
|---|---|---|---|---|---|---|---|---|
| 1 | 2 | 3 | 4 | 5 | 6 | 7 | 8 | 9 |

ADRESSE (numéro, voie) :

Lieu-dit ou suite de l'adresse :

Commune :

Code postal :

Bureau distributeur :

PERSONNE MORALE (Nom du représentant légal ou statutaire) :

PROMOTEUR IMMOBILIER (le cas échéant) NOM ou DÉNOMINATION :

ADRESSE :

TÉLÉPHONE :

## 2 TERRAIN (le terrain est l'îlot de propriété constitué par la parcelle ou par l'ensemble des parcelles contiguës appartenant à un même propriétaire ou à une même indivision)

### 21. DÉSIGNATION DU TERRAIN

ADRESSE DU TERRAIN (Numéro, voie, lieu-dit) :

Commune :

Code postal :

Bureau distributeur :

NOM ET ADRESSE DU PROPRIÉTAIRE DU TERRAIN (s'il n'est pas le demandeur) :

### 22. CADASTRE ET REMEMBREMENT

SUPERFICIE de la parcelle constituant la propriété : _____ m²

Le terrain est-il inclus dans le périmètre d'une opération de REMEMBREMENT RURAL en cours de réalisation ? OUI NON

Le terrain a-t-il DÉJÀ fait l'objet d'une opération de REMEMBREMENT RURAL ? OUI NON

SECTIONS CADASTRALES, et pour chaque section le(s) numéro(s) de la ou des parcelles :

### 23. SITUATION JURIDIQUE DU TERRAIN

1. Le terrain est-il situé dans un lotissement ? OUI NON — NOM DU LOTISSEMENT OU DU LOTISSEUR :

NUMÉRO(S) DU OU DES LOTS : — LOTISSEMENT autorisé le — Surface hors-œuvre nette constructible sur le lot

2. Le terrain est-il issu depuis MOINS DE DIX ANS d'une plus grande propriété ? OUI NON — SI OUI, DATE D'ACQUISITION :

3. UN CERTIFICAT D'URBANISME a-t-il été délivré pour le terrain ? OUI NON — DATE DU CERTIFICAT : — NUMÉRO DU CERTIFICAT

4. S'agit-il d'un terrain provenant de la DIVISION d'une propriété bâtie ? OUI NON

### 24. OCCUPATION ACTUELLE DU TERRAIN

1. Existe-t-il des bâtiments sur ce terrain ? OUI NON (dans l'affirmative, remplir la rubrique 341)

2. Parmi ces bâtiments, y en a-t-il qui sont destinés à être démolis à l'occasion de la réalisation du projet ? OUI NON (dans l'affirmative, remplir la rubrique 342)

3. Y a-t-il eu sur le terrain des bâtiments qui ont été démolis depuis le 1-4-1976 ? OUI NON (dans l'affirmative, remplir la rubrique 344)

## DEMANDE (FACULTATIVE) D'ARRÊTÉ D'ALIGNEMENT

Aucune construction ou installation ne peut être élevée en bordure d'une voie publique sans être conforme à l'alignement.
L'arrêté d'alignement permet au demandeur de connaître avec exactitude la ou les limites actuelles ou futures du domaine public routier en bordure du terrain sur lequel la construction ou l'installation est envisagée.
Cette demande sera transmise au(x) service(s) compétent(s) ; le ou les arrêtés d'alignement seront adressés directement aux demandeurs.

Je demande que me soit délivré le ou les arrêtés d'alignement en application de l'article L. 112-1 du Code de la construction pour la ou les voies bordant le terrain désigné ci-dessus au cadre 2.

DATE :

SIGNATURE :

## 3 PROJET

### 31. ANTÉRIORITÉ

Si le projet a fait l'objet d'une demande de permis de construire antérieure, indiquer ci-contre son numéro :

Cachet et signature de l'architecte ou de l'agréé en architecture

### 32. AUTEUR DU PROJET ARCHITECTURAL

| Architecte | Agréé en architecture | Maître-d'œuvre indépendant | Demandeur | Autres | QUALITÉ : |
|---|---|---|---|---|---|

NOM, PRÉNOMS ET ADRESSE (dans le cas où l'auteur du projet architectural n'est pas le demandeur) :

### 33. CARACTÉRISTIQUES DU PROJET

#### 331. NATURE DES TRAVAUX (cocher l'une des cases numérotées de 1 à 6, puis répondre aux questions complémentaires pour les cas 1, 4, 5, 6)

| Construction neuve | Extension ou surélévation d'un bâtiment existant | Création de niveaux supplémentaires à l'intérieur d'un bâtiment existant | Changement de destination des locaux à l'intérieur d'un bâtiment existant | Installation de locaux dépourvus de fondations | Autres travaux |
|---|---|---|---|---|---|
| 1 | 2 | 3 | 4 | 5 | 6 |

Nombre de bâtiments créés

ANCIENNE DESTINATION ◁

| Logement | Hébergement hôtelier | Commerces ou artisanat | Entrepôts commerciaux | Locaux industriels | Bâtiments agricoles | Bureaux | Aires de stationnement | Combles ou sous-sols non aménagés |
|---|---|---|---|---|---|---|---|---|

Autre

NATURE DES TRAVAUX

| Maison mobile | Chalet démontable | Autre (à préciser au dessous) |
|---|---|---|

Nature (à préciser ci-dessous)

#### 332. DESTINATION DES LOCAUX PROJETÉS

| Logement | Hébergement hôtelier | Commerces ou artisanat | Entrepôts commerciaux | Locaux industriels | Bâtiments agricoles | Bureaux | Locaux relevant du régime des installations classées | Autre (à préciser) |
|---|---|---|---|---|---|---|---|---|

Dans le cas où la construction projetée concerne un ouvrage de production, de transport, de distribution ou de stockage de l'énergie, cette énergie est-elle destinée principalement à une utilisation directe par le consommateur ? OUI NON

#### 333. ASPECT EXTÉRIEUR (Pour les projets complexes, compléter, si besoin est, cette rubrique par une notice descriptive ou des indications sur les plans de façades)

| ENDUITS EXTÉRIEURS ▷ | MATÉRIAUX APPARENTS EN FAÇADE | MENUISERIES EXTÉRIEURES EN FAÇADE | TOITURES | CLÔTURE |
|---|---|---|---|---|
| NATURE | | | | |
| COULEUR | | | | |

#### 334. AIRES DE STATIONNEMENT (En cas d'impossibilité de réaliser les aires de stationnement, voir rubrique B 15 dans le tableau des pièces complémentaires à joindre éventuellement)

| AIRES DE STATIONNEMENT AFFECTÉES AUX BESOINS DE L'OPÉRATION ▷ | (a) DANS LES BÂTIMENTS | (b) EN SURFACE | (c) EN DEHORS DU TERRAIN AFFECTÉ À L'OPÉRATION | TOTAL | DANS LE CAS (c), INDIQUER CI-DESSOUS L'ADRESSE DES AIRES DE STATIONNEMENT |
|---|---|---|---|---|---|
| NOMBRE D'EMPLACEMENTS | + | + | = | | |
| SURFACE EN MÈTRES CARRÉS (y compris les circulations) | + | + | = | | DISTANCE À PARCOURIR ENTRE LE TERRAIN DE LA CONSTRUCTION PROJETÉE ET L'AIRE DE STATIONNEMENT AFFECTÉE À L'OPÉRATION _____ m |

#### 335. ESPACES VERTS

| NOMBRE D'ARBRES DE HAUTE TIGE À ABATTRE | NOMBRE D'ARBRES DE HAUTE TIGE À CONSERVER | NOMBRE D'ARBRES DE HAUTE TIGE À PLANTER | SUPERFICIE TOTALE DES ESPACES VERTS PLANTÉS OU GAZONNÉS _____ m² |
|---|---|---|---|

336. DENSITÉ DE LA CONSTRUCTION

246

**34. DENSITÉ DE LA CONSTRUCTION** *(Voir notice explicative)*

| | NIVEAUX | SURFACES HORS ŒUVRE BRUTES (S) | SURFACES DÉDUITES — Combles et sous-sols non aménageables (a) | Terrasses, balcons et loggias surfaces non closes au rez-de-chaussée (b) | Stationnement des véhicules dans la construction (c) | Logement des récoltes, des animaux, du matériel agricole, serres de production (d) | SURFACES HORS ŒUVRE NETTES S − (a + b + c + d) (Sn) |
|---|---|---|---|---|---|---|---|
| 341 BÂTIMENTS EXISTANTS AVANT L'OPÉRATION | TOTAL | (S) m² | (a) m² | (b) m² | (c) m² | (d) m² | (Sn) m² |
| 342 BÂTIMENTS DESTINÉS A ÊTRE DÉMOLIS | TOTAL | (S) m² | (a) m² | (b) m² | (c) m² | (d) m² | (Sn) m² |
| 343 BÂTIMENTS CONSERVÉS | | SURFACE HORS ŒUVRE NETTE DES BÂTIMENTS CONSERVÉS   Sb = Sb1 − Sb2 | | | | | (Sn) m² |
| 344 | | (Éventuellement) SURFACE HORS ŒUVRE NETTE AVANT LE 1-4-1976 | | | | | m² |
| 345 SURFACES CRÉÉES A L'OCCASION DU PROJET | | | | | | | |
| | TOTAL | (S) m² | (a) m² | (b) m² | (c) m² | (d) m² | (Sn) m² |
| 346 VALEUR DU TERRAIN | | (A NE REMPLIR QUE POUR LES CAS PRÉVUS DANS LA NOTICE EXPLICATIVE)   VALEUR DU TERRAIN « NU ET LIBRE » | | | | | F/m² |

*NE RIEN INSCRIRE DANS LE CADRE CI-DESSOUS RÉSERVÉ A L'ADMINISTRATION*
DATE DOSSIER COMPLET — JOUR MOIS AN
DATE DÉCISION — JOUR MOIS AN
ZONAGE POS — 1 POS publié / POS approuvé / 2 Pas de POS — Zone — 01 En zone agglo. / 02 Hors zone agglo.
NATURE DE LA ZONE — Zone opérationnelle 3 / Lotissement 4 / Hors zone 5
RÉSERVE FONCIÈRE — ZIF 1 / ZAD 2 / Pré-ZAD 3 / Hors zone 4
AUTORITÉ COMPÉTENTE — Maire au nom de la commune 1 / Président EPCI 2 / Maire au nom de l'État 3 / Commission de la République 4
NATURE DE LA DÉCISION — Favorable / Défavorable 2 / Sursis 3 / 4

**35. HAUTEUR DE LA CONSTRUCTION**
LA HAUTEUR MAXIMALE DE LA CONSTRUCTION EST DE [   ] MÈTRES

**36. PARTIE HABITATION**

361. COMPOSITION — Agrandissement de logements existants ▷ Nombre de pièces créées [   ] — Construction ou création de nouveaux logements ▷ Nombre de logements créés [   ] — Réalisation de chambres (hôtels, foyers, centres) ▷ Nombre de chambres créées [   ]

362. FINANCEMENT ENVISAGÉ — Nombre de logements dont le financement principal relève de ▷ Prêt à l'accession à la propriété (PAP) [   ] / Prêt locatif aidé (PLA) [   ] / Prêt conventionné [   ] / Autre financement [   ]

363. UTILISATION PRINCIPALE ENVISAGÉE — Occupation personnelle 1 / Vente ou location vente 2 / Location vide ou meublée 3 / Mixte à majorité location 4 / Mixte à majorité accession 5 / Logement de fonction ou occupation à titre gratuit 6

364. DESTINATION — Résidence(s) principale(s) 1 / Résidence(s) secondaire(s) 2

365. HABITATION COMMUNAUTAIRE — Foyer de personnes âgées 1 / Foyer de jeunes travailleurs 2 / Foyers de travailleurs immigrants 3 / Foyers pour enfants ou adolescents 4 / Logement de transition 5 / Cité universitaire 6 / Casernement (gendarmes ou sapeurs pompiers) 7 / Autre 8

366. HÉBERGEMENT (commercial ou non) — Pension de famille 1 / Colonie de vacances 2 / Motel 3 / Hôtel 4 / Autre (dont internat) 8

367. HABITATION : NOMBRE PRÉVU DE — Chambres indépendantes [   ] / Logts 1 pièce [   ] / Logts 2 pièces [   ] / Logts 3 pièces [   ] / Logts 4 pièces [   ] / Logts 5 pièces [   ] / Logts 6 pièces et + [   ]

368. TYPE DE CONSTRUCTION — HABITAT INDIVIDUEL Nombre de logts [   ] / HABITAT COLLECTIF Nombre de logts [   ] / Nombre maximum de niveaux prévus [   ]

**37. PARTIE A USAGE AUTRE QU'HABITATION** Indiquer les trois principales destinations prévues ainsi que leur surface hors œuvre nette en mètres carrés (brute pour les locaux agricoles)

| | Destination | Surface hors œuvre | Cadre réservé | | Destination | Surface hors œuvre | Cadre réservé |
|---|---|---|---|---|---|---|---|
| 1. | | | | 2. | | | |
| 3. | | | | SURFACE TOTALE DE LA PARTIE A USAGE AUTRE QU'HABITATION = | | | |

**4  ENGAGEMENTS DU DEMANDEUR** *(Voir notice explicative)*

Je soussigné, auteur de la présente demande : CERTIFIE exacts les renseignements qui y sont contenus et m'engage à respecter les règles générales de construction prescrites par les textes pris en application de l'article L. 111-3 du Code de l'Urbanisme. (L'attention du demandeur est appelée sur les articles L. 480-1 à 480-12 du Code de l'Urbanisme relatifs aux sanctions pénales applicables en cas de violation des règles de construction prescrite.)

Pour les personnes physiques, je déclare édifier ou modifier la construction projetée [ ] POUR MOI-MÊME [ ] POUR AUTRUI

NOM :        SIGNATURE
DATE :

**Plate 9   The application form for a *permis de construire***

Plate 9 is the form for the application for a *permis de construire* and it illustrates the standardised, detailed information which is required about the applicant (1); the plot of land (2) its *cadaster* and legal ownership (22–3); the architect (32); the nature of the project (33), floorspace (34), height (35) and housing composition (36). The form allows the applicant to request the local authority to provide details of road alignments that affect the property. All permissions and preliminary declarations have similarly standardised and comprehensive forms.

247

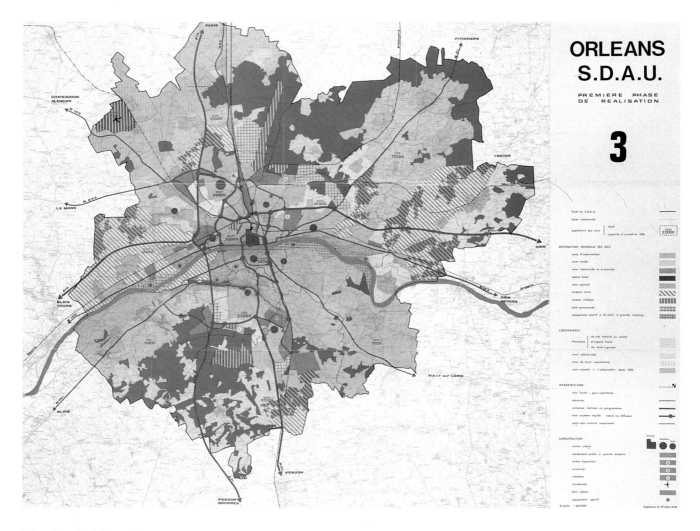

**Plate 10**  A *Schéma Directeur d'Aménagement et d'Urbanisme*
for Orléans (the 10 year, first phase of realisation,
map)

Plate 10: The content of the SDAU and its successor the SD
are closely prescribed by the *Code de l'Urbanisme* so that each
follows a standard format with maps possessing a standard
legend. This is the ten-year projection map which portrays
Orléans in 1984, and provides a convenient reference point to
assess its contemporary relevance.

This particular SDAU was approved in 1974 and embraced
40 communes in and around Orléans. As such it relied upon
projections of growth based upon data from the 1960's and it
predicted a population growth rate of about 3 per cent from
1968 to 1985, a significant over-estimate. This map reveals the
projected situation in 1985 with an elaborate network of
roads largely unrealised, and major residential expansion
which will not take place. In the meantime, industrial
developments or science parks in the communes on the edge
of Orléans unimagined by the plan emphasise the SD's
essential irrelevance as a planning document; an obstacle to,
rather than an instrument of, sound planning.

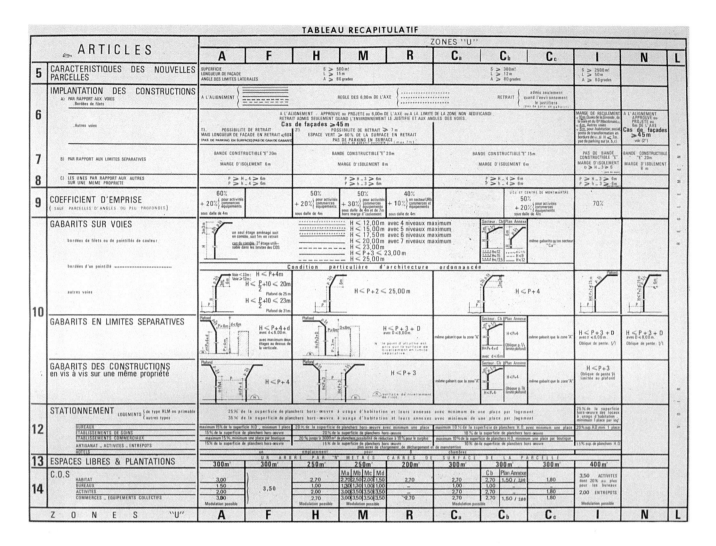

**Plate 11  Recapitulative table of POS regulations for 'U'
Zones in Paris**

Plate 11 shows the summary table taken from the Paris POS.
It illustrates clearly the technical nature of the POS
regulations as they apply to building form, envelope, and
space around the building. Along the horizontal axis the
different Urban Zones are denoted while on the vertical axis
the different articles of the regulation are listed (nos. 5–14).

Minimum parcel sizes (5), siting regulations (6–8), coverage of
the plot (9) with incentive schemes for including special uses,
building envelope (10), amounts of parking (12), tree planting
and landscaping (13) and the COS (14) are spelt out as a set
of basic rules ensuring certain maximum amounts of
development and certain minimum amenities.

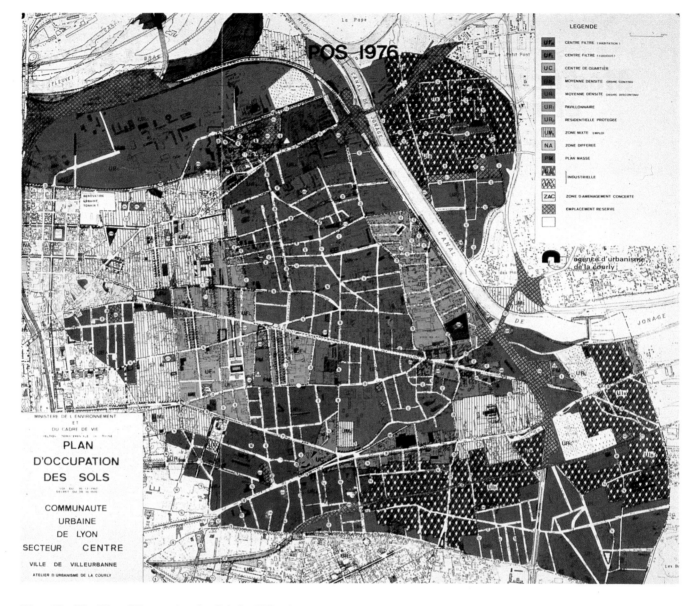

**Plate 12    The *Plan d'Occupation des Sols* for Villeurbanne, Isère 1976**

These two POS maps in Plates 12 and 13 provide an illustration of the basic zoning map which is the key graphic document in the POS. The legend reveals the different zones as well as additional designations for public infrastructure, redevelopment zones, road widenings, etc.

Villeurbanne is a late nineteenth, early twentieth century suburb north east of Lyon. Land use and building form are very mixed, although one would not get that impression from the broad brush zoning map of 1976 which rigidly segregates land uses (Plate 12). Elaborated in a period of high growth in the early 1970's, the plan was clearly inappropriate for the development context of the late 1970's and the process of revision began immediately.

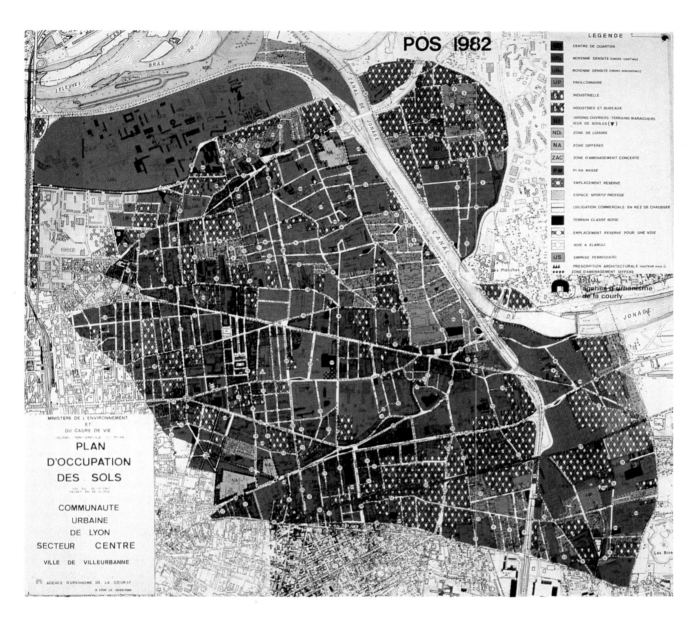

**Plate 13   The *Plan d'Occupation des Sols* for Villeurbanne, Isère 1982**

In the 1982 plan (Plate 13) the main aim was to preserve the fragile equilibrium of land use, employment and social mix, and the zoning explicitly rejects the segregation of uses proposed in 1976. The most striking contrast between the 1976 and 1982 maps is the distribution of industrial zoning, so concentrated in 1976, so dispersed in 1982. This is a direct attempt to counter de-industrialisation by encouraging the landowners to retain their industrial land uses, and offering them the incentive of a certain proportion of office uses in return. Other innovations included architectural prescriptions and the protection of shops in certain streets. All in all, the 1982 POS of Villeurbanne epitomises the attempts to make the POS more sophisticated, flexible, and responsive to local conditions.

251

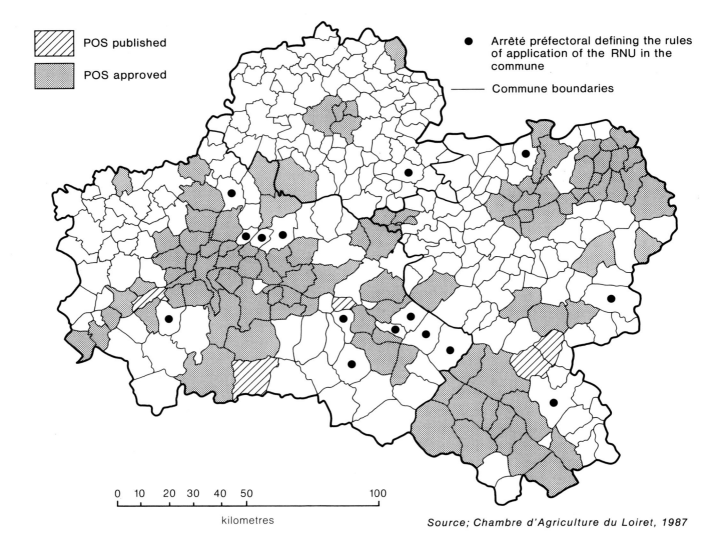

POS published

POS approved

● Arrêté préfectoral defining the rules of application of the RNU in the commune

—— Commune boundaries

0  10  20  30  40  50                    100

kilometres

*Source; Chambre d'Agriculture du Loiret, 1987*

**Plate 14   Coverage of the POS for the department of Loiret (1987)**

Plate 14 reveals the patchy coverage of POS in the largely agricultural department of Loiret. Even around Orléans itself coverage is not entirely consistent with the pressure for development, while by contrast in the north east and south east there is good coverage in very rural communes. The map also reveals POS that have reached the publication stage, and those communes where a prefectoral *arrêté* has produced a guide to the RNU as a first step on the road towards prescribing a POS (see paragraph F4.101).

# FEDERAL REPUBLIC OF GERMANY

AJ Hooper

# G1   THE BACKGROUND TO PLANNING AND CONTROL

## Introduction

G1.1   Any study involving international comparison in the field of public policy inevitably raises intellectual problems relating to the 'systems' to be compared, and to the identification of their special characteristics. A necessary first step, given the focus of public policy, is to establish the legal concept of the state in the various countries to be compared, in order to identify the relevant institutions which exercise legitimate authority and which comprise the state apparatus. This is particularly necessary when (as is the case in this present research) the countries to be compared all have a multi-tier structure of government. Hence there is a need to establish the precise nature of central-local state relationships, and the context of the legal status of the respective government authorities, in order to provide a structure for comparative purposes.

G1.2   However, the articulation of such structures carries with it associated dangers relating to the artificial presentation of 'fossilised' or 'ideal' systems of government and policy formulation/ implementation. The complex flux of political interests which provides the bond linking central and local governments in all countries can easily be lost in the search for accurate description of government institutions and policy channels. In particular, given structures and institutions may evolve over time and operate in quite different ways in response to changing economic, political and social realities, whilst superficially presenting a static or enduring form. In other words, the processes of public policy are as relevant as the structures through which they take place, in any comparative international analysis.

G1.3   Perhaps the greatest danger in comparative analyses of this type is the temptation to identify the phenomena to be compared through the blinkered perspective of one country (an almost inevitable difficulty given the cultural problems associated with international studies). Having identified a policy field in one country, its 'equivalents' are sought out in other countries and policy instruments compared directly in terms of their effectiveness in achieving expressed goals. Such analyses inevitably distort the universe under observation, and can too readily descend to the simple parade of prejudice.

G1.4   Accordingly, in the account which follows, an attempt has been made to present the basic outline of the structure and process of the planning system and practice for the control of development in the Federal Republic of Germany (FRG) as dynamically evolving in response to external and internal pressures. The views of both academic commentators on, practitioners in, and consumers of, this 'system' are presented where appropriate in order to provide an informed critical perspective on contemporary developments.

G1.5   Whilst most German political and legal institutions have a long and complex history, the foundation of the Federal Republic of Germany in 1949 radically altered most of these, and created many which were new. The basis of the new German state was to be parliamentary democracy, legal security, civic liberties, human rights, private property and private enterprise. A Basic Law (*Grundgesetz*) was drawn up by a Parliamentary Council of delegates from the *Länder* parliaments from September 1948 and adopted in May, 1949. The choice of the term 'Basic Law' rather than 'constitution' was intended to signify the provisional nature of the measures rather than the final constitution of a separate state. With the coming into force of the Basic Law, the Federal Republic of Germany came into existence, sanctioned by the election of its first parliament, the *Deutsche Bundestag* on August 14, 1949. The provisional constitution has become permanent for the forseeable future, providing a viable foundation for a stable, democratic system of society (Bulka et al, 1986, pp.70-71, 92).

## The Basic Law

G1.6   The control of development and the framework system of physical development planning in the FRG can be properly understood only within the context of the constitutional provisions established in the *Grundgesetz* (GG - Basic Law) of 1949. The Basic Law (Article 20, paragraph 1) establishes the FRG as a democratic and social federation, comprising 11 *Länder* (States) (Schmidt-Assmann, 1986, p.335; David, 1986, p.16).

G1.7   The principle of federalism ensures that legislative, administrative and judicial powers are distributed between the Federal and *Länder* parliaments and administrations, local government authorities and the law courts. This principle is enshrined in a written constitution, protected by the constitutional law courts (David, 1986, p.16).

G1.8   The legal structures of this federal form of government thus ensure

(i) "a constitutional division of powers between the levels of government, so that each level has an autonomous right of decision in some areas of government; neither derives its authority from the other, and neither can change the legal relationship between them unilaterally

(ii) each of the political authorities also possess the quality and the apparatus of state within the sphere of responsibilities assigned to it". (Blair, quoted in Uppendahl, 1985, p.22).

G1.9    The legal system enshrines fundamental principles laid down in the Basic Law

(i)   guarantees of civil rights (*Grundrechte*)

(ii)  a guarantee that any person affected by public decisions may have recourse to the law courts (*Rechtsweggarantie*, Article 19, paragraph 14, GG). This right may not be restricted by any law. (Schmidt-Assmann, 1986, p.339)

(iii) the establishment of the 'rule of law' as a constitutional principle (*Rechtsstaatsprinzip*), with the judiciary subject to that law (ie in particular to the written law) (Article 20, paragraph 3, GG) (David, 1986, pp.19, 43)

## Parliamentary Institutions and the Constitution

G1.10    The *Bund* (Federation) and *Länder* are mutually independent entities having their own parliaments and constitutions (Schmidt-Assmann, 1986, p.335). Since both bodies have legislative powers, a relationship of 'competitive legislation' exists between the two (Ahrens et al, 1980, p.3), though the *Länder* laws may not contradict those of the *Bund*. Hence Federal Law is superordinate to *Länder* Law, and the Federal government is charged with ensuring that *Länder* laws are compatible with each other, do not adversely affect other *Länder*, or infringe the constitutional "principle of equality of living conditions" (*Gleichwertigkeit der Lebensbedingungen*; Article 72, paragraph 2N.3 GG) between and within *Länder* (Ahrens et al, 1980, p.1).

G1.11    The legislature at the Federal level comprises three institutions.

(i)   The *Bundestag* (House of Representatives) - elects the *Bundeskanzler* (Federal Chancellor) and is responsible for enacting Federal legislation. The government is solely responsible to this body, which is elected for four years by a system of 'personalised proportional representation.'

(ii)  The *Bundesrat* (Federal Council) - has to agree all acts of the *Bundestag* which affect *Länder* affairs or influence the 'competitive legislation' of the *Länder* and the Federal government. There is no statutory definition of the scope of these powers but if there is disagreement between the two bodies a mediation committee of both chambers seeks a compromise. In

the absence of such a compromise, the *Bundestag* can overrule the objections of the *Bundesrat*. The *Bundesrat* is not elected, but consists of members of the *Länder* governments or their delegates.

(iii) The *Bundesversammlung* (Federal Convention) - whose sole function is to elect the *Bundespräsident* (Federal President), the representative of the republic who nominates the chancellor and signs all Federal Acts. (Ahrens et al, 1980, p.3; Bulka et al, 1985, pp.97-104).

G1.12    Thus legislation typically involves both main parliaments or houses on the Federal level, and the influence upon Federal legislation by the *Länder* gives them significant power in central affairs. Furthermore, other than in the special fields defined within the Basic Law (Article 30 GG) as Federal responsibilities (foreign service, defence, customs, post and telecommunications, railways, Federal waterways and shipping administration), the *Bundesregierung* (Federal government) has no administrative structure with which to execute legislation. Hence the *Länder* act as the executive arm of the state in many fields. The essence of the constitutional division of powers established by the Basic Law thus relates to the provision whereby the Federal government and parliament retain the majority of legislative powers (either exclusively or concurrently with the *Länder*) whilst the *Länder* are responsible for the greater part of administration. This has been termed the "secret" of the German constitution (Uppendahl, 1985, p.22).

G1.13    As the administrative laws and ordinances (whether Federal or *Länder*) relating to planning and the control of development are not defined as special fields within the Basic Law, all authorities acting in this domain are *Länder* or local government authorities. Hence the German system of planning and control of development mainly operates at *Länder* level or below, with the proviso that both *Länder* and local government authorities have to comply with the 'framework regulations' established under the *Raumordnung* (regulation of space) provisions of the Basic Law (Article 75, No.4 GG). The scope of these provisions has been defined by the Federal Constitutional Court as comprising "the comprehensive, superior planning and ordering of space, superior in the sense that it is above the local level combining and harmonizing the various special planning activities." These Federal framework rules establish a general pattern ensuring a basic consistency in the planning legislation of the different *Länder* and producing a broad similarity in the structure of planning authorities and their operational procedures. (Kimminich, 1986, pp.183-189).

## Local Government Authorities and the Constitution

G1.14    The Basic Law (Article 28, Section 2 GG) also guarantees, however, the principle of local autonomy to local government authorities, in that they are given "the inalienable right to direct all affairs of the local community as their own responsibility, regulated solely by public law." This constitutional right operates within the context of the legal framework established by Federal and *Länder* legislation. In a recent ruling of major constitutional significance (1983 *Rastede* decision) the Federal Administrative Court stated that Article 28 of the Basic Law guarantees to all local government authorities a "core domain of functions which must not be infringed or removed by any law." The range of matters encompassed by this relative autonomy is, however, unspecified.

G1.15 As the administration of Federal and *Länder* laws is usually delegated to local government authorities under Article 28 of the Basic Law, the latter may be seen both to represent 'higher' tiers of government locally and constitute communal self-administration. Besides being an issue of major constitutional importance, it has direct practical relevance since the implementation of the control of development in the FRG takes place largely at this local level.

## The Judiciary and the Constitution

G1.16 The legal system in the FRG is not based upon a distinction between common and civil law, but is founded upon ancient germanic legal traditions and Roman law (David, 1986, p.19). The highest court in the FRG is the Federal Constitutional Court (*Bundesverfassungsgericht*), which is the guardian of the Basic Law, and which rules in disputes between the Federation and the *Länder*, or between individual Federal bodies. It examines Federal and *Länder* legislation as to its conformity with the Basic Law. If it rules a law unconstitutional it can no longer be applied. The Federal Republic's law is predominantly written law, most of it Federal. With the exception of the Federal Constitutional Court, no judge is bound by any superior court's ruling on a similar case, and therefore the interpretation of the written law is of fundamental importance. The Federal German court system consists of five branches

(i) the 'ordinary courts', responsible for criminal law and civil proceedings

(ii) the labour courts, responsible for employer/ employee disputes

(iii) the administrative courts - responsible for the resolution of disputes relating to administrative law, particularly challenges by individuals against public administrative actions

(iv) the social courts - which deal with disputes relating to social insurance

(v) the fiscal courts - dealing with taxation etc.

These courts are hierarchically organised, usually on three levels and there are numerous possibilities for legal review, with in principle two appeal courts. There is a tendency for increasing resort to the courts and the invoking of appeal possibilities, with a consequent over-burdening of the courts and associated delays (Bulka et al, 1986, pp.104-110).

G1.17 The law is a major influence upon the process of administration in Germany. The Basic Law gives a constitutional right of access to the law courts for a citizen aggrieved by the actions of a public authority (see paragraph G1.9 (ii) above), and hence potentially every administrative action may be the subject of consideration by the administrative courts. Preliminary appeal procedures and the procedure of administrative courts are regulated by the *Verwaltungsgerichtsordnung* (1960 and subsequent amendments) (law regulating the law courts procedure in administrative affairs -VwGo). The VwGo applies equally in all *Länder*, but subsidiary *Länder* regulations, not required by the VwGo, may produce variability in decisions between *Länder* in relation to apparently similar cases. Since the Federal Administrative Court has ultimate authority, close attention is paid to its rulings.

G1.18 In addition, the administrative high courts may review legal provisions with a legal rank inferior to a formal law enacted either by a *Land* or the Federal government authority or as a local statute by a municipality. The specific judicial procedure is *Normenkontrollverfahren* (control of statutory rules) (David, 1986, p.41).

G1.19 It follows, therefore, that the courts which have most immediate relevance for the control of development and physical development planning are the administrative courts. Except for the Federal Administrative Court, these courts are established by each *Land* and comprise

(i) the Local Administrative Courts (*Verwaltungsgericht*)
- where an action begins, irrespective of whether the public action giving rise to the complaint emanates from a local, *Länder* or Federal authority

(ii) the Higher Administrative Courts in the *Länder* (*Oberwaltungsgericht* in northern FRG, *Verwaltungsgerichtshof* in southern FRG)
- primarily courts of fact and law

(iii) the Federal Administrative Court in Berlin (*Bundesverwaltungsgericht* - BVerwG)
- a court of legal revision, considering violations of Federal law. The Federal Constitutional Court and a *Land's* own Constitutional Court have the power to declare *Länder* laws unconstitutional.
(Schmidt-Assmann, 1986, p.340).

G1.20 Decisions of the administrative courts are subject to a right of appeal to the competent higher administrative court. Further appeal to the Federal Administrative Court in Berlin is allowed only if it has been admitted in the decision of the higher administrative court (but it must be so admitted when the case is of fundamental importance and when the decision deviates from a relevant decision of the Federal Administrative Court). In the event of the non-admittance of the further appeal, the applicant may raise a complaint with the Federal Administrative Court (Kimminich, 1986, pp.204-205).

G1.21   The *Länder* have their own constitutional courts. But the supreme court in the FRG for considering consistency with the Federal law, and in particular with the Federal Constitution, is the Federal Constitutional Court. Cases can be brought into this court by the Länder and Federal governments and by individuals. The administrative law courts have the jurisdiction to interpret a law only in a manner that is in accordance with the Basic Law. If they reach the conclusion that it is not, they have to submit the matter to the Constitutional Court for consideration (David, 1986, pp.42-43).

## The Constitution and Private Property Rights

G1.22   The German Civil Code relating to land law is based upon principles which are to a large extent alien to English law. The types of rights that a person is permitted to have in land are provided in the Code, and constitute a 'closed' catalogue in that an owner may have only those rights listed therein (Cohn, 1973, p.56). These are registered in a Land Register, which in practice is available to the interested public's inspection. In general, only by registering land can absolute proprietary land ownership rights be established.

G1.23   The development of German constitutional law relating to property rights has tended to extend the guarantee of such rights to a fundamental general constitutional principle, providing a more explicit statutory protection than in many other western parliamentary democracies. Article 14 of the Basic Law guarantees private property (*Eigentumsgarantie*), both in relation to public expropriation and in terms of legal regulation affecting private property rights. However, this guarantee is not without limitation, since the Basic Law also states (Article 14, paragraph 2 GG) that (private) property rights have to serve the public welfare (Schmidt-Assmann, 1986, p.337; David, 1986, p.71).

G1.24   The ability of the state to impose legal restrictions and obligations upon private property rights with regard to the social commitment of such rights (*Sozialbindung des Eigentums*) without the necessity for compensation is of fundamental importance for the control of development and physical development planning in the FRG. In principle nearly all the restrictions imposed by plans binding on private individuals can be defined as consequences of the social commitment of property rights. The state may thus legitimately set limits to the exercise of private property rights in relation to a wide range of considerations such as public health and safety, neighbours' rights, conserva-

tion, ecological protection and land use planning (Schmidt-Assmann, 1986, p.337; David, 1986, pp.71-72). Indeed, this may be considered to be one exemplification of the founding principles of the constitution in terms of a social federation.

G1.25   Whether the state exercises such rights in a constitutional manner is subject to redress by the courts, and may involve the civil courts (in the case of expropriation, Article 14, paragraph 3, GG) as well as the constitutional and administrative courts. The complexities of such issues has produced a large body of case law (David, 1986, pp.72-73).

G1.26   Nevertheless, the constitutional guarantee of private property leads to the 'principle of freedom of building' within the framework of the law. In principle, a person's right to build on his/her own land is guaranteed by the constitution, it is not a right vested in the state. There is a difference of opinion among legal commentators as to whether the building permit has merely an ascertaining function (in which the permit does not possess the status of a legal title but is only an administrative act establishing the fact that the projected development is in full accordance with the relevant laws) or is a constitutive act (opening the way for building operations which would otherwise not be permitted). In any event, the outcome is that "in German law building is not an activity generally forbidden and allowed only exceptionally by building permits. Rather it is an activity generally allowed governed by a host of regulations, the fulfillment of which is controlled by the building permit" (Kimminich, 1986, p.195).

G1.27   The 'principle of freedom of building' with its connotations of owners of building sites having a right to obtain a building permit improves the position of owners in relation to public authorities in case of remedies, and is a fundamental feature in the operation of the control of development. As will be seen below (section G3), it establishes a relationship between development and planning context. The whole system of land use planning in the FRG may thus be seen as a means of facilitating the development of private property rights in a socially responsible manner.

G1.28   Above all else, the significance of written law for the operation of the system of the control of development in the FRG is clearly of paramount importance. Its central objective is the establishment of certainty, both in relation to constitutional rights and duties and with regard to the use and development of land. The consequent development of a highly complex and codified system of law inevitably carries the danger of legalism (*Verrechtlichung*), which some commentators allege the German planning system suffers from (David, 1986, p.40; Schmidt-Assman, 1986, p.341).

## The Federal Administration

G2.1   The Federal government (*Bundesregierung*) is led by the *Bundeskanzler* (Federal Chancellor), who is elected by the *Bundestag* and who nominates the *Bundesminister* (Federal ministers). In administrative undertakings the power of the Federal government is more circumscribed than its legislative role (See G.1, G1.7 above). The most significant ministries related to planning are *Wirtschaft* (economy), *Finanzen* (finance), *Verkehr* (transportation) and the Federal Ministry for *Raumordnung, Bauwesen und Städtebau* (Regional Planning, Building and Urban Development - BMBau) (see Figure G1).

G2.2   This last Ministry is responsible for supervising the 'framework regulations' associated with all planning measures concerning the use of land for any purpose (*Raumordnung* - the regulation of space, Article 75, no.4, GG), and has jurisdiction for land law (*Bodenrecht*) relating to local urban planning and urban renewal (*Städtebaurecht*) (Article 74, no.18, GG). In addition, it oversees Federal housing programmes.

G2.3   In none of these duties, however, does it have direct executive jurisdiction, and there are specific constitutional restrictions upon responsibilities for controlling the *Länder* implementation of Federal and *Länder* legislation (David, 1986, p.17). There are, for example, no regional offices of the BMBau. The main powers of co-ordination/ initiation with respect to planning at the Federal level are therefore of two types.

(i)   The enactment and supervision (with recourse to the courts, if necessary) of 'framework regulations' which the *Länder* and local government authorities have to respect when setting up their own regulations within the framework. Hence the important principle operates in the FRG that all planning activities at one level are in conformity with policies and programmes operating at a higher-ranking level.

(ii)   Financial imperatives emanating from Federal aid in the implementation of planning programmes (Article 91, a,b,GG, giving joint responsibility for specific fields (*Gemeinschaftsaufgaben*); Article 104 a) paragraph 4 GG, giving Federal aid to specific *Länder* investments (*Bundesinvestitionshilfen*) (David, 1986, p.17).

G2.4   The shared tasks include

(i)   the construction of new university and college buildings and the extension of those existing

(ii)   the improvement of regional economic structure

(iii)   the improvement of agricultural structure

(iv)   coastal protection

Programmes involving Federal aid to the *Länder* include

(i)   the financing of local traffic projects

(ii)   grants towards housing programmes

(iii)   grants towards urban renewal
(Ahrens et al, 1980, p.14)

G2.5   Given the great complexity and legal formalities associated with the *Bundesraumordnung* 'framework regulations', much of the day-to-day activity of the BMBau concerns advice to the *Länder*, local government authorities, public authorities and the private sector on the interpretation of such regulations. One of the aims of current legislation in this field is to clarify, simplify and streamline these regulations (BMBau, 1987, pp.44-46).

G2.6   As well as the *Raumordnung* 'framework regulations', the Federal government has attempted, in co-operation with the *Länder*, to provide a measure of Federal and inter-*Länder* spatial planning through programmes of *Raumordnung und Landesplanung* (comprehensive regional planning). Under Article 75, No.4 GG the *Bundestag* may provide a legal framework (*Rahmengesetzgebungsbefugnis*), which the individual *Länder* make concrete in specific programmes. The goals defined in these programmes then become binding for sectoral and local authorities (David, 1986, p.56). An inter-governmental Ministerial Conference for Spatial Planning (the standing conference of Federal and state cabinet members responsible for supra-local planning), with an advisory status, is the instrument for securing Federal-*Länder* co-operation in this regard. This body has become moribund over the last decade, and recent developments have led to a less significant role for the Federal government in this field, with an increased role for the *Länder* and local government authorities. (Indeed, there has been some speculation that the BMBau itself may be abolished as part of current reforms to the planning system.)

G2.7   In addition to responsibilities relating to *Bundesraumordnung*, at the federal level are the *Fachplanungen* (sector programmes), eg planning for railways, air transport, autobahns. The Federal government's sectoral

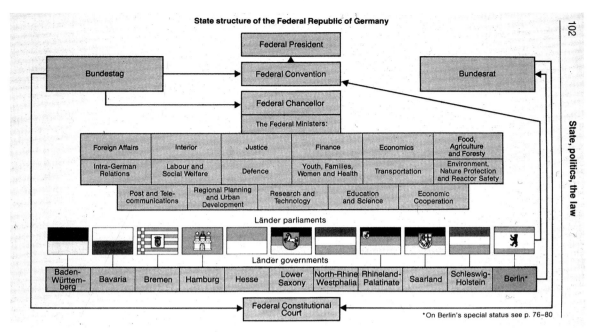

**Figure G1   State Structure of the Federal Republic of Germany**

Figure G1 illustrates the basic relationships between the legislative bodies, the executive and the judiciary in the Federal Republic of Germany. Together these institutions embody the democratic, federalistic law-abiding state order prescribed in the Basic Law. The Bundestag (House of Representatives) is the chief legislative body, but the Bundesrat (Federal Council), representing Länder interests, also participates in legislation relating to Länder matters. The Bundesregierung (Federal Government) consists of the Federal Chancellor (elected by the Bundestag) and the Federal Ministers (nominated by the Chancellor). The Federal Chancellor holds a strong position, being the only member of government elected by Parliament and responsible to it. The tasks of the Federal President are mainly representational, associated with the role of head of state. He is elected by the Federal Convention, consisting of the Bundestag deputies and members elected by the Bundesrat.

He nominates the Federal Chancellor to the Bundestag, and appoints Federal Ministers on the nomination of the Chancellor.

It should be noted that, within the Federal Government, only certain ministries have executive power under the Basic Law. For other functions (including planning and the control of development) the Länder act as the executive arm of the state. The Länder are not merely provinces, but states in their own right with their own sovereign powers. Each Land has its own constitution, parliament and government. The principal field of lawmaking of the Länder is in cultural affairs, and their major activity is administration (implementing federal legislation).

Within the whole field of legislative and executive action, the rule of law is supreme, protected ultimately by the Federal Constitutional Court.

planning is one means of influencing comprehensive regional planning, though this implies considerable inter-sector co-ordination, both at the Federal level and with the *Fachplanungen* of the *Länder*. On a strictly legal interpretation, *Fachplanungen* (sectoral planning) is of a subordinate legal rank to *Raumordnung und Landesplanung* (comprehensive regional planning coordinating supra-local and sectoral planning) but political and administrative realities tend to produce an equivalence of powers between the two, since sectoral programmes are associated with financial means for implementation. Nevertheless, the legal requirements relating to *Raumordnung und Landesplanung* establish a procedural framework of plans, policies and programmes which have in principle to be acknowledged by sectoral planning agencies (David, 1986, p.55).

G2.8   In addition to the abstract principles and procedures of the 'framework regulations' are the various

environmental laws, the most significant of which in the field of planning of land use are

(i) the *Bundes-Immissionsschutzgesetz* - BImschG (1974, p.82) - Federal legislation relating to pollution control from noise, smell etc.

(ii) *Wasserhaushaltsgesetz* (1976) - a Federal framework legislation relating to water resources

(iii) *Atomgesetz* (1976; 1980) - Federal legislation relating to nuclear activities

(David, 1986, pp.51-52)

### The *Länder* Administrative Framework

G2.9   The Federation comprises nine 'space-*Länder*' (though *Bayern* terms itself a 'Free-State' rather than a '*Land*-State'), two city-states and West Berlin, a city state though not technically part of the Federal area, being under Four-Power status (see Figure G2).

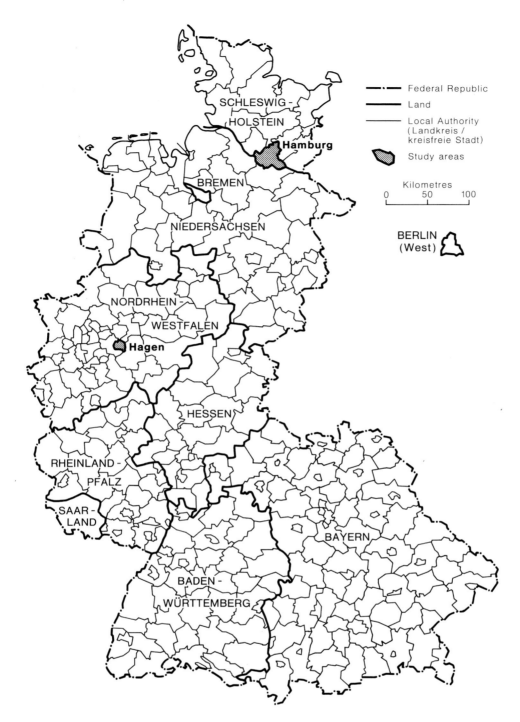

**Figure G2    The Federal Republic Länder and Local Authority Areas (1981)**

Figure G2 shows the eleven Länder (or states) comprising the Federal Republic of Germany. Nine of these are 'space-Länder' and three are 'city-states', the latter each essentially comprising a single metropolitan area. West Berlin, however, is technically not part of the Federal area, being under Four-Power status.

The Länder vary tremendously in area and size of population – though North Rhine Westphalia is only the third largest Land in terms of area, it has by far the largest population, indicating the very dense coalescence of the cities comprising the Ruhr conurbation. One of the case study areas considered in the study (Hagen) is located on the south-east edge of this conurbation. The other case study area is Hamburg.

The 'city-state' structure gives to respective Länder a more

'unitary' system of government than in the 'space-Länder'.

The figure also shows the distribution of Kreis (county) and Stadt (county-free) local authority areas, and also shows the location of the two case study municipalities. Notwithstanding the extensive reform of local government areas and functions from 1968–787 (see Tables G2.1, G2.2), there are approximately 8,500 municipality areas in the F.R.G., differing widely in area and population size.

Hagen is a Kreisfreie Stadt (county-free town) in the Land of North Rhine Westphalia, on the edge of the Ruhrgebiet (Ruhr region).

Hamburg is a 'city state' located in a strategic position in the Kustenländer, the four coastal Länder comprising the North German Plain.

261

| Land | Area (sq.km.) | Population (1986) (millions) |
|------|---------------|------------------------------|
| *Baden-Wurttemberg* | 35,751 | 9·2 |
| *Bayern* (Bavaria) | 70,553 | 11·0 |
| Berlin | 480 | 1·9 |
| Bremen | 404 | 0·7 |
| Hamburg | 755 | 1·6 |
| *Hessen* (Hesse) | 21,114 | 5·4 |
| *Niedersachsen* (Lower Saxony) | 47,447 | 7·2 |
| *Nordrhein-Westfalen* (North Rhine Westphalia) | 34,062 | 16·8 |
| *Rheinland-Pfalz* (Rhineland Palatinate) | 19,848 | 3·6 |
| *Saarland* | 2,571 | 1·0 |
| *Schleswig-Holstein* | 15,721 | 2·6 |

G2.10 Each of the 'space-*Länder*' is governed by a *Länder* government with a *Ministerpräsident* (prime minister). The latter is elected by the *Länder* parliament (*Landtag*). He/she has his/her own office (*Staatskanzlei*), and in many *Länder* either this office or the Ministry of the Interior fulfils the functions of the supreme planning authority. Some *Länder* (Bavaria, Saarland, North Rhine-Westphalia) have established specific ministries to administer *Landesplanung* (*Land* planning).

G2.11 The situation is slightly different in the case of the city states, and can be illustrated with reference to Hamburg. The city state has its own parliament (*Bürgerschaft*), elected for four years. Its executive is collegiate, the Senate Council (*Senat*), which elects its chief political/administrative officer (*Bürgermeister*) who appoints the *Senators* heading each administrative department (building, economic development etc).

G2.12 Typical *Länder* departments include internal affairs, economy, finance, transport etc. The *Länder* have extensive legal responsibilities, and have residual powers for matters not otherwise allocated within the Basic Law. Article 30 of the Basic Law establishes that the exercise of executive powers and the discharge of executive functions rests with the *Länder* insofar as the Basic Law does not prescribe or permit other arrangements. Hence the *Länder* have responsibility for executive action in most subjects, including many in which legislative power lies with the *Bundestag*. Articles 84 and 85 of the Basic Law entitle the *Länder* to administer Federal laws either 'as their own affairs' or as agents of the Federal government (Uppendahl, 1985, p.22).

G2.13 Important legislative fields retained by the *Länder* include the supervision of local government within their own areas, aspects of environmental protection and education. Whether administering Federal or *Länder* laws, however, the *Länder* usually delegate executive actions to their respective local government authorities.

G2.14 In North Rhine Westphalia a co-ordinated system of planning for the development of the *Land* territory is the responsibility of the *Ministerium für Stadtentwicklung, Wohnen und Verkehr* (Ministry for Urban Planning, Housing and Transport - MSWV). In Hamburg this function is provided by the *Baubehörde* (Building Department). The main functions of the MSWV are the preparation and continuous monitoring of *Länder* planning laws, the development of comprehensive *Länder*-wide development programmes and plans and the supervision of the local planning system.

G2.15 In order to achieve the proper administration and implementation of Federal and *Land* legislation, six of the *Länder* have set up *Regierungsbezirke* (administrative districts) headed by *Regierungspräsidenten*, high ranking chief executives appointed by the *Ministerpräsident* for the *Land*.

G2.16 An important aspect of the work of the *Regierungsbezirk's* office concerns the co-ordination of Federal, *Land* and local government authorities' land use planning activities. To this end, in North Rhine Westphalia the *Regierungspräsidenten* meet regularly. The relevant legislation (see below, paragraph G5.26(ii)) requires that the development plans of all local government authorities conform to the provisions of regional and/or *Land* plans, but the legislation also provides for local government authorities to participate in regional planning. Hence advisory regional planning councils (*Bezirksplanungsrat*) are elected by the local government authorities of their respective areas, and facilitate a degree of effective influence upon the *Regierungsbezirk* (besides those informal links between officers and elected representatives on the different levels of government). However, in the field of land use planning, the office of the *Regierungsbezirk* retains considerable reserve powers to ensure conformity of planning activities at the different levels (see below), and the role of this component of the FRG administrative framework has been hitherto underestimated (Ahrens et al, 1980, p.6; Uppendahl, 1985, p.33; David, 1987, p.31).

G2.17 In order to ensure that all activities involving planning for the use of land above the local level maintains the principle (backed by legislation) of conformity with higher-ranking plans, a special procedure known as *Raumordnungsverfahren* can be invoked by the *Länder* or *Regierungsbezirke* authorities to ensure such conformity. This procedure is appropriate only where the *Länder* and *Regierungsbezirke* development plans are diagrammatic in form (as, for example, in Bavaria) and hence require interpretation. In North Rhine Westphalia, *Länder* and *Regierungsbezirke* plans are detailed and therefore the *Raumordnungsverfahren* procedure is not needed.

G2.18 North Rhine Westphalia is divided into five *Regierungsbezirke* (administrative districts) - Köln, Düsseldorf, Münster, Detmold and Arnsberg (see Figure G3). Each of these administrative districts constitutes a

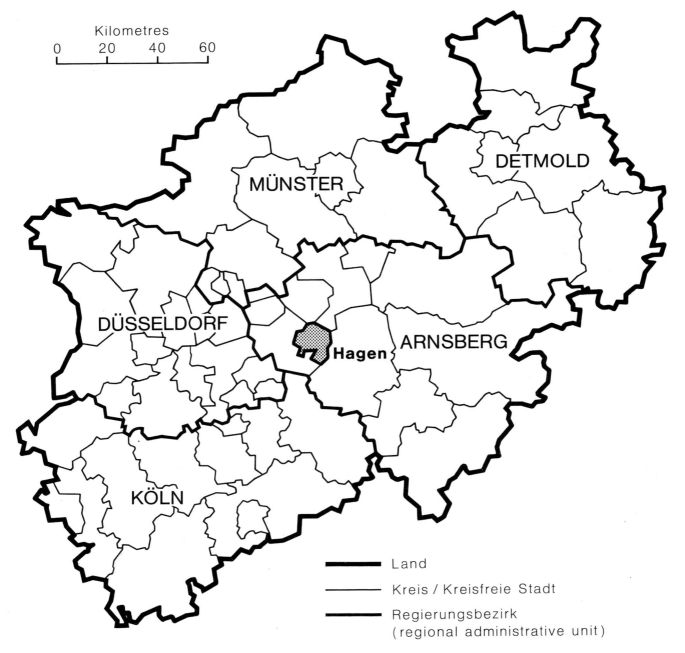

**Figure G3   Regional Administrative Units in North Rhine Westphalia**

Figure G3 shows the administrative sub-units of the Land of North Rhine Westphalia. The Land is divided into five Regierungsbezirke (administrative districts), each of which comprises a planning region for which is prepared a regional development plan consistent with plans and programmes of the Land. The preparatory land use plans (Flächennutzungspläne) and the binding land use plans (Bebauungspläne) of the cities (Städte) and counties (Kreise) must confirm with the provisions of these regional development plans, which in turn must comply with the Land development plans and programmes.

planning region for which is prepared a regional development plan consistent with the plans and programmes of the *Land*.

G2.19   A similar administrative district organisation (but at a different scale) exists for the 'city-states'. In Hamburg, for instance, the *Baubehörde* (Building Department) acts as the equivalent of the relevant Ministry at the 'space-*Länder*' level, with seven *Stadt Bezirke* (city district offices of the central Department).

G2.20   The *Regierungsbezirke* thus constitute an important intermediate tier of government. During the 1960s and 1970s they evolved as part of a centralist trend within the Federal constitutional structure, acting as crucial vehicles for the co-ordination of Federal and *Länder* programmes for the promotion of structural change. Local administration was increasingly integrated into *Länder* and Federal administration, co-ordinated by an extended system of development programmes and plans and creating a growing financial dependency of local government authorities on Federal and *Länder* resources. Given the large size of some *Länder* territories, deconcentration of implementation of programmes inevitably produced a streamlined system of state functions on a hierarchical basis, with the *Bezirksregierung* (District Administrative Office) playing a pivotal role (Uppendahl, 1985, pp.30-32; Konukiewitz & Wollmann, 1982, pp.83-95).

G2.21   However, the *Regierungsbezirk* also has an important role in the ever-growing involvement of local government authorities in the implementation of regional, *Länder* and Federal policies, which may act as a limitation to centralisation trends. The *Rastede* decision of the Federal Administrative Court in 1983 (see paragraph G1.14, above) has confirmed the constitutional right of local government authorities to a core domain of functions, so that they are not merely the agents of Federal and *Länder* government. Current changes in Federal legislation in the field of land use planning have as one of their explicit aims an increase in the sovereignty of local government authorities with regard to planning (BMBau, 1986, p.42), and procedurally this will affect the role of the *Regierungsbezirk* in approving local land use plans (see below, paragraph G5.9). Since local planning is a local government authority activity, and is not developed out of regional planning, the role of the *Regierungsbezirk* in approving locally-produced plans is of key importance in mediating the relationship between Federal, *Länder* and local planning (David, 1987, p.31) (see Figure G4).

## The Local Administrative Framework

G2.22   Local government (*Kommunale Selbstverwaltung*) in the FRG comprises three types of *Kommune* (commune)

(i)   *Gemeinden* (municipalities)

(ii)   *Landkreise* (counties)

(iii)   *Kreisfreie Städte* (county-free towns or cities).

G2.23   All Western Germany is divided into *Gemeinden* which are constitutionally sanctioned corporate entities comprising a spatial unit with an effective organisation for political decision-making and planning at the lowest level of the political system. Typical functions include

(i)   general administration

(ii)   finance

(iii)   law, order and safety

(iv)   schools and culture

(v)   social and health policy

(vi)   building and town-planning

(vii)   local facilities

(viii)   commerce and transport

(Nassmacher & Norton, 1985, p.106; Uppendahl, 1985, pp.23-24; BMBau 1976, p.24).

G2.24   There are two types of *Gemeinde* (municipality)

(i)   those belonging to a county (*Kreisangehörige Gemeinden*)

(ii)   those not belonging to a county (*Kreisfreie Städte*) (David, 1986 (6), p.39).

G2.25   *Landkreise* 'duties' include

(i)   those which cannot be performed by one *Gemeinde*, such as, for instance, the construction of roads

(ii)   supplementary duties such as the maintenance of hospitals which small authorities cannot possibly afford

(iii)   equalising small duties such as payment for subsidies to weaker local authorities

(iv)   general responsibilities such as, for example, the support of secondary schools
(Uppendahl, 1985, p.24).

The *Landkreise* thus form a second local government level, with subsidiary functions for the *Gemeinden* (municipalities) of their territory (David, 1986 (6), p.39). If *Gemeinden* lack the financial and/or organisational means to provide tasks, these affairs can be taken over by this next higher level of government.

G2.26   The *Kreisfreie Städte* possess both *Gemeinde* and county functions. They may be sub-divided for administrative purposes into *ortsteile* (districts), but these do not constitute corporate self-governing entities as do the *Gemeinden* within the *Landkreise*.

G2.27   Geographically, a *Kreisfreie Stadt* may be found

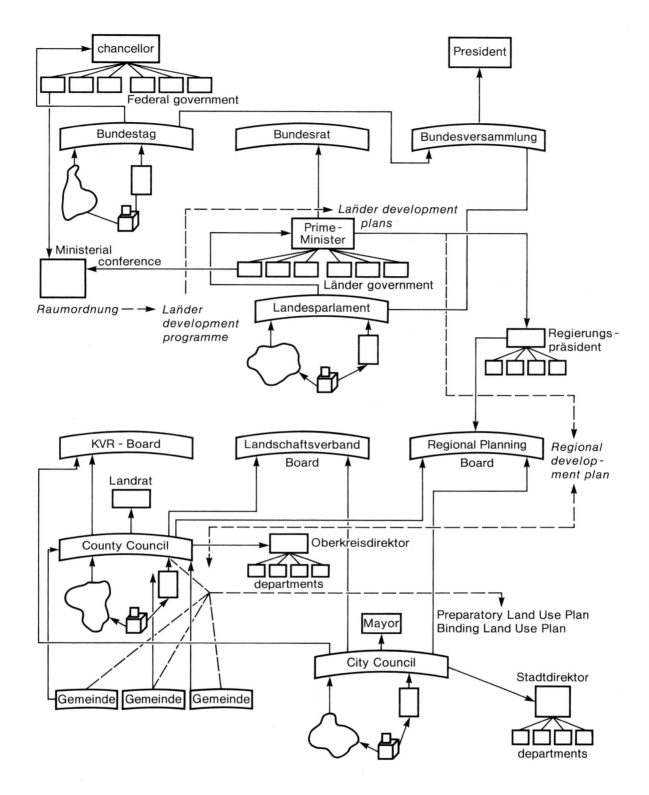

Source : from Estermann, 1980, p. 2

**Figure G4   Administration, Planning Authorities and Plans in the FRG**

Figure G4 demonstrates the complex interrelationships between government and administration comprising the three tiers of government in the F.R.G. (Federal, Länder and Municipality) as these relate to land-use planning and the control of development. In one sense, there is a hierarchical relationship, descending from Federal to Municipality level, in terms of a comprehensive system of statutory development plans. But the relationship is not strictly hierarchical, being generally largely supervisory on the part of the Federal Government, and with both Länder and Municipalities having strong sovereign powers in this field, so that the relationship should more properly be considered a partnership. In certain Länder (for instance, North Rhine Westphalia) the administrative districts (Regierungsbezirke) play a crucial role in mediating between 'higher' and 'lower' tiers of the government.

encompassed on all sides by a *Landkreis*. In North Rhine Westphalia all *Gemeinden* are known as *Städte* (towns), though again being divided between *Kreisfreie Städte* and *Kreisangehörige Städte* (Ahrens et al, 1985, p.10).

G2.28 It has been noted above (paragraph G2.12) that Federal and *Länder* tasks may be administered by local government authorities as agents of the *Länder*. The role of the *Regierungsbezirk* has already been noted in this respect (paragraph G2.15), but at the lower administrative tier these duties may be performed by the *Kreisfreie Städte*, *Gemeinden* or *Landkreise*. Functions of this kind given to *Landkreise* include

(i) the regulation of commerce

(ii) the administration of public health and educational grants

(iii) the supervision of the *Gemeinden* within their respective areas (Uppendahl, 1985, p.24).

G2.29 A central feature of government administration in the FRG is thus the close co-operation necessary between Federal government, Land goverment, administrative district government and *Kreis* and *Gemeinde* officials. This has been termed a "vertical departmental companionship" which characterises the German form of Federal goverment (Uppendahl, 1985, p.24).

G2.30 In the fields of self-government (which is guaranteed by the Basic Law) local authorities are subject only to legal control by the *Land*, ie the *Land* is only allowed to ensure that the law is kept; the purpose of its actions is determined by each community itself. For *Kreisfreie Städte* this legal supervision is provided by the *Regierungsbezirk*; for *Kreisangehörige Gemeinden* it is provided by the *Landratsamt*, the county administration in the *Landkreis*. When municipalities and counties are charged with the execution of Federal and *Land* laws falling outside the fields of self-government, they are subject not only to legal control by the *Land* but in some cases are given detailed directives to work to (Bulka et al, 1986, p.125).

G2.31 All *Länder* planning laws permit or order the creation of 'regions' for planning purposes on a level between the *Regierungsbezirk* (district) boundaries. The legal entities thus combined form a planning community, syndicate or association (Kimminich, 1986, pp.189-190). *Gemeinden* can thus form a *Verband* (association), either within or beyond the *Kreis* by

(i) compulsion from the *Regierungspräsident*, through a Land statutory instrument

(ii) by agreement between *Gemeinden*.

Such *Verbände* (associations) may include the association of one or more *Kreisfreie Stadt* with adjacent *Gemeinden* (which themselves may be, under *Land* law, part of a *Landkreis*). The association of *Gemeinden* in a *Verband* does not necessarily result from a lack of resources or organisational means - *Gemeinden* have a right to associate in this manner, though conditions are specified in the respective *Federal* and *Land* legislation. The association of *Gemeinden* for planning purposes (ie the *Verband*) then has legal jurisdiction in the field(s) for which the *Gemeinden* have become associated.

G2.32 In North Rhine Westphalia the *Kommunalverband* Ruhr (KVR) and the *Landschaftsverbände* for *Westfalen-Lippe* and for *Rheinland* respectively cover the whole area of North Rhine Westphalia, but constitute only restricted specific jurisdictions in particular forms of association providing municipal partnership, and also requiring municipal funding (see Figures G5 and G6). The functions of such associations have been severely affected by local government reforms in 1975. Formerly the KVR area was under the planning authority of one body, the *Siedlungsverband Ruhrkohlenbezirk* (SVR), traditionally one of the earliest regional planning authorities in the world (Kunzmann, 1980, p.49). Since 1977 responsibility for planning for the Ruhr area has been transferred to three *Regierungsbezirke*. The KVR now has only residual functions in relation to waste disposal, recreation and open space provision and promotion of the Ruhr area. The *Landschaftsverbände Rheinland* and *Westfalen-Lippe* provide central facilities for the *Gemeinden*, run hospitals for the mentally handicapped, administer museums and build and maintain roads (Ahrens et al, 1980, p.11).

## Local Government Reform

G2.33 From 1968-1978 sweeping administrative reforms of local government authorities were made in all *Länder* except Bremen, Hamburg and Berlin, which ultimately reduced the number of *Gemeinden* in the FRG from 24,282 in 1968 to 8,505 by 1983 (Siedentopf, 1980, p.87; BMBau, 1986, p.131; Uppendahl, 1985, p.28) (see Tables G2.1, G2.2 and Figures G7 and G8).

G2.34 The basic structure of the administrative system defined by the constitutional allocation of government pow thus comprises five tiers

(i) 1 *Bund* (federation)

(ii) 11 *Länder* (states)

(iii) 26 *Regierungsbezirke* (*Länder* administrative districts) a 12 *Regionalverbände* (regional associations)

(iv) 237 *Kreise* (counties) and 91 *Kreisfreie Städte* (county-f towns)

(v) 8,505 *Gemeinden* (municipalities)

Thus, for example, in North Rhine Westphalia there are *Regierungsbezirke*, 54 *Kreise*, 23 *Kreisfreie Städte*, 31 *La kreise* and 396 *Gemeinden* (see Table G2.1).

Kilometres
0    20    40    60

LANDSCHAFTSVERBAND
WESTFALEN - LIPPE

Hagen

LANDSCHAFTSVERBAND
RHEINLAND

——— Land

——— Kreis / Kreisfreie Stadt

——— Landschaftsverband
(Advisory Planning Association)

**Figure G5    Landschaftsverbände (Advisory Planning
Associations) North Rhine Westphalia**

There are two Landschaftsverbände (advisory planning associations) in North Rhine Westphalia, though the range of functions they administer is small (hospitals for the mentally handicapped, museums, advice on historic buildings and areas etc.).

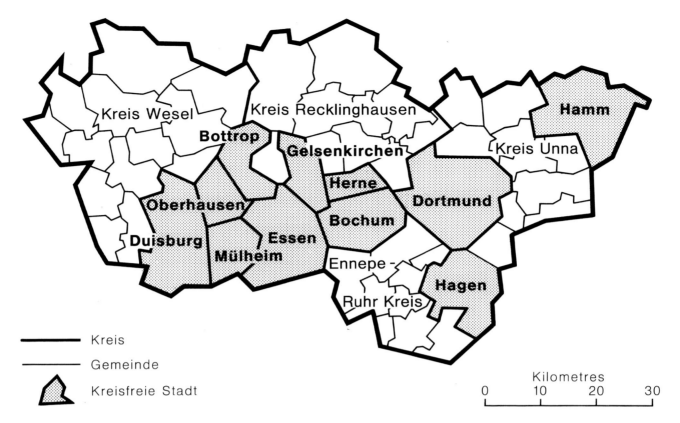

Kreis

Gemeinde

Kreisfreie Stadt

Kilometres
0    10    20    30

**Figure G6   Communal Association of Ruhr Local Authorities**

The Kommunal Verband Ruhrgebeit (Communal Association of Ruhr Local Authorities) is a voluntary association of local authorities seeking to provide a regional voice representing the Ruhr area, but its powers are strictly circumscribed and relate mainly to the preparation of advisory plans and environmental and recreational policies. Its predecessor, the SVRC Siedlungsverband Ruhrkohlenbezirk) was a strong regional planning association, but the reform of local government in 1977 transferred the regional planning functions to the three Regierungsbezirke (see Figure G4). The Ruhr now comprises 42 gemeinde (communities) falling within four kries (counties), and a further 11 gemeinde having the status of kreisfreie Städte (county-free towns). The complexity of German local government is one of the features militating against any conception of a simple hierarchical planning system.

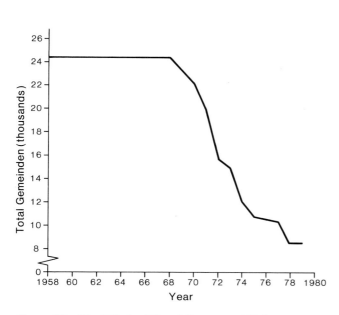

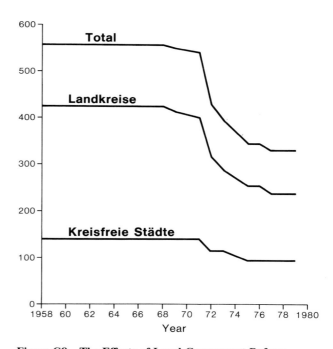

| Figure G7 | The Effects of Local Government Reform 1958–1979: (a) *Gemeinde* | Figure G8 | The Effects of Local Government Reform 1958–1979: (b) *Landkreise* and *Kreisfrei Städte* |

**Table G2.1   Administrative Areas Comprising the Federal Republic (1983)**

| Land | Regierungsbezirk | Kreise insgesamt | Kreise Kreisfreie Städte | Kreise Land kreise | Gemeinden |
|---|---|---|---|---|---|
| Schleswig-Holstein | – | 15 | 4 | 11 | 1,131 |
| Hamburg | – | 1 | 1 | – | 1 |
| Niedersachsen | 4 | 47 | 9 | 38 | 1,031 |
| Bremen | – | 2 | 2 | – | 2 |
| Nordrhein-Westfalen | 5 | 54 | 23 | 31 | 396 |
| Hessen | 3 | 26 | 5 | 21 | 427 |
| Rheinland-Pfalz | 3 | 36 | 12 | 24 | 2,303 |
| Baden-Württemberg | 4* | 44 | 9 | 35 | 1,111 |
| Bayern | 7 | 96 | 25 | 71 | 2,050 |
| Saarland | – | 6 | – | 6 | 52 |
| Berlin (West) | – | 1 | 1 | – | 1 |
| Bundesgebiet (total) | 26 | 328 | 91 | 237 | 8,505 |

\* Außerdem: 12 Regionalverbände

**Table G2.2   Local Government Reform 1970, 1981, 1983**

| Gebiet | 1970 Anzahl | 1970 durch-schnittl. Bevölkerung | 1970 durch-schnittl. Flächen (km²) | 1981 Anzahl | 1981 durch-schnittl. Bevölkerung | 1981 durch-schnittl. Flächen (km²) | 1983 Anzahl | 1983 durch-schnittl. Bevölkerung | 1983 durch-schnittl. Flächen (km²) |
|---|---|---|---|---|---|---|---|---|---|
| Bundesgebiet insgesamt | – | 60,650,599 | 248,575.84 | – | 61,712,689 | 248,691.92 | – | 61,306,669 | 248,706.21 |
| Bundesland | 11 | 5,513,691 | 22,597.80 | 11 | 5,610,245 | 22,608.36 | 11 | 5,573,334 | 22,609.66 |
| Regierungsbezirk | 30 | 2,021,687 | 8,285.86 | 26 | 2,373,565 | 9,565.07 | 26 | 2,357,949 | 9,565.62 |
| Kreis | 542 | 111,901 | 458.63 | 328 | 188,148 | 758.21 | 328 | 186,911 | 758.25 |
| Gemeinde | 22,510 | 2,694 | 11.04 | 8,504 | 7,257 | 29.24 | 8,507 | 7,207 | 29.24 |

## Local Administrative Structures

**G2.35** All local government authorities make a clear distinction between the deliberative body, which is a council elected by universal franchise, and the executive or administration, headed by a political executive.

The council and the chief executive are legally the policy-making institutions of local government. An important function of the council is to decide on the functions to be performed by the municipality in its own right (Nassmacher & Norton, 1985, p.112). Constitutionally, however, local government authorities are part of the *Länder* and subject to the *Länder's* right of supervision under legal aspects. (BMBau, 1982, p.4).

**G2.36** In the FRG the function of political executive fuses the representation of the municipality and the chairmanship of the *Rat* (council). It is carried out in different forms in different states (Norton, 1985, pp.43-44). It can be undertaken

(i) by a single person, a full-time salaried *Bürgermeister* or *Oberbürgermeister* directly elected

(ii) by a collegiate executive board (*Magistrat*), including the *Bürgermeister*, elected by councils, with an honorary council chairman. The city states of Berlin, Hamburg and Bremen also have a collegiate executive (*Senat*)

(iii) by an honorary *Bürgermeister* or *Oberbürgermeister*, together with a chief executive officer who is a salaried municipal official, the *Gemeindedirektor*, *Stadtdirektor* or *Oberstadtdirektor*, elected by the council.

(Nassmacher & Norton, 1985, p.113).

**G2.37** The *Bürgermeister* may be directly elected or appointed by the council. They have an initial tenure of six, eight or twelve years, with the possibility of renewal. Honorary *Bürgermeister* are appointed for the tenure of their councils, the latter being elected at one time for periods of four to six years (Nassmacher & Norton, 1985, p.114).

**G2.38** In the *Kreise* (counties), councils (*Kreistag*) are elected for four or six years. They elect an executive committee which administers the county together with a salaried chief executive official (*Landrat* or *Oberkreisdirektor*) who also has chief executive responsibilities for the *Land* Government (Nassmacher & Norton, 1985, p.114).

**G2.39** The chief executive official in a municipality is elected by the *Rat* (council), and thus combines political accountability with responsibility for administration. In Hagen, a *Kreisfreie Stadt* in North Rhine Westphalia, this official is elected for eight years at a time by the *Rat*, which is itself elected every five years. (If after the term of office expires the *Rat* wishes to dispense with the services of this official his/her salary must be paid for a further period of office in lieu of renewing the post).

**G2.40** Below this post of political executive, officers are appointed on the basis of technical qualification and are public servants rather than political appointments. At the municipal level, chief officers will typically have three administrative responsibilities

(i) administering the local authority's own policies and programmes

(ii) administering Federal and *Länder* tasks, where the municipality has no legal competence

(iii) administering tasks under direction from the *Länder* (via the *Regierungspräsident*).

In this last capacity, officers may experience role-tension, in that potentially competing obligations to different authorities may have to be reconciled in particular policy fields.

**G2.41** Variability in German local government arises from the principle of general competence. German, unlike British local authorities, can undertake all services and measures of local interest not reserved to other public bodies, without having to obtain permission from higher levels of government (Nassmacher & Norton, 1985, p.108).

## The Central/Local Relationship

**G2.42** In the field of land use planning and the control of development, the role of the Federal government is largely confined to the provision and supervision of 'framework regulations' - written principles contained in a law (Section 2, BROG). In general terms, the *Länder* prepare development programmes, enact legislation relating to the control of development and supervise the operation of the local planning system in the various local government authorities. (Article 74 No.18 GG & BBauG; Article 75 No.4 GG & BROG). Operationally, the *Regierungsbezirke* (administrative districts) play a significant role in both these fields, making concrete the development programmes in terms of more detailed policies and plans, ensuring conformity of local development plans and dealing with appeals relating to development permissions (*Bauaufsichtsbehörden*) falling within the jurisdiction of *Kreisfreie Städte* (see below, paragraph G6.8). The *Bund* (federation) and *Länder* thus provide national and regional policy contexts and the legal framework of development controls within which the local government authorities engage in *Bauleitplanung* (local development planning) (see Figure G9).

**G2.43** The local government authorities are in many senses the prime agents in land use planning and the control of development in the FRG. They are responsible for preparing the development plans which impinge most directly upon the land use and development process, and for operating the system of development permissions. Whilst these powers are exercised within the context of *Länder* supervision, such supervision relates largely to legal aspects (and it is here that the system of written law has its greatest impact and significance). Given the great variety between *Gemeinden*, and the variable relationships between *Gemeinden* and higher tier authorities among the different *Länder*, the essence of land use planning and development control in the FRG is a partnership of co-operation between authorities among and between the different tiers in the Federal system. A common feature of most policy fields, including land use planning, is that of implementation at the *Gemeinden* level, and it is at this local level that the rich complexity of the control of development in the FRG is to be found.

G2.44    The current position of local government in the FRG has been nicely summarised by Pauley (1985, p.8):

> "... within the existing administrative and constitutional framework local government in West Germany is fit and strong, understands its role and purpose and is both accepted and valued by the Federal Government for its contribution to the social and economic fabric of the nation. That is not to say that there is no tension about the division of powers and responsibilities and no anxieties about the abilities of the revenue sources to meet the revenue demands. But it is to say that there is no crisis."

It is in this context of a "fit and strong" system of local government that the current operation of the control of development in the FRG must be examined.

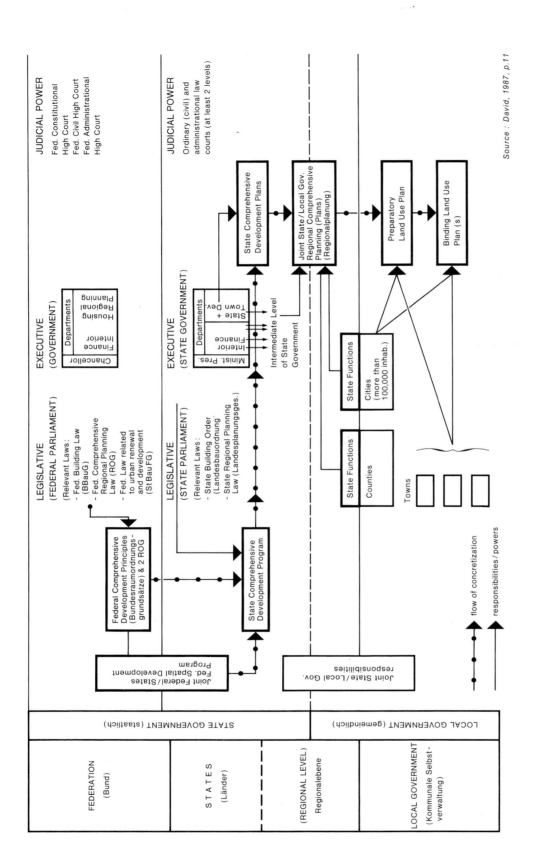

**Figure G9   Land Use Planning Responsibilities in the F.R.G.**

This diagram illustrates the complexity of the relationships between the preparation and approval of land use plans among the three tiers of government. Essentially, it seeks to demonstrate how rather abstract spatial goals and principles are concretised into comprehensive regional plans and finally into detailed local plans, with a progressive emphasis upon the detailed land use components of the relationships and upon their legally-binding character. At each level, the respective roles of legislature, executive and judiciary are clearly specified in the Basic Law, though the actual working relationships (given the variety between and among Länder and municipalities) are complex.

Labels within the figure:

JUDICIAL POWER
Fed. Constitutional High Court
Fed. Civil High Court
Fed. Administrational High Court

JUDICIAL POWER
Ordinary (civil) and administrational law courts (at least 2 levels)

EXECUTIVE (GOVERNMENT)
Chancellor
Departments
Finance
Interior
Housing
Regional Planning

EXECUTIVE (STATE GOVERNMENT)
Minist. Pres.
Departments
Interior
Finance
State + Town Dev.
Intermediate Level of State Government

LEGISLATIVE (FEDERAL PARLIAMENT)
(Relevant Laws :
- Fed. Building Law (BBauG)
- Fed. Comprehensive Regional Planning Law (ROG)
- Fed. Law related to urban renewal and development (StBauFG)

LEGISLATIVE (STATE PARLIAMENT)
(Relevant Laws :
- State Building Order (Landesbauordnung)
- State Regional Planning Law (Landesplanungsges.)

Federal Comprehensive Development Principles (Bundesraumordnungs-grundsätze) & 2 ROG

State Comprehensive Development Program

Joint Federal/States Fed. Spatial Development Program

Joint State/Local Gov. responsibilities

State Comprehensive Development Plans

Joint State/Local Gov. Regional Comprehensive Planning (Plans) (Regionalplanung)

Preparatory Land Use Plan

Binding Land Use Plan (s)

State Functions — Counties
State Functions — Cities (more than 100,000 inhab.)
Towns

FEDERATION (Bund)
STATES (Länder)
STATE GOVERNMENT (staatlich)
(REGIONAL LEVEL) Regionalebene
LOCAL GOVERNMENT (gemeindlich)
LOCAL GOVERNMENT (Kommunale Selbst-verwaltung)

flow of concretization
responsibilities/powers

*Source : David, 1987, p.11*

# G3    THE SCOPE AND DEFINITION OF DEVELOPMENT

## Problems of Definition

G3.1    In the FRG the distinction between planning and development control is less clear than it is in Britain. This arises from the fundamentally different legal relation between private landownership and the planning system in the two countries (see paragraph G1.26, above). In the FRG, the drawing-up and adoption of binding land use plans has direct legal consequences for the use of land, and thus the understanding of development control in West Germany covers both the granting of a development permission and the adoption of binding land use plans. Hence the binding land use plan (where it exists) is the instrument for development control, and the development permission realises the binding land use plan (David, 1987, pp.16-17; Albers, 1973,p. 6). Applications for development which are in accordance with approved binding land use plans leave no discretion to the authority - they must be approved, and failure to do so can result in an action by a developer to the administrative courts to request a positive decision by the court to permit development in accordance with the binding land use plan. This legal constraint derives directly from the constitutional guarantee of private property rights (see paragraph G1.27) (David, 1986, pp.33-34; 1987, pp.16-18; Kimminich, 1986, p.195).

G3.2    "Development" is itself a problematic term in German. Its broadest translation is *Entwicklung*, but the connotations of this term refer to the comprehensive development planning activities of municipalities with no specific legal significance in terms of land use and development, rather than the narrower definition inherent in British town and country planning legislation. Another meaning of the term 'development' (*Erschliessung*) refers only to the construction of highways, sewerage systems and utilities for new residential or industrial areas, and is a municipality responsibility (Kimminich, 1986, p.194). The German expression "*Baugenehmigung*" covers the review of a development proposal under all the relevant legal requirements, including planning, building regulations, infra-structure requirements, etc. It thus includes both planning permission and building permission, and the development permission encompassing both these aspects is administered by *Bauaufsichtsbehörden* (building control departments) (David, 1987, pp.16-17).

G3.3    Since development permissions relate to binding land use plans (which in turn relate to a whole hierarchy of plans and programmes at the several government tiers) and to a wide range of regulations of public law at the local, *Länder* and Federal level, the definition of development itself has to be derived from a complex range of legislation.

G3.4    The focus of this legislation is urban development legislation (*Städtebaurecht*), of which urban planning legislation (*Stadtplanungsrecht*) is a major component in the context of urban physical development policy (*Städtebaupolitik*) and comprehensive urban development policy (*Stadtentwicklungspolitik*) (David, 1986, p.21).

## Legislation Relating to Land Use Planning and the Control of Development

G3.5    The main Federal law in this field is the *Bundesbaugesetz* (BBauG - Federal Building Law) of 1960 (amended 1969, 1976, 1979 and amended and retitled *Baugesetzbuch* - BauGB December 1986 to take effect from July 1987). This replaced and consolidated the earlier construction and building laws of individual *Länder* (Schmidt-Assmann, 1986, p.334; David 1986, pp.25-28). It establishes the basis of land use planning and control, by providing a description of the task of urban land use planning (*Bauleitplanung*) which is

"to prepare and to control the land use, - by constructing buildings or otherwise -, of the sites in the municipality according to the regulations of this law". (Section 1, Paragraph 1 BBauG, cited in David, 1986, p.29).

This task is assigned in the BBauG to the *Gemeinden*.

The term 'control' (*leiten*) is thus to be found in the *Bundesbaugesetz*, though it is refined in other legislation (see below).

G3.6    Since the Basic Law entitles the Federation (Article 75, no.4 GG, *Rahmengesetzgebungsbefugnis*) to enact 'framework regulations' which the *Länder* and municipalities have to respect when setting up their own regulations, including all planning measures concerning the use of land for any purpose, the Federation enacted the *Bundesraumordnungsgesetz* (BROG -Federal Comprehensive Regional Planning Law) in 1965 (amended 1976, 1980, 1986). In German law, land use controls derive not only from urban land use planning (*Bauleitplanung*) but also from comprehensive development planning, concerned with coordinating all planning and implementation activities of public authorities (Federal,

*Länder* and municipal), and which has a hierarchical legal structure. *Raumordnungsplanung* thus refers to the central planning for the use of land at the Federal or *Länder* level, and is concerned with the co-ordination of activities of other authorities which implement their (sectoral or local) plans themselves. This *Raumordnung* activity (comprehensive regional planning) has been defined by the Federal Constitutional Court as

"the comprehensive, superior planning and ordering of space, superior in the sense that it is above the local level combining and harmonising the various special planning activities" (Kimminich, 1986, p.188; David, 1986, pp.53-57).

G3.7 *Raumordnung* procedures involve an important distinction in the German legal system between two types of planning with spatial consequences.

(i) Comprehensive, integrated planning (*zusammenfassende Planung* or *Gesamtplanung*). Though municipalities' land use planning is of this nature, it does not fall within the compass of *Raumordnung* procedures which are by legal definition supra-local. Comprehensive regional planning is thus supra-local, supra-sectoral and co-ordinating.

(ii) Sectoral planning (*Fachplanung*) concerns all kinds of institutionalised and sometimes legally regulated planning activities which pursue specific policies with spatial impacts (eg road construction, landscape protection etc). Some of these activities may directly consume land, others may have more indirect land use implications. Some may be embodied in binding land use plans, others not (David, 1986, p.54).

G3.8 Under the BROG legislation (Section 5 paragraph 4) supra-local, comprehensive planning has a higher legal status than sectoral planning (and local comprehensive planning). Whilst administrative and political realities often tend to produce rather an equivalence of powers between sectoral and regional comprehensive planning, the legal framework has in principle to be acknowledged by sectoral planning agencies and municipalities in exercising their land use planning functions (David, 1986, p.55; David, 1987, p.13; Schmidt-Assmann, 1986, p.335).

G3.9 The various *Länder* have consequently modified their old planning laws, or have enacted new laws, to accommodate the provisions of the BROG (in 1979 in North Rhine Westphalia, replacing its old law of 1962). The main instrument of *Raumordnung* and *Landesplanung* is that the goals defined in the *Länder* programmes and plans are legally binding for the sectoral and local authorities (David, 1986, p.56).

G3.10 In addition to the Federal and state laws, subordinate legislation and statutory instruments (*Rechtsverordnung* - legal ordinances) contain more detailed provisions than the law itself. They are detailed by the Federal or *Länder* governments, or by the *Länder* ministries with a formal legal basis in the law, binding on

municipalities. In addition, and of practical relevance, are internal administrative guidelines (*Richtlinien, Verwaltungsvorschriften*) of general importance or for single cases (*Erlasse*). They are all subordinated to legislation passed by the parliaments (of importance in relation to the control effected by the law courts) (David, 1986, p.23).

G3.11 Whilst there is, then, no comprehensive overall Federal land use planning, the abstract principles of the BBauG and BROG, together with environmental laws, create a legal context for land use planning which formulates general principles for land development and control. Individual *Länder* enactments may supplement this framework, but Federal law is exclusive within its own ambit.

G3.12 In addition to the BBaug and BROG legislation, urban renewal law constitutes an integral component of planning legislation. Urban renewal (*Sanierung*) is the subject of separate legislation, the *Städtebauförderungsgesetz* (StBauFG - Urban Renewal and Development Act) of 1971, amended 1976, though the German legal system includes urban renewal law as part of town and country planning and land legislation. The StBauFG may be used only in special circumstances relating to

(i) new independent suburbs or new towns in priority development areas

(ii) the expansion of existing towns

(iii) urban renewal projects

(Pfeiffer, 1973, p.83).

G3.13 This Act does not therefore cover all aspects of legislation necessary to development, being restricted spatially (as indicated above) and temporally in so far as general development regulations become valid once again for the defined development/renewal area after execution of the special measures under the Act (BMBau, 1976, p.31).

G3.14 The Act provides instruments for more positive and comprehensive planning of development and renewal schemes. Its prime purpose was to facilitate the expansion and physical adjustment of regional commercial centres to the needs of expanding service industries and central place functions (Konukiewitz & Wollmann, 1982, p.88), through the provision of a legal framework for Federal matching grants (covering up to a third of total costs) in partnership with the *Länder* governments, to local renewal and development projects (see paragraph G2.4).

G3.15 Special powers introduced in the Act extended the scope of positive planning by municipalities within the designated areas. In general, the urban renewal programme was to preserve private property or recreate private ownership as far as possible, and to work through co-operation with existing proprietors, who would undertake the bulk of new construction and modernisation work. Such restoration and development was to be carried out according to the provisions of the BBauG. In order to ensure the progress of such programmes, additional regulations of land law were introduced to ensure the participation of those private landowners not able or unwilling to engage in the development procedure (see below), (BMBau, 1976, pp.31-34; David, 1986, p.23).

G3.16 Over the last decade, the emphasis in the Urban Renewal and Development Act programmes has changed from wholesale urban redevelopment with a strong regional development component towards inner urban area renewal and rehabilitation. In January 1985 the StBauFG was substantially amended and simplified, by the modification of some of the positive planning measures and the removal of the absolute obligation of municipalities to prepare binding land use plans for such areas.

G3.17 These changes were effected partly in anticipation of the measures to be introduced as a consequence of the new *Baugesetzbuch* (Building Law Code) to come into force in July 1987. In this new Code, the BBauG (Federal Building Law) and the StBauFG (Urban Renewal and Development Act) are being combined in order to make their provisions mutually compatible in a simplified legislative form. The new Code has four sections as described below.

(i) Section 1 - General Urban Development Law
Essentially retains the regulations of Parts One to Six of the Federal Building Act (ie master planning, the safeguarding of master planning, regulation of building and other uses, compensation, land regulation, expropriation and development).

(ii) Section 2 - Specific Urban Development Law
Contains regulations for urban renewal, rehabilitation and development measures which formerly came under the StBauFG, together with legislation on the preservation of townscapes, on urban development ordinances, on the social plan, on financial equalisation in case of hardship, and tenancy regulations. This section also contains regulations relating to urban development measures in connection with agricultural structural measures. Furthermore, the provisions for urban renewal are made compatible with the rest of urban development law, provided that the peculiarities of urban renewal do not require any independent regulations.

(iii) Section 3 - Residual Matters
This section contains provisions relating to the determination of land values, general and administrative regulations and those relating to the court procedures dealing with building land matters.

(iv) Section 4 - Regulation Procedures
This section regulates both the transitional and final provisions.

(BMBau, 1986, pp.24-44)

## Physical Development Planning

G3.18 'Development' in the FRG, as noted in paragraph G3.1 above, is controlled simultaneously through land use planning and the control measures. The development permission (usually termed a 'building permit') unites these several regulatory provisions in a single administrative act which establishes with full legal certainty that the projected development is in full accordance with all the relevant laws (see paragraph G1.26).

G3.19 Under the Federal legislation outlined in the preceding section, the *Bundesbaugesetz* (BBauG) is the main Federal law concerned with land use control and establishes the basic provisions for *Bauleitplanung* (physical development planning). The BBauG makes provision for two types of land use plans.

(i) The *vorbereitender Bauleitplan, Flächennutzungsplan* (pre-paratory land use plan) which has no direct legal effect on the land use rights of private landowners, but which is legally binding for the public authorities which participated in the adoption procedure (including the municipality which adopts the plan).

(ii) The *verbindlicher Bauleitplan, Bebauungsplan* (legally binding land use plan), which directly determines the legal land use characteristics of a site and affects every present or future landowner.

G3.20 The *Flächennutzungsplan* (preparatory land use plan) is intended to cover the whole area of a municipality, and must conform with all higher-order plans. It forms the basis for any subsequent binding land use plans, which must in turn comply with its provisions (David, 1987, p.19). It is essentially a zoning plan (Kimminich, 1986, p.191), designating areas for specific land uses (see Figure G10a). The BBauG (Section 5) specifies what is to be indicated (see below), and the *Baunutzungsverordnung* (BauNVO- Federal Land Utilisation Order)(1977 amended 1986) designates the general land use classifications (*Bauflachen*) as follows

(i) housing areas (W)

(ii) mixed building areas (M)

(iii) manufacturing/industrial areas (G)

(iv) special areas (S)

(Article 1, paragraph 1) BauNVO Von Klitzing, 1975, pp.90,93; David, 1986, p.32).

G3.21 The *Bebauungsplan* contains the legally binding provisions for all types of land use, especially for building purposes (see Figure G10b. The BBauG (Section 9) determines what can be regulated in a binding land use plan with regard to its content and designated land uses, with the BauNVO providing in detail (Article 1 paragraph 2) the different specific types of land use (*Baugebiete*):

(i) small residential estates (WS)

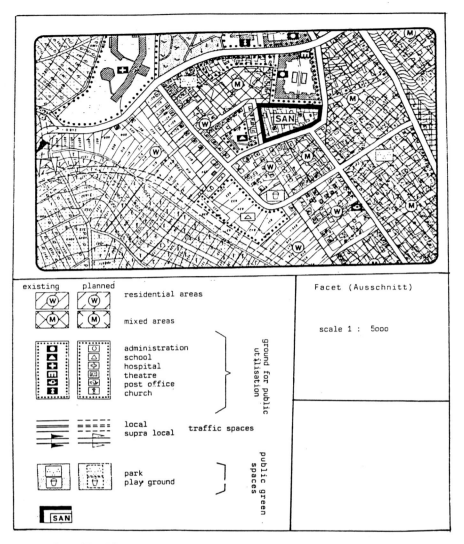

Legend:

| existing | planned | | |
|---|---|---|---|
| (W) | (W) | residential areas | |
| (M) | (M) | mixed areas | |
| | | administration | ground for public utilisation |
| | | school | |
| | | hospital | |
| | | theatre | |
| | | post office | |
| | | church | |
| local | | traffic spaces | |
| supra local | | | |
| | | park | public green spaces |
| | | play ground | |
| SAN | | | |

Facet (Ausschnitt)

scale 1 : 5ooo

**Figure G10   (a)  Preparatory Land-Use Plan
(Flächennutzungsplan) – extract**

This figure illustrates diagrammatically the scale and content of a typical preparatory land-use plan. The form in which such plans may be produced is specified in the Federal Graphic Representation in Urban Land-Use Planning Order (1981) (see Figure G14). Essentially the plans are land-use zoning plans differentiating between four broad urban land uses (housing areas; mixed building areas; manufacturing/industrial areas; special areas) on a map base, in terms of existing and proposed (planned) development. The plan makes specific provision for land required for public uses, including open space. The functional road hierarchy and urban renewal areas are also clearly demarcated.

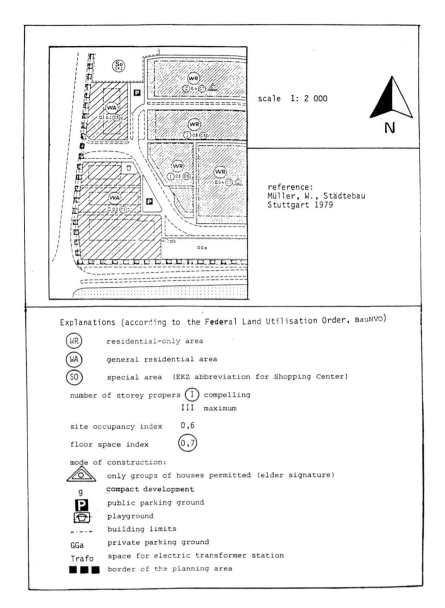

scale 1: 2 000

N

reference:
Müller, W., Städtebau
Stuttgart 1979

Explanations (according to the Federal Land Utilisation Order, BauNVO)

WR    residential-only area

WA    general residential area

SO    special area   (EKZ abbreviation for Shopping Center)

number of storey propers (I) compelling

III    maximum

site occupancy index    0,6

floor space index    (0,7)

mode of construction:

        only groups of houses permitted (elder signature)

g       compact development

P       public parking ground

        playground

-..-    building limits

GGa     private parking ground

Trafo   space for electric transformer station

■ ■ ■   border of the planning area

**Figure G10    (b)  Binding Land-Use Plan (Bebauungsplan)**

This figure illustrates diagrammatically the scale and content of a typical binding land-use plan. The specification of land uses is more detailed than in the preparatory land-use plan (up to ten land uses may be detailed), and includes measures of land intensity regulation such as height of buildings; site occupancy index; floor space index; mode of building construction; and building lines. Detailed provision for car-parking and infrastructure is also indicated.

(ii) residential only areas (WR)

(iii) general residential areas (WA)

(iv) special residential areas (WB)

(v) village areas (MD)

(vi) mixed areas (MI)

(vii) central city areas (MK)

(viii) manufacturing areas (GE)

(ix) industrial areas (GI)

(x) special areas (SO)

(Von Klitzing, 1975, p.93; David, 1986, p.32).

G3.22   Under the provisions of the BBauG (Section 30) the binding land use plan must have a minimum content relating to

(i) type and extent of land use

(ii) the amount of land to be covered with building

(iii) areas required for local traffic purposes.

Besides the type of land use, the scale and intensity of use is controlled by measures such as floor space indexes, total structural volume limits, site occupancy indexes and controls over number of storeys. In addition, controls over type of construction, alignment and building limits are usual.

G3.23   The BBauG (Section 5 and 9) and the BauNVO (Section 1, Paragraphs 1 & 2) thus provide for general and specific land use designations 'to the extent necessary' in approved *Bauleitpläne* (land use plans) at the local (ie municipality) level. Both types of plan are designed to regulate the use of land, and because land is such an important element of a local authority's activities, both of these types of comprehensive land use plan are very important local authority instruments (Schmidt-Assmann, 1986, p.346).

G3.24   All land within the Federal Republic is under the jurisdiction of a local authority, even areas belonging to the Federation or *Länder*, and hence local authorities occupy an important position in land use planning - at the local level, they are the bearers of planning power (Schmidt-Assmann, 1986, p.347). As such, it is a form of public law, subject to constitutional and administrative law (see above, paragraphs G1.23-1.27). Within this process of administrative rule making only the *Bebauungsplan* has the legal sanction to establish or change the land use designation of a site. As David has noted (1986, p.30) this results in the "extraordinary importance in every day planning practice" of the existence of a binding land use plan for all those interested or affected by the legally admitted land use contained therein.

G3.25   For each of the specified land use types in a binding land use plan, the BauNVO regulates what kind of development is admitted or can be admitted as an exception (David, 1986, p.32).

G3.26   If development does not comply with the content of a binding land use plan, the authority has to consider whether the plan itself, or a legal regulation which interprets the plan, explicitly allows an exception to be granted (Section 31, paragraph 1, BBauG). If no such exception can be granted, the authority has further to consider whether a specific dispensation may be granted for individual cases (David, 1986, p.33; Kimminich, 1986, p.198 ). Such dispensations are subject to the requirements that they must be necessary for reasons of public welfare; are required for planning reasons and do not affect the essential structure of urban land use plans; are required in order to overcome an unforeseen obstacle to the implementation of the development plan; and if the proposed change, even if interests of the neighbourhood are taken into account, can be reconciled with public interests. The granting of an exception or a dispensation according to Section 31 of the BBauG can be applied only in connection with an existing *Bebauungsplan* (David, 1986, p.36).

## Regulations which Substitute for a Binding Land Use Plan

G3.27   However, the adoption of a binding land use plan is only one instrument for municipalities to control land use. The BBauG provides two other crucial regulations which apply in the absence of such binding land use plans, and this is of great significance given the fact that, in general, only part of a municipality's territory is covered by such plans. These regulations relate to

(i) built-up areas for which no binding land use plan exists (Section 34 BBauG)

(ii) peripheral development, outside built-up urban areas (Section 35 BBauG) with no binding land use plan (Schmidt-Assmann, 1986, p.338; Kimminich, 1986, p.198; David, 1986, pp.34-35; David, 1987, p.22).

G3.28   In the case of built-up areas for which no binding land use plan exists, a development project is permissible if the type and extent of building use, the manner of construction and size of built-up area are in conformity with the character of adjacent development and if the provision of infrastructure is guaranteed. The standards for healthy living and working conditions must be complied with; there must be no injurious effect upon amenity (BBauG Section 34, paragraph 1).

G3.29   In the case of peripheral development, outside built-up urban areas and where no binding land use plan exists, development projects can be admitted if no conflicting public interests exist, and if the provision of infrastructure is guaranteed. Special conditions attach to development projects classed respectively as privileged and general cases.

G3.30   Privileged cases (Section 35, paragraph 1, BBauG) include

(i) minor development serving agricultural purposes, such as agricultural dwellings

(ii) development serving the purposes of public utilities (tele-communication, electricity, gas, heating, water), waste disposal or an industrial plant

(iii) buildings which because of their harmful effects on their environment or because of their special purpose have to be erected outside populated areas, including development for the purpose of research, development or use of nuclear power for peaceful purposes or the disposal of nuclear waste

(Kimminich, 1986, p.198).

G3.31 General cases under BBauG, Section 35, paragraph 2 are those which may be permitted provided the public interest does not forbid it, with the public interest deemed to be especially negatively affected in the cases involving development projects which

(i) contradict the provisions of the *Flächennutzungsplan*

(ii) create pollution

(iii) create expenditure for road building; the provision of other traffic facilities; facilities for the disposal of rubbish and sewage; security measures; measures to maintain public health or other purposes which are uneconomic

(iv) create dangers for water supply and regulation

(v) are detrimental to the interest of nature protection, landscape conservation and the conservation of historic monuments

(vi) are injurious to townscape and landscape

(vii) injuriously affect the natural characteristics of the landscape as a recreational resource

(viii) lead to scattered residential development or the filling in or extension of an existing area of this kind.

Measures to improve the structure of agriculture have to be taken into special account.

G3.32 Development projects which are spatially significant according to the privileged and general cases of Section 35 BBauG are not to contradict the objectives of spatial organisation expressed in national and *Länder* planning programmes.

G3.33 In applying the provisions of Section 35 BBauG it should be noted that the requirements relating to paragraph 1 are cumulative, in that if a proposed development project falls within one of the privileged categories the authority must still consider whether it is contrary to the public interest and whether the provision of infrastructure is guaranteed. Unless both of these conditions are complied with, the development proposal will not be permitted (Kimminich, 1986, p.198). If it is not a privileged case, any aspect of the proposal which hinders the grant of permission would constitute a legal reason for a refusal of a development permission. Hence

in practice (unless a development proposal falls within a strictly controlled privileged category) there is little chance of general development projects obtaining permission under the provisions of this section of the BBauG, though exceptions have occurred in the past.

G3.34 The provisions of Sections 34 and 35 of the BBauG therefore have the practical effect of replacing the *Bebauungsplan* since they alleviate the necessity to prepare such a plan. The crucial sections of the BBauG (Sections 29-37) apply to all development projects which involve the erection, alteration or change of use of buildings, mineral extraction and large scale excavation and drainage of land (Section 29, BBauG). Whereas the provisions of the *Bebauungsplan* establish with great precision the scope, scale, intensity and type of admitted development, there is considerably more flexibility concerning the admittance of development proposals invoking Section 35 and particularly Section 34 of the BBauG. Nevertheless there are cases which inevitably require the adoption of a *Bebauungsplan*.

## The Rural Environment

G3.35 The land use zoning principles established by *Flächennutzungspläne* (preparatory land use plans) do not apply only to urban areas. The BBauG does not recognise a difference between urban and rural municipalities in this respect - zoning applies to the whole territory of the FRG. In smaller rural communities not subject to land development pressures, the division between agricultural land and building land has been established by time-honoured practice, sometimes on the basis of *Länder* legislation predating the BBauG, and does not require a new *Flächennutzungsplan* (Kimminich, 1986, p.192). But any proposal to permit the building of a residential or industrial annex would require the prior enactment of a *Flächennutzungsplan* (unless falling within the scope of Section 35 of the BBauG, unlikely for major developments).

G3.36 In addition, *Länder* legislation relating to open space preservation, complementary to the *Bundesnaturschutzgesetz* (BNatG, Federal Nature Protection Act, 1976) makes provision where appropriate for the designation of protected areas of different, intensive, protection. These comprise

(i) *Naturschutzgebiete* - areas of outstanding beauty

(ii) *Landschaftsschutzgebiete* - areas of protected landscape

(iii) *Naturparke* - protected parks

(iv) National Parks.

These designations severely restrict building activity, but can also affect the intensity of farming and forestry activity (David, 1986, p.49). The relevant legislation in North Rhine Westphalia is the *Landschaftsgesetz*, 1980.

G3.37 There is a legal assumption (BNatG, Section 8, paragraph 7) that agricultural activities abiding by the rules of orderly agricultural or forestry practice do not constitute an illegal infringement upon the natural environment. Other legislation, in particular the *Flurbereinigungsgesetz*, 1976 (Land Redistribution Law) relating to agricultural re-allocation and consolidation of land holdings has an impact upon agricultural land development (David, 1986, p.50).

## The Definition of Development

G3.38 As well as the provisions of Federal and *Länder* legislation relating to land use planning stemming from the BBauG (Federal Building Law), development proposals have to comply with the provisions of *Landesbauordnung* (State Building Orders - BauO) contained in *Länder* legislation. These relate to the securing of public safety, welfare and environmental protection. Though each *Land* has separate legislation, the *Länder* voluntarily based their enactments on a model draft (*Musterbauordnung*) prepared by a commission of experts from the respective *Länder* ministries headed by a representative of the Federal Ministry (BMBau)(David, 1986, p.28). The relevant legislation for North Rhine Westphalia is the *Landesbauordnung Nordrhein-Westfalen* (BauONW), 1984 (see Appendix A). The legal norms to be applied when a building permit is issued fall within two headings, therefore: material building law (*materielles Baurecht*) - relating to building regulation approval - with statutory authority at the Land level, and formal building law (*Bauleitplanung*) - relating to land use planning - with statutory authority at the Federal and Land level (Kimminich, 1986, p.194). In both cases, it is the *Baugenehmigungsbehörde* (Building Department) of the *Landkreis* and the *Kreisfreie Stadt* which is responsible for the issuing of building permits (Section 57, BauONW), and hence all aspects deriving from public law are considered in a single procedure (*Baugenehmigungsverfahren*)(David, 1986, p.36).

G3.39 All building statutes make building permits obligatory for "the construction, alteration, change in use, demolition or removal of any building" (Section 60, BauONW). But all of them allow a list of exceptions (Section 62, BauONW), the more important being

(i) buildings unfit for habitation with a total volume of less than 45 cubic metres (outside villages 5 cubic metres) and without rest rooms or heating devices; this exception, however, does not apply to garages, showrooms and shops

(ii) one-storey agricultural buildings without heating devices, if their ground space is less than 70 square metres and if they are only for storage or for temporary shelter of animals

(iii) greenhouses without heating devices and up to a height of 4 metres for professional gardeners

(iv) manure containers up to 2.5 metres ... Swimming pools of less than 100 cubic metres, provided they are not outside a village ... 'Insignificant structures', eg terraces, arcades, dovecotes, hunters' rests, garden porches and structures for the airing of carpets.

(Kimminich, 1986, p.196).

G3.40 The BauO, covering aspects of safety, life, health and welfare considerations (eg minimum standards for public rooms and living quarters) make provisions for the standards of construction of buildings and the general technical demands which have to be met in the development project. The following subjects are treated in detail (see Appendix A for full coverage)

(i) Provisions for public access including Fire Department and ambulance requirements; determination of distances between boundaries and buildings under fire regulations; adequate lighting and ventilation of living quarters (Part 2, BauONW).

(ii) Technical requirements for the construction of buildings, including:

(a) matters of stability, fire protection, insulation against excess noise and heat loss as well as measures relating to public health. The demands made on the individual parts of the building depend to some extent upon the use and to some extent upon the manner of construction.

(b) Heating installations, water supply and drainage, minimum conditions for dwellings and single living rooms, the external form of the building construction and extra-mural advertising display.

(Part 3, BauONW).

The respective *Länder* Building Ordinances are complemented by legislative ordinances which partly contain technical details of construction but also make settlements on certain subjects wherever the Building Ordinance itself consists only of blanket injunctions. Furthermore, wherever details of technical construction are not specified in detail in other legislative ordinances, the development must be in accordance with generally recognised rules of building science (established by the *Deutsches Institut für Normung* (DIN - German Standards Institute)) (BMBau, 1976, 26-27).

G3.41 It should be noted that the *Landesbauordnung* includes specific provision for aesthetic control of building activities (Part 3, Section 1, paragraph 12(1) BauONW). Furthermore, a municipality may add its own controls relating to building regulations (material building law) in addition to those of the Land (Section 81 BauONW).

G3.42 In principle, therefore, the level of detailed building control is very great. For example, in the *Kreisfreie Stadt* of Hagen a building permit would not be required for a small garden shed within a domestic curtilage, but it would be required for the change of use of a garage within a domestic curtilage to residential use. Most domestic extensions require permission, as do means of enclosure for plots (walls, fences) with differing provisions relating to situation *vis-à-vis* adjacent highways. In addition, a *Bebauungsplan* (binding land use plan) can specify the most comprehensive and detailed provisions requiring building permits, including regulations incorporating roof pitch (to +/- 5°), materials, colour of external finish etc.

G3.43   The number of detailed building regulations is very great. Fire prevention in buildings is regulated in 35 laws, ordinances, guidelines and decrees. In addition there are more than 30 extremely detailed DIN standards, with around 660 specific standards applied in the narrow field of the building trade. These are supplemented by an additional 1,700 technical standards and regulations which directly affect the building trade, eg for ventilation, heating or sanitation. A decision by the *Länder* Ministers for Building in 1985 should reduce the number of standards and guidelines which apply to the supervision of construction work from its present level of around 215 to 120. This will lead to a noticeable simplification of approval procedures (BMBau, 1986, pp.45-47).

## Special Controls

### Licensing
G3.44   With regard to the control of possible undesirable activities (eg amusement arcades, sex shops), a system of licensing exists but in relation only to:

(i)   the licensing of an individual to trade (ie whether competent and not disqualified by criminal activity)

(ii)   the licensing of machines (but not premises) for amusement arcades.

Such licensing provisions do not derive from land law or planning regulations, and apply irrespective of possible effects on the urban environment. To prevent particular uses, a *Bebauungsplan* (binding land use plan) could be prepared specifically excluding such uses. In general, regulations of the Federal Building Law and respective *Länder* guidelines refer only to land use requirements, not to the user, and as yet there has been no final decision by the Federal Administrative Court as to the legality of such measures.

### Advertisements
G3.45   Advertisement control is strict, and is regulated in the building orders of the *Länder* without substantial variation. The statutes adopt a wide definition of advertising so as to include all signs, letterings, show boxes, illuminations, display areas for posters, etc. Only public notices for religious services are excluded. Outside developed areas advertising is prohibited, as is the case at the edge of developed areas if it is visible outside the limits of this area. Exceptions include

(i)   advertisements on the premises of the plant or business

(ii)   highway signs at the city limits announcing the names and types of business of certain types of firms in the city (usually hotels)

(iii)   signs at crossroads in the countryside pointing to industrial plants

(iv)   advertisements at airports, stadia and playgrounds, provided they are not visible across the landscape

(v)   advertising at exhibitions.

So far as advertising is not prohibited altogether, the following rules have to be respected. Advertising equipment which falls under the definition of 'buildings' presupposes a building permit for their construction. All other kinds of advertising equipment must be constructed in such a way as

(a)   not to disfigure existing buildings, streets, towns or landscapes

(b)   not to impair the safety and facility of traffic

(Kimminich, 1986, p.200).

### Historic Buildings, Monuments and Archaeological Sites
G3.46   The protection of historic buildings is a matter of *Länder* law, covered by *Denkmalschutzgesetz* (1980 - NW) (legislation for the protection of ancient monuments and buildings of historic interest). Usually a permit is required for the following operations

(i)   demolition or alteration of historic buildings

(ii)   removal or alteration of parts of historic buildings

(iii)   construction, alteration or demolition of new buildings.

The quality of a building as an historic building is established by inclusion in a register prepared and kept either by the relevant authority at the *Land level (Landesamt fur Denkmalspflege)* or by the municipality itself (as in the case of Hagen, NW) which then liaises with the responsible *Land* authority (in North Rhine Westphalia this is the responsibility of *Landschaftsverbände* - see paragraph G2.31). Since the listing of buildings has legal consequences for the landowner, the list is approved by the municipality, and listed building consent is administered by the municipality as an integral element of the development permit (Kimminich, 1986, pp.200-201; David, 1986, pp.50-51).

G3.47   Special permission is required for development proposals involving disturbance of archaeological sites. Protected areas may be designated by ordinance, and these must be shown in the *Flächennutzungsplan* (preparatory land use plan)(Kimminich, 1986, p.201) (see Plate 15).

### Tree Preservation Orders
G3.48   Municipalities may enact local statutes requiring the preservation of specified trees, and many *Gemeinden* have utilised such powers (eg Hagen, North Rhine Westphalia in 1984).

## Environmental Laws

G3.49   Whilst there is no single specific environmental law, environmental aspects are contained in a wide variety of laws and fields of legislation. These can be summarised in terms of

(i) Federal legislation, such as the *Atomgesetz* (nuclear activity law) and the *Immissionsschutzgesetz* (air pollution and noise) - see G2.8(i)

(ii) Federal legislation relating to Water Management (*Wasserhaushaltsgesetz*), detailed by *Länder* legislation (*Landeswassergesetz* NW 1979) - see G2.8(ii)

(iii) certain types of sectoral legislation, where the BNatG (see G3.35 above) requires that sectoral plans (such as road construction or agricultural reallocation of land holdings) which may have land consumption and aesthetic destruction implications be accompanied by supplementary plans to safeguard the natural environment and the landscape (*Landespflegerische Begleitpläne*) prepared by the sector authorities in charge of implementing the development (David, 1986, p.50).

Where special permission procedures apply, as for large industrial plants or nuclear facilities falling under (i) above, the land use aspects relating to locational suitability covered by the BBauG still apply as usual (David, 1986, p.37). Permission is technically granted by the *Länder*, but in practical terms the actual processing of the application is carried out by the municipalities.

G3.50   In 1975 the Federal Cabinet approved the application of a 'model procedure' for subjecting all Federal actions to environmental compatibility testing. Individual *Länder* have frequently enacted more specific legislation or regulations on similar environmental matters. In certain cases, they have adopted general procedures for environmental compatibility testing which apply where existing procedures are considered deficient (Wood & Lee, 1978, p.25). At the local level environmental impact assessments can only be enforced by including specific provisions in a *Bebauungsplan*. As Gravells notes (1986, p.404) the implementation of EEC Council Directive 85/337 in 1988 will mean a shift from a discretionary to a mandatory application of such environmental assessment practices for relevant development proposals, and the current changes in the BBauG incorporate the requirements of Federal and EEC environmental legislation.

## Conclusions

G3.51   It is important to stress that in the FRG the term development control encompasses both the granting of a development permission and the adoption of binding urban land use plans. Essentially this arises from the requirement that the authority responsible for granting the development permission which is formally required by law for the legal commencement of development on any site is legally bound by the content of the binding land use plan. This in turn stems from the constitutional guarantee of property rights (David, 1987, pp.17-18).

G3.52   However, the existence of a binding land use plan does not obviate the need to obtain a development permission, and the building permit involves issues going beyond considerations of planning and land use to encompass public safety, welfare and environmental protection. In cases where no binding land use plan exists (Sections 34 & 35 BBauG - see paragraph G3.27-G3.34 above), the formal building law and material building law together in their respective provisions constitute the relevant control instruments.

G3.53   The building permit thus unites in one procedure the relevant requirements of public law, both planning and building regulations. The scope of these requirements is wide and of great complexity and detail - currently regulations determining building law can be found in approximately 250 laws, ordinances and general administrative provisions of the Federation alone, which rises to 1,400 if the provisions of the *Länder* are included (BMBau, 1986, p.45). The building permit is administered by the relevant municipality, so that the implementation of control of development in the FRG is a responsibility of local government.

# G4 THE CONTROL OF DEVELOPMENT

## Building Permits

G4.1   The main agency in the operation of the control of development in the FRG is the *Kreisfreie Stadt* (County-free town) or the *Kreis* (County), since these authorities are responsible for the administration of both the formal building law and the material building law (ie building regulations) through the *Baugenehmigungsverfahren* procedure (development permission, usually referred to as building permit). As David notes (1986, p.29) this establishes a key feature of the land development system in the FRG as being the dependent position of the landowner/developer in relation to the municipality, rather than a public-private partnership in the field of development.

G4.2   The granting of building permissions is the responsibility either of *Kreisfreie Städte* or, in the case of municipalities belonging to a county (*Kreisangehörige Gemeinden*) the *Landkreise* (counties) (see paragraphs G2.22-G2.28) (David, 1986, p.39; 1987, p.13; Kimminich, 1986, p.146). However, the counties are not in charge of adopting urban land use plans, which are within the municipalities' jurisdiction. An application for a building permit has to be presented to the municipality for the territory where the building site is situated. If the municipality which has received the application is not competent to deal with the application (ie if it is not a *Kreisfreie Stadt*) it will be forwarded together with the municipality's views to the county (*Kreis*) administration (Kimminich, 1986, pp.196-197). The county administration will consult with the affected *Gemeinde* before making a final decision in order not to contradict the municipal planning jurisdictions (David, 1986, p.34). Without the consent of the *Gemeinde*, the *Baugenehmigungsbehörde* (Building Permit Department) of the *Landkreis* is legally forbidden to issue the permit in the case of Sections 31,33 to 35 BBauG.

G4.3   In order to enable the authority to process the application for a building permit, each application has to include all blueprints and technical calculations. Applications are normally made on standard proforma (see Figures G11 and G12), but may be made by letter provided all the relevant details specified in the BauO are given. Applications are almost always submitted by professional architects, who typically consult the *Bebauungsplan* (binding land use plan) and work up the detail of the development proposal from its provisions, certifying on the plan and the application that it is in precise conformity with the plan. In Hagen all plans and

forms are placed on microfiche, with a paper file supporting each fiche entry. Any owner of land can, for a small charge, obtain copies of all previous building permits on the site, complete with full details of development proposals (in addition, full details of proprietary land interests are available from the Land Registry).

G4.4   Data from each building application is coded by type of development, square metres floorspace etc in very great detail and sent

 (i)  to the *Länder* statistical department

 (ii) to tax and rating authorities.

A charge is made for each building permit application, which is based on a fixed percentage of the development cost calculated in terms of the external skin of the building, excluding internal finish (BauO).

G4.5   The municipality of Hagen, North Rhine Westphalia has, for the whole of the area of the municipality, full map coverage to a scale of 1:50,000 and 1,5000 in bound form, with aerial photographs facing the respective map on each page to assist in the processing of building permit applications.

G4.6   In processing applications for building permits, municipalities undertake a wide range of formal and informal consultations. Formal consultees are specified in the BauO, and differ according to the regulations under which the development is to be admitted (Sections 30, 34 & 35 BBauG) (David, 1986, p.40). The full range of consultation is, however, at the discretion of individual municipalities. The approving authority has, however, to verify whether the proposed development complies with the regulations of public law unconnected with building legislation, and for this purpose other competent authorities must be granted a hearing (BMBau, 1976, p.27).

G4.7   In those cases where the Federal Building Law (BBauG) makes specific provision for exceptions to normal control (eg Section 31, paragraph 2, see above, G3.30) the approval of the *Regierungsbezirk* is required prior to the issue of the consent. The role of the *Regierungsbezirk* is, however, strictly limited to the supervision of legal procedural regulations, protecting the *Länder* interest in the planning process (David, 1986, p.40).

Stadt Hagen
Bauordnungsamt

**5800 H A G E N**
Friedrich-Ebert-Platz
Verw.-Hochhaus

☐ Antrag auf Vorbescheid ☐ Bauantrag nach § 60 BauO NW ☐ Bauantrag nach § 64 BauO NW

Eingangsvermerk der Bauaufsichtsbehörde

Antragsdatum    Aktenzeichen

| I | Bauherr | Entwurfsverfasser | Bauleiter[1] |
|---|---|---|---|
| Zuname | | | |
| Vorname | | | |
| Stellung (im Beruf) | | | |
| Wohnung, Str. Nr. | | | |
| (PLZ) Wohnort | | | |
| Fernruf Nr. | | | |

**II Baugrundstück**
Straße, Haus-Nr.:
Gemarkung:                    Gemeinde
                             Flur        Flurstück

**III Errichtung**
☐ Einfamilienhaus ☐ gewerbl. Vorhaben ___ Wohneinheiten
☐ Mehrfamilienhaus ☐ Werbeanlage ___ Garage(n)
☐ Wohn- und Geschäftshaus ☐ Warenautomat ___ Stellplätze
☐ landwirtschaftl. Vorhaben ☐ sonstige Vorhaben ___ Einfriedigung
Genaue Bezeichnung

**IV Änderungen**
☐ Abbruch ☐ Aufstockung ☐ Änderung der äußeren Gestaltung
☐ Umbau ☐ Erweiterung ☐ Nutzungsänderung
☐ Instandsetzung ☐ Dachgeschoßausbau ☐ sonstige Vorhaben
Genaue Bezeichnung

**V Haus- und Grundstückseinrichtungen**
☐ Grundstücks-entwässerung mit ☐ Anschluß an Sammelkanal ☐ Kleinklär-anlage ☐ Sonstige Entwässerung
☐ Feuerungsanlage für ___ kW Nennwärmeleistung ☐ feste Brennst. ☐ Heizöl ☐ Gas
☐ Lagerbehälter für ___ l Rauminhalt ☐ Heizöl ☐ unterirdisch ☐ Flüssiggas ☐ oberirdisch ☐ Sonstiges ☐ im Gebäude
☐ Einfriedigung
☐ Sonstige Anlage
Genaue Bezeichnung

**VI Genaue Fragestellung für den Antrag auf Vorbescheid gem. § 66 BauO NW: (ggf. auf bes. Blatt)**

**VII Bindungen für die Beurteilung des Vorhabens**
☐ Bodenverkehrsgenehmigung ☐ Heimstätte ☐ Bauherr ist nicht Eigentümer · oder Erbbauberechtigter des Baugrundstückes
☐ Baulast ☐ Kleinsiedlung ☐ sonstige Bindungen
☐ Vorbescheid ☐ Wohnungsbauförderungsmittel wurden/werden beantragt
Genaue Bezeichnung

**VIII Kosten**
Rohbaukosten:
Herstellungs-kosten f. Anlagen und Einrichtungen:
                    Herstellungskosten (reine Baukosten)

[1] Angabe gem. § 2 BauStatG v. 27.7.78 (BG Bl I S.1118)
[2] Sofern Bauleiter noch nicht angegeben werden kann, ist er in der Baubeginnanzeige zu benennen

---

**2. Städtebaurecht (BBauG) für das Baugrundstück**

**2.1 Beurteilungsgrundlage**
☐ § 30 BBauG  Bebauungsplan Nr.
☐ § 33 BBauG  Bebauungsplan Nr.
☐ § 34 BBauG  Nicht verplanter Innenbereich
☐ § 35 BBauG  Außenbereich    ☐ Absatz 1    ☐ Absatz 2

**2.2** Im Bebauungsplan festgesetzte (§ 30 BBauG) oder nach der vorhandenen Bebauung zu ermittelnden (§ 34 BBauG) städtebaurechtlichen Merkmale

**2.21 Baugebiete**
☐ WS ☐ WR ☐ WA
☐ MD ☐ MI ☐ MK
☐ GE ☐ GI
☐ SW ☐ SO

**2.22 Bauweise**
☐ offen    ☐ geschlossen

**2.23** Zahl der Vollgeschosse (Z):
**2.24** Grundflächenzahl (GRZ):
**2.25** Geschoßflächenzahl (GFZ):
**2.26** Baumassenzahl (BMZ): " ___ m²

**2.3** Anrechenbare Grundstücksgröße:
(nur Flächen hinter d. Straßenbegrenzungslinie)

**2.4** Nachweis der Einhaltung der städtebaurechtlichen Merkmale
**2.41** Geschoßzahl (Z)
   $Z$ geplant ___  v. $Z$ zulässig
**2.42** Grundflächenzahl (GRZ)
   Neubau ___ m²
   Vorh. Bebauung: ___
   $GRZ_{vorh.} = \frac{Grundflächen}{Grundstücksgröße}$ = ___  v. $GRZ_{zul.}$
**2.43** Geschoßflächenzahl (GFZ)
   Neubau ___ m²
   Vorh. Bebauung: ___
   $GFZ_{vorh.} = \frac{Geschoßflächen}{Grundstücksgröße}$ = ___  v. $GFZ_{zul.}$
**2.44** Baumassenzahl (BMZ)
   Neubau ___ m³
   Vorh. Bebauung: ___
   $BMZ_{vorh.} = \frac{Umbauter\ Raum}{Grundstücksgröße}$ = ___  v. $BMZ_{zul.}$

(i.Verfahren)

---

**Figure G11    Building Permit Application Form – Hagen, North Rhine Westphalia**

This six-page application form represents the 'standard' application form for an outline, normal or simplified (ie small-scale) building permit in a Kreisfreie Stadt (county-free town). Sections I–VIII on the first page related to the names and addresses of the owner, designer and builder, together with location, type of proposed development, services/infrastructure, legal requirements and construction costs. Pages 2–6 require more detailed information on these and related topics where appropriate.

284

Die Seiten 4 und 5 sind auszufüllen, wenn eines der nachfolgend aufgezählten Merkmale zutrifft:

    a) Wohnungsbau
    b) Nichtwohnbauvorhaben über 350 cbm umbauten Raum
    c) Nichtwohnbauvorhaben mit reinen Baukosten über 25.000,- DM

**Anleitung:**

Als Zugang werden bei Neubauten die Summe der ermittelten Gebäudemassen, Wohn- und Nutzflächen, Anzahl der Wohnungen und Zimmer, nach den einzelnen Merkmalen aufgegliedert, eingetragen. Als Zimmer im Sinne dieser Statistik werden ausschließlich Aufenthaltsräume gezählt.

**Die Küche gilt als Aufenthaltsraum!**

Bei An- und Umbauten sind unter Zugang die Summe der Werte der alten und neuen Baumaßnahme einzutragen, während unter Abgang die vorhandenen Gebäudemassen vor dem An- bzw. Umbau zu berücksichtigen sind.

**Rechtsgrundlage**

Die Statistik der Bautätigkeit im Hochbau ist durch das zweite Gesetz über die Durchführung von Statistiken der Bautätigkeit und die Fortschreibung des Gebäudebestandes (2. BauStatG) vom 27. 7. 1978 BGBl. I S. 1118) angeordnet.

Auskunftspflichtig sind gem. § 3 des 2. BauStatG für die Hochbaustatistik die Bauherren bzw. die mit der Baubetreuung Beauftragten (Architekten).

---

**3. Bauordnungsrecht (BauO NW)**

**3.1 Berechnung der Anzahl der Stellplätze (Stpl.) für Kraftfahrzeuge (§ 47 BauO NW)**

**3.11 Wohnungen**
Zahl der Wohnungen _____ x 1,0 = _____ Stpl.

**3.12 Gewerbliche Flächen** = _____ Stpl.

    Gesamtzahl der Stellplätze = _____ Stpl.

Davon sind/werden

☐ _____ Stellplätze auf dem Baugrundstück angelegt.

☐ _____ Stellplätze durch Eintragung einer Baulast nach § 78 BauO NW auf einem fremden Grundstück öffentlich-rechtlich gesichert.

☐ _____ Stellplätze gem. § 47 (5) BauO NW abgelöst.

**3.2 Berechnung der Größe des Spielplatzes für Kleinkinder und Angaben über seine Ausstattung.**

(§§ 9 und 81 (3) BauO NW i.V.m. der Satzung über die Beschaffenheit und Größe von Spielplätzen für Kleinkinder in der Stadt Hagen vom 25.2.1972)

**3.21 Flächenbedarf**

1. bis 5. Wohnung = 25 m²

6. " Wohnung (je Wohnung 7 m²) = _____ m²

    = _____ m²

**3.22 Ausstattung mit Spielgeräten und Bänken.**

☐ Der Standort des Kinderspielplatzes auf dem Baugrundstück sowie die Ausstattung mit Spielgeräten und Bänken ist aus dem beigefügten Lageplan 1 : 200 zu entnehmen. (§ 1 (6) der Verordnung über bautechnische Prüfungen (BauPrüfVO) vom 6.12.1984 – GV NW S. 774)

☐ Die Anlegung eines Kinderspielplatzes entfällt, da der Neubau von Wohnungen nicht geplant ist oder die Zahl der geplanten Wohnungen mit Kinderzimmern nicht über zwei liegt.

---

**4.1 Art des Gebäudes** (bitte künftige Nutzung angeben)

Bitte ankreuzen

| | | Anzahl | |
|---|---|---|---|
| | | neuer Zustand | alter Zustand*) |
| Wohngebäude (ohne Wohnheim) | | | |
| ohne Eigentumswohnungen | 1 | 12 | 20 |
| mit Eigentumswohnungen | 2 | 13 | 21 |
| Wohnheim für | | | |
| Studenten | 3 | 14 | 22 |
| Pflegepersonal | 4 | 15 | 23 |
| Andere Berufstätige | 5 | 16 | 24 |
| Alte/r Menschen | 6 | 17 | 25 |
| Sonstige Gruppen | 7 | 18 | 26 |
| Nichtwohngebäude (bitte Art angeben) _____ | 21 | 19 | 27 |

(z. B. Bürogebäude, Werkshalle, Küche, Scheune)

Wohnungen (nach der Zahl der Räume einschl. Küchen)

| | neuer Zustand |
|---|---|
| mit | |
| 1 Raum | |
| 2 Raumen | |
| 3 Raumen | 28 |
| 4 Raumen | 29 |
| 5 Raumen | 30 |
| 6 Raumen | 31 |
| 7 und mehr Raumen | |
| Zahl der Räume in Wohnungen mit 7 und mehr Räumen | |

Von den Wohnungen sind vom Bauherrn bzw. künftigen Erwerber eigengenutzt Wohnungen

| | | |
|---|---|---|
| | | 24-28 |
| Wohnungen mit | | |
| Kochnische/Kochgelegenheit | | |
| Zweitem Bad bzw. getrenntem Duschraum | | 26 |
| Zweit-WC | | |

| | neuer Zusta. | alter Zustand*) |
|---|---|---|
| Sonstige Wohneinheiten | 32 | 34 |
| Räume in sonstigen Wohneinheiten | 33 | 35 |

**4.2 Art der Bautätigkeit**

Bitte ankreuzen

| | |
|---|---|
| Errichtung eines neuen Gebäudes | |
| in konventioneller Bauart | 1 |
| im Fertigteilbau | 2 |
| Baumaßnahme an einem bestehenden Gebäude | 3 |

**4.3 Größe des Zugangs**

Bei Errichtung eines neuen Gebäudes

| | Anzahl |
|---|---|
| Grundstücksfläche (§ 19 Abs. 3 BauNVO) | 01 |
| Grundfläche (§ 19 Abs. 2 u. 4 BauNVO) | 02 |
| Geschoßfläche (§ 20 Abs. 2 u. 3 BauNVO) | 03 |
| Rauminhalt – Brutto (DIN 277) | 04 |
| Zahl der Vollgeschosse (nach LBO) | 05 |

Bei allen Baumaßnahmen

| | neuer Zustand | alter Zustand*) |
|---|---|---|
| Nutzfläche in m² (DIN 277, o. Wohnfläche) | 06 | 09 |
| Wohnfläche in m² (DIN 283) der Wohnungen | 07 | 10 |
| der sonst. Wohneinheiten | 08 | 11 |

**4.4 Größe der Räume**
in Wohnungen und sonstigen Wohneinheiten

| | Anzahl | |
|---|---|---|
| | neuer Zusta. | alter Zustand*) |
| Küchen | | |
| bis unter 6 m² | 36 | 45 |
| 6 bis unter 12 m² | 37 | 46 |
| 12 m² und mehr | 38 | 47 |
| Zimmer (über 6 m²) | | |
| bis unter 10 m² | 39 | 48 |
| 10 bis unter 15 m² | 40 | 49 |
| 15 bis unter 20 m² | 41 | 50 |
| 20 bis unter 25 m² | 42 | 51 |
| 25 m² und mehr | 43 | 52 |
| Einzelzimmer über 6 m² außerhalb von Wohneinheiten | 44 | 53 |

**6 Veranschlagte Kosten des Bauwerkes** (siehe DIN 276)

| | 1000 DM |
|---|---|
| | 54 |

*) Alten Zustand bitte nur bei Baumaßnahmen an bestehenden Gebäuden angeben.

**(Allgemeine Angaben)**

KA 3  Lsp I

**Der Bauherr zählt zu den**   Bitte ankreuzen

Bitte ankreuzen bzw. zutreffende Ziffer einsetzen

**Öffentlichen Bauherren**

| | |
|---|---|
| Bund | 01 |
| Länder | 02 |
| Gemeinden und Gemeinde-verbände | 03 |
| Sozialversicherung | 04 |

**Unternehmen**

| | |
|---|---|
| Gemeinnützige Wohnungs- und landliche Siedlungsunternehmen | 05 |
| Sonstige Wohnungsunternehmen | 06 |
| Immobilienfonds | |
| Sonstige Unternehmen (ohne Wohnungsunternehmen) | 07 |

| | |
|---|---|
| Land- und Forstwirtschaft, Tierhaltung, Fischerei | 08 |
| Produzierendes Gewerbe | 09 |
| Handel, Kreditinstitute und Versicherungs-gewerbe, Dienstleistungen | 10 |
| Verkehr und Nachrichtenübermittlung (ohne Bundesbahn und Bundespost) | 11 |
| Bundesbahn und Bundespost | 12 |

12-13

**Privaten Haushalten**

| | |
|---|---|
| Selbständige | 13 |
| Beamte, Angestellte | 14 |
| Arbeiter | 15 |
| Rentner, Pensionäre | 16 |
| Sonstige private Haushalte | 17 |

Organisationen ohne Erwerbscharakter

Der Bauherr ist Sanierungsträger (§ 33 StBauFG)   ja / nein  14

**Das Baugrundstück liegt** (Bei allen Baumaßnahmen)

in einem förmlich festgelegten Sanierungsgebiet (§ 5 StBauFG)
in einem Ersatz- bzw. Ergänzungsgebiet (§ 11 StBauFG)
in einem städtebaulichen Entwicklungsbereich (§ 53 StBauFG)
außerhalb der genannten Gebiete   [1][2][3][4]  15

**(Nur bei Errichtung eines neuen Gebäudes)**

a) im Geltungsbereich eines qualifizierten Bebauungsplans (§§ 30, 33 BBauG)
innerhalb von im Zusammenhang bebauten Ortsteilen (§ 34 BBauG)
mit einfachem Bebauungsplan
ohne Bebauungsplan
im Außenbereich (§ 35 BBauG)   [1][2][3][4]  16

b) sofern ein Bebauungsplan mit Baugebiets-festsetzung nach BauNVO vorliegt, in einem

| | | |
|---|---|---|
| Kleinsiedlungsgebiet | (WS) | 01 |
| reinen Wohngebiet | (WR) | 02 |
| allgemeinen Wohngebiet | (WA) | 03 |
| besonderen Wohngebiet | (WB) | 04 |
| Dorfgebiet | (MD) | 05 |
| Mischgebiet | (MI) | 06 |
| Kerngebiet | (MK) | 07 |
| Gewerbegebiet | (GE) | 08 |
| Industriegebiet | (GI) | 09 |
| Sondergebiet für Erholung | (SO, § 10 BauNVO) | 10 |
| sonstigen Sondergebiet | (SO, § 11 BauNVO) | 11 |

17-18

oder sofern kein Bebauungsplan mit Baugebiets-festsetzung nach BauNVO vorliegt. Die Eigenart der näheren Umgebung entspricht einem

| | | |
|---|---|---|
| Wohngebiet | (W) | 1 |
| Dorfgebiet | (M) | 2 |
| Mischgebiet | | 3 |
| Gewerbegebiet | (G) | 4 |
| Sondergebiet | (S) | 5 |

19

**Zu 4.1 (Art des Gebäudes)**

Bei Errichtung eines neuen Wohngebäudes
Das Wohngebäude dient Ferien-, Wochenend-, Erholungszwecken   ja / nein   20

Haustyp des Wohngebäudes
Einzelhaus (1)   Gereihtes Haus (3)
Doppelhaus (2)   Sonst Haustyp (4)   21

Bei allen neu zu errichtenden Gebäuden
Art der Konstruktion
Skelettbau (1)   Massivbau (2)   22

Überwiegend verwendeter Baustoff
Stahl (1)   Sonst Mauerstein (4)
Stahlbeton (2)   Holz (5)
Ziegel (3)   Sonstiges (6)   23

Unterkellerung
keine (1)   mit 1 Untergeschoss (2)
mit 2 und mehr Untergeschossen (3)   24

Abwasserablauf direkt in
Öff. Kanalisation mit Klärwerk (1)
Öff. Kanalisation ohne Klärwerk (2)
Kleinkläranlage (DIN 4261) (3)
Sonstige Abwasserbehandlungsanlage (4)
Grube (Behälter u.a.) (5)
ohne Abwasseranschluß (6)   25

Art der Beheizung
Fernheizung (1)   Etagenheizung (4)
Blockheizung (2)   Einzelraumheizung (5)
Zentralheizung (3)   keine Heizung (6)   26

Vorwiegende Heizenergie
Koks/Kohle (1)   Fernwärme (5)
Gas (2)   Wärmepumpe (6)
Strom (3)   Solarenergie (7)
(4)   Sonstige (8)   27

Klimaanlage   ja / nein   28

**Zu 4.2 (Art der Baulichkeit)**

Bei Baumaßnahmen am bestehenden Gebäude
Ändert sich die Nutzungsart des ganzen Gebäudes?   ja / nein   29

Wenn ja, bitte frühere Nutzung angeben

**Bei Wiederaufbau, Ersatzbau, Wiederherstellung**
In welchem Jahr wurde das Gebäude (Gebäudeteil) abgebrochen, zerstört o.ä.?   19

**Zu 4.3 (Größe des Zugangs)**

Nachgewiesene Pkw-Stellplätze (alle Stellplätze sind nach Art und Lage anzugeben)

| | | |
|---|---|---|
| Art | Garagen | 32-35 |
| | offene Stellplätze | 36-39 |
| Lage | auf dem Baugrundstück | 40-43 |
| | auf einem getrennten Grundstück | 44-47 |
| | durch Ablösung bei der Gemeinde | 48-51 |

---

Folgende Unterlagen, die der Verordnung über bautechnische Prüfungen (BauPrüfVO) vom 6.12.1984 (GV NW S. 774 / SGV NW S. 232) entsprechen, sind beigefügt.

Die Klammerwerte für die Zahl der Ausfertigungen gelten, wenn der Kreis untere Bauaufsichtsbehörde ist. Weitere Ausfertigungen sollen zur Beschleunigung des Verfahrens eingereicht werden, wenn andere Behörden oder Dienststellen zu beteiligen sind (§ 1 Abs. 3 BauPrüfVO).

### A. Allgemeine Bauvorlagen

| Nr. | | |
|---|---|---|
| 1. | 2-(3)-fach | Lageplan Maßstab 1:500 (§ 2 BauPrüfVO)  [ ] amtlich beglaubigt oder angefertigt |
| 2. | 2-(3)-fach | Übersichtsplan Maßstab 1: |
| 3. | 2-(3)-fach | Berechnungen des Maßes der baulichen Nutzung (§ 2 Abs. 5 BauPrüfVO) |
| 4. | 2-(3)-fach | Bauzeichnungen Maßstab 1:100 (§ 3 BauPrüfVO) |
| 5. | 2-(3)-fach | Baubeschreibung (§ 4 BauPrüfVO) |
| 6. | 2-(3)-fach | Berechnung des Rauminhaltes nach DIN 277 |
| 7. | 2-(3)-fach | Berechnung der Rohbaukosten [ ] Herstellungskosten für Anlagen und Einrichtungen |
| 8. | 2-(3)-fach | Berechnung der Wohnfläche/Nutzfläche nach DIN 283 |
| 9. | 2-( )-fach | Statische Berechnung [ ] Bewehrungs- und Konstruktionszeichnungen (§ 5 BauPrüfVO) |
| 10. | 1-( )-fach | Erklärung zur Bauantrag nach § 3 WärmeschutzÜVO |
| 11. | 2-( )-fach | Nachweis des Schallschutzes |
| 12. | 2-( )-fach | Nachweise des baulichen Brandschutzes (§ 5 Abs. 3 BauPrüfVO) |
| 13. | -fach | |

### B. Bauvorlagen für Behälteranlagen  > 5.000 l

| Nr. | | |
|---|---|---|
| 14. | 2-(3)-fach | Lageplan mit Standort der Anlage (bei gesondertem Antrag) |
| 15. | 2-(3)-fach | Beschreibung für Behälter [ ] Zulassungsbescheid 1-fach |
| 16. | 2-(3)-fach | Grundriß und Querschnitt Lagerraum |
| 17. | 2-(3)-fach | Berechnung der Herstellungssumme |
| 18. | -fach | |

### C. Unterlagen für die Eintragung einer Baulast, Erteilung einer Befreiung oder Vereinigung von Flurstücken

| Nr. | | |
|---|---|---|
| 19. | -fach | unbeglaubigter Grundbuchauszug neuesten Datums |
| 20. | -fach | Lageplan neuesten Datums von ÖbVI oder Katasterbehörde |
| 21. | -fach | Befreiungsantrag mit Begründung |
| 22. | 1-fach | Einverständniserklärung des (der) Grundstücksnachbarn |
| 23. | 1-fach | Veränderungsnachweis über die Vereinigung der Flurstücke |
| 24. | -fach | |

### D. Unterlagen für die straßenbaurechtliche Ausnahmegenehmigung bzw. Zustimmung bei Anbau an Kreis-, Land- oder Bundesstraße

| Nr. | | |
|---|---|---|
| 25. | 1-fach | Flurkartenauszug (Übersichtsplan) |
| 26. | 1-fach | Meßtischblatt mit Kennzeichnung des Baugrundstücks |
| 27. | 2-fach | Lageplan 1:500 |
| 28. | 2-fach | Straßenquerprofil mit Angabe der Kilometrierung |
| 29. | -fach | |

### E. Unterlagen für Vorhaben mit besonderen Anforderungen gem. §§ 50 und 51 BauO NW

| Nr. | | |
|---|---|---|
| 30. | 2-fach | Übersichtsplan mit Eintragung vorhandener Nutzungen und planungsrechtlicher Festsetzungen |
| 31. | 3-fach | Maschinenaufstellungsplan mit Rettungswegen und Notausgängen |
| 32. | 3-fach | Betriebsbeschreibung (§ 4 Abs. 3 BauPrüfVO) |
| 33. | 3-fach | Besondere bauliche Vorkehrungen für Behinderte |
| 34. | -fach | |
| 35. | | |

Weitere Ausfertigungen sind beigefügt zu Nr.: ...   Unterlagen werden nachgereicht zu Nr.: ...

Der Standsicherheitsnachweis kann durch einen Prüfingenieur zu Lasten des Bauherrn geprüft werden.

Vollmachtserklärung für Entwurfsverfasser (Architekt, _____ )

Mir/Uns ist bekannt, daß die Bauaufsichtsbehörde den Bauantrag gem. § 67 BauO NW i.V. mit der Tarifstelle 2 AVwGebO NW gebühren-pflichtig zurückweisen kann, wenn die Bauvorlagen erhebliche Mängel aufweisen.

Mir/Uns ist bekannt, daß gem. § 70 (5) BauO NW vor Zugang der Baugenehmigung nicht mit der Bauausführung begonnen werden darf, und daß die Zuwiderhandlung mit einer Geldbuße bis zu 100.000,-- DM (§ 79 BauO NW) geahndet werden kann.

Gebühren für evtl. erforderliche Baulasteintragungen werden von mir/uns übernommen (§ 13 (1) Ziffer 2 GebG NW).

[ ] [ ] [ ]

_____   _____
Bauherr (Datum, Unterschrift)   Entwurfsverfasser (Datum, Unterschrift)

An das Bezirksamt / Ortsamt

Bauprufabteilung / Bauprufung

**ANTRAG** auf

1 ☐ Baugenehmigung
2 ☐ Nutzungsgenehmigung
3 ☐ Abbruchgenehmigung
4 ☐ Werbegenehmigung
5 ☐ Vorbescheid

nach § 60 bzw. § 65 Hamburgische Bauordnung (HBauO) vom 1. Juli 1986.

**Wird von der Behörde ausgefüllt**

Geschäftszeichen ____________

Sachbearbeiter / in ____________

Telefon ____________

Eingangsstempel

Bauherr / Antragsteller

Name ____________

Straße, Hausnr. ____________

Ort ____________

Baugrundstück

Straße, Hausnr. ____________

Grundbuch ____________

Flurstück ____________

Blatt ____________

Baublock-Nr. ____________

Bezeichnung des Bebauungsplanes

Bezeichnung des Vorhabens

Bei baulichen Anlagen nach § 63 Absatz 3 HBauO:
Die Prüfung der bautechnischen Nachweise wird beantragt.

nein ☐ ja ☐, dann ____________ Datum / Unterschrift

Für die Bearbeitung von Anträgen werden Gebühren nach der BauGebO erhoben

☐ Herstellungskosten (DIN 276) ____________ DM

☐ Es besteht Gebührenfreiheit wegen ____________ (Ggf. Nachweis)

Als Entwurfsverfasser, Sachverständiger nach § 55 HBauO / Bauvorlageberechtigter nach § 64 HBauO

Name ____________

Straße ____________

Telefon ____________ ____________ Datum / Unterschrift

Als Grundstückseigentümer*, wenn nicht zugleich Bauherr

Name ____________

Straße ____________

Telefon ____________ ____________ Datum / Unterschrift

*) nicht erforderlich bei Antragsart 4 und 5

Als Bauherr (auch bei Abbrucharbeiten)

Telefon ____________ ____________ Datum / Unterschrift

**Figure G12 Building Permit Application Form – Hamburg**

This application form is for a building permit, change of use permit, advertisement or application for preliminary discussion. It requires details of the type and location of the proposed development, names and addresses of owner, designer and builder and relationship to appropriate binding land use plans.

G4.8 Neighbours are involved in the procedure of issuing building permits, though the provisions laid down in the BauO of the *Länder* vary. In Baden-Württemberg consultation with neighbours is a general rule, but in North Rhine Westphalia it is discretionary. In Hagen, for example, consultation with neighbours would take place only in the case of contentious proposals, and would involve only immediately adjacent properties for building regulation matters, extending to the wider neighbourhood for planning matters. Where such consultation is obligatory, the plans filed with the application for the permit have to be signed by the owners of all plots adjacent to the projected building site. The refusal of a signature by a neighbour does not prevent the authority from issuing the permit, but does enable that neighbour to litigate against the permit (see below paragraphs G4.26 and G6.9) (Kimminich, 1986, p.199).

G4.9 The period of time within which a decision must be given by the municipality is not regulated by law, though if an application for a building permit has not been considered within 'adequate time' the applicant is entitled to file an action with the administrative court. The term 'adequate time' is interpreted by legal commentaries to mean 'a good period of time within which a decision may be expected by experience'. Since the filing of an action to the administrative courts is precluded within three months after the filing of an application, this is the period of time which at least has to be tolerated for the total inactivity of the authority (Kimminich, 1986, p.205).

G4.10 Most applications for building permits are determined by officers. However, those of 'extraordinary importance' are referred to a special sub-committee of the full council of the municipality by the elected director in consultation with respective heads of section. The elected director gives general policy directives to officers, but in general not specific instructions in relation to individual building permits. Individual officials may find themselves subject to conflicting obligations in relation to *Land* or *Gemeinde* responsibilities in relation to specific development proposals (eg listed buildings conflicting with urban proposals, involving the *Landeskonservator* in the respective *Landschaftsverband*, see paragraph G2.39). Without a specific directive from the *Regierungspräsident*, which is rare, the municipality's priorities prevail.

G4.11 The authority responsible for issuing the building permit must apply both the formal and the material building law. In practice, the formal building law (ie 'planning') provisions are considered first, because the detailed scrutiny of building regulation provision is superfluous if the proposed development contradicts approved urban land use plans (Kimminich, 1986, p.197).

G4.12 A proposal may, consequently

(i) comply fully with both the formal and material building law, in which case a permit must be issued

(ii) deviate from the legal norms, in which case the permit may be issued under the condition of effecting the changes prescribed by the authority. In this case the permit is valid only if the building is constructed according to the altered plans. Otherwise the building is illegal and may be demolished on the order of the authority. Changes to the approved plan may require a new application provided the permission has not already come into legal force, and work may be resumed only after the permit for the amended proposal has been obtained (Kimminich, 1986, p.197; BMBau, 1976, p.27).

(iii) deviate from the legal norms to such an extent that permission is refused.

A building permit ceases to be valid after two years following its legal incontestability if the project has not been begun or if it has been interrupted for a period of one year. The period of validity of the permit may be extended for a year at a time.

G4.13 The refusal rate is very low, especially in areas where there is extensive coverage of binding land use plans. For example in the *Bezirk Hamm-Mitte* in Hamburg, with a full coverage of binding land use plans, the figures for building permits in 1986 were:

| | |
|---|---|
| Rejected | 0.4% |
| Permitted without conditions | 20.0% |
| Permitted with conditions or by exemptions | 79.6% |

G4.14 One of the reasons for the low refusal rate is the extensive consultation that takes place with prospective developers prior to the submission of a building permit application. In addition, before an application for approval to build is made, an interim decision may be requested or given, valid for two years, often relating to the reparcellation and acquisition of land, prior to a detailed development proposal. The charge for such interim building permits is lower than the full permit charge. In the event of the preparation of a binding land use plan contradicting such an interim building permit, the municipality would be liable for compensation. Furthermore, work on the construction site and on individual preliminary stages for individual components of a building may also be approved before the building permit itself has been granted (partial building permit)(B-MBau, 1976, p.27). To indicate the overall significance of these provisions, the municipality of Hagen grants approximately 600 building permits per annum, but handles about 3000 files per annum for all development proposals, many of which are applications prior to a full building permit application.

G4.15 In addition to the guidance and advice provided by municipalities, the *Länder* ministries issue many guidance documents or 'circulars' relating to formal and material building law, which interprets the provisions of legislation. In the FRG, such a 'circular' is just an interpretation of the law. If a municipality or individual does not wish to accept this interpretation they are obliged to show that their interpretation is in accordance with the law. The administrative law courts refer directly to the law, not to the interpretation offered by such circulars. However, in practice there are so many legal cases and the law is so complex that the 'circulars' act as practical guides to both municipalities and prospective developers. Often containing considerable detail, such 'circulars' are specifically intended to inform the public of the *Länder* interpretation of the law.

G4.16 Effective control over building operations once under way is secured by the requirement that one set of approved plans and one copy of the building permit are required to be available at all times on the building site, which may be inspected by municipality officials upon demand. In order to ensure this control when there is no activity on the site, the proprietor is obliged to post a notice outside the building site announcing his/her name and full address, which may be removed upon permanent occupation of the building (Kimminich, 1986, p.197) (see Figure G13).

G4.17 Until 1984 all buildings for which a building permit was required had to pass two comprehensive inspections at different stages of the operations:

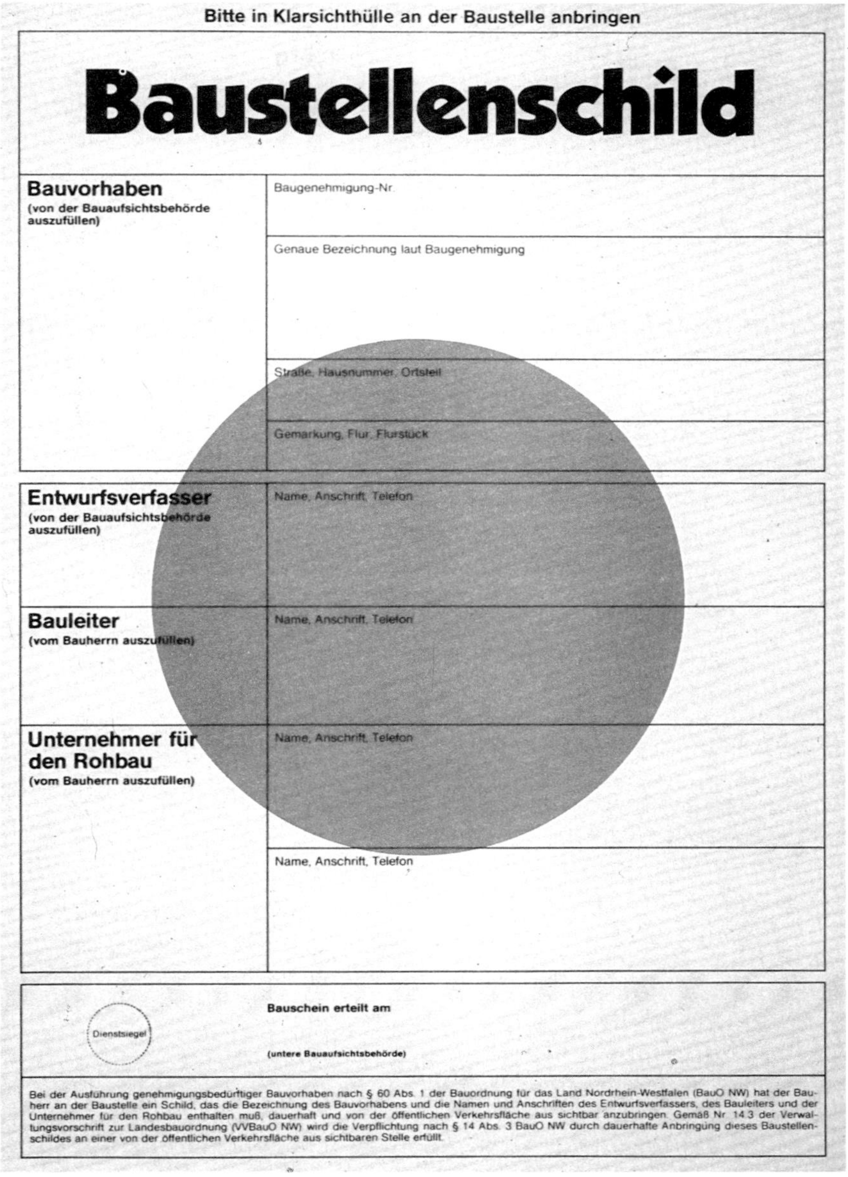

Bitte in Klarsichthülle an der Baustelle anbringen

**Baustellenschild**

| | |
|---|---|
| **Bauvorhaben** (von der Bauaufsichtsbehörde auszufüllen) | Baugenehmigung-Nr. |
| | Genaue Bezeichnung laut Baugenehmigung |
| | Straße, Hausnummer, Ortsteil |
| | Gemarkung, Flur, Flurstück |
| **Entwurfsverfasser** (von der Bauaufsichtsbehörde auszufüllen) | Name, Anschrift, Telefon |
| **Bauleiter** (vom Bauherrn auszufüllen) | Name, Anschrift, Telefon |
| **Unternehmer für den Rohbau** (vom Bauherrn auszufüllen) | Name, Anschrift, Telefon |
| | Name, Anschrift, Telefon |

Dienstsiegel

**Bauschein erteilt am**

**(untere Bauaufsichtsbehörde)**

Bei der Ausführung genehmigungsbedürftiger Bauvorhaben nach § 60 Abs. 1 der Bauordnung für das Land Nordrhein-Westfalen (BauO NW) hat der Bauherr an der Baustelle ein Schild, das die Bezeichnung des Bauvorhabens und die Namen und Anschriften des Entwurfsverfassers, des Bauleiters und der Unternehmer für den Rohbau enthalten muß, dauerhaft und von der öffentlichen Verkehrsfläche aus sichtbar anzubringen. Gemäß Nr. 14.3 der Verwaltungsvorschrift zur Landesbauordnung (VVBauO NW) wird die Verpflichtung nach § 14 Abs. 3 BauO NW durch dauerhafte Anbringung dieses Baustellenschildes an einer von der öffentlichen Verkehrsfläche aus sichtbaren Stelle erfüllt.

**Figure G13 Building Site Notice – Hagen, North Rhine Westphalia**

This notice has to be displayed at all times on the building site, and specifies a brief description of the permitted development together with name, address and telephone number of designer, builder and site owner.

(i) the 'raw brick inspection', necessary for all types of buildings. The interior finishing may be commenced only after this inspection

(ii) the final inspection.

A certificate was issued for each inspection (Kimminich, 1986, p.199).

Since 1984 in North Rhine Westphalia this detailed inspection requirement has been modified, because municipalities had been held liable to claims in law because they had inspected and certificated a building. Now the authority's certification relates only to matters 'in the public interest', and the authority must be informed by a certain stage of the development's progress when it is free to carry out an inspection or not (Section 77, BauONW). In addition, there are simplified types of authorisation for small-scale residential development (Section 64, BauONW).

## Enforcement

G4.18 The enforcement of the control of development is strict. Any operation contravening public law regulations may be immediately stopped by the competent authority (including all violations of public law occurring on the occasion of any kind of operation, whether requiring a building permit or not. A building lacking a permit or deviating from approved plans may be demolished if lawful conditions cannot be secured otherwise (Kimminich, 1986, p.199). In addition, fines of up to 100,000 DM can be imposed for the largest types of development, with 6,000 DM being the norm for unauthorised minor development (such as an individual dwelling). The building department may also issue a stop notice, requiring the cessation of unauthorised building works.

## Flexibility and Control

G4.19   Under the provisions of the BBauG detailed in paragraphs G3.23-G3.35 above the entire area under the jurisdiction of municipalities is divided into three kinds of zone

(i) zones with a binding land use plan (Section 30 BBauG) where development is permitted which complies with the content of the plan

(ii) zones where no binding land use plan exists but where surrounding areas have already been built on (Section 34 BBauG) - in such zones development is permitted if it is in keeping with the existing land use pattern and is 'fitting' to the environment

(iii) zones having open land, very often at the periphery of urban areas, without a binding land use plan. In such zones, development can be permitted if no conflicting public interest exists (Section 35, paragraph 2 BBauG). There are strictly defined special and general allowances (see paragraphs G3.30-G3.32)

(Schmidt-Assmann, 1986, p.338; Kimminich, 1986, p.198; David, 1986, pp.34-35; 1987, p.24).

These provisions permit different degrees of flexibility in the operation of the control of development in the FRG, depending upon whether or not a binding land use plan is in force (or in preparation).

G4.20   In the case of zones with a binding land use plan, control is in principle total (within the provisions of public law). The *Bebauungsplan* (binding land use plan) may be used as a positive instrument of control, with the intention of providing a comprehensive land use plan for the future development of land. Since the passing of the BBauG in 1960, most development has been peripheral urban development, so that many *Bebauungspläne* deal with the co-ordination of 'greenfield' development for town expansion and new housing. Large cities usually have numerous binding land use plans, each one covering a small area of the city (they are known colloquially as 'postage stamp plans'). Some cities, such as the city-state of Hamburg, may have a total coverage of *Bebauungs-pläne* (hence rendering Section 34 of the BBauG irrelevant for the control of development), but for most cities the coverage is approximately 25 per cent by area. Small municipalities may manage to fit all regulations into one *Bebauungsplan*. If a municipality does not plan any building activity, it is not obliged to draw up a binding land use plan (Kimminich, 1987, p.193).

G4.21   The *Bebauungsplan* may also be used as an instrument of negative control, designed to prevent development (eg new retail or warehouse development at the urban periphery), but a *Bebauungsplan* containing nothing more than the exclusion of a specific kind of land use would not be legal. There needs to be a particular reason justifying an exclusion, and a minimum positive indication of the permitted legal use must be provided. Where a *Bebauungsplan* is intended to change the land uses already permitted by the operation of Sections 34 & 35, the BBauG (Section 44 and Section 42 BauGB 1987) makes provision for compensation when the new use provided in the *Bebauungsplan* results in a depreciation of the potential development value which the owner could be expected to obtain from development in accordance with Section 34/35 provisions. Whereas previously development proposals may have triggered the preparation of a *Bebauungsplan* in urban renewal areas, in future the charges in Section 34 of the 1987 BauGB may reduce the legal pressure to prepare a plan. This may be interpreted as a reduction in the degree of control of development, or alternatively as the introduction of a further degree of flexibility into a system of potentially legally constricting development plans.

G4.22   For development proposals falling in areas for which a binding land use plan is in preparation, the application for a building permit will not be considered (Section 14) until the plan comes into force, provided the municipality has passed a by-law. But the *Baugeneh-migungsbehörde* (Building Permit Department) may issue a building permit before the plan comes into force, with the consent of the municipality (Section 33, 36, BBauG). Compensation for the losses incurred during this transitional period is paid only if the time exceeds four years (Section 18 BBauG) (Kimminich, 1986, p.194; David, 1987, p.26). This 'freezing' of development previously required the consent of the *Regierungsbezirk*, but now current changes permit such a 'freeze' by by-law.

G4.23   Where a binding land use plan is not in force (or in preparation) the provisions of Section 34 of the BBauG, especially in its amended 1987 form, provide the greatest degree of flexibility within the operation of formal building law. This Section applies only to developed areas, 'developed' in this context being legally defined as parts of a municipality with houses in close neighbourhood. In this case the building permit must be issued if the proposed development is in conformity with the adjacent land uses (Kimminich, 1986, p.198). The municipality may define sites at the periphery which are considered to be part of 'built-up areas' (Section 34, paragraph 2, BBauG).

G4.24   In the case of mixed areas of residential and industrial development falling within Section 34 BBauG jurisdiction in North Rhine Westphalia, the *Land* Ministry (MSWV, see paragraph G2.13) have until recently required the separation of residential and industrial uses. The application of such planning regulations to new development poses few problems, but in 'mixed areas' their strict application is impracticable. Consequently the *Land* Ministry has introduced a 'circular' suggesting how a compromise may be reached between existing and proposed uses, requiring the reduction of noise/smell emission but also some endurance of a degree of emission by affected areas. The intention is to produce an improvement in the existing situation, rather than the achievement of an abstract ideal. The 'circular' brings together the binding regulations and administrative advice of the MSWV, the Economic Ministry and the Emission/environmental Inspectorate, and this circular is equally applicable to the Emissions Inspectorate and municipalities. Up to 1980 such problems were solved only by the removal of industrial premises to other locations, caused by the requirements of emission legislation and the requirements of the Administrative High Court protecting the rights of property owners. The impossibility of financing the removal of industry and the employment consequences of such a policy have thus resulted in the production of the 'circular' seeking a degree of compromise. This may be seen, once again, either as a reduction in effective planning control or the achievement of a degree of practical flexibility. In any event, although a circular may interpret the law, it cannot change it.

G4.25  Some expert commentators have expressed the view that current modifications to the BBauG will have the practical effect of extending the scope of Section 34 regulations, and reducing the likelihood that *Bebauungs-pläne* will be prepared for such areas. In the context of a federal policy focus upon urban renewal (BMBau, 1986, 27), this is likely to be a major change in the future operation of the control of development in the FRG.

G4.26  Third parties, mostly owners of adjacent sites, may seek to oppose the granting of development permission based on Section 34 BBauG by the authority. This requires that they can prove that the legal rights protecting neighbours have been violated. But Section 34 is not generally recognised by the law courts as having this legal character and therefore it requires additional grounds of objection by the neighbouring owner to win the case (David, 1987, p.25).

G4.27  For zones of open land (often at the urban periphery), with no binding land use plan, the authority is granted in practice a wide degree of discretion as to whether it considers public interests would be violated by the development for which permission is sought. Since the majority of Section 35 cases involve new development at the fringes of a municipality on land in existing agricultural use where a binding land use plan does not exist, and the authority has only limited powers to refuse a building permit for privileged cases, most of the practical problems involve legal dispute over the definition of privileged development (David, 1987, p.26).

G4.28  Hence the existence of Sections 34 and 35 of the BBauG may be viewed favourably or unfavourably by

municipalities, depending upon their attitude to development proposed under these regulations. The municipality may, of course, draw up a *Bebauungsplan* for such areas to obviate any problems.

## Related Powers to Control Land Use and Development Infrastructure Charges

G4.29  Municipalities possess a significant influence over development under the provisions of Section 123 BBauG, which specifies that landowners have to pay up to 90 per cent of the development costs relating to the provision of streets, parking places and other public facilities, and 100 per cent of the cost of facilities for the supply of gas, water, electricity and sewerage (Kimminich, 1986, pp.207-208). Even if a developer is willing to carry all the costs, the municipality must carry at least 10 per cent. This is intended to oblige municipalities to be economical with land and with its capital investment and is therefore a practical constraint on municipalities' planning policies, especially with regard to the release of land for residential development.

## Safeguarding Measures for Urban Planning

G4.30  As well as the freezing of development activities for up to four years (Section 14 and 17 BBauG) and the postponement of the granting of a building permit for one year (Section 15 BBauG) in areas for which a binding land use plan is in preparation (see above, paragraph G4.22), Section 19 of the BBauG provides powers authorising municipalities to approve the sub-division of land in such areas. The granting of such authorisation establishes a legal presumption in favour of the use of land for which the subdivision took place (and cannot therefore be refused within the following three years on planning grounds, Section 21 BBauG) (David, 1986, p.38).

G4.31  In addition, municipalities have pre-emption rights under Section 24-25a BBauG and Sections 17 & 57 StBauFG, which allow them to enter a contract for the purchase of land under the same conditions as negotiated between a private vendor and buyer (David, 1986, pp.38-39). Previously, municipalities could generally reduce the price of such transactions if they were deemed to be above the current market price, ultimately sanctioned by the Administrative Courts on appeal (see Section 28a BBauG). With the amalgamation of urban renewal law with formal building law in the *Baugesetzbuch* of 1987, such reductions will be permissible only for purchases of land situated in areas covered by binding land use plans which are required for public land uses (roads, parks, hositals, etc), and if an expropriation would be legally admissible. In practice, pre-emption is of minor significance for municipalities because of lack of funds to purchase land.

## Municipality Land Policy Instruments and Binding Land Use Plans

G4.32 Many German municipalities own a considerable part of their territory, generally agricultural land outside the built-up area. There is a long tradition of the purchase of municipal land stocks, by agreement in the open market (though sometimes against a background of reserved compulsory purchase powers). More recently, such acquisition policy has declined as local authority finances have dwindled. Municipalities can sell land at prices below market value in order to attract commercial or industrial development or to subsidise housing. As a general rule, municipalities convey the freehold title in land to private purchasers (David, 1986, p.73).

G4.33 There are three types of urban land policy instrument available to municipalities to ensure the development of land in accordance with binding land use plans

(i) expropriation/compulsory Purchase (*Enteignung*)-(Section 85-122 BBauG)

(ii) *Bodenordnung* (reorganisation of land holding ordinances), comprising

    (a) reallocation of land (*Umlegung*) (Section 45-79 BBauG)

    (b) minor adjustments of site boundaries (Grenzregelung)(Section 80-84 BBauG)

(David, 1986, p.75).

G4.34 The right of preemption exists in areas with a binding land use plan; in areas covered by specific by-laws, such as urban renewal; or fixed areas for reallocation of land (*Umlegung*). Approval of sub-division of land exists in areas with a binding land use plan, and in areas covered by Sections 34 and 35 of the BBauG.

G4.35 Expropriation in the field of urban planning can occur where there is a valid binding land use plan, defining the specific land use for which the land is needed (Kimminich, 1986, p.206; David, 1986, p.76). Under Section 85 of the BBauG, specific authorisation is given for the following

(i) to use a plot in accordance with the stipulations of the alignment plan or to prepare it for such use

(ii) to build houses on hitherto unused plots which are outside the area of an alignment plan but within developed areas, so as to close gaps

(iii) to procure land for the purpose of compensation for expropriation.

(Kimminich, 1986, p.207).

Land acquired under such powers (by the *Regierungspräsident* - Section 104 BBauG) is compensated at market value (Section 142, BBauG).

G4.36 Once again, municipalities tend to avoid using expropriation powers as it is a time consuming and costly process, and is not received favourably by the public or private landowners. The removal of the requirement in urban renewal areas to prepare a binding land use plan will reduce the likelihood of expropriation in such areas (and entail the use of Section 34 BBauG provisions).

G4.37 Reallocation of land (*Umlegung*) is a compulsory exchange of sites within a binding land use plan area in order to adapt the ownership pattern of land to the requirements of the content of the plan (Section 45, BBauG). It is not legally construed as expropriation, even though the municipality may deduct a certain amount of land for the requirements of local infrastructure provision. Under the procedure, every participating landowner receives a site of equivalent value either in the same location or in a location of equivalent value in the land transfer area. In exceptional cases the landowner may be compensated with money or with land outside the land transfer area. As participating owners must not gain or lose through the implementation of the scheme, and as land values in such areas tend to rise, such measures tend in practice to facilitate the municipalities, own land use requirements in designated areas at low or zero cost. Hence this measure facilitates bargaining and negotiation with participating private landowners, particularly in inner city areas involved in urban renewal (Farrell & Jones, 1987, pp.20-21); David, 1986, pp.78-81). Similar powers exist in relation to the reallocation and consolidation of agricultural land holdings (*Flurbereinigungsgesetz*, 1953).

G4.38 Disputes relating to expropriation and reallocation of land decisions are resolved by the ordinary law courts with special juries for building matters (Section 157 BBauG, Section 217 BauGB)(David, 1986, p.41).

G4.39 Section 80 paragraph 1 of the BBauG also permits compulsory minor adjustments of lot boundaries, for instance to make neighbouring plots more easily developed. This procedure is possible, however, even where a binding land use plan is not applicable (Trieb, 1973, p.70).

## The Administration of the Control of Development: the example of Hagen

G4.40 The *Kreisfreie Stadt* of Hagen in North Rhine Westphalia (1984 population, 213,215) may be taken as an example of a typical middle-size city in the FRG with many of the problems and opportunities relating to urban land use planning commonly experienced in the FRG. These are (see Plate 22)

(i) urban renewal - given the age of the housing stock, much is ripe for rehabilitation of the social and physical fabric. Such problems are typical of large areas of the FRG (BMBau, 1986, p.27) where, since the supply and demand of housing is now largely balanced, attention is focussing upon condition of the stock

(ii) large derelict industrial sites - the city has recently experienced the closure of a large steel processing plant, with the loss of 12,000 jobs. The city has designated an extensive area for development of new industrial employment under the 1971 Urban Renewal Act provisions. This is a common problem in the economic restructuring of North Rhine Westphalia

(iii) environmental and ecological issues are of increasing political prominence, both nationally and locally.

G4.41    The organisation of the planning department is headed by the Principal, elected for an eight-year term of office, responsible to the city council (*Stadtrat*) whose members are elected for five year periods of office. The Principal appoints his deputy, who is a civil servant. The deputy is responsible for the day-to-day running of the department, which is divided into two sections.

(i) The Plan Section, comprising 60 persons with responsibilities for

(a) the preparatory land use plan

(b) binding land use plans

(c) other non-statutory plans and programmes

(d) sectoral plans.
    The operating costs of this section are as follows

| | |
|---|---|
| personnel | 3,250,000 DM p.a. |
| materials | 175,000 DM p.a. |
| incidentals | 175,000 DM p.a. |
| Total | 3,600,000 DM p.a. |

(ii) The Development Control Section, comprising 25 persons in four sections covering a city area (though not in area offices), responsible for

(a) building permits

(b) negotiation and advice on development proposals

(c) enforcement and appeals

The operating costs of this section are as follows

personnel & materials 1,900,000 DM p.a.
fees for building permits amounted to
approximately          1,500,000 DM p.a.

G4.42    On average, the Development Control Section processes 3,000 files a year, and issues 600 permits, though many of the files are applications prior to a full building permit, enquiring whether planning permission is likely to be forthcoming. Charges for such applications are lower than for full permits. Until recently, the processing of a complex building permit took on average approximately eight months to complete. A new management procedure has been instituted, under which in the first week all internal and external consultations, processing etc are set in operation, followed by a special meeting of all relevant heads of section (roads, surveying, sewerage etc) of the municipality to facilitate the rapid processing of applications, including the arrangements for infrastructural payments. This has resulted in a reduction of the average time to process applications to approximately three months, and is considered a successful innovation by the Principal.

G4.43    The most frequently experienced problems in the operation of the control of development were stated to be

(i) change of use of buildings without a permit (since these are more difficult to observe than new construction)

(ii) building more than allowed by a permit

(iii) weekend houses/camping

It is felt that the system works well, and that there are no real problems with the enforcement of regulations.


## Conclusions

G4.44    The control of development in the FRG is effected by a complex range of legislation at the Federal, *Länder* and municipality level. However, all aspects of formal building law and material building law are united in the grant of a building permit, which is executed at the local level (*Kreis* or *Kreisfreie Stadt*).

G4.45    Land use planning and control of development has three integral components

(i) preparation and adoption of land use plans

(ii) safeguarding measures for land use planning (freezing of development activities; postponement of building permits; subdivision controls; preemption powers etc)

(iii) regulation and control of building and land use

The regulation and control of building and land use differs according to whether the development

(a) falls within a designated area of a binding land use plan

(b) is situated within built-up urban areas without a binding land use plan

(c) is situated on the periphery, outside built up areas and lacking a binding land use plan.

G4.46 Whereas formerly most development and development control was 'plan led', a recent tendency has been to substitute alternative regulations of the formal building law (*Bundesbaugesetz*) - especially Sections 34 and to a lesser extent 35 - for binding land use plans. This tendency will be increased with the relaxations permitted by the 1987 changes to the *Bundesbaugesetz*, which will particularly affect urban renewal programmes of cities. This may be interpreted either as a diminution of planned control of development, or as the introduction of a degree of practical flexibility in the planning process. Whether or not a *Bebauungsplan* exists, in virtually all cases there will be a *Flächennutzungsplan*.

G4.47 In general, it is believed by most participants and users that the operation of the control of development in the FRG works efficiently and equitably. The current Federal initiative to integrate comprehensively all legal principles of urban development into a single framework, to produce legal and administrative simplification through the removal of unnecessary regulations in order to simplify the development process, and to increase the sovereignty of local authorities with regard to planning is widely welcomed at the *Länder* and municipality level. This policy focus reflects changing development pressures in the light of significant economic, political and social changes over the last decade or so.

G4.48 The *Bundesbaugesetz* applies to the whole territory of the FRG, except for Free Port areas (such as Hamburg), where the formal building law applies only to the regulation of the impact of development on adjacent areas (eg industrial emissions). Within such areas, only the material building law (building regulations) is applicable.

## Federal Planning Policy Instruments

G5.1   Besides the regulations of the formal planning law (*Bebaugesetz* 1960) the Federal Government provides a legal framework for comprehensive regional planning at the national and *Länder* level in the Federal Regional Planning Act (*Bundesraumordnungsgesetz*, (BROG) 1965). All three levels of government - Federal, *Länder* and local - are charged with its implementation, but since the Federal Government has no executive powers in the field of planning, the 'skeleton' law provided by BROG can contain only guiding and binding statements of principle in the matter of regional planning.

G5.2   After six years of discussions between the Federal Government and the *Länder*, agreement was reached on a Federal Regional Planning Programme in 1975, which

(i)   established 38 planning regions (revised to 75 in 1982) as a basis for sophisticated regional analysis

(ii)   identified problem regions in relation to employment structure and infrastructure, including the 'Zonal Fringe Region' or frontier zones - a 30km-50km belt abutting the GDR and Czechoslovakia

(iii)   established a framework for the co-ordination of Federal programmes with a direct bearing on regional development (highway construction, regional economic policy, urban renewal and development etc).

The strategy emphasised development centres and development axes to connect urban agglomerations, both within the Federal territory and with neighbouring countries (BMBau, 1976, pp.14-18; BMBau, 1977; Konukiewitz & Wollmann, 1982, pp.89-90). Because of lack of political consensus over the programme within the Federal system, no development centres in problem regions were designated, and the programme was approved not by the Federal Government but by the 'Ministerial Conference for Spatial Planning', the standing conference of Federal and state cabinet members responsible for supra-local planning. The programme remains essentially a statement of principle, and the Federal Ministerial Conference for Spatial Planning has been moribund for a decade.

G5.3   With the intended end of joint funding arrangements between the Federal Government and the *Länder* for urban renewal projects accompanying the changes proposed in the *Baugesetzbuch* of 1987, the Federal

influence in housing and urban renewal policy is likely to decline. In these circumstances it will be for the *Länder* Ministers responsible for regional and *Länder* level planning to co-ordinate their programmes and plans within the conception of the joint *Länder* Federal Programme. A Federal Regional Planning Report is produced every four years (the most recent in 1986), reviewing trends and setting out objectives and principles of inter-regional planning (Section 11, BROG). A subsuming binding political goal, widely acknowledged, is that of "creating and maintaining equivalent living conditions" in all subregions of the territory of the FRG (BMBau, 1982, p.8), though the objective of maintaining the basic natural living conditions (ie ecological conservation) has assumed increased prominence in the present decade.

## *Länder* Planning Policy Instruments

G5.4   All *Länder* are obliged (Section 5(1) BROG) to set up statewide plans for the development of the *Land*. These plans may include provisions for inter-*Länder* planning, such as is practised by the four coastal *Laender* (*Kustenländer*) comprising the *Konferenz Norddeutschland* (Conference of North Germany) set up in 1969, which resulted in a regional plan (*Entwicklungsmodell* - scale 1:100,000) for Hamburg and its hinterland (1969) and a regional policy report for the four *Länder* in 1975.

G5.5   Each *Länder* prepares a Comprehensive Development Programme (*Landesentwicklungsprogramm*) out of which are developed Comprehensive Development Plans (*Landesentwicklungspläne*). These do not follow a uniform pattern and are highly diagrammatic (being illustrated at a scale of 1:1,000,000 maximum or 1:200,000 minimum). Such plans consist of a text, maps and explanations. Their contents include population developments, the system of central places (upper and middle centres), the division of the *Länder* into regions, the larger priority areas (eg water conservation, recreation), the development axes, and locations of large-scale industrial and atomic undertakings (Schmidt-Assmann, 1986, p.345). After preparation, these Comprehensive Development Plans are approved by the *Land* cabinet, not the *Landtag* (Parliament).

G5.6   North Rhine Westphalia has six such *Landesentwicklungspläne*, covering specific sectors of policy (spatial and settlement structure; open space; water management and recreation; noise protection; mining and mineral resource development; land for major development projects) (see Plate 23).

G5.7 These *Landesentwicklungsprogramm* and *Landesentwicklungspläne* are concretised at the administrative district level (*Regierungsbezirk*) in the *Gebietsentwicklungspläne* (regional development plans) which are considerably more detailed than the higher level land use plans of the *Länder*. Such plans appear as a combination of text and maps to a scale of 1:50,000, available in some cases to a scale 1:25,000). The text consists of binding declarations and accompanying explanations, the maps also in part containing binding statements (binding on all public agencies and authorities, including the municipalities within the respective areas of the plans). The binding provisions usually relate to central places (small and lower centres); the subdivision of land in development axes; priority areas for rest and recreation, environmental protection, water conservation, agricultural and forestry matters; the location of infrastructural plants of regional importance; and the desirable directions of development in the distribution of residential and employment locations among the municipalities (Schmidt-Assmann, 1986, pp.345-346) (See Plate 24).

G5.8 The Land Comprehensive Development Plans and the *Gebietsentwicklungspläne* (regional development plans) correspond very closely. The *Regierungsbezirk* officials prepare the *Gebietsentwicklungspläne*, but the *Land* Ministries have considerable influence in their formal and substantive content. The plans are approved by the Regional Planning Board (comprising representatives of the *Land* and the constituent municipalities of the administrative district) and by the *Land*. If the *Land* Ministry is not satisfied with a *Gebietsentwicklungspläne* it is not able to replace such a plan with one of its own; it may only require a new plan to be prepared, and, of course, it will refuse approval of the plan to which it objects.

G5.9 The provisions of the *Gebietsentwicklungspläne* once approved become binding on municipalities. This is secured by the requirement that municipalities' land use plans must be submitted to the *Regierungsbezirk* for approval. In this sense, local land use plans are not developed out of regional plans, rather the two types of plans have to be in conformity with each other, following the principle established in the BROG that all activities have to be in conformity with higher-ranking plans. In the 1987 amendments to the *Bebaugesetz* an important change will take place in this approval process. Once the binding land use plans of municipalities are submitted to the *Regierungspräsident* there will be a period of three months in which to raise legal objections against the plans, after which they are deemed to be approved if there has been no objection from the *Regierungspräsident*. Hence a negative act will replace the positive act of approving every binding land use plan. However, *Flächennutzungspläne* will still require 'full' approval.

G5.10 The preparation of both Land Comprehensive Development Plans and *Gebietsentwicklungspläne* (regional development plans) is a lengthy process, which can take from two to ten years. The plans look forward 10-15 years, and are periodically updated. Until a revised plan is approved, deviations from the old plan are permitted in individual instances by ministerial dispensation, ensuring a degree of flexibility (Schmidt-Assmann, 1986, p.346).

G5.11 As an example of the resource implications of such plan preparation, the *Regierungspräsident* Dusseldorf (North Rhine Westphalia) office comprises 60 planning staff, with all plans prepared in-house.

G5.12 Since according to the BBauG the preparatory land use plans and the binding land use plans of municipalities have to be consistent with the goals of the *Land* Comprehensive Development Programme (Section 1, paragraph 4), as set out in the *Landesentwicklungspläne* and the *Gebietsentwickslungspläne*, the role of the *Regierungsbezirk* is clearly pivotal in ensuring the compatibility of *Land* and local planning. This is achieved in practice through the provisions of the *Gebietsentwickslungspläne*, and the approval by the *Regierungspräsident* of local land use plans. If local land use plans do not conform, the *Regierungsbezirke* may not prepare plans themselves, but may approve the plans under the condition that the municipality makes the necessary changes in the plan (Sections 6 & 11, BBauG). During the last decade in the *Regierungsbezirk* Mnster (North Rhine-Westphalia) on only 10 occasions has the *Regierungspräsident* refused approval of local land use plans, with some 500-700 binding land use plans submitted annually.

G5.13 It is to be noted that the BROG does not contain any provisions for public participation by citizens or landowners at the level of *Land* or regional planning (Kimminich, 1981, p.276), but it does require intensive participation and co-operation of local government authorities for all comprehensive regional planning activities.

## Local Planning Policy Instruments

G5.14 Comprehensive urban land use plans at the local level provide the framework for control with a basis in the *Bundesbaugesetz* (BBauG) (Section 1, paragraph 2). The Federal Urban Land Use Planning Graphic Representation Order (*Planzeichenverordnung* 1981) (see Figure G14) ensures a common cartographic representation in all urban land use plans. The *Baunnutzungsverordnung* (Federal Land Utilisation Order, 1977) defines the land use types which can be provided by such plans. Together these measures provide common principles and rules for the exact presentation and interpretation of urban land use plans in all the municipalities throughout the FRG (David, 1986, p.64; Schmidt-Assmann, 1986, p.346).

G5.15 These comprehensive urban land use plans form a hierarchy of their own, and comprise two types of plan

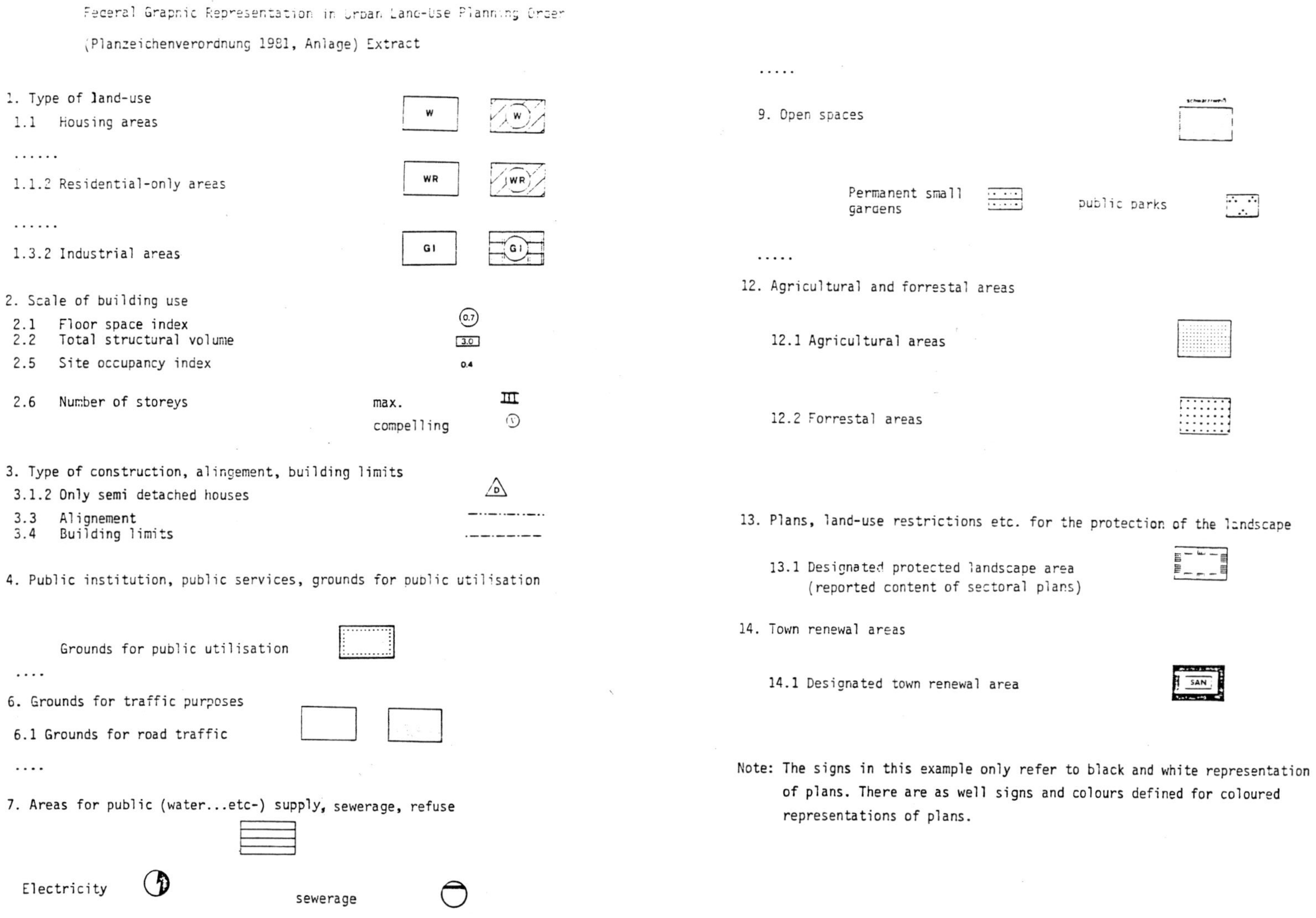

Federal Graphic Representation in Urban Land-Use Planning Order

(Planzeichenverordnung 1981, Anlage) Extract

1. Type of land-use
1.1 Housing areas
......
1.1.2 Residential-only areas
......
1.3.2 Industrial areas

2. Scale of building use
2.1 Floor space index
2.2 Total structural volume
2.5 Site occupancy index
2.6 Number of storeys                    max.
                                         compelling

3. Type of construction, alingement, building limits
3.1.2 Only semi detached houses
3.3 Alignement
3.4 Building limits

4. Public institution, public services, grounds for public utilisation

    Grounds for public utilisation

....
6. Grounds for traffic purposes
6.1 Grounds for road traffic
....
7. Areas for public (water...etc-) supply, sewerage, refuse

Electricity          sewerage

......
9. Open spaces

                        Permanent small
                        gardens           public parks

......
12. Agricultural and forrestal areas

    12.1 Agricultural areas

    12.2 Forrestal areas

13. Plans, land-use restrictions etc. for the protection of the landscape

    13.1 Designated protected landscape area
         (reported content of sectoral plans)

14. Town renewal areas

    14.1 Designated town renewal area

Note: The signs in this example only refer to black and white representation
      of plans. There are as well signs and colours defined for coloured
      representations of plans.

**Figure G14   Federal Graphic Representation in Urban Land
Use Planning Order**

This figure provides an extract of the detailed specifications
provided in Federal planning legislation for the graphical
representation to be adopted by municipalities in drawing-up
and approving urban land-use plans. It should be noted that,
as well as a black and white representation of plans, detailed
guidance is provided for the colour representation of plans.

297

(i) the *Flächennutzungsplan* - preparatory land use plan

(ii) the *Bebauungsplan* - binding land use plan

(See Plates 18 and 25 and Plates 21 and 26 respectively for examples).

## The *Flächennutzungsplan*

G5.16   The *Flächennutzungsplan* (preparatory land use plan) is a single plan which has to be prepared for the whole territory of a municipality, providing the land use designations for development needs 10-15 years ahead, though the law does not prescribe any time horizon. The scale of presentation is not specified in regulations, but ranges in practice from 1:50,000 to 1:5,000. The documents consist of an official plan containing the most important designations, with a key facilitating interpretation of the plan, together with a supporting text setting out further details and the reasoning contained in the plan. Only the graphic representation (ie the plan or plans) is a legal document, the explanatory documents being interpretative only.

G5.17   The regulations (BBauG, Section 5 paragraph 2) provide that 'to the extent necessary' the following information shall be shown

(i) areas zoned for building, according to general and particular type of building (further specified in the *Baunutzungs-verordnung*)

(ii) public buildings

(iii) major transport facilities

(iv) land for public utilities

(v) green areas

(vi) bodies of water

areas for tipping and mineral extraction          (vii)

areas for agriculture and forestry.          (viii)

In addition, other special requirements (such as urban renewal areas - *Sanierungsgebiete*) may be indicated (Von Klitzing, 1980, p.92).

G5.18   The *Flächennutzungsplan* is therefore essentially a zoning plan, having the appearance of a master plan (see Figure G10(a) and for examples, Plates 18 and 25). Its provisions must conform with regional plans. The provisions of the *Flächennutzungsplan* are legally binding upon the municipality adopting the plan and for the public authorities taking part in the adoption procedure. As the provisions of these land use plans are not binding upon individuals, landowners or citizens, they do not directly create legal land use rights. However, the *Flächennutzungsplan* forms the basis for subsequent binding land use plans, which have to conform with the preparatory land use plan(s). If there is no conformity, the preparatory land use plan has to be amended, and in

many large cities it is common to find preparatory land use plans amended a hundred times since adoption (David, 1986, p.59; 1987, pp.19-21).

G5.19   Some municipalities, particularly in metropolitan areas, have chosen not to prepare a *Flächennutzungsplan*, because of its self-binding character, preferring to work on the basis of older plans adopted before the enactment of the *Bundesbaugesetz* in 1960. The *Länder* have exerted pressure on such authorities to accelerate the preparation of new plans, but some continue to work with pre-1960 plans or with legally non-effective *Flächennutzungsplan* drafts (David, 1986, p.64), sometimes for up to 17 years.

G5.20   All municipalities have jurisdiction for adopting both *Flächennutzungspläne* and *Bebauungspläne* plans - this is their constitutional prerogative. The plans for small rural *Gemeinden* (*Kreisangehörige*) may be drawn up by the county (*Kreis*) or by private planning consultants, but have to be approved by the council of the *Gemeinde*. Though the adoption of land use plans is a constitutional prerogative of the municipalities, Section 6, paragraph 1 of the BBauG regulates that a land use plan comes into force only after the approval of the respective district administrative office of the *Länder* (ie in North Rhine Westphalia, (the *Regierungspräsident*). Such approval, however, relates only to legal issues (David 1986, p.40).

## The *Bebauungsplan*

G5.21   The *Bebauungsplan* (binding land use plan) constitutes the second level of the local plan hierarchy, and the effective basis for the implementation of detailed development control (see Figure G10(b) and, for examples, Plates 21 and 26). Binding land use plans are small in scale, being adopted for selected parts of a municipality's area where development is likely to take place. They have the status of local statutes (Satzung, Section 10, BBauG) and their provisions are legally binding upon all parties, both public and private individuals and landowners. They are both instruments of public control and the legal framework for private development.

G5.22   The documents consist of a plan and written statement, both of which constitute legal documents. The scale of presentation is not specified in regulations, but ranges in practice from 1:2,000 to 1:500. The form of presentation is set out in the *Baunutzungsverordnung* (Federal Land Utilisation Order) and the content which must by law be included is set out in Section 30 of the *Bundesbaugesetz*. The matters which may be included are set out in Section 9 BBauG.

For this purpose the binding land use plan may contain - to the extent necessary - the following legal provisions

(i) the limits of the planning area, ie the extent of local statutes

(ii) the permitted land uses - following the detailed categories established in the *Baunutzungsverordnung* (paragraph 4 permits ancillary uses to a dominant use providing no detriment is caused to that prevailing use)

(iii) the densities for land use, determined for each different use zone by the number of permitted storeys, site coverage limits and plot ratio (also laid down in the *Baunutzungsverordnung*) The Federal Administrative Court has ruled that these detailed plot ratios should be interpreted as a range, not as a strictly interpreted average

(iv) limits on height and mass, establishing building envelopes through building lines and height restrictions which can be fixed as compulsory (of permissive maxima)

(v) land required for public purposes - such as roads, footpaths, open spaces, public buildings, public utilities

(vi) anti-noise and clean air zones, and flood areas.

The binding land use plan may prescribe minimum lot sizes and, from July 1987, the maximum lot sizes for plots designated for housing for reasons of economic use of land (Section 9, paragraph 1, No.3 BauGB); public rights of way, common access roads and parking areas; preservation of trees or new planting. The external appearance of buildings (ie design control) is permitted by an enabling BBauG clause, but is authorised in *Länder* planning acts and municipality statutes (Trieb, 1983, pp.66-68).

G5.23    The binding land use plan is the most important basis in German law for

(i) planning control (Section 29, BBauG)

(ii) sub-division of plots (Section 19, BBauG)

(iii) allocation of land uses (Section 45, BBauG).

There is a tendency to use binding land use plans only in very specific cases, the current changes in the BBauG extending the scope of the substitute regulations (Sections 34, 35 BBauG). But binding land use plans are mandatory

(a) where development activities in inner city areas conflict with the urban development policies of the municipality, even if in accordance with the surrounding environment

(b) where Sections 34,35 BBauG do not admit a development and only a binding land use plan can resolve the problem

(c) if a municipality wishes to use instruments which legally must be based on a binding land use plan. (David, 1986, p.35)

G5.24    If a municipality does not plan any building activity, it is not obliged to draw up a binding land use plan. There are some municipalities which lack binding land use plans, but these are generally exceptional and confined to small rural *Gemeinden*.

## Other Local Planning Instruments

G5.25    Besides the two statutory local plans detailed above, municipalities may prepare the following

(i) *Stadtentwicklungspläne* (urban development plans) as comprehensive programmes for land use, infrastructures, economic and demographic development involving budgetary strategies for their entire area. The preparation of such plans has declined somewhat over the last decade. Smaller area plans may be drawn up for parts of towns (*Stadtteile*) reflecting these themes (see Plates 19 and 20 for examples).

(ii) In addition, *Standortprogramme* (location programmes) are an instrument for local planning, necessary if grants from the *Länder* government are required for large infrastructure development programmes (Ahrens et al, 1980, p.16). These and other plans do not have the legal status of the *Flächennutzungsplan* and the *Bebauungsplan* as regards land use. In 1976 (Section 1, paragraph 5) BBauG introduced an amendment requiring decisions on urban development policy contained in formal resolutions of the councils of municipalities to be observed when urban land use plans are adopted, but the legal relevance of these resolutions was very restricted (David, 1986, p.64) and this requirement is abolished in the BauGB 1987 (Section 1, paragraph 5).

## Urban Renewal

G5.26    Until 1987, urban renewal (*Sanierung*) was subject to separate legislation, the *Städtebauförderungsgesetz* (Urban Renewal Act - StBauFG, 1971). The provisions of this legislation are concerned with building rights and land ownership in designated improvement and development schemes, regulating the planning, sub-division and implementation phases as well as financing (including the recoupment of development values) and organisational arrangements. The Act increased municipality control over real estate markets and development activities by creating special legal powers for municipalities, applying only in the designated areas. The financing of such schemes included Federal grants, channelled through the *Länder* budgets. The joint funding arrangements have now ceased, replaced by block grants to the *Länder*. The requirement that binding land use plans for such areas must always be drawn up was abolished in 1985, and expropriation associated with binding land use plans is likely to be of reduced significance in future as policy moves in the direction of rehabilitation and renewal rather than development.

299

## The Process of Plan Preparation and Approval

G5.27 The preparation and adoption of urban land use plans (ie the *Flächennutzungspläne* and the *Bebauungspläne*) is a process regulated by the *Bundesbaugesetz* and local government legislation relating to the rules and practices for the internal municipal decision-making processes. It involves both officers and elected representatives in a two-phase process (for full details see Appendix B).

(i) Planning Phase Involving an informal preparatory period, followed by a formal resolution of the municipality to prepare a plan, and the preparation of a draft plan by the planning department which is approved by the council as a basis for public and private representation for a period of one month, after which the plan may be amended before subsequently being adopted by a resolution of the council as a local statute.

(ii) Approval Phase Under the *Bundesbaugesetz*, all preparatory and binding land use plans require approval by the relevant state authority, (but see paragraph 5.9 above) usually at the intermediate (*Regierungsbezirk*) level. This approval is restricted to an examination of legal procedures only. The plan then becomes operational upon public notification of the *Länder* authority's approval (David, 1986, pp.68-69).

G5.28 The period of plan preparation is lengthy. Preparatory land use plans may take three to eight years to prepare, binding land use plans from less than a year to three and more. The *Gebietsentwicklungspläne* (regional development plans) and *Flächennutzungspläne* (preparatory land use plans) are usually continuously amended - for example, the *Flächennutzungsplan* for Hamburg dating from 1953 and most recently revised in 1984 has been amended in minor form on 90 occasions since that latter date. A change in either the *Flächennutzungsplan* or the *Bebauungsplan* requires a change in the other. This lengthy period for preparation, approval and amendment is one reason why some municipalities have adopted non-statutory or draft plans, or have sought to substitute alternative (ie Sections 34, 35 BBauG) regulations. Another reason is that once a municipality has adopted a binding land use plan it has the legal obligation to provide the infrastructure needed in cases where landowners have already made an advance payment of the infra-structure levy required by the municipality (see G4.29) (David, 1987, p.35).

## Public Participation in Land Use Plans

G5.29 Public participation is provided for in the German planning system only on the local level. Both the *Flächennutzungsplan* and the *Bebauungsplan* are required (under Section 2a BBauG) to allocate periods of one month for public representations (though not for minor alterations which do not affect the basic outlines of the plan). In the case of a *Bebauungsplan*, the BBauG allows (Section 13) an exception to the rule of public participation if the proprietors of adjacent plots do not object to changes proposed in the plan, and this is considered by one commentator to constitute a legal loophole (Kimminich, 1981, p.276). The revised Section 13 provides for limited public participation of those who are affected by the intended changes. Current changes in the BauGB are intended to strengthen participation as compared with the BBauG.

G5.30 The most extensive provisions for public participation are contained in the Urban Renewal Act, with detailed provisions for public consultation in designated areas. Though one commentator has alleged that the regulations are vague and consist of consultation rather than effective participation (Kimminich, 1981, p.277), they do represent an extension of the principle of participation and have been widely deployed in successful urban renewal projects.

G5.31 Thus the preparation of *Gebietsentwicklungspläne* (regional development plans) permits no private objections, the preparation of *Fachpläne* (sector plans for public development) permits only those with private rights adversely affected to object, whilst in the case of the *Flächennutzungsplan* or *Bebauungsplan* anyone may object but the municipality and *Regierungsbezirk* are not compelled to heed the objections. In the case of a permission granted under Section 34 BBauG there is not even the provision for public representations which the *Bebauungsplan* affords, because such an act involves only *Länder* Building Orders, not an act of municipal planning.

G5.32 A large number of laws through which planning for specific purposes is regulated in the FRG (but which extend beyond the scope of land use planning) require public hearings (eg *Bundesfernstrassengesetz*, 1976 - law for construction of highways; *Atomgesetz*, 1957 - law for the construction of nuclear power plants). The provisions relating to these hearings allow observations to be made by the public. The only individuals entitled to start a legal action against plans of which they do not approve are those who are potentially affected in their own rights (such as property, health etc) (see paragraph G6.9) (Kimminich, 1981, p.278).

## The Administration of Development Plans: the example of Hamburg

G5.33 Hamburg, a city-state, has a population of 1.61 million persons (1984) (see Plate 16). In its administrative and planning structure it demonstrates significant differences with the situation in 'space-*Länder*', notably with regard to the system of development plans and the operation of development control. The system of development plans is more 'unitary' in character, and development control more centralised than in the 'space-*Länder*'.

G5.34 For the purposes of the BBauG, Hamburg has also the status of a *Gemeinde*. The chain of command in administering planning and control of development is

(i) *Senator* (elected), responsible for each administrative department

(ii) Vice-*senator* (*Staatsrat*) (non elected, appointed by *Senator*)

(iii) *Oberbaudirektor* - elected for 9 years, responsible for the Building Department

(iv) *Baubehörde* (Building Department) administrative structure, comprising career 'civil servants'.

G5.35 The *Baubehörde* provides the central direction of policy, with seven *Stadtbezirke* (district level offices) (see Plate 17). There is a third level (*Ortsaemter*) which has consultative political status but not full executive powers (eg the *Ortsamt* gives building permissions but has no planning department). The *Baubehörde* prepares and amends plans for the whole territory of Hamburg. The *Bebauungspläne* are prepared at the *Stadtbezirk* (district) level, but upon initiation by the *Baubehörde*. There is thus strong central planning direction, and a detailed hierarchy of plans, among which are the following

(i) The *Entwicklungsmodell* (1969) or regional development plan (comparable with the *Landentwicklungspläne* of the 'space-*Länder*), to a scale of 1:100,000 drawn up with the co-operation of the adjacent *Länder*, indicating a policy of growth corridors with satellite growth centres. (The Hamburg Chamber of Commerce (*Handelskammer Hamburg*) considers this to be a useful tool for the regional co-ordination of planning).

(ii) The *Flächennutzungsplan* (1973, revised in parts when necessary) which in style is between the extremes of detail and generalisation which such plans are permitted and is to a scale of 1:50,000 (see Plate 18).

(iii) *Stadtteilentwicklungsplanung* (STEP), district development plans providing corporate proposals bridging the scale between the *Flächennutzungsplan* and the *Bebauungsplan*. These STEPs are to a scale of 1:5,000, and are supplemented by Programme Plans dealing with specific details and land use requirements. (These latter are considered very useful by the private sector in dovetailing the requirements of statutory land use plans and in adapting to change) (see Plates 19 and 20).

(iv) *Bebauungsplan*, of which there is a complete coverage in Hamburg, in combination with the *Baustufenplan* (the binding land use plan preceding the 1960 BBauG) where no more recent *Bebauungsplan* has been approved (see Plate 21).

G5.36 In addition the *Baubehörde* prepares *Dichtemodell* (Density Model) 'regulation plans' (1980) which specify for each public transportation node concentric zones with specified permitted densities. These are not legally binding, but are used as bargaining instruments. The *Baubehörde* also prepare *Fachbereichspläne* (subject plans) for such topics as design guidance, and *Funktionspläne* (non-statutory plans) for general guidance. Every 15 years a publication is produced on Hamburg, its buildings and planning system, last revised in 1984. Like the STEPs, these plans do not derive from the BBauG, but are used as planning and negotiating/bargaining tools. Only the *Flächennutzungsplan* and the *Bebauungsplan* relate to formal planning law. Over a thousand *Bebauungspläne* have been approved in Hamburg since 1960.

G5.37 For such an extensive programme of plan preparation a large staff is required, with the *Baubehörde* planning staff dealing with *Flächennutzungsplan* and Programme Plans comprising 120 persons, together with a further 15 dealing with procedures and preparation of binding land use plans. In the *Stadtbezirke* (districts), of which there are 7, are a further 20 staff per district. The *Baubehörde* central office prepares a manual or handbook for Districts advising on the preparation of *Bebauungspläne*, which is revised every 6 months.

G5.38 The procedure for all plan preparation and approval is that the initiative for plan preparation comes via the *Baubehörde* from the *Senator*, with a draft plan prepared by the *Stadtbezirk* (District) and submitted to the *Baubehörde* where all interested parties are brought together to consider the plan (*Abstimmung* - conference of interested parties). This includes the important Chamber of Commerce. The plan is then made available at the district level for one month for public representations, with two routes to its subsequent adoption

(i) approval by the executive (*Senat*) - plan becomes an ordinance (*Verordnung*). If the local council (*Ortsamt*) wishes to reject the plan by a vote of more than 25 per cent, the plan must be referred to the *Bürgerschaft* (parliament)

(ii) if the plan involves changes to a *Bebauungsplan* which has been a real law, then approval must be sought from the *Bürgerschaft*. The amended plan then has the status of a real 'law'.

G5.39 It is considered by the planning authority that plan preparation and approval is faster in Hamburg than in 'space-*Länder*' because of the unitary system of development plans and unitary governmental system. However, the Hamburg Chamber of Commerce considers that the typical two-year plan preparation period is too long in terms of the requirements of commercial companies to adjust their investment decisions to short-term market changes. In addition, the Chamber of Commerce feels that the increasing participation of the public in plan making is prolonging the planning process unnecessarily.

## Conclusions

G5.40 The control of development in the FRG takes place in the context of a hierarchy of statutory plans and regulations, which provide a successively increasing degree of policy direction and detailed land use control as one proceeds from the Federal through the *Länder* to the municipal level. Land use plans are the responsibility of the *Länder* and municipalities, and represent a symbiotic relationship between the levels of public authorities entitled by the *Bundesbaugesetz* to control land use and development. At the level of detailed development control, the municipalities are responsible for preparing and implementing the two local land use plans (*Flächennutzungsplan* and *Bebauungsplan*) which impinge most directly upon development processes, subject only to legal ratification (which applies to all applicable regulations, substantive as well as procedural) by the intermediate level of the *Länder* (*Regierungsbezirk*).

G5.41 The only plan which is legally binding upon public and private agencies alike is the *Bebauungsplan* (binding land use plan), a small scale plan prepared wherever development proposals are expected or intended. Federal regulations ensure a common presentation of such plans throughout the FRG and a minimum content, but the precise content and degree of detailed control varies at the discretion of the preparing municipality (typically less detail is prescribed for public than for private land uses). Usually the degree of control specified is very precise, amounting to a detailed development brief for implementation, though some limited dispensations may be made under the provisions of Federal regulations.

G5.42 The preparation of such plans is a complex and lengthy process, and there is an emerging trend to confine their preparation to specific development requirements (such as 'greenfield' development or inner city renewal), or to substitute other regulations (Sections 34,35 *Bundesbaugesetz*) where appropriate. The German planning system may be in a process of transition from a highly detailed legally-binding plan-led system towards a more indicative system of plans and case-oriented planning and development control (David, 1987, p.39).

G5.43 The economic and political changes currently favouring the role of municipalities in planning procedures are associated with a reduced role for national land use development planning at the Federal level. At the same time, the increasing conflicts between development needs and those of environmental conservation have resulted in an increased resort to law by both private landowners and the public at large, perhaps indicating a reduced toleration of public planning decisions without testing the opportunities to block plans by taking cases to court. In reviewing *Bebauungspläne*, some commentators consider that the limits of the courts' discretion have perhaps been exceeded, resulting in a restriction of the planning discretion of municipalities, though this is a complex issue.

## Administrative Law and Land Use Control and Development

G6.1   Article 19(4) of the Basic Law (GG) contains a wide-reaching guarantee of legal protection against the state. Anyone believing their rights to be violated by the administrative decision of a public authority may have recourse to the courts. Since land use plans and control regulations are official acts, they fall within the domain of public law and may be contested in the administrative courts (see paragraph G1.9(ii)).

G6.2   Decisions falling within the administrative autonomy of municipalities may be checked by higher administrative authorities only for their legality not their expediency, ie to confirm that the municipality has respected all laws and especially control of whether a proper balancing of interests took place (Section 1, paragraph 7 BBauG). In addition, the Federal Administrative Court (*Bundesverwaltungsgericht*) has ruled (1969, 1981) that in addition the municipalities have to respect the limits of their 'planning freedom'. Within these limits the municipality's planning decisions are not subject to judicial examination, but the court may determine whether a municipality, in exercising this freedom, has exceeded the limits established in the principles of Section 1, part 6 BBauG. This section provides that preparatory and binding land use plans have to serve the social and cultural needs of the population and must give due consideration to the requirements of business and industry, agriculture, youth activities, traffic and national defence as well as the conservation of nature and the shaping of cities and landscapes (Kimminich, 1986, p.202).

## Administrative Courts and Land Use Plans

G6.3   A preparatory land use plan (or any other plan prepared by the *Land*) cannot be the direct object of appeal to the court since its provisions are not legally binding against private individuals.

G6.4   Actions against binding land use plans are possible according to Section 47 of the Federal Law of Administrative Court Procedure (*Verwaltungsgerichtsordnung* - VwGO) which permits the higher administrative courts the examination of legal rules below the rank of formal laws (ie ordinances, statutes etc). Since a binding land use plan is a legal rule it can be subjected to such a procedure (*Normenkontrolverfahren*). The procedure may be initiated by any person who has suffered or is expected to suffer a disadvantage by the application of the stipulation, as well as any government agency (Kimminich, 1986, p.203). Unlike the judicial decisions in other cases, which are of legal relevance only between those who participate in the procedure, binding land use plans which are reviewed in this manner and considered null and void no longer have any legal validity (David, 1987, p.33; 1986, p.41).

G6.5   Section 1, paragraph 7 of the *Bundesbaugesetz* establishes a very important obligation upon municipalities - the rule of fairly balancing and weighing interests (private and public). This is rigorously tested in the courts, and municipalities have often been considered to have disproportionately favoured certain interests or obviously disregarded essential aspects.

G6.6   Under Section 155(a) BBauG (1979) a landowner must bring an action against the provisions of a plan within one year (in relation to procedural/formal mistakes). If he/she does not do so, it is very difficult to bring an action before the administrative courts (involving other mistakes, Section 155(b) BBauG). The current revisions to the BBauG tighten this provision, in that after 7 years even important cases of violation of the principle of fairly balancing and weighing interests cannot be brought before the law courts (the result of pressure from the powerful association of local government authorities).

G6.7   In the *Regierungsbezirk* Münster, an average *Flächennutzungsplan* can generate 13,000 objections to the *Regierungsbezirk*, whereas a *Bebauungsplan* might generate up to 800 objections on average, some of which may proceed to the administrative courts.

## Administrative Courts and Building Permits

G6.8   Most of the cases brought before the administrative courts in the field of planning concern individual administrative decisions, usually relating to building permits. Before an appeal against the refusal of a building permit or against the imposition of unfavourable conditions can be made to the administrative court, the applicant has to file an administrative appeal to the authority at the next highest level (called *Widerspruchsverfahren*). This higher authority will be either the

*Regierungsbezirk* in the case of *Kreisfreie Städte* (county-free towns), counties in certain cases, or to the *Land* Minister if no other higher authority exists. After the final decision in this administrative appeal procedure, the appellant has to bring the case to court within one month after being properly informed of the result, or, if not informed about the right of suing in court, one year (Kimminich, 1986, pp.203-204; David, 1986, pp.42-43). There is a charge for this appeal procedure. In the *Regierungsbezirk* Münster there are approximately 600 appeals per annum under this *Widerspruchsverfahren* procedure referred from *Kreisfreie Städte*.

G6.9   To bring a successful action in the administrative courts it is not sufficient that a decision be shown to be illegal - the claimant must also establish that his legal rights have been violated by that decision. This effectively restricts third-party rights of appeal (David, 1987, p.33; 1986, p.42), but does permit an appeal against the grant of a building permit by an aggrieved person.

G6.10   Decisions of the administrative courts are subject to a right of appeal to the competent higher administrative court.

G6.11   In addition to an appeal to the next highest government authority and to the courts, a person aggrieved by a local authority decision may submit his/her case to the Petition Committee of the *Land* Parliament (a political, not a legal, mechanism). The Petition Committee is not confined to planning issues. This appeal route is open only to individual citizens, not development companies. The Committee consults with the municipality, *Regierungspräsident* and Minister before reaching a conclusion on the issue. The procedure often favours applicants, and is free. The *Petitionsausschuss* can be used after the *Widerspruchverfahren* procedures. The *Petitionsausschuss* operates on a political basis; it has no right to abolish decisions of other bodies of government, and cannot overturn a legal decision.

G6.12   The *Petitionsausschuss* will rarely if ever be appropriate in the case of a neighbour opposing a building permit. This is due to the legal effect of the rule precluding access to the court beyond one month after the decision (see paragraph G6.7 above), for after this date the right of the permit holder to start building cannot be contested.

## Judicial Issues

G6.13   Although the West German planning system is very strictly governed by written law, the wide scope of the relevant regulations leaves a considerable discretion to the authorities (David, 1987, p.38). The limits of this discretion are, under the German Basic Law, open to challenge in the administrative courts. This raises the vexed question of the separation of procedural and substantive issues in planning decisions. It has been pointed out that the German system of deriving unambiguous administrative principles for the implementation of plans at the local level from goals deriving from higher order political structures is fraught with contradictions (Grauhan, 1973, p.303) which inevitably brings judicial decisions into the arena of political policy making. Legal commentators have noticed a tendency for the Federal Administrative High Court to develop independent principles applied by the judiciary, adapted from blanket clauses in legislation, and thus to widen the control functions of the judiciary over the executive (David, 1986, pp.43-44). Such considerations raise fundamental issues in relation to the Basic Law. That planning issues are at the heart of such issues is perhaps not surprising at a time of such rapid economic, social and political change. The adaptability and stability of the federal system established by the Basic Law is likely to ensure that the changes which take place in the field of planning and the control of development will be evolutionary and based upon a complex coalition of interests.

## Introduction

G7.1 The Federal Republic of Germany (population, 61.2millions (1984); territory 248,706 sq.km) is a democratic and social federation with a written constitution under which government is exercised at three levels; the Federal; the eleven states (*Länder*)(population, 0.7millions to 16.8millions) and approximately 8,500 municipalities (minimum population approximately 8,000). At each level, the higher tier has to a certain degree a supervisory power over the lower; but the lower tiers, in particular the local government authorities, have independent sovereign powers and can pursue their own policies provided they do not conflict with those of the higher.

G7.2 Thus for planning, the Basic Law establishes (Article 83) a separation of powers, whereby, as a rule, the *Länder* execute Federal law, with the municipalities being responsible for the implementation of the control of development. The Federal government formulates broad goals and principles for the spatial distribution of population and employment and the environment, which are made concrete in the development plans and programmes of the *Länder*. Based on these, the municipalities prepare urban land use plans. The hierarchy of planning from Federal through *Länder* to municipality level is not one of central direction but of mutual cooperation.

## Planning Legislation and Instruments

G7.3 The main planning legislation is found in three Federal Acts.

(i) The Federal Building Law (1960 with various major amendments) which establishes the regulations for land use planning and the control of development.

(ii) The Federal Comprehensive Regional Planning Law (1965) which regulates the position of public authorities (Federal, *Länder* and municipal) within the comprehensive development planning system.

(iii) The Urban Renewal and Development Act (1971) which provides instruments for the comprehensive planning of defined urban renewal and development areas.

The Federal Building Law and Urban Renewal and Development Act have recently been integrated in a comprehensive Federal Act (*Baugesetzbuch*) which will provide a unified system of regulation for land use planning, control of urban development and urban renewal from July 1987.

G7.4 Whereas the aforementioned Federal legislation is final and obligatory for the *Länder* (Federal law being exclusive in its ambit), each of the *Länder* enacts a comprehensive regional planning legislation within a broad Federal framework legislation. Further, each of the *Länder* controls development through Building Orders, which regulate the provisions for the grant of building permits, an activity carried out by the relevant municipalities and counties (which relates only to aspects of safety, health, security and aesthetics, a distinctly separate area of law from land use planning).

G7.5 There is a wide range of Federal and *Länder* legislation covering pollution, environmental protection, major public sectoral planning (eg roads, power etc), most of which are linked to planning control.

## The Control of Development

G7.6 The basic system of control is fundamentally linked to the constitutional protection of private property rights. There exists a freedom to build, but only within the framework of the law. Under the Basic Law, these rights may legitimately be constrained by land use planning and other requirements for specific purposes laid down in written law. Hence development is normally allowed within the host of regulations controlled by the issuing of a building permit. This is an administrative act, and can be contested in the administrative law courts. Written law, of great complexity and detail, provides the context for the operation of land use planning and control of development.

G7.7 Under the Federal Building Law, the clearest plan provision for land use is provided by the *Bebauungsplan* (binding land use plan) which establishes with precise legal certainty the intended use and built form of land within the area regulated by the plan. The central elements of this provision include the following

(i) The adoption of a binding land use plan is a municipality responsibility and discretion (controlled only under legal aspects by the *Land*), and once approved is binding in its provisions on all private and public agencies - it is the only urban land use plan in the FRG with this legal status.

(ii) The binding land use plans, created under the Federal urban land use planning legislation, are small in scale and apply usually only to parts of a municipality's area. They are drawn up where development is anticipated or intended, and thus have been used in the main for new development at the urban periphery of redevelopment of urban areas. Many urban areas, particularly in the inner parts of a large metropolis, are not covered by such plans.

(iii) A major purpose of planning control is to safeguard and 'realise' (implement) the uses and provisions shown in the plan.

(iv) Most development (in particular for housing) is carried out by the private sector and non-municipal agencies. Municipalities often own building land on the urban periphery, but compulsory purchase and municipality involvement in the direct implementation of development for private purposes is the exception rather than the rule.

G7.8   The method of control - besides the preparation of binding land use plans - is through the issue of a building permit, which combines the provisions of planning and building regulations.

(i) The municipality's building department receives the application and
- first checks the provisions of statutory local plans involving

  (a) the *Bebauungsplan* (binding land use plan), if one exists

  (b) the relevant substitute regulations if no binding land use plan exists
  - then checks the building regulation requirements
  - consults other departments of the municipality or *Länder*, together with other relevant public sector agencies.

(ii) The decision is made, and the permit issued/refused, with or without conditions, by the chief municipal official (or municipal civil servants signing on instruction by him/her) and not by the municipal council. Thus

  (a) if in accordance with the plan and regulations, a permit cannot be refused

  (b) if not in accord, a permit can be granted only if based on specific exemptions provided for in the plan, or under very restricted legally defined conditions if not provided for in the plan.

(iii) There is no statutory period provided for the issue of a permit, but the applicant has to wait at least for a three- month minimum period before he/she can have the case considered by the administrative courts.

(iv) An applicant refused a permit can appeal to the next highest authority and then the administrative courts.

## The Definition of Development and the Scope of Control

G7.9   Development requiring permission is defined in the *Bundesbaugesetz*, which regulates land use control. In addition, *Länder* Building Orders specify building and other land use activities requiring permission or formal information to the building authority. These latter relates to matters of public health, safety, security, aesthetics etc; the former relates to urban land use matters. Broadly, the control of development can cover

(i) all building construction including site planning and layout; the density and intensity of development; the height and mass of buildings; and aesthetic control

(ii) demolition of buildings

(iii) agricultural buildings (though some are of a privileged character) but not agricultural land use other than in special areas

(iv) mineral working and extraction (mining activities requiring specific authorisation)

(v) open spaces, parks, green areas

(vi) all changes of use (as defined in the regulations) for areas covered by the *Flächennutzungsplan* (preparatory land use plan) and the *Bebauungsplan* (binding land use plan). Federal and *Länder* legislation establish the relevant land use classes

(vii) car parking

(iii) advertisements

(ix) all infrastructure and utilities.

G7.10   The level of detail varies depending on the required level of control. But in particular the binding land use plans, if they exist, are usually very precise in terms of layout, building disposition, mass, density and type.

*Flexibility in Control*
G7.11   The system does provide room for flexibility and negotiation, using three principal methods

(i) exemptions already specifically provided in binding land use plans (but which must be sanctioned by the respective *Länder* authority, if applied in a specific case).

(ii) where 'substitute' regulations for binding land use plans apply, or a binding land use plan does not exist. These regulations derive from

  (a) Section 34 of the Federal Building Law, providing that, in developed areas where a binding land use plan does not exist, development shall conform to the dominant character of the adjacent areas

(b) Section 35 of the Federal Building Law, providing special and general categories of development which may be permitted outside developed areas

The requirements of Section 35 are strict, but Section 34 provides considerable latitude to municipalities for flexible control without a binding land use plan.

*Control Workload*
G7.12   There are no comprehensive statistics available, but the situation in an average municipality where binding land use plans exist indicates that the norm is for a very high proportion of building permits to be approved (over 90 per cent), though most involve negotiation, exemptions etc (70 per cent). Figures are not available for the situation outside binding land use plan areas.

G7.13   On average, applications are processed and decisions issued within three months, though this may extend to eight months or more for complex developments.

## Policy and the Preparation of Plans

G7.14   The binding legal land use plan is the *Bebauungsplan*, which has to be in conformity with the provisions of the *Flächennutzungsplan* (preparatory land use plan). Both these plans are prepared by municipalities, and have to be approved by the relevant *Länder* administrative authority (usually the *Regierungsbezirk* - administrative district of the *Land*). This approval is procedural and substantive, intended to ensure conformity with all applicable laws. The *Flächennutzungsplan* is binding on public authorities, the *Bebauungsplan* as well on all private and public agencies and individual landowners. These plans are

(i)   prepared for the municipalities by the planning sections of building departments with public consultation

(ii)  placed on deposit for one month for objections to be considered by the municipal council

(iii) approved by the council

(iv) sent to the *Regierungsbezirk* for checking regarding legal aspects.

G7.15   The process of plan preparation can be lengthy, taking from less than one year up to five years or more, and can involve hundreds of objections to municipalities to be submitted to the *Regierungsbezirk* for consideration within the approved procedure for urban land use plans, if not fixed by the municipality.

G7.16   Each of the *Länder* has its own Development Plan and Programme setting out a spatial strategy, made concrete in the Regional Development Plans of the *Regierungsbezirk*. All land use plans prepared by municipalities must conform with the goals defined in these plans, and with the Federal regional planning objectives set out in the Federal Regional Development Law. The role of the *Regierungsbezirk* in coordinating both the plan and control activities and responsibilities of the *Länder* and municipalities is significant. The *Länder* planning ministries issue circulars and advice interpreting policy and legislation. There is no Federal land use plan. But the Federal planning ministry prepares a Federal comprehensive regional planning report every four years, covering more general aspects of overall national spatial development.

## Objections and Appeals

G7.17   The most significant aspect of German law affecting planning derives from the administrative law courts. The Basic Law establishes a constitutional right for citizens to challenge the legality of administrative decisions. The function of the courts is to interpret the law by reference to the legislation and the constitution.

G7.18   A person who is refused permission for a building permit or who wishes to dispute the conditions attaching to a permission can appeal to the next highest authority and after this to the administrative law courts. Such procedures are complex, lengthy and, because of the delay, expensive. In order to bring a successful action a decision must not only be shown to have been illegal, but to be a violation of the legal rights of an applicant. These rights include the constitutional protection of private property, and hence effectively limit third party rights of appeal.

G7.19   A preparatory land use plan cannot be the direct object of appeal to the administrative courts. It is controlled by the *Regierungsbezirk* during the state approval procedure for the preparatory land use plan. Actions to the administrative courts against binding land use plans are permitted, and concern both procedural and material issues (eg the role of appropriately and fairly weighing the different interests involved).

G7.20   However, there is a growing tendency for the borderline between procedural and substantive issues to become unclear, and the role of the courts in planning and control issues is increasing in significance.

## The Private Sector and Planning

G7.21   Although the system of land use regulation and control of development in the FRG is one in which the public sector prepares a range of development plans which rely for their successful implementation upon the actions of a predominately private sector development process, to a remarkable extent this system is not one of conflict between the two sectors. Land use planning is widely seen by the private sector and the general public to be a useful and necessary component of a democratic and social federation. In a sense, the rather dependent position of the private landowner/developer in relation to the municipality (with regard to the operation of the urban land use planning system) is widely accepted to be a legitimate relationship, since it is protected by appropriate constitutional guarantees. The necessity for all public authorities fairly to weigh and balance competing interests ensures a strong voice for the private sector in plan-making and implementation, and the close relationship between physical development plans and the operation of the control of development lends a strong element of certainty to the operation of the planning system. Where criticism is made of the planning system by the private sector, it relates more to the tendency for delay to set in with extensive public participation; for an alleged over-rigidity and inflexibility in the interpretation of regulations; and an emerging concern that a growing interest in environmental and ecological issues amongst the general public may unduly influence the planning system in relation to the needs of development.

## Conclusions

G7.22    The chief characteristics of the system of control in the FRG are that it

(i) links building and planning control in a single instrument, ie the building permit

(ii) provides a measure of public consultation in the preparation of plans, and opportunity for legal redress (by directly affected) persons in relation to the issuing of building permit by recourse to the administrative courts

(iii) provides legal certainty for developers and their neighbours in approved plan areas

(iv) provides an increasing measure of municipal discretion in the operation of control

(v) provides a policy framework at *Länder* level.

G7.23    However, the system is open to criticism in that it

(i) involves potentially very lengthy procedures for the preparation and revision of plans

(ii) may be overly detailed and inflexible

(iii) is more appropriate for negative planning and the control of major new developments as extensions to built-up areas, and may be less appropriate for the renewal of existing urban areas

(iv) can lead to a lack of certainty and clear guidance in relation to regulations which substitute for a binding land use plan.

G7.24    Some of those difficulties are being addressed in the recent revision of planning and control regulations, which are directed towards simplification of procedures, reduced regulation, less state involvement in the procedures run by the municipalities but a greater autonomy for municipalities in devising and implementing planning policies and control of development in their areas. These current changes having been welcomed in principle by the private and public sectors, though some in the private sector remain to be convinced that all parts of the public sector (in a system as strongly decentralised as that in the Federal Republic of Germany) will respond fully to the opportunities provided by the new reforms in planning legislation.

# GLOSSARY

**Abstimmung**
conference of interested parties

**Atomgesetz**
federal Nuclear Act 1976; 1980

**Bauaufsichtsbehörde**
building authority/department, which issues building permits (relating to both land use planning and building regulation matters). A generic term. (see also *Baubehörde* and *Baugenehmigungsbehörde*)

**Baubehörde**
building authority/department, which issues building permits (relating to both land use planning and building regulation matters). In common use referring to the main department concerned with planning and building matters, particularly for larger cities (eg Hamburg). (see also *Bauaufsichtsbehörde* and *Baugenehmigungsbehörde*)

**Bauflachen**
building land area, building land, building space

**Baugebiet**
specific land use areas.

**Baugenehimigungsbehörde**
building authority/department, which issues building permits (relating to both land use planning and building regulation matters). A generic term. (see also *Baugufsichtsbehörde* and *Baubehörde*)

**Baugenehmigung**
building permit, construction permit

**Baugenehminungsvesfahren**
procedure to obtain a building permit

**Baugesetzbuch**
federal Act consolidating and unifying land use planning and urban renewal, combining and revising the federal acts relating to planning (the Federal Building Law) and urban renewal (the Urban Development Assistance Act).

**Bauleitplanung**
physical development planning, town and county planning, area development planning

**Baunutzungsverordnung**
land use zoning order

**Bauordnung**
building code

**Bebauungsplan**
binding land use plan, which directly determines the legal land use characteristics of a site, binding upon all private and public agents

**Bezirksplanungsrat**
district planning office - the *Bezirk* is an administrative area of a *Land*

**Bezirksregierung**
district administration

**Bodenordnung**
land code/order

**Bodenrecht**
land law

**Bund**
federation

**Bundesbaugesetz**
federal Building Act, Federal Building Law - until the *Baugesetzbuch* of 1986/87, the main Federal Act regulating land use planning in the Federal Republic of Germany. (Though see also *Bundesraumordnungsgesetz*)

**Bundesfernstrassengesetz**
federal Trunk Road Act/Law

**Bundes-Immissionsschutzgesetz**
federal Control of Pollution Act (noise, smell etc) 1974; 1982. Act relating to Protection Against Environmental Hazards and Pollution

**Bundesinvestitionshilfen**
federal grant aid to specific *Länder* development projects

**Bundeskanzler**
federal Chancellor

**Bundesminister**
federal minister

**Bundesministerium für Raumordnung, Bauwesen und Stadtebau**
federal Ministry for Regional Planning, Building and Urban Development

**Bundesnaturschutzgesetz**
federal Nature Protection Act/Law, 1976

**Bundesrat**
upper House of Parliament

**Bundesraumordnung**
federal planning

**Bundesraumordnungsgesetz**
federal Comprehensive Regional Planning Act/Law,
(1976, 1976, 1980, 1986) - provides Federal framework
regulations for comprehensive development planning at
the *Länder* level, co-ordinating all Planning and
implementation activities of local authorities

**Bundesregierung**
federal government

**Bundesverfassungsgericht**
federal Constitutional Court

**Bundesversammlung**
federal Assembly

**Bundesverwaltungsgericht**
federal Administrative Court

**Bïrgermeister**
the political executive of a municipality

**Bürgerschaft**
lower House of Legislature (Bremen; Hamburg; Berlin)

**Dankmalschutzgesetz**
preservation of Monuments Act/Law

**Deutsche Bundestag**
german Federal Parliament

**Deutches Institut für Normung**
german Standards Institute

**Dichtemodell**
density model (used in Hamburg to regulate population
densities near major transportation/activity nodes)

**Eigentumsgarantie**
constitutional guarantee concerning private property
rights

**Enteignung**
expropriation

**Entwicklung**
development, in the sense of the comprehensive
planning activities of municipalities and public agencies

**Entwicklungsmodell**
development model (in Hamburg, a regional framework
plan)

**Erlass**
dispensation, exemption

**Erschliessung**
development, in the sense of building land
improvements (Facilities, usually public utilities, such
as pavements or sewers, which added to land, increase
its usefulness)

**Fachbereichsplan**
sector or subject plan (see *Fachplanungen*)

**Fachplanungen**
sector planning (see *Fachbereichsplan*)

**Flächennutzungsplan**
preparatory land use plan, essentially a zoning plan for
the entire area of a muunicipality. Binding on public
agencies

**Flurbereinigungsgesetz**
(agricultural) land redistribution act/law (1976)

**Funktionsplan**
non-statutory plan

**Gebeitentwicklungsplan**
territorial development plan, area development plan -
for a sub-unit of a *Land* (such as a *Bezirk*)

**Gemeinde (plural Gemeinden)**
community, commune (obsolete) municipality (a city or
town chartered to govern itslf)

**Gemeindirektor**
chief executive officer of a *Gemeinde*

**Gemeinschaftsaufgaben**
joint responsibility for specific fields of action in
relation to Federal aid for the implementation of
planning programmes

**Gesamtplanung**
comprehensive, integrated planning

**Gleichwertigkeit der Lebensbedingungen**
constitutional principle of equality of living conditions

**Grenzregelung**
adjustment of plot boundaries

**Grundgesetz**
basic Law, comprising the written constitution of the
Federal Republic of Germany

**Grundrechte**
constitutional guarantee of civic rights

**Kommunale Selbstverwaltung**
community/local self-government

**Kommunalverband Ruhr**
union of Ruhr communities

**Kommune**
community, commune (obsolete)

**Konferenz Norddeutschland**
conference of North Germany, an advisory
inter-*Länder* Planning organisation for the four
north-coast *Länder* (Schleswig-Holstein, Hamburg,
Bremen and Lower Saxony)

**Kreis (plural Kreise)**
county

**Kreisangehörige Gemeinde**
*Gemeinde* belonging to a county

**Kreisfrie-Stadt**
*Gemeinde* not associated with a County, having
sovereign municipal rights, notably in relation to land
use planning

**Kreistag**
county council

**Kustenländer**
coast *Länder* (in North Germany - Schleswig-Holstein, Hamburg, Bremen, Lower Saxony)

**Land (plural Länder)**
state, of which there are eleven in the Federation

**Landesamt für Denkmalspflege**
preservation of monuments office/board

**Landesbauordnung**
land building code

**Landesentwicklungsplan**
land development plan

**Landesplanungsgesetz**
land development plan

**Landesentwicklungsprogramm**
land development programme

**Landespflegerische Begleitplan**
land conservation plan

**Landesplanungsgesetz**
*Land* (State) planning law

**Landeswassergesetz**
Land Water Management Act/Law (1979 in North Rhine-Westphalia)

**Landkreis**
county (see *Kreis*)

**Landrat**
chief executive official of a county

**Landschaftsgesetz**
Landscape Protection Act/Law (1980 in North Rhine-Westphalia)

**Landschaftsschutzgebiete**
landscape preserve

**Landschaftsverband**
advisory planning association

**Landtag**
*Land* Parliament

**Magistrat**
municipal council, board, government

**Materielles Baurecht**
material building law - building regulations

**Ministerium für Stadtentwicklung, Wohnen und Verkehr**
Ministry for Urban Planning, Housing and Transport, North Rhine-Westphalia

**Ministerpräsident**
Prime Minister of a *Land* government

**Munsterbauordung**
model building code

**Naturpark**
nature recreational area

**Naturschutzgebiet**
nature reserve

**Normenkontrollverfahren**
control of statutory rules - procedure whereby administrative high courts may review legal provisions with a rank inferior to a formal law enacted either by a *Land* or the Federal government authority or as a local statute by a municipality

**Oberbaudirektor**
executive head of the building department of a local authority

**Oberbürgermeister**
political executive of a municipality

**Oberstadtdirektor**
chief executive officer of a town or city

**Oberwaltungsgericht**
superior administrative court

**Ortsamt**
local board

**Ortsteil**
part of a locality

**Petitionsausschuss**
Petition Committee of the *Land* Parliament in North Rhine-Westphalia

**Planzeichenverordnung**
Federal Urban Land Use Planning Graphic Representation Order, 1981

**Rahmengesetzgebungsbefugnis**
provision of the Federal Basic Law entitling the Federation to enact 'framework regulations' which the *Länder* and municipalities have to respect when setting up their own regulations

**Rat**
council

**Raumordnung**
public task to provide a certain spatial order by co-ordinating and initiating planning, implementation and public investment (not run by the co-ordinating and initiating authority). A legal term. It relates in principle to all levels of administration and government and might be run there under specific designations and within a specific legal framework and scope (eg as *Landesplanung* on the *Land* level, or as urban land use planning on the municipal level). In particular it is used in relation to the political task of the Federal Government to provide comparable living conditions throughout the Federal Republic

**Raumordnung und Landesplanung**
national, regional and *Land* (State) planning - umbrella terms for all types of space-related planning relating to physical, economic, social or environmental aspects

**Raumordungsplanung**
regional policy planning

**Raumordnungsverfahren**
special procedure to ensure all planning activities are in conformity with higher-ranking plans (applies to all

development activities above the local level), in order to conform with the principles of the comprehensive regional planning system

**Raumplanung**
spatial planning, relating to physical, economic, social or environmental aspects

**Rechtsstaatsprinzip**
establishment of the 'rule of law' as a constitutional principle

**Rechtsverordnung**
ordinance

**Rechtsweggarantie**
constitutional guarantee that any person affected by public decisions may have recourse to the law courts

**Regierungsbezirk**
government district

**Regierungsprsident**
executive of the *Land* government at district level

**Richtlinien**
administrational guidelines

**Sanierung**
redevelopment

**Sanierungsgebiete**
redevelopment area

**Satzung**
by(e)-law, statute

**Senat**
senate Council, Lower House of Legislature (Bremen, Hamburg, Berlin)

**Siedlungsverband Ruhrkohlenbezirk**
settlement Association for the Ruhr District - regional planning organisation preceding the Kommunalverband Ruhr

**Sozialbinding des Eigentums**
constitutional social obligations relating to the exercise of private property rights

**Staatskanzlei**
Prime Minister's office of a *Land*

**Staatsrat**
Vice-Senator (of Hamburg *Länder* government)

**Stadt (plural Städte)**
town, city

**Stadtbezirk**
city district, town district

**Stadtdirektor**
chief executive officer of town or city

**Städtebauförderungsgesetz**
Urban Renewal and Development Act, 1971, 1976 -

from 1987, integrated into *Baugesetzbuch*, a unifying Federal Planning Act

**Städtebaupolitik**
urban policy

**Städtebaurecht**
urban building law

**Stadtentwicklungsplan**
urban development plan, town development plan, city development plan - of a comprehensive nature

**Stadtentwicklungspolitik**
urban development policy

**Stadtteilentwicklungsplanung**
development planning for an urban district.

**Stadtplanungsrecht**
urban planning law, town planning law, city planning law

**Stadtrat**
city council

**Standortprogramm**
location programme

**Umlegung**
reallocation of land

**Verband (plural Verbände)**
union, association - often for planning purposes

**Verbindlicher Bauleitplan**
detailed local development plan

**Verrechtlichung**
legalism

**Verwaltungsgericht**
administrative court

**Verwaltungsgerichtshof**
higher administrative court

**Verwaltungsgerichtsordnung**
code of the administrative court.

**Verwaltungsvorschriften**
administrative rules

**Vorbereitender Bauleitplan**land use plan

**Wasserhaushaltsgesetz**
Federal Water Act 1976

**Widerspruchsverfahren**
pre-appeal proceeding, whereby a person wishing to appeal to an administrative court must first submit the case to an authority at the next highest level of the government system

**Zusammenfassende Planung**
comprehensive, integrated planning

# REFERENCES

ALBERS, G **Notes on the German Planning System**, Paper delivered to Conference on Comparative Studies of European Cities and City Regions, 5th-8th June, 1973, Lanchester Polytechnic (mimeo)

ALBERS, G (1986), "Changes in German town planning: a review of the last sixty years", **Town Planning Review**, vol 57, no 1, pp 17-34

ARDAGH, J (1987), **Germany and the Germans**, London: Hamish Hamilton

ARNDT, H-J (1966), **West Germany: Politics of Non-Planning**, Syracuse: Syracuse University Press

BACH, L C (1980), **Urban and Regional Planning in West Germany: A Bibliography of Source Material in the English Language**, Department of Urban and Regional Planning, University of Dortmund, Federal Republic of Germany, Council of Planning Librarians No 30, Chicago: Illinois, U S A

BAESTLEIN, A, HUNNIUS, G, JANN, W, KONNUKIEWITZ, M and WOLLMANN, H (1978), "State grants and local development planning in the FRG'", in Hanf, K and Scharpf, F W (editors), **Interorganisational Policy-Making**, London: Sage

BALFOUR, M (1982), **West Germany: a contemporary history**, London: Croom Helm

BENNETT, R J (1983), "The finance of cities in West Germany", **Progress in Planning** 21(1) pp 7-62

BERENTSEN, W H (1985), "Regional planning in Central Europe: Austria, the Federal Republic of Germany, the German Democratic Republic, and Switzerland", **Environment and Planning C: Government and Policy** 3(3) pp 319-339

BORCHARD, K (1973), **German Experiences of the Planning Problems of Cities and City Regions**, Paper delivered on Conference on Comparative Studies of European Cities and City Regions, 5th-8th June 1973, Lancester: Lancaster Polytechnic (mimeo)

BOWDEN, P (1980), **North Rhine-Westphalia: North-West England - Regional Development in Action**, London: Anglo-German Foundation for the Study of Industrial Society

BROADBENT, T A and REIDENBACH, M (editors) (1987), **Urban Infrastructure in Britain and Germany**, London: Anglo-German Foundation for the Study of Industrial Society

BULKA, H D, MICHEL, H-G and WULLENKORD, C (editors) (1986), **Facts about Germany - the Federal Republic of Germany**, Verlag GmbH. Presse-und Informationsamt der Bundesregierung, Bonn: (Press and Information Office of the Government of the FRG)

BURTON, I, WILSON, J and MUNN, R O (1983), "Environmental impact assessment: national approaches and international needs", **Environmental Monitoring and Assessment**, 3(2) pp 133-150

BURTENSHAW, D (1974), **Economic geography of West Germany**, London: Macmillan

BURTENSHAW, D, BATEMAN, M and ASHWORTH, G J (1981), **The City in Western Europe**, Chichester: Wiley

CHILDS, D and JOHNSON, J (1981), **West Germany - Politics and Society**, London: Croom Helm

CLEMENT, K (1984), "The nuclear inquiry - German style", **Town & Country Planning**, vol 53 nos 7-8 pp 214-215

COPRIAN, W and KUNZMANN, K R (1975), **English Language Literature on Urban and Regional Planning in the Federal Republic of Germany**, 1955-75, Dortmund: University of Dortmund

COUCH, C (1985), **Housing conditions in Britain and Germany**, London: Anglo-German Foundation

DARIN-DRABKIN, H (1977), **Land policy and urban growth**, Oxford: Pergamon

DAVID, C-H (1986) "The German Planning and Land Use System" forthcoming in Harr, C M and Kayden, J **International Land Use Law Treatise**, Harvard Law School (mimeo)

DAVID, C-H (1987), **German Planning and the development control system: a comparative report**, Occasional Paper No.14, Centre for Environmental Studies, Department of Architecture and Planning, Belfast: The Queen's University of Belfast

DAVIDSON, A W and LEONARD, J E (editors) (1973), **Urban development in France and West Germany**, Property Studies in the United Kingdom and Overseas, no 2, Centre for Advanced Land Use Studies, College of Estate Management, Reading: esp. Cohn, E **The Land Market in Germany**, pp 56-63, Trieb, M **Land use planning and control in Germany**, pp 64-75, Pfeiffer, U **Government initiative in urban development in Germany**, pp 76-86, Wittwer, G **Experiences with the financing of planned communities in the Federal Republic of Germany**, pp 87-98, Wittwer, G **Experiences in financing a new town: the new town of Wulfen**, pp 99-108, Wick, R **The urban development business in Germany**, pp 109-114

DAVIES, R L (editor) (1979), **Retail planning in the**

**European Community, Saxon House, Germany:** pp 135-151

DORING, H and SMITH, G (editors) (1982), **Party government and political culture in West Germany**, London Macmillan

ENVIRONMENTAL RESOURCES LTD. (1976), **Law and practice relating to pollution control in Germany**, London: Graham and Trotman

ESTERMANN, H (editor) 1980, **Planning Organisation and planning problems in the Federal Republic of Germany: a reader**, Dortmund: Institute of Urban and Regional Planning, University of Dortmund, vols 2,3,4

EVERSLEY, D E C (1974), "Britain and Germany - Local Government in Perspective", in Rose, R (editor), **The Management of Urban Change**, London: Sage

FARRELL, M R and JONES, G A (1987) "'Umlegung' - the thinking man's three-dimensional monopoly", **Property Management** vol 5 no 1, pp 19-34

FEDERAL MINISTRY OF REGIONAL PLANNING, BUILDING AND URBAN DEVELOPMENT (1975), **Regional Planning Programme for the General Spatial Development of the FRG (Federal Regional Planning Programme)**, Regional Planning Series no 06-002, Bonn-Bad Godesberg: Federal Ministry of Regional Planning, Building and Urban Development

FEDERAL MINISTRY OF REGIONAL PLANNING, BUILDING AND URBAN DEVELOPMENT (1976), **Habitat: National Report for the United Nations Conference on Human Settlements - Federal Republic of Germany**, Bonn:

FEDERAL MINISTRY OF REGIONAL PLANNING, BUILDING AND URBAN DEVELOPMENT (1977), **Regional Planning and Urban Development : an Overview of the Federal Regional Planning Programme, the 1974 Regional Planning Report and the 1975 Urban Development Report**, (Bonn-Bad Godesberg: Federal Ministry of Regional Planning, Building and Urban Development)

FEDERAL MINISTRY FOR REGIONAL PLANNING, BUILDING AND URBAN DEVELOPMENT (1982), **Regional Planning Report 1982** - Summary (Bonn-Bad Godesberg: Federal Ministry of Regional Planning, Building and Urban Development)

FEDERAL MINISTRY FOR REGIONAL PLANNING, BUILDING AND URBAN DEVELOPMENT & FEDERAL MINISTRY FOR ECONOMICS (1982), **Federal Republic of Germany: Monograph on the Human Settlements situation and Related Trends and Policies**, September 1982, for United Nations Economic Commission for Europe, Committee on Housing, Building and Planning

FEDERAL MINISTRY FOR REGIONAL PLANNING, BUILDING AND URBAN DEVELOPMENT (BMBau) (1986), **Renewal of housing policy and urban development policy**, Bonn-Bad-Godesberg

FRANZEN, L F (1982), "Environmental compatibility assessment of nuclear facilities in the FRG", **Environmental Impact Assessment Review** 3 (2-3) pp 271-288

GARNER, J F (editor), (1986), **Planning Law in Western Europe**, Oxford: North Holland Publishing Co., 1st edition 1975, 2nd edition 1986, esp. Kimminich, O, Town and Country Planning law in the Federal Republic of Germany, Chapter 8, pp 188-215, Gravells, N P, Town and Country planning law: the European dimension, Chapter 15, pp 389-407

GOSS, A (1975)," Planning in West Germany", **Town and Country Planning**, January 1975, pp 26-30

GRAFIN zu CASTELL RUDENHAUSEN, A (1982), "Population change and regional development in the FRG", pp 325-348 in Eversley, D and Killmann, W. (editors), **Population change and social planning**, London: Edward Arnold

GRAUHAN, R-R (1973), "Notes on the structure of planning administration" in Faludi, A, **A Reader in Planning Theory, Oxford: Pergamon, pp 297-316**

GRIEVES, F L (1984), "Environmental protection in the Federal Republic of Germany: focus on the Saarland", **Environmental Reviews**, 8(3) pp 252-269

GUDE, S et al "Urban policy in the FRG", in Schwartz, G G (editor), **Advanced Industrialization and the Inner Cities,** Lexington Books Lexington, Mass: (pp 99-140)

GUNSTEREN, H R (1976), **The Quest for Control: A critique of the rational-central-rule approach in public affairs**, esp. Chapter 3 London: Wiley

HAJDU, J G (1979), "Phases in the post-war German urban experience", **Town Planning Review**, vol 50 no 3, pp 267-86

HAJDU, J, "Postwar development and planning of West German cities", pp 16-39 in Wild, T (editor), **Urban and rural change in West Germany**, London: Croom Helm

HALL, P and HAY, D (1980), **Growth centres in the European urban system**, London: Heinemann

HALLETT, G (1976), **The Social Economy of West Germany**, London: Macmillan

HART, D A (1980), **Urban Economic Development: lessons for British cities from Germany and America**, Reading: Occasional Papers POZ, School of Planning Studies, University of Reading

HAY, D and WEGENER, M (1985), "The Dortmund Region", **Scandinavian Housing and Planning Research**, vol 2, pp 225-230

HASS-KLAU, C, **Planning research in Germany** London: SSRC

HAYWOOD, D J and NARKIEWICZ, O (editors), (1978), **Planning in Europe**, London: Croom Helm

HEINZE, G W (1985), "Regional policy and transport in the Federal Republic of Germany", **Environment and Planning C: Government & Policy**, 3(3), pp 269-284

HELLEN, J A (1974), **North Rhine-Westphalia**, Problem Regions of Europe series, London: Oxford University Press

JUNG, H U (1982), "Regional policy in the FRG", **Geoforum** 13(z), pp 83-96

KENNEDY, D (1984), West Germany, pp 55-74, in Wynn, M. (editor) **Housing in Europe**, London: Croom Helm

314

KIMMINICH, O (1981), "Public participation in the Federal Republic of Germany", **Town Planning Review**, vol 52, No 3, July pp 274-9

KIMMINICH, O (1986), Town and country planning law in the Federal Republic of Germany, in: Garner, J.F. (editor), (1975), **Planning Law in Western Europe**, Oxford: North-Holland Publishing Co., Oxford - 2nd edition Garner, F.J. & Gravells, N.P. (editors), Elsevier Science Publishers B.V. (North Holland)

KIMMINICH, O (1981), "Public participation in the Federal Republic of Germany", **Town Planning Review**, July, vol 52, no 3, pp 274-279

KOLINSKY, E (1984), **Parties, opposition and society in West Germany**, London: Croom Helm

KLOSS, G (1976), **West Germany: an introduction**, London: Macmillan

KNAPP, V (editor), **International Encyclopaedia of Comparative Law**, vol 1, National Reports, J C B Mohr -P Siebeck - The Hague

KONUKIEWITZ, M (1983), "Can regional planning work? experiences from a research project on the implementation of spatial policies in the FRG", **Policy and Politics**, 11(1), pp 87-98

KONUKIEWITZ, M and WOLLMANN, H (1982), "Physical planning in a Federal system: the case of West Germany", in McKay, D.H. (editor), **Planning and politics in Western Europe**, London: Macmillan

KRUMME, G (1974), "Regional policies in West Germany", pp 103-135 in Hansen, N M (editor), **Public policy and regional economic development. The Experience of Nine Western Countries**, Cambridge, Mass: Ballinger

KUNZMANN, K R (1985), "The Federal Republic of Germany", pp 8-25 in: Williams, R.H. (editor), **Planning in Europe**, Allen & Unwin, Urban & Regional Studies 11.

KUNZMANN, K R, ESTERMANN, H and ROJAHN, G, "Development trends in the regional and settlement structure of Federal Germany", **Built Environment**, vol 7, no 3-4, pp 243-254

KUNZMANN, K R and ROJAHN, G (1980), **The Ruhr within the context of European Regional Planning Objectives**, (mimeo) Institute of Town and Country Planning, University of Dortmund, September 1980

KUPPER, U I (1984), "Decentralising the city: attempts and experiences in Cologne Koln), FR Germany", **Geo Journal**, 9 (4), pp 407-419

LEWIS, J R and HUDSON, R H (1982), **Regional Planning in Europe**, London Papers in Regional Science 11, London: Pion

LITTLECHILD, M (1982), "Regional policy in Germany", **Town and Country Planning**, 51(6), pp 155-159

MASSER, I (1984), "Gross national comparative planning studies: a review", **Town Planning Review**, 55(2), pp 137-160

MAYHEW, A (1969), "Regional planning and the development areas in West Germany", **Regional Studies**, 1969, vol 3, pp 73-79

MAYNTZ, R and SCHARPF, F W (1975), **Policy-making in the German Federal Bureaucracy**, Amsterdam: Elsevier

MCKAY, D H (editor), (1982), **Planning and Politics in Western Europe**, London: Macmillan, esp. Chapter 4, Konukiewitz, M and Wollmann, H, Physical planning in a federal system: the case of West Germany

MELLOR, R E (1978), **The Two Germanies**, New York: Harper & Row

MICHAEL, R (1979), "Metropolitan concepts and planning policies in West Germany", **Town Planning Review**, vol 50, no 3, pp 287-312

MINNS, R and THORNLEY, J (1976), **Public control of companies in West German city planning**, Working Note no 429, London Centre for Environmental Studies

MOOR, N (1983), **The Planner and the Market**, London: George Godwin

MORGAN, J F W (1976), "Planning law and practice in the FRG", **Estates Gazette** 1976, vol 238, 10th April

MULLER-IBOLD, K (1974), "Administration and Urban Change in Germany", in Rose, R (editor), **The Management of Urban Change in Britain and Germany**, London: Sage, pp 171-183

NORMANN, P (1966), "The Federal Planning Framework in West Germany", **Journal of Town Planning Institute**, vol 52, no 3, March 1966, pp 91,93

NORTON, A (editor), (1985), **Local government in Britain and Germany**, Institute of Local Government Studies/ Anglo-German Foundation

NORTON, A (1985), "A short review of local government systems in other western democracies", **Local government studies**, 1985, vol 11 no 5, September/ October, pp 35-48

PARKER, G (1979), **The countries of community Europe: a geographical survey of contemporary issues**, London: Macmillan

PATERSON, W I and SMITH, G (1981), **The West German model: perspectives on a stable state**, London: Cass

PAULEY, R (1985), "Local Government in Britain and West Germany: Its present condition and future role", **Local Government Studies**, vol 11, 1 Jan/Feb 1985, pp 6-16

PERRY, N (1966), "The Federal planning framework in West Germany", **Journal Town Planning Institute**, vol 52, pp 91-93

PFEIFFER, U, "Market forces and urban change in Germany", in Rose, R. (editor), **The Management of Urban change in Britain and Germany**, London: Sage

PUNTER, L (1981), **The inner city and local government in Western Europe**, Reading: College of Estate Management, Centre for Advanced Land Use Studies

REICHSTEIN, J (1984), "Federal Republic of Germany", pp 37-47, in Cleere, H, **Approaches to the archaeological heritage**, Cambridge University Press

REISSERT, B (1980), 'Federal and State transfers to local government in the FRG: a case of political immobility', in Ashford, D F (editor), **Financing urban government in the welfare state**, London: Croom Helm

RIDLEY, F F (editor), (1979), **Government and administration in Western Europe**, Oxford: Martin Robertson

ROSE, R (editor), (1974), **The Management of Urban Change in Britain and Germany**, London: Sage

ROSNER, R (1975), "Town and Regional Planning in Germany", **The Planner**, December 1975, vol 61, no 10, pp 375-378

ROWAT, D C (editor), (1980), **International handbook on local government reorganisation**, London: Aldwych, esp. Sidentopf, H West Germany: Chapter 7, pp 85-91, Doeker, G West Germany: North Rhine-Westphalia, Chapter 27, pp 321-331, Wagener, F West Germany: A Survey, Chapter 28, pp 332-340

SANT, M (editor), (1974), **Regional Policy and Planning for Europe**, Farnborough: Sage

SCHMIDT, K (1987), "German property: Legal formalities, taxation and succession rules", **Estates Gazette**, vol 282, May 9, 1987, pp 666-667

SCHMIDT-ASSMANN, E, **German Federal Republic**, Chapter 13 in Patricios, N N (editor), (1986) **International Handbook on Land Use Planning**, London: Greenwood Press

SMITH, E O (1983), **The West German Economy**, London: Croom Helm

SMITH, G (1982), **Democracy in West Germany: parties and politics in the Federal Republic**, London: Heinemann

SOELL, H (1985), "Basic questions concerning the law of nature protection", **Environmental Policy and Law**, 14 (2-3), pp 58-67

SOLESBURY, W (1968), "Local planning and the German Bebauungsplan", **Journal of the Town Planning Institute**, vol 54, no 3, pp 117-122

STILES, R (1983), "An outline of landscape planning in West Germany", **Landscape Research**, 8(3), pp 23-27

STUMPF, M (1984), Land consolidation in the Federal Republic of Germany, pp 335-357, in Steiner, F R and Van Lier, H N (editors) **Land conservation and development**, Elsevier: Developments in Landscape Management and Urban Planning

UPPENDAHL, H, Intergovernmental relations in the Federal Republic of Germany: an overview, in Norton, A (editor), (1985), **Local government in Britain and Germany**, Anglo-German Foundation for the Study of Industrial Society/Joint Centre for Regional Urban and Local Government Studies, University of Birmingham

WATERHOUSE, A (1978), **City Planning in a Social Market Economy: Rationality and Evolution in German Planning Legislation**, Papers in Planning and Design, No.17, University of Toronto, 1978

WATERHOUSE, A (1979), "The advent of Localism in Two Planning Cultures: Munich and Toronto", **Town Planning Review**, vol 50, no.3 pp 313-324

WHITTICK, H (editor), (1974), **Encyclopaedia of Urban Planning**, New York: McGraw-Hill, see esp. Albers, G, 'Germany, West (Federal Republic)', pp 459-483

WILD, M T (1979), **West Germany, a Geography of its People**, Folkestone: W. Dawson & Son, (and (1981) Harlow: Longman)

WILD, M T (editor), (1983), **Urban and rural change in West Germany**, London: Croom Helm

WILLIAMS, R H (1978), "Urban planning in Federal Germany", **The Planner**, March 1978, vol 64, no 2, pp 46-7

WILLIAMS, R H (1984), **Planning in Europe: urban and regional planning in the EEC**, London: Allen & Unwin

WOLMAN, H (1983), "Central government policies towards declining urban economies in Western Europe and the United States of America", **Environment & Planning C: Government & Policy**, 1(3), pp 299-315

WOOD, C and LEE, N (1978), **Physical planning in the Member States of the European Economic Community**, Occasional Paper no 2, Department of Town and Country Planning, University of Manchester.

ZIELINSKI, H (1983), "Regional development and urban policy in the Federal Republic of Germany", **International Journal of Urban and Regional Research**, 7(1), pp 72-91

# APPENDIX A    LAND BUILDING ORDER NORTH RHINE WESTPHALIA 1984

*Landesbauordnung Nordrhein-Westfalen* 1984
(20th edition, January 1985)

# APPENDIX B    PROCEDURE FOR THE ADOPTION OF A LEGALLY BINDING LAND USE PLAN

Requirements under the planning legislation

I.   Planning phase
   *1. Informal phase*
   - planning initiative taken by the council or the city administration
   - preparations of sketches or drafts
   - information, contacts, participation
   - internal contacts between the different branches-departments of the city administration
   - contacts to the district councils (if existing)
   *2. Formal phase*
   - resolution of the city council (existing district councils, Bezirksvertretungen, have no powers to take planning decisions, but have only advisory functions of considerable political importance), that for a certain designated area a binding land use plan is to be prepared (§ 2 Abs.1, S. 2 BBauG)

   - information to the general public respectively the objectives and the purpose of the plan in order to initiate a public discussion to be continued during the formal planning phase (§§ 2 Abs. 5, 2a Abs. 2 BBauG)

   - preparation of a draft by the city administration (planning department)

   - resolution of the city council (based on preparatory recommendation of the planning committee) that this draft is to be laid out for inspection by bodies representing public interests and by private individuals during one month, giving the opportunity of making suggestions and representing during this time. General notice is given to the public and supplementary notices to the affected authorities.
   (Vorbingen von Anregungen und Bedenken, vgl. § 2 Abs. 6 BBauG)

   - consideration of the suggestions and representations made by public bodies and individualsa by the city council (based on a report worked out by the city administration)

   - an important legal constraint on the content of the plan is exercised by the legal requirement, that all interests, public and private, must be included in the consideration and must be appropriately weighted and fairly balanced, §1 Abs. 6, 7 BBauG, Abwägungsgebot)

   - if the draft is subjected to considerable alterations, the procedure of laying out for inspection for a month giving the possibility to make suggestions and representations is to be repeated

   - final adoption of the plan by a resolution of the city council as a local statute (BeschluB des Bebauungsplans als Satzung, § 10 BBauG), including the consideration of the representations made during the procedure.

II   Approval phase
   The plan needs the approval of the relevant state authority, in general of state authority of the intermediate level (staatliche Mittelinstanz: Regierungspräsident, § 11 Abs. 2, 6 Abs.2 BBauG). The approval is restricted to an examination only under legal aspects.

III   Coming into force
   The plan comes into force by public notification of the state authority's approval (§ 12 BBauG).
   Public display of the plan with the written statement giving the reasons for the content of the plan.

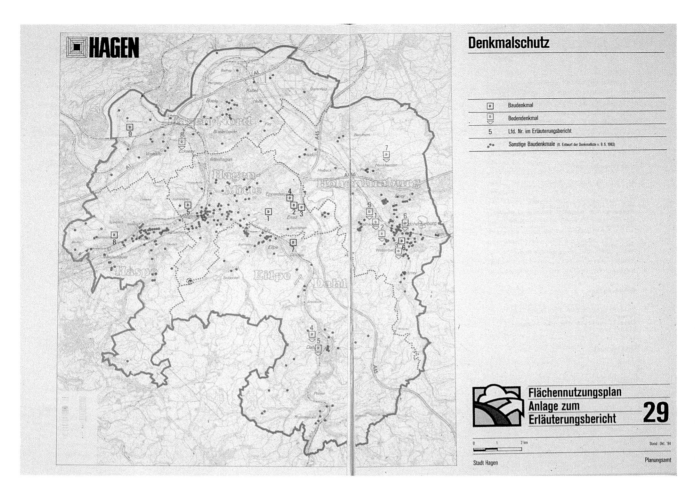

**Plate 15    Denkmalschutz (Historic Conservation) – Hagen,
North Rhine Westphalia**

Plate 15 shows the distribution of historic structures and
monuments in Hagen, North Rhine Westphalia. Policies for
these structures and monuments comprise a specific section of
the current Flächennutzungsplan (preparatory land use plan)
for the municipality.

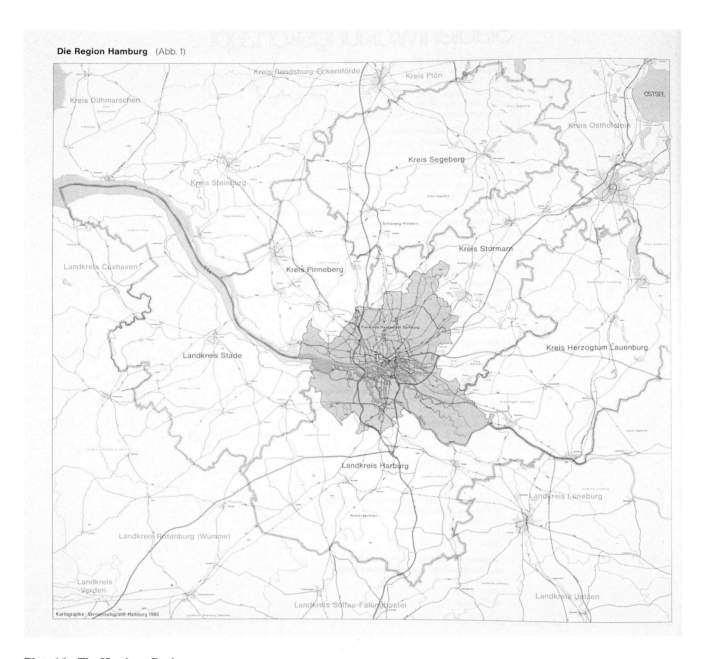

Die Region Hamburg  (Abb. 1)

**Plate 16   The Hamburg Region**

Plate 16 sets Hamburg in its strategic context with respect to
major transportation routes and its relationship with adjacent
Kreise (counties) in the metropolitan fringe.

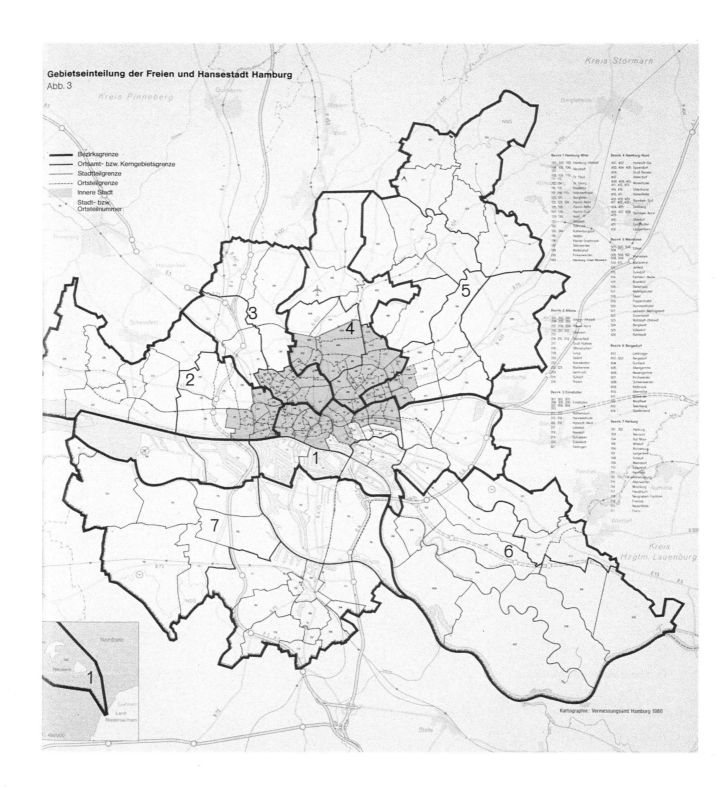

**Plate 17    Administrative Districts of Hamburg**

Plate 17 illustrates the administrative sub-divisions of
Hamburg in terms of the seven Bezirke (districts) and their
constituent components (core-zone; city quarter; locality –
these latter having only administrative or 'parish-council'
functions).

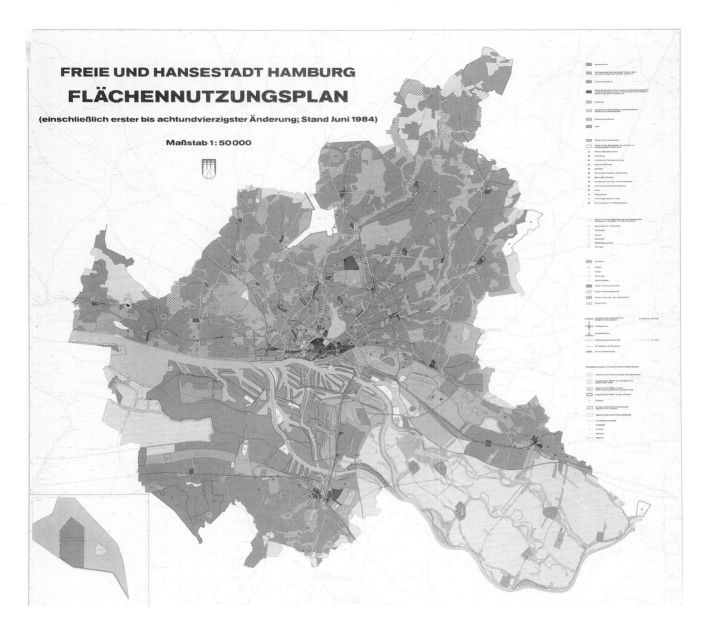

**Plate 18   Preparatory Land Use Plan – Hamburg 1984**

Plate 18 represents the basic land-use zoning plan for
Hamburg – its provisions are binding upon all public-sector
agencies, and it provides the framework for the more detailed
Stadtteilentwicklungsplan (comprehensive land use plans for
city districts), Programmplan (programme plans) and
Bebauungsplan (binding land use plans). Compare with Plate
25, and Figure G10(a).

327

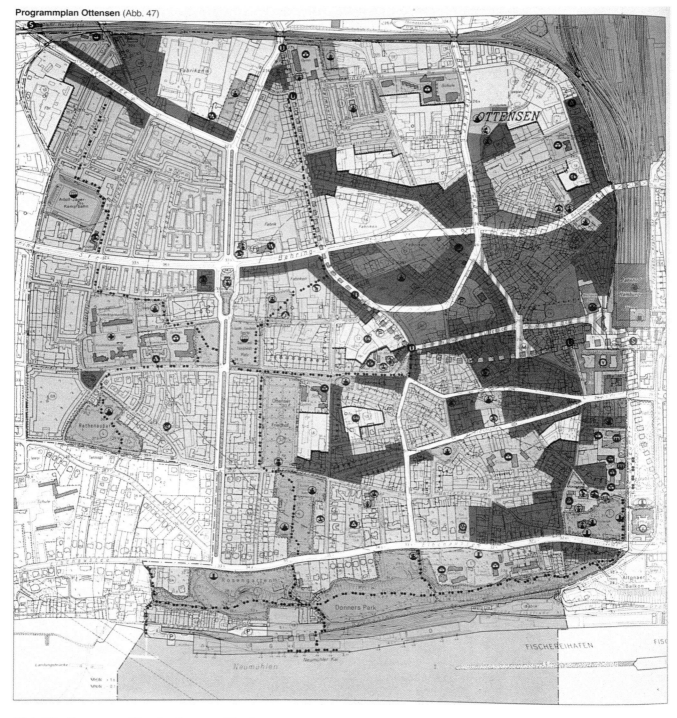

**Plate 19    Programme Plan for Ottensen, Hamburg**

Plate 19 indicates one of the non-statutory plans devised to
concretise the provisions of the Flächennutzungsplan
(preparatory land use plan) on the scale of a city district.

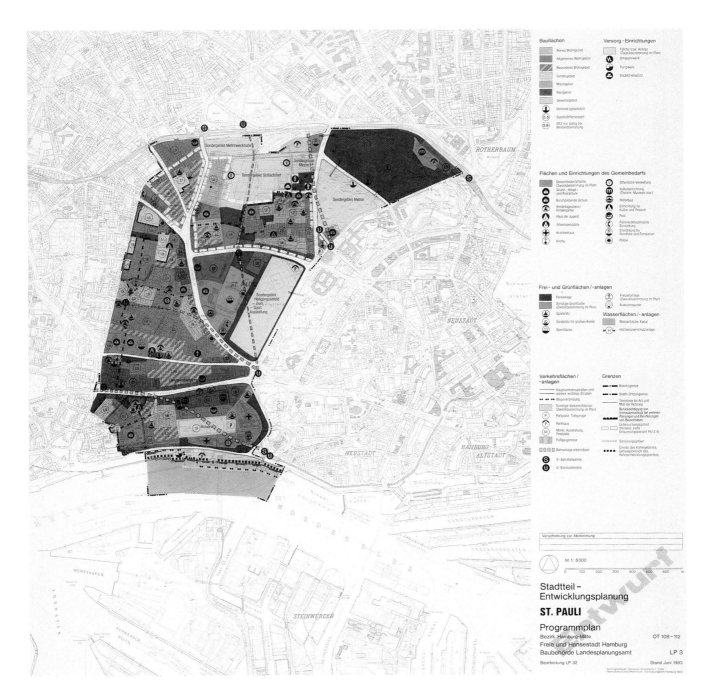

**Plate 20    Stadtteil-Entwicklungsplan (City District Comprehensive Plan) for St Pauli, Hamburg**

Plate 20 provides a further indication of the programme plan approach to specifying details of the Flächennutzungsplan (preparatory land use plan) on the scale of a city district noted in Plate 19.

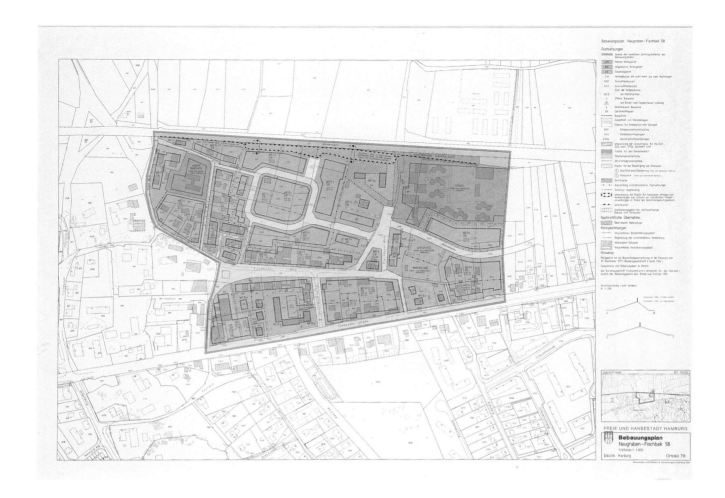

**Plate 21     Bebauungsplan (Binding Land Use Plan) for
Neugraben-Fischbek, Hamburg**

Plate 21 provides a concrete example of the level of detail
provided for in binding land-use plans which is shown
diagrammatically in Figure G10(b). Detailed land uses,
intensity of land use, building lines, car parking and

infrastructure provisions are clearly specified. The provisions
are legally binding on all public and private agencies.
Compare with Plate 26.

330

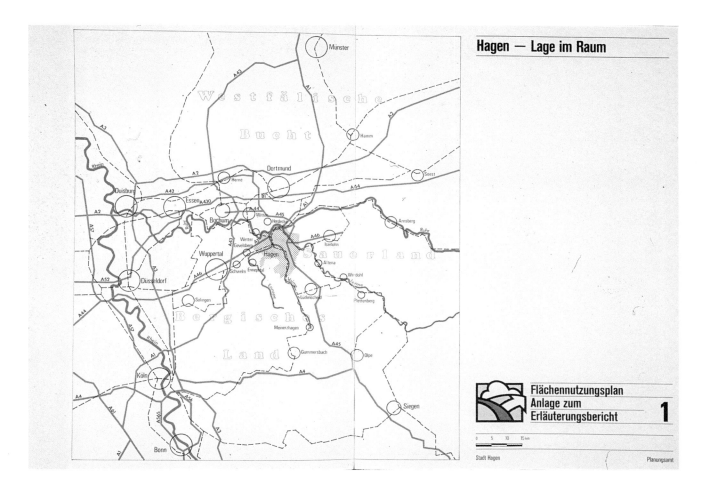

Flächennutzungsplan
Anlage zum
Erläuterungsbericht

1

0   5   10   15 km

Stadt Hagen

Planungsamt

**Plate 22   Hagen – Regional Location**

Plate 22 indicates the location of the Kreisfreie-Stadt
(county-free town) of Hagen, North Rhine Westphalia in
relation to adjacent centres and transportation routes within
the Ruhr region.

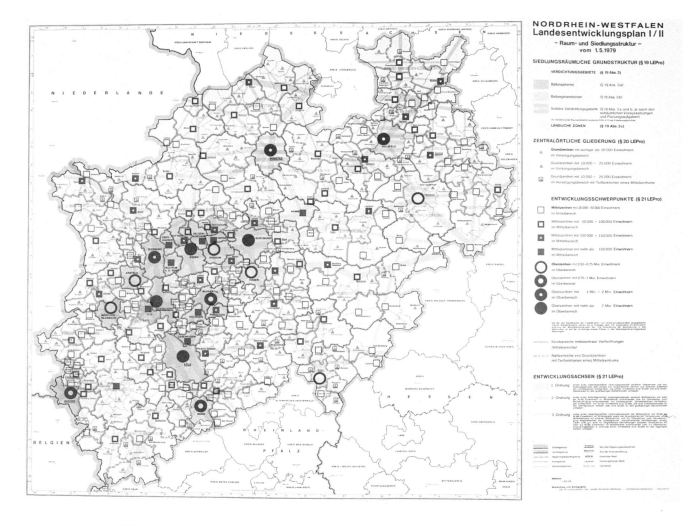

**Plate 23   North Rhine Westphalia – Comprehensive
Development Plan I/II**

Plate 23 illustrates the strategic settlement structure
provisions of the Comprehensive Development Plan for
North Rhine Westphalia, indicating the planned urban
hierarchy and accessibility corridors. This plan represents the
closest approximation to a regional physical plan in the
F.R.G.

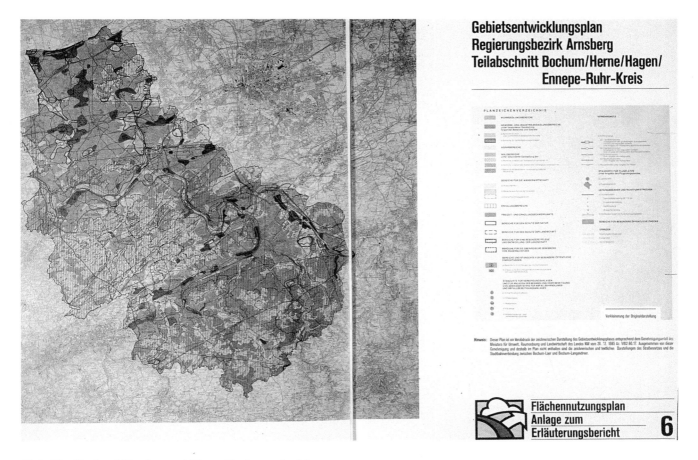

**Plate 24   Regional Development Plan – Regierungsbezirk (administrative district) Arnsberg, North Rhine Westphalia**

Plate 24 illustrates the concretisation of the provisions of the Comprehensive Development Plan for North Rhine Westphalia (indicated in Plate 23) for that part of the administrative district of Arnsberg (see Figure G3) comprising Bochum/Herne/Hagen/Ennepe – Ruhr Kreis. It is thus a sub-regional physical development plan, setting out the main land-use zones to which preparatory and binding land use plans must conform. It is prepared by the Regierungsbezirk (administrative district of the Land).

**Plate 25   Flächennutzungsplan (Preparatory Land Use Plan) –
Hagen, North Rhine Westphalia**

Plate 25 illustrates the detailed land use zoning provisions of a preparatory land use plan. These provisions must confirm with those of the Land Comprehensive Development Plan (see Plate 23) and the Gebietsentwicklungsplan (regional development plan) (see Plate 24). The provisions shown in the plan are binding on all public sector agencies, and provide the framework for the preparation of binding land use plans (Bebauungsplan). The Flächennutzungsplan is prepared and approved by the municipality, but the Regierungsbezirk ensures its confirmity with higher-level plans and policies. Compare with Plate 18 and Figure G10(a).

334

**Plate 26    Bebauungsplan (Binding Land Use Plan) –
Hovestadtstrasse, Hagen, North Rhine Westphalia**

Plate 26 illustrates the detailed specification of land uses,
intensity of land use, building lines, car parking, open space
and infrastructure shown diagrammatically in Figure G10(b).
The provisions are legally binding on all public and private
agencies. Compare with Plate 21.

# THE NETHERLANDS

HWE Davies

# N1.  THE BACKGROUND TO PLANNING AND CONTROL

## Introduction

N1.1  The Netherlands (population, 14 million, see Figure N1) is a decentralised unitary state with a written constitution under which all executive power is centralised in government, that is the Crown comprising King and Ministers, but in which authority is delegated to provincial and municipal governments through legisla-tion and legal instruments enacted by the States-General (parliament). It is a small, densely populated, highly urbanised country, allegedly the "most planned country" in Europe (Dutt & Costa, 1985 p. 1).

N1.2  The first planning legislation was in the Housing Act 1901 which started the process of legally binding plans and building permits. As its name implies, planning

**Figure N1   The Netherlands**

Figure N1 shows the twelve provinces. The three western provinces (North and South Holland, Utrecht) contain Randstadt Holland, the ring of towns comprising the four largest cities and other smaller towns. National planning policies have sought to direct growth away from the west to the north eastern provinces and Limburg (see Plate 29). The map also shows the two principal land reclamation and flood protection areas around Ijsselmeer (now Flevoland Province) and the Delta of the Rhine/Maas/Scheldt in Zeeland and South Holland.

was an offshoot of housing policy. Planning and the control of development involve all three levels of government and extend beyond housing into many other matters of land use and the environment. Today, the current legislation comprises (Brussaard, 1986; Staatsblad, 1986; De Vries-Heijnis et al, 1985)

(i) the Physical Planning Act 1962 (amended 1985). This creates the administrative framework for planning at all three levels of government; and the statutory procedures for the preparation and approval of national key planning decisions by government; *streekplannen* (regional plans), by the provinces; *struktuurplannen* (general framework, or structure plans), by the municipalities; and *bestemmingsplannen* (legally binding plans), also by the municipalities

(ii) the Housing Act 1962 (amended 1982). This, *inter alia*, places a duty on the municipalities to adopt their own *bouwverordening* (building regulations), and to issue *bouwvergunningen* (building permits) for all construction

(iii) the Town and Village Renewal Act 1984. This gives municipalities authority to prepare *stadsvernieuwingsplannen* (urban renewal plans) and *leefmilieuverordeningen* (residential environment regulations). They provide additional powers for the control of development in existing built-up areas as well as more positive powers for redevelopment and environmental improvement

(iv) the Monuments and Historic Buildings Act 1961. This gives central government power to list buildings of historic interest, and to designate conservation areas both of which powers become important instruments of control

(v) a lot of environmental legislation by which development may be controlled, either directly, by licensing, or through control through the housing and planning acts.

N1.3 The essential character of the system therefore is that the Planning Act sets up a structure of national, regional and local planning in which the higher tiers can impose administratively binding directives on the lower tiers, and in which the lowest tier is a plan legally binding on all citizens and government, implemented by the building controls of the Housing Act. Its purpose is to create the opportunities by which the 'destination' (that is, the buildings and uses) shown in the *bestemmingsplan* may be 'realised', that is, implemented within the controls imposed by government.

## The Constitutional Context of Planning

N1.4 The planning system thus has its origins and basis in several key features of Dutch history and society. In the first place, it is guided by the principle of *rechtstaat* (the constitutional state) embodying the concept of *rechtszeterheid* (legal certainty) which is fundamental to all Dutch law and administration (van der Cammen, 1984). This derives from Roman law transmitted through the Napoleonic code into the modern constitution of the country. It means that the state guarantees, by legislation, judicial control and public enforcement, a framework of rules which ensure for the citizen his basic rights. It means that the statutes are inviolable, and that there is a basic separation of powers: administrative authority is based exclusively on enacted law; an independent judiciary ensures due process and compliance with the law. The law is found only in the constitution, the legislation and the legal instruments, but it is interpreted and elaborated through jurisprudence.

N1.5 The implication of this for planning and the control of development is that the focus of control is actually on plan-making with reliance on detailed plans and regulations all of which have statutory force. In principle, proposals for development in accordance with a *bestemmingsplan* and the building regulations cannot be refused a building permit; those not in accord cannot be granted a permit.

N1.6 A second guiding principle is that of democratic accountability. The constitution as such does not guarantee the rights of property. They are found rather in the civil codes, but public administrative law is superior to private law. Thus individual rights are safeguarded through administrative law by building into the system opportunities for objection to draft plans and appeal against administrative decisions such as the grant or refusal of building permits.

N1.7 Furthermore, there is a very clear separation of powers under the constitution and the legislation even though this results in a very complex machinery of government. At each level of government there are separate bodies with sharply defined responsibilities: the legislature (parliament, provincial or municipal council); and the executive (cabinet, provincial or municipal executive). It means that there is a multitude of different departments or agencies at each level responsible for the various sectors of public policy, each deriving its authority from legislation. It means too that for judicial purposes there is a distinction between administrative law (which governs virtually all planning matters) and the civil and criminal law. The former is the responsibility of the very old-established, powerful Council of State, and the latter, the Supreme Court. The only court superior to either is the European Court of Human Rights.

N1.8 Physical planning therefore is very much a matter of vertical coordination between, and horizontal coordination within, the different levels of government. In practical terms, this means that the preparation and the implementation of planning policies are the responsibility of the executive at each level of government, whereas the approval of policy, and the enactment of legislation, decrees and regulations or plans is the responsibility of the elected bodies at each level. The interests and rights of the individual are then safeguarded by appeal to the municipal or provincial council and ultimately to the Council of State (Brussaard, 1979).

N1.9 A third, important characteristic of Dutch society and politics is the extent to which policy, at least in the field of housing and planning, has been built on a very broad consensus. Traditionally, Dutch society divided sharply into several distinct social groups which found their political voice in the various catholic, protestant, liberal and labour parties, none of which had an overall majority. This alignment of the parties has shifted but, over most of the modern period, coalition governments have been in power and the result has been the so-called politics of "accommodation" between different points of view in which a compromise is sought (Hamnett, 1985a). Combined with the constitutional procedures, legislation thus takes a very long time to reach the statute book from its first inception, usually in a government-appointed expert committee and the emergence of a consensus. Thus, for instance, the 1962 Physical Planning Act had its origins in the Van den Bergh State Commission of 1950; and the first moves which culminated in the Town and Village Renewal Act 1984 were started in government reports in the late 1960s with the first bill before the States General in 1976 (Amsterdam, 1983).

## Planning, Development and Land Policy

N1.10 Another key factor, crucial to an understanding of the planning system, is the geography of the Netherlands. It is the most densely populated country in Europe, but much of it is at or below sea-level, on land reclaimed from the sea, including the most heavily urbanised western part of the country. The preparation of land for urban development over most of the country is exceptionally expensive, to compensate for the poor ground conditions and the need for complex land drainage and protection from the sea (Thomas et al, 1983).

N1.11 As a consequence, throughout its history, there has been strict control of urban development at what is now the local, municipal level, with lots of direct involvement by the state. At its most intense, this involvement by the state can be seen in the special planning and development organisations set up for the reclamation of the Zuyder Zee and the Delta project. The former, the *Rijksdienst voor de IJsselmeer*, has now been replaced by the elected Flevoland province with six municipalities.

N1.12 Today state involvement means that many municipalities have an active land policy. The usual process for new urban development, whether for housing, industry or commerce, is for the municipality to acquire the land (compulsorily or by agreement, roughly at existing use value plus compensation); to prepare it for development including drainage, roads and infrastructure; and to sell or lease the land to private developers or

housing associations (Thomas et al, 1983). Thus, for example, the City of Amsterdam owns 75 per cent of all land in the municipality, including all the built-up areas outside the historic core, having retained the freehold; and 70 per cent of all Dutch urban development in 1973 was on land first acquired by municipalities (Amsterdam, 1983; Thomas et al, 1983). It is comparatively rare for major developments to be undertaken exclusively by the private sector, with the state's role confined to control. In so far as it does occur, it is likely to be on freehold land, for redevelopment.

N1.13 The involvement of the state in urban development is intensified by Dutch housing policy since 1901 (Hetzel, 1985). Dwellings fall into three categories (Thomas et al, 1983), namely

(i) those built under the Housing Act by municipalities or, more usually, housing associations, with a high level of state contribution to the cost of development (38 per cent of all new construction, between 1971 and 1984)

(ii) *premie* housing built for sale or rent by housing associations, financial institutions and pension funds and the private sector, with a moderate level of state subsidy (43 per cent, between 1971 and 1984)

(iii) free housing, built for sale or rent by the private sector, with no subsidy for the cost of construction, though still eligible for mortgage tax relief (19 per cent, between 1971 and 1984).

N1.14 The *bestemmingsplan* thus has a vital, formal role in the urban development process, going beyond its regulatory function. It is the trigger for the compulsory purchase and eventual disposal of the land. As such, it has to include a financial appraisal. It may also be accompanied by land development regulations which lay down the financial contribution which a private developer may have to make towards the cost of drainage and infrastructure. All of these have to be agreed by the responsible levels of government before building permits can be issued. Central and provincial governments are thus also involved. Municipalities were formerly given annual quotas of subsidised housing by central government, as part of its national housing and planning policies, and are now given financial allocations for construction under the housing acts. These crucially affect the financial viability of housing development and therefore the mix of free and state subsidised housing in a locality, and therefore the final content of the *bestemmingsplan*. Finally, the Netherlands has a system of land registration, with special land registry offices to record all land systematically. Land registrations are then shown on *bestemmingsplannen*.

N1.15 The consequence of this involvement by the state is that at every level, but especially the municipal level, much of Dutch planning is positive, oriented towards creating development opportunities and initiatives, especially for the extension of towns. It is done by the municipality acting chiefly in its own interest as far as possible, accepting national policies where it has no choice or they accord with local interests. The municipality is both promoter and controller, working closely with the private sector. Its key instrument, the *bestemmingsplan*, lays down a binding legal framework for development which has to be achieved, and which then remains in force until altered by due process. It can be argued that planning and development are therefore, in theory, hindered by a rigid and inflexible system linking plans and decisions which creates legal certainty at the price of inflexibility (Thomas et al, 1983). Nevertheless, this is not the whole story. Planning and, in particular, *bestemmingsplannen* are also used restrictively as the guardian of the existing environment, preventing change especially in rural areas.

## The Evolution of Planning

N1.16   Planning in the Netherlands started in 1901 in a very simple form, concerned only with planning the extensions to towns. Since 1945 it has gone through two upheavals, in each case a response to changing problems and priorities in national policy. Three themes in particular have dominated post-war Dutch planning (Van den Berg et al, 1982).

(i) Economic policy initially was concerned with post-war economic recovery focused especially on port and industrial developments in the western provinces; but increasingly became a matter of counteracting the economic imbalance between the congested, highly urbanised western provinces, the northern provinces, with relatively high unemployment as their agricultural base declined, and the southern provinces with a collapsing coal mining economy and other obsolete industries including textiles.

(ii) Housing and urbanisation policy concentrated at first on the absolute housing shortage, especially in the big four cities of Randstad Holland, and on the problem of urban congestion, urban sprawl and the need to conserve the remaining open land of the west, the so-called green heart; and, since the 1970s, the problems of inner city urban decline and the need for urban renewal.

(iii) Environmental concerns have risen in importance with atmospheric and water pollution, ecological balance and related matters, especially those generated by the industrial and urban development of Randstad Holland and the reclamation of the IJsselmeer and Delta projects.

N1.17   These concerns found expression in a series of three national planning reports in 1960, 1966 and 1973/1977. Each, but especially the third, gave formally approved guidelines for the national policy for urban settlements and the rural hinterland, identifying growth cities and centres with target populations, buffer zones and the like. These were intended to provide the framework for the provincial regional plans which, in turn, would be the criteria for the provincial approval of municipal plans (Hazelhoff, 1981).

N1.18   The first response to these problems came in the 1950s and eventually culminated in the two key acts, the Physical Planning Act 1962 and the Housing Act 1962, giving the modern system of planning and control. They created the structure, as well as the instruments, of control, covering land use as well as development. Within the constitutional constraints, and the Dutch tradition of state intervention, they were a response to the massive, post-war shortage of housing, the need for dispersal from the congested urban centres, and the need to conserve open land. These were the classic land use problems of growth, and the response was equally classic: strict regulation within a framework of certainty provided by legally binding plans.

N1.19   The second response came during the later 1970s and early 1980s and is only now beginning to see expression in the Town and Village Renewal Act 1984, the 1985 amendments to the Physical Planning Act, and the working parties now active on a revision of the building regulations. In physical terms, it was a recognition of changing priorities. Urban renewal and arresting the decline of inner city areas required different instruments from the straightforward *bestemmingsplan*, partly to ensure that funds would be forthcoming for the implementation of policies; partly to prevent unwelcome uses as much as to create the opportunities for realisation. In social terms, the changes in 1984 and 1985 provided the first steps towards ensuring wider, formal opportunities for public participation in the preparation of plans, going beyond the already existing rights of objection.

N1.20   Above all, in political terms, the present response is part of the trend towards deregulation, by giving more opportunities for a flexible implementation of plans. It follows the statement presented jointly by the ministers of physical planning and economic affairs to the States General in 1983 on the deregulation of physical planning and environmental management (Staatsblad, 1986: *Nota van Toelichting*). It forms part of a wider batch of measures for deregulation also affecting environmental hygiene, housing and building, and monuments.

N1.21   Taken together, these developments mark a potential shift in the planning and control of development away from its original guiding principles. The declared aim is still to maintain a degree of legal certainty but it is proving a very difficult concept to maintain in the face of demands for greater administrative discretion which in any case is essential for the efficient and effective operation of a planning system (van Gunsteren, 1976).

N1.22   In the opinion of at least one developer, active in residential and commercial development and used to working in partnership with municipalities, the planning system reflects the Dutch desire for detailed, democratic control. It is lengthy, but "a good plan takes time". The *bestemmingsplan* is the necessary condition for building to start, by giving developers as well as government a degree of certainty. It means that the limits of competition are known, and that is the essential, desirable characteristic of the system. But a partnership must be based on negotiation and compromise. The eventual, approved plan at its best should encapsulate the agreement between developer and municipality.

N1.23 Arguably, as will be shown in the report, the 1962 system was too rigid, the evidence being the very high proportion of all building permits for developments of all sizes which were issued either through the procedures for altering a plan, or virtually by ignoring the plan; in either case, the idea of guaranteed certainty was being lost. The new 1985 system is intended to bring back certainty in a modified form by creating more opportunities for controlled flexibility; whether it will do so, it is too soon to say.

## Outline of the Report

N1.24 This account is based largely on literature and experience of the 1962 system of the planning and control of development. Details are given of the new 1984/85 alterations to that system, but so far very few if any of the new-style plans have been prepared, let alone implemented.

The report continues as follows.

N2 describes the administrative framework of planning, identifying the various organs of the state at each level of government.

N3 tackles the problems of finding a definition of the use and development of land and buildings which are subject to control, and thus identifying the scope of that control.

N4 describes the instruments and procedures for control of development at the municipal level, including those for the alteration and elaboration of a plan, as these have become part of the control mechanism.

N5 reviews the preparation of policy as the source of authority for primary control at the municipal level. This means going down through the hierarchy, from national planning policy instruments, through regional plans and structure plans, down to the *bestemmingsplan*. The *bestemmingsplan* is the basic instrument of control but its content must not conflict with the directives of higher levels of government and thus primary control cannot be understood in isolation from those higher instruments of policy.

N6 discusses the challenges to control possible through appeal, ultimately, to the Council of State but only after earlier stages have been gone through. The very minor role of the civil courts is briefly mentioned.

N7 summarises the description of the system for the control of development in the Netherlands, and suggests some general conclusions.

**Appendix A** summarises those parts of a set of building regulations of most direct relevance to planning control; and **Appendix B** summarises the regulations in two contrasted *bestemmingsplannen*, an inner city area, and a semi-rural area on the fringe of a conurbation.

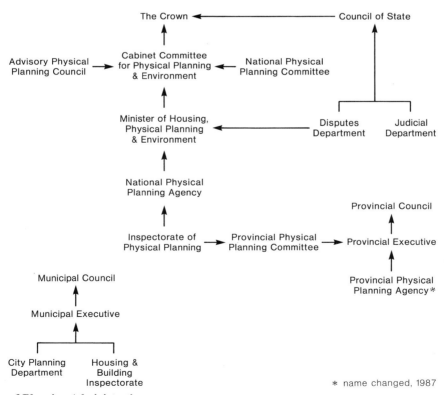

**Figure N2   Structure of Planning Administration**

Figure N2 shows the basic structure of planning administration involving the three elected tiers of government (national, provincial, municipal) and the Council of State, responsible for administrative justice.

## The Central/Local Relationship

N2.1 Government in the Netherlands is exercised at three levels: national, provincial and municipal (see Figure N2). The country has been described as "a decentralised, unitary state" (Brussaard, 1986). The different levels of government in the various parts of the country have independent legislative and administrative powers. At each level, the higher authorities have a supervisory power over the lower, thus ensuring a degree of national unity. Nevertheless, the supervisory power is not strictly hierarchical and *dirigiste*. The lower tier authorities can pursue their own policies and draw up their own regulations, and these need not necessarily be in accord with those of the higher tier; they can have their own policies provided they do not conflict with those of the higher tiers. But, if necessary, the higher tier authorities and, ultimately, national government do have the power to enforce compliance with their policies.

N2.2 The relationship between the different levels of government is therefore complex. In principle, the formal autonomy of each level of government is guaranteed by the constitution, with the precise relations in any particular field of government being defined in the legislation. The Council of State then provides a safeguard for ensuring the legal status and independence of local from central government, as well as that of the citizen in regard to government (Raad van State, 1987). In practice there is a much greater degree of supervision than is at first apparent, through the national government's monopoly of financial control over all three levels, through its advice, and through its informal political and administrative relations with provincial and municipal government (Thomas et al, 1983).

N2.3 The administrative structure is also complex and highly formalised, involving roughly similar structures at each level of government, but with greater complexity at the higher tiers. In particular, the legislative and executive functions are kept separate. People appointed as ministers cannot continue as members of parliament, though at lower levels of government the executive is headed by elected members of the provincial or municipal council. At the same time, the judicial function is split between administrative law, with its own tribunals and procedures, and the civil and criminal codes, through the various courts, the courts of appeal and the Supreme Court.

N2.4 In accordance with these relations, the administrative framework for planning is defined precisely in the

Physical Planning Act 1985, for the most part following the pattern established by the original Physical Planning Act in 1962.

## National Level

N2.5 The national level is as follows, much of its work in planning being on the horizontal coordination of sectoral policies between departments, and the vertical integration between the levels of government (Brussaard, 1986).

N2.6 *Staten-Generaal* (States-General), is the legislative body, comprising

(i) the First Chamber (elected by the provincial councils)

(ii) the Second Chamber (direct elections every four years). The elections have usually resulted in a coalition government, which since 1982 has been led by the Christian Democratic Alliance, the membership of the chamber being as follows (van Hezik & Verbeijen, 1983)

|  | 1983 | 1986 |
|---|---|---|
| Labour Party | 47 | 52 |
| Christian Democratic Alliance | 45 | 54 |
| People's Party for Freedom & Democracy | 36 | 27 |
| Others (8 parties) | 22 | 17 |
| Total | 150 | 150 |

The Standing Committee for Housing and Physical Planning oversees planning activities of government, especially through the debate on the budget for which an annual report is prepared by the ministry.

N2.7 The government is the executive body.

(i) The Crown, constitutionally the King and Ministers.

(ii) The Council of Ministers, (the Cabinet) which includes

(a) *Raad voor Ruimtelijke Ordening en Milieuhygiene*, (the Cabinet Committee for National Physical Planning and Environment). It is the chief coordinating body of ministers, chaired by the Prime Minister and with a membership involving nearly all ministers in the cabinet (Crince le Roy, 1975)

(b) many other overlapping committees including a new one on urban renewal.

(iii) *Ministerie van Volkhuisvesting, Ruimtelijke Ordening en Milieu-beheer*, (Ministry of Housing, Physical Planning and Environment). Ministerial responsibility for planning is defined in the Physical Planning Act. The Minister has to prepare a national policy on physical planning, submit an annual report on physical planning to the Second Chamber, and prepare and decide upon plans to be approved by the States-General after consultation in the Council of Ministers. The Ministry was originally concerned only with housing and planning, but more general responsibility for the environment was added in 1982, following legislation on environmental protection (Jones, 1984), and a general reorganisation of ministerial responsibilities.

To assist the Minister, the following are created by the Physical Planning Act.

(a) *Rijksplanologische Dienst*, (National Physical Planning Agency created under section 52 of the Act). This civil service department comprises four directorates, namely physical planning, physical development, social and economic development, and general administration. In addition, there are five regional inspectorates which act as the 'eyes and ears' of the Ministry in the provinces and municipalities, with formal, statutory powers to be consulted and to offer advice at the provincial level of government. Finally, the agency also contains a special advisory bureau (*Bureau van de Adviseeur t.b.v. de Raad van State*), with a special role in investigating objections to *bestemmingsplannen*.

(b) *Rijksplanologische Commissie* (the National Physical Planning Committee created under section 51 of the Act). This is an inter-departmental committee of officials with responsibility for coordinating the policies of different ministries as they affect physical planning.

(c) *Raad van Advies voor de Ruimtelijke Ordening* (Advisory Physical Planning Council created under section 54 of the Act). This is a very important part of the machinery of government, "a point of concentration for democratic participation of the community together with the government in the sphere of physical planning" (Brussaard, 1986). Its membership is defined by statute, and includes the representatives of different public interest and sectoral organisations ranging from housebuilders to nature conservation; technical and academic experts; and persons with experience of provincial and municipal government (Crince le Roy, 1975). Its duties chiefly are advisory and consultative, but are guaranteed by the legislation.

In addition, the Minister has other directorates within his Ministry, responsible for Housing, Environmental Protection, Government Buildings, and Land Registration.

(iv) *Ministerie van Welzijn Volksgezondheid en Cultuur* (Ministry of Welfare, Health and Culture). This Ministry is responsible for historic buildings and conservation areas. It contains

(a) *Rijksdienst voor de Monumentenzorg* (National Monuments Agency), the civil service department for this function

(b) *Rijsdienst Oudheidkundig Bodemonderzoek* (National Agency for Archaeological Surveys).

N2.8 The judiciary includes the Council of State, responsible for administrative law, and the Supreme Court and its subordinate courts, responsible for civil and criminal law.

(i) *Raad van State* (Council of State). This was founded in 1531 and is the highest advisory organ of state responsible for administrative justice and law. Its 24 members are nominated by the Minister of the Interior in consultation with the Minister of Justice, and are appointed by the King. It has three main duties (Raad van State, 1985).

(a) The Plenary Council is responsible for commenting on all legislation introduced in the States-General before its enactment.

(b) *Afdeling Geschillen* (Disputes Department) is responsible for hearing objections to the Crown provided for in the legislation including objections to *bestemmingsplannen*. Its decisions take the form of recommendations to the Crown.

(c) *Afdeling Rechtspraak* (Judicial Department) was created in 1976 by *Wet Administratieve Rechtspraak Overheids-beschikkingen* AROB (Administrative Tribunal on the Decisions of Public Authorities Act). This hears appeals against all forms of administrative decisions at every level of government including, for instance, the grant or refusal of a building permit.

(ii) The civil and criminal courts are involved in planning primarily in the enforcement of planning law, as opposed to the challenges provided for in the legislation which are dealt with by the Council of State. It includes (Koopmans, 1971)

(a) Hoge Raad (Supreme Court)

(b) the five Courts of Appeal

(c) the 19 District Courts

(d) the 62 Peace Courts.

## Provincial Level

N2.9 Until recently there were eleven provinces, of which the largest, Zuid Holland, has a population of 3.2 million, including Rotterdam and The Hague. A twelfth province, Flevoland with a population of 180,000, was recently created for the reclaimed *polders* of IJsselmeer; it is the smallest (see Figure N1).

N2.10 The provinces are the level at which much of the vertical and horizontal coordination of government is concentrated. They receive the policies and directives of the national government and ensure that they are carried forward for implementation by the municipalities. In similar fashion, they act as a channel through which the interests of the municipalities are passed up to the national government. But, above all, it is in the provinces that the policies of the different sectors of government are coordinated, between for instance housing, physical planning and transportation.

N2.11 In line with the separation of powers, each province comprises

(i) *provinciale state* (provincial council), the elected body of deputies, chaired by a provincial governor who is a political appointment by the Crown. It is responsible for approving all provincial policies, legislation and regulations

(ii) *gedeputeerde staten* (provincial executive), deputies nominated by the provincial council from among their number and chaired by the provincial governor. It is responsible for the executive work of the province, including preparing provincial level plans, and approving municipal level plans

(iii) *provinciale planologische commissie* (provincial physical planning committee), which combines the roles of the national committee and national advisory council with a similar structure and membership including the provincial heads of the different government departments and lay experts. It is thus able to ensure that the different sectoral interests are properly coordinated through physical planning.

N2.12 In most provinces, the technical civil service department for planning formerly was the *provinciale planologische dienst* (the provincial planning agency). However, the planning structure has recently been reorganised in Zuid Holland (Zuid Holland, 1987) and also has been, or is about to be, similarly reorganised in all of the other provinces except Noord Holland. Since 1 January 1987, planning has been the responsibility in Zuid Holland of *Dienst Ruimte en Groen* (literally the Spatial and Green Agency). It includes five departments, namely

(i) local planning, responsible for all work in advising on, and approving or revising, *structuurplannen* (structure plans) and *bestemmingsplannen* prepared by municipalities. It is thus the department chiefly concerned with planning and the control of development. It has an establishment of 12 planners, 15 lawyers (reflecting the legalistic basis of plans), five accountants (reflecting the financial aspects of plans) and 28 support staff

(ii) regional planning, responsible for all the work on preparing *streekplannen* (regional plans)

(iii) housing

(iv) recreation and conservation

(v) agriculture.

N2.13 The first aim of the reorganisation is to bring into one agency the responsibility for technical planning and policy, giving advice directly to the provincial executive. The provincial planning committee then reports separately to the provincial executive; thus ensuring that all matters affecting different provincial departments are properly coordinated before decisions are finally made by the provincial executive.

## Municipal Level

N2.14 The third tier is the municipalities of which there were 842 in the 1960s (Macmullen, 1979), but the number had reduced to 714 by 1986, including the three largest cities each with a population of over 400,000.

N2.15 The organisation of municipalities under the Municipalities Act 1851 comprises

(i) *gemeenteraad* (municipal council), the elected body of *raadsliden* (councillors) chaired by a *burgemeester* (burgomaster) who is a political appointment by the Crown. It is responsible for the approval of policies, plans and regulations, and appeals against the decisions of the executive. Cities such as Amsterdam and Rotterdam have 45 members of council

(ii) *college van burgemeester et wethouders* B&W (municipal executive) comprising the burgomaster and a number of *wethouders* (aldermen) appointed from amongst their numbers by the municipal council, and responsible for all executive decisions including the preparation and implementation of policies and plans. Nine of the councillors in Rotterdam become the municipal executive.

2.16 The municipality is responsible for all local government functions, organised in departments or agencies with a member of the municipal executive responsible for one or more departments, and various committees of council overseeing their functions. The range of functions and organisation of the city of Rotterdam, for instance, comprises (Rotterdam, 1984)

(i) town clerk and secretariat

(ii) city treasurer, with audit department, accounting office

(iii) executive departments
- fire                               - youth work
- police                           - arts
- health                          - libraries
- social services            - recreation
- education                    - zoo
- traffic
- planning and urban renewal
    - city development
    - building and housing inspectorate

- urban renewal project administration
- public housing
- housing affairs (registers, rents etc)
- public works department
  - construction
  - parks and gardens

(iv) operating departments
  - Port of Rotterdam Authority
  - Rotterdam Transit Authority
  - City Water Utility
  - City Energy Utility.

N2.17 In addition, some of the larger municipalities, including Amsterdam and Rotterdam, have directly elected borough or district councils. At first, in 1947, Rotterdam had only two, purely advisory councils. But under the Local Government Act 1961, borough councils can be given more power. Since 1973 Rotterdam has had nine such councils, with a number of delegated responsibilities including building permits, traffic regulations, local welfare services, neighbourhood open spaces (Rotterdam, 1984).

N2.18 Urban renewal is usually organised in decentralised project groups, with responsibility for the planning and implementation of urban renewal projects including the preparation of *bestemmingsplannen* and the allocation of housing within city-wide guidelines. In Rotterdam the groups have equal representation of council officials and local residents (Rotterdam, 1981).

N2.19 Municipalities rely very heavily on central government finance. The system in its present form had its origins in the 1960 Financial Relations Act, which established the Municipal Fund, generated as a fixed proportion of national tax revenues, and the formulae for its distribution by government (van den Berg et al, 1982). Subsequent legislation covers municipal borrowing rights, and the recent Town and Village Renewal Act for instance allocates funds for housing and urban renewal. Thus the very high proportion of municipal finance which comes through grant, and the control of capital borrowing for which ceilings are approved by the provinces on behalf of central government, give central government a very powerful lever over local government. It affects very directly the control of development by the municipalities as financial approval is part of the planning process.

**Table N2.1   Local Government Finance, 1960, 1975.**   (%)

|  | 1960 | 1975 |
|---|---|---|
| **Municipal revenues** | | |
| Local taxes and income | 11.4 | 7.1 |
| Specific grants | 44.1 | 54.8 |
| General Municipal Fund grant | 42.7 | 38.7 |
| Total (rounded) | 100.0 | 100.0 |

| **Local taxes and income** | | |
|---|---|---|
| Municipal supplement on property tax | 26.7 | 13.1 |
| Municipal supplement on personal tax | 12.1 | 9.4 |
| Taxes on streets, sewerage etc | 22.9 | 38.9 |
| Other taxes | 13.1 | 4.5 |
| Proceeds from municipal services | 25.0 | 34.0 |
| Total (rounded) | 100.0 | 100.0 |

*Source:* Thomas et al, *1983*

N2.20 The make-up of revenues in 1975 is shown in Table N2.1 and remains roughly the same in 1986. It comprises

(i) a general, or basic, grant from the municipal fund, allocated chiefly on the basis of population, area, density etc, with a refinement element for special local circumstances such as ground conditions, historic buildings etc

(ii) special, or specific, grants from central government departments for particular services including police, education, social welfare, roads, public transport etc

(iii) local property and other taxes

(iv) net revenues from municipal operations etc.

N2.21 Thus municipal responsibility for planning and the control of development under the municipal executive is usually as follows.

(i) Structure plans and *bestemmingsplannen* are prepared by the municipality or, in small municipalities, by town planning consultants. In Rotterdam, for instance, they are prepared by the city development division of the planning and urban renewal department, with a staff of 180.

(ii) The various urban renewal plans are prepared by project groups, and a coordinated urban renewal project administration in Rotterdam.

(iii) Building and construction permits are issued by the municipal executive on the recommendation of the *bouw-en-woningtoezicht* (building and housing inspectorate), though there may be some delegation to elected borough or district councils within the city. The Rotterdam department, which is part of the planning and urban renewal department, celebrated its centenary in 1986. It has a staff of 180. But the department in, say, Albrandswaard (a small semi-rural municipality south of Rotterdam with a population of 13,600) has a staff of only two inspectors.

(iv) In addition, a number of specialised agencies and advisory bodies have to be consulted at various stages in the planning and control process. Their functions and memberships are defined in national legislation or municipal regulations. The larger municipalities appoint their own organisations, but smaller municipalities may combine to establish joint arrangements. Key organisations in the Rotterdam area, for instance, include

(a) *welstandscommissie* (aesthetic committee). The requirement for such committees is contained in the model building regulations issued by *Verening van Nederlandse Gemeenten* (the Association of Netherlands Municipalities) and adopted by virtually all municipalities (Verening van Nederlandse Gemeenten, 1986). Rotterdam and Capelle aan/den IJssel (population 50,000) have their own committees but the other 14 municipalities in the area have a joint committee. The committees are made up of architects and others nominated by the municipality. Their function is to comment and give independent advice on all applications for building permits

(b) *agrarische adviescommissie bouw- en aanlegvergunning* is a committee of agricultural experts offering independent advice on applications for building and construction permits for agricultural enterprises, according to regulations in the *bestemmingsplan*, (Albrandswaard, 1986)

(c) *landilijke gebeiden en kwaliteitzorg* is a committee of landscape experts with a similar function to (b), on all applications in specified areas of landscape quality and natural environment

(d) *Dienst Centraal Milieubeheer Rijnmond*, (Central Environmental Control Agency) was originally part of Rijnmond council but is now supported by the 16 municipalities in the area. It is responsible for monitoring atmospheric pollution under the Air Pollution Act and other nuisances under the Nuisance Act, and issuing licenses in both connections.

## Administrative Reform

N2.22    Three issues have dominated recent discussion about the administrative reform of local government. The first has been the number and size of the provincial councils, with proposals for an increase in number from 11 to anywhere between 17 and 25. The demand for reform has come chiefly from the provinces themselves and academic commentators. Its motivation lies in the enlarged scale of urban problems, suburbanisation and the inner cities, and financial redistribution. The proposals got as far as a bill on administrative regions in 1971, in which a system of five regions for the country was proposed, but failed to win support (Hamnett, 1975). The City of Amsterdam has also proposed the idea of a city province covering all municipalities within about 20 km, to be responsible for the preparation of a structure plan, land allocation, housing, environment, traffic and transport and economic development (Amsterdam, 1983).

N2.23    The only supra-municipal authority actually to be created was the Rijnmond council, for Rotterdam and the 23 surrounding municipalities, later reduced to 15, with responsibility for supervising and coordinating policies in the vicinity of Europoort and the Nieuwe

Waterweg especially on waste disposal, water matters and pollution. It was created in 1964 and lasted twelve years, being abolished in 1986, chiefly on the grounds of its ultimate lack of power over the planning policies of the smaller municipalities. The Second Chamber finally rejected the idea as it was clear that Zuid Holland was not to be divided into two provinces, and Rijnmond would therefore create a permanent fourth tier of government.

N2.24    The second issue is that local government is currently being reorganised by the amalgamation of smaller municipalities to give a minimum size of 10,000 population. The effect will be to reduce the number of municipalities in Zuid Holland for instance from 160 to 100. Nationally the effect so far has been to reduce the number of municipalities from 842 in the 1960s to 714 by 1986. The main period of reorganisation was between 1968 and 1978 but even by 1986 nearly half of the municipalities have a population of less than 10,000 as the following figures show

| | |
|---|---|
| below 2,000 | 27 |
| 2,000 - | 126 |
| 5,000 - | 194 |
| 10,000 - | 190 |
| 20,000 - | 125 |
| 50,000 - | 35 |
| 100,000 - | 10 |
| 150,000 - | 4 |
| over 250,000 | 3 |

At the same time as amalgamation reduces the number of very small municipalities, the four largest cities of Randstad Holland are losing population at a rapid rate, with many implications for planning policy.

| | 1971 | 1986 |
|---|---|---|
| Amsterdam | 820,000 | 679,000 |
| Rotterdam | 679,000 | 571,000 |
| The Hague | 538,000 | 444,000 |
| Utrecht | 278,000 | 230,000 |

N2.25    The third matter has been that of decentralisation within the larger municipalities. Both Amsterdam and Rotterdam have used powers under the Local Government Act 1961 to create directly elected borough councils, with delegated powers as noted above. However, Amsterdam has wished to go further, giving its borough councils full responsibility for the preparation and approval of *bestemmingsplannen* within the framework of a *structuurplan* prepared by a proposed city province (Amsterdam, 1983). The intended scope of decentralisation is very great, involving about three-quarters of all municipal activities other than finance and major regulations. The aims are greater efficiency and bringing local government closer to the people.

N2.26    Local government reorganisation is progressing but so far there has been no progress in increasing the number of provinces, and the decentralisation within cities remains something specific for each municipality within the power of the Local Government Act.

## Problems of Definition

N3.1   There is no single, universal, comprehensive definition of development for planning purposes. The Physical Planning Act talks about, but does not precisely define, physical planning. Historically, since 1901, it has been shifting beyond its original boundaries of the built environment into land use, and, later, socio-economic and cultural fields. Thus the key government memorandum on physical planning which preceded the 1962 Act defined it as "The deliberate state control of the process due to the interaction between the land and those living on it ... aimed at making the given environment more subservient to the promotion of human prosperity, and beyond it to human happiness" (quoted in Crince le Roy, 1975).

N3.2   Although there is no all-embracing definition, in practical terms physical planning in the Netherlands focuses on the promotion and control of the use of land and buildings, and construction in the built environment. The legal basis for the definition of development is found in a wide range of planning and housing instruments, rather than in a single source. Furthermore, there is no universal definition. The national legislation gives responsibility for the exercise of legally binding control over development to the municipalities. Each municipality can adopt its own definition and, even within a municipality, the definition will vary depending on the legal status of the plans and instruments covering different areas.

N3.3   The two principal types of instrument which every municipality is required to prepare and adopt and which thus contain the definitions of development are

  (i) *bouwverordening* (building regulations) for the whole of the municipality. Section 2(1) of the Rotterdam building regulations under the Housing Act 1962, for instance, states that building covers

> "any construction of sizeable proportions made of wood, stone, metal or other materials which is either directly or indirectly linked with or based on the ground in its location"

  (ii) *bestemmingsplannen*, literally the destination plans covering most of the municipality, and comprising the plan itself and regulations. Section 10(1) of the Physical Planning Act states that the *bestemmingsplan* shall

> "deal with the allocation of the land covered by the plan to the degree required, consistent

with good physical planning, and where necessary, in view of such allocation, provisions shall be made regarding the use of the ground covered by the plan and regarding the buildings erected thereon. Only for urgent reasons may such provisions include restrictions on the most efficient use of the land and they may not contain any requirements regarding structure of the agricultural industry. Water shall be considered as land covered by water".

N3.4   Development therefore is what the *bestemmingsplan* or building regulations say it is, separately in each municipality or, indeed, in each separate area covered by a particular plan. However, the Housing Act 1962 does indicate what the building regulations are intended to cover; and the Physical Planning Act and the supplementary Physical Planning Decree 1985 give very general guidance on the *bestemmingsplan*. In both cases challenges to the courts (discussed in paragraph N6.10) have either confirmed or, very much more rarely, dismissed certain items from effective control. But in such challenges, the test has been the strict, legal interpretation of the acts and decrees, as the constitution of the Netherlands makes the legislation supreme and does not guarantee property rights.

N3.5   The definitions of development in the building regulations and *bestemmingsplannen* are therefore mandatory and legally enforceable. They define what is legally permitted, whether it be the building standards and planning criteria which have to be satisfied according to the regulations in both documents, or whether it be the precise pattern of land uses shown on the plan. They are therefore very rigid in their legal certainty according to basic constitutional principles. In practice, they are workable only in the extent to which the regulations and the plans also contain provisions giving some degree of flexibility, and allowing some discretion in their application, namely

  (i) the scope and procedures for the 'elaboration' of certain, *globaal*, or more general, regulations which are expressed deliberately in terms of *hoofdlijnen* (headlines) needing to be amplified in detail in order to be effective as legally binding definitions

  (ii) the arrangements under which exemptions may be granted from the strict interpretation of the regulations.

N3.6 The plans and regulations at first sight are not only very rigid, but their definitions of development are also potentially very varied as they are the primary responsibility of 714 individual municipalities. Nevertheless the variety is more potential than real. In the vast majority of municipalities, the plans are actually prepared by a relatively small group of planning consultants. The professional education of planners, and their networks, is another force for consistency. More importantly, the legal procedures for the approval of plans and regulations by the provinces and, if challenged, the Council of State and the Ministry of Physical Planning, also ensure consistency of definition.

N3.7 There are, as well, strong tendencies towards a unification, or harmonisation of definitions, partly for administrative efficiency, partly in the trend towards deregulation. Thus ever since 1965 the Association of Netherlands Municipalities has been devising model building regulations which to a large extent are now followed by virtually all municipalities. The government is now reinforcing the harmonisation thus achieved by deregulation. Its aim is to introduce by 1990 a new building decree which will specify in global terms the scope of the building regulations, with less reliance on detailed specifications (Verening van Nederlandse Geeenten, 1986). However, the new draft decree has been heavily criticised and work is still in progress on trying to arrive at an agreed formula.

N3.8 Similarly, following the 1985 amendments to the Physical Planning Act, the government is preparing a memorandum on the form and content of *bestemmingsplannen*. One of its aims is the simplification and standardisation of definitions of development as part of the process of deregulation (Ministerie VROM, 1986). The 1985 Act itself started the process, by extending the opportunities for flexibility in the drafting and implementation of *bestemmingsplannen*.

N3.9 The rest of this chapter therefore gives an indication of the scope and definitions of development by reviewing (a) the building regulations, as illustrated by those of the City of Rotterdam and the model building regulations; (b) *bestemmingsplannen*, chiefly drawing on plans for an inner urban area in The Hague and for a small, semi-rural municipality, Albrandswaard, on the edge of Rotterdam (see Appendix).

N3.10 In addition, three other matters need to be discussed for an understanding of the potential scope of development control. They are

(i) listed buildings and conservation plans, prepared under the Monuments Act, 1961

(ii) urban renewal area plans and environmental amenity regulations, prepared under the Town and Village Renewal Act, 1984

(iii) other licensing and control systems chiefly covering environmental questions about the use of land and

buildings, under a long list of environmental legislation.

## *Bouwverordening* (Building Regulations)

N3.11 Building regulations were approved by an increasing number of municipalities during the nineteenth century until the Housing Act 1901 made their adoption obligatory on all municipalities. Currently, the requirement is contained in the Housing Act 1982, which amends the major 1962 Act (Ministry of Housing, 1982b). Under section 2(1) of the Act, every municipality has to have a set of building regulations covering its entire area, and approved by the provincial executive. The Act states that the regulations may cover the construction of new buildings and the maintenance of existing buildings; the use of buildings; and their demolition, wholly or in part. They must also stipulate the procedures for control and enforcement.

N3.12 The purpose of the regulations is to ensure the erection and maintenance of technically sound buildings, and to control general and detailed aspects of the physical environment. A full summary, concentrating on planning matters, is given in Appendix (a).

N3.13 The building regulations include a wide range of matters relating to physical planning, as well as those more purely concerned with building construction and maintenance. In summary, the planning matters are

(i) the design and layout of buildings

(ii) demolition

(iii) conversion for residential use

(iv) existing and planned use of buildings and open spaces

(v) matters for further elaboration in supplementary regulations
  - car-parking
  - advertisements.

### Design and Layout of Buildings
N3.14 All new building construction is controlled. The planning considerations for which there are, for the most part, quite specific and precise criteria or standards include

(i) building lines, front and rear; the space about buildings in relation to each other on and off the site; the height of buildings. That is, a building must not interfere with existing or planned surrounding buildings and their compliance with regulations (articles 32,33)

(ii) the on-site design, layout and construction of roads and open spaces

(iii) the siting of a building in relation to the type of road frontage and access

(iv) the 'outward appearance', or aesthetic control. A building "in itself and in conjunction with the surrounding existing or planned development ... (must comply with) ... reasonable aesthetic requirements" (article 34). This can be very detailed, down to the colour of external painting. However, it is a matter of judgement, as to how, and how far, control is exercised (see paragraph N4.4).

N3.15 These regulations apply to all building construction wherever it is located. The only qualifications are

(i) certain building is exempted from needing permission, though it still has to comply with the regulations. This covers

   (a) fences, aerials, small non-residential buildings

   (b) non-commercial greenhouses, buildings for public utilities

   (c) on-site roads built by specified public authorities

(ii) other, minor construction within specified limits may be given a permit despite contravening the regulations, at the discretion of the municipal executive and subject to various procedures for objection by neighbours.

*Demolition*
N3.16 The demolition of buildings without the permission of the municipal executive is prohibited, and in general is likely to be permitted only if actually so ordered, or if it is also covered by a permit for a new building. This is probably a standard regulation in the larger cities and historic municipalities but may not be in widespread use elsewhere. However, the municipal executive does have power to grant a permit for a minor demolition.

*Conversion for Residential Use*
N3.17 There are special regulations in certain larger cities selected by government controlling the conversion of a dwelling into apartments, and the conversion of non-residential buildings into residential use. In each, the primary legislation is in the housing acts. There are also regulations under the Empty Property Act 1981, giving certain municipalities the power to require all vacant dwellings to be registered and, if necessary, requisitioned and allocated to people on the housing waiting list (Ministry of Housing, 1981g).

*Existing and Planned Use of Buildings and Open Spaces*
N3.18 This is a very complex matter, further illuminating the definition of what might be controlled under the building regulations. The rules are laid down in the building regulations, article 352, although in any dispute the *bestemmingsplan* takes precedence over the building regulations.

(i) Any use of land or buildings, and associated construction not being buildings (eg, a surfaced car park), contrary to their *bestemmings* (destination) in an earlier plan prepared under the 1901 Act or to that

implied by their construction, unless superseded by a later *bestemmingsplan*, is prohibited. The prohibition extends in rural areas to non-agricultural use of land, and all agricultural buildings. The only general exemption to this can be for various very short-term, temporary uses.

(ii) However, this prohibition can be relaxed, and an exemption given from the regulation under certain circumstances including if the prohibition would prevent, without good reason, the most efficient use; if the building or land can no longer be sensibly used for its original destination; or if the exemption would not have far-reaching and irreversible consequences for the original destination. Examples which have been tested in the courts include the continued use of a factory to prevent it becoming vacant, or the use of a disused barn for car repairs. Conversely, the continued use as a surgery of part of a dwelling in a predominantly residential street was refused permission, as being contrary to the policy of the municipality, and the refusal was upheld on appeal.

(iii) In general the prohibition under the building regulations relates only to uses which require construction. A change of use which does not require construction, and which involves a use which does not conflict with the original design of the building, may take place without a permit unless expressly prohibited by a subsequent *bestemmingsplan*. It was by this means, for instance, that in one case a supermarket came to occupy a factory building in an area shown for commercial purposes on the *bestemmingsplan*, exposing a serious loophole in the control of out-of-town shopping centres, since rectified by provincial directives (Borchert, 1986). However, some local regulations have been made more strict. For instance, the Rotterdam building regulations prohibit all changes of use in certain parts of the inner city near the central area.

(iv) Specifically, the use of vacant land for industrial purposes is prohibited, as is the use of vacant land in rural areas for non-agricultural purposes.

(v) Existing uses may continue, provided that they were lawful when this particular regulation came into force and the use continues unchanged. But existing uses may not be used to prevent proper realisation of the destination given in the plan. However, this regulation can be waived, and permission given if there is good reason.

(vi) The only general exceptions to these rules are that space within the curtilage of a building may be used without permission for parking, loading and unloading; and permission may be given by the municipal executive for a limited period for exemption from the requirement to comply with the *bestemmingsplan*.

*Matters for Further Elaboration*
N3.19 The model building regulations specify two other

matters which are subject to control, but on which the municipal executive has to formulate and approve local regulations. They are

(i) car-parking standards, which are usually in a *bestemmingsplan* but for which Rotterdam for instance has local regulations defining ratios for different uses in different parts of the city

(ii) advertisements, for which Rotterdam has devised a classification of different types of streets and highways, and set standards for each. The only exemption to this control is for small, unilluminated advertisements.

N3.20 In summary, the discussion shows that there is no easy or general answer to the question of how development subject to control on physical planning grounds may be defined according to the building regulations. The answer can be found only by close study of the regulations, including the model regulations and the appeal decisions on their interpretation. But basically the definition covers all building construction, and its scope extends to the site planning, layout and design of buildings. The standards and criteria are specified in the regulations, in most cases with numerical precision allowing for value judgement only in aesthetic matters. If the regulations are not satisfied precisely, a building permit strictly speaking will not be issued, and any contravention may be subject to enforcement. The definition furthermore may extend to the demolition of buildings. The only exceptions not requiring permission are for very insignificant matters, with a further set of minor divergences from the regulations which may be exempted from the need to comply, though in the latter case a permit is nevertheless necessary.

N3.21 The problem comes with the use of land and buildings, including uses and changes of use. These may be controlled under the building regulations, but control is exercised mainly through *bestemmingsplannen*. Thus the definitions and scope of control over use are best understood through the *bestemmingsplan*.

## Bestemmingsplan (Destination Plan)

N3.22 The *bestemmingsplan* is the legally binding plan. It gives a detailed description of the planned use and physical form of new development or redevelopment, in effect thus defining the permitted development as any non-conforming development is illegal, and liable for enforcement action. That is, the plan describes in whatever is the desired level of detail the ultimate 'destination' of land use and built form to be 'realised'. It follows that it is the plan and regulations themselves which uniquely define development in any particular case.

N3.23 To a large extent, the control of development is through the procedures for the preparation, approval, implementation and alteration of a *bestemmingsplan*, although the actual instruments of control are the

*bouwvergunning* (building permit) and the *aanlegvergunning* (construction permit), for development not being actual building construction.

N3.24 The municipality is responsible for the preparation of its own plans, working through the procedures under the Physical Planning Act described in section N5. It is obliged to prepare *bestemmingsplannen* for its entire non-built-up area, and may, if it wishes, prepare plans for its existing built-up areas. The effect of this is that, in theory, the entire built-up area developed since about 1901 and all undeveloped land outside the built-up areas is covered either by a *bestemmingsplan* or its predecessor, the *uitbreidingsplan* (extension plan). The built-up area existing in 1901 is not necessarily so covered. In fact, virtually the entire area of at least the western, heavily urbanised provinces, is covered by plans.

N3.25 Each municipality prepares its own plans and therefore there is no uniformity of scope or content. In the beginning, under the Housing Act 1901, the scope of the *uitbreidingsplan* was usually limited to defining street and canal lines. The Housing Act 1921 extended the scope to include land uses, and the later 1931 Act to cover building lines, front and rear, and the height of buildings (Thomas et al, 1983).

N3.26 The actual definitions and scope of development subject to control can be found by looking at the *voorschriften* (regulations) which accompany the plan. The plan itself then shows the various planned destinations in whatever detail is required. The plan may typically be to a scale of 1/1,000 in a heavily built-up area, or 1/5,000 in a non-built up area. The level of detail, through inset maps and diagrams which form part of the plan, can go down to distinguishing between a house and its garden, with separate regulations for each, or can identify individual floors in a multi-storey building.

N3.27 The precision of the definitions hinges on two factors.

(i) The desired level of detail of the eventual 'realised destination' of the plan which may be very precise or, in line with the 1985 Act, may be more generalised.

(ii) The degree to which the municipality wishes to retain some flexibility either to give a measure of discretion to the municipal executive in permitting a divergence from the plan; or to establish at first only the broad principles for development to be worked out in detail at a later date, closer to the time of implementation.

In both instances, the key words which recur in discussion about the Dutch planning system are *globaal*, meaning rough, or generalised plans; and *hoofdlijnen*, meaning 'headlines' or generalised statements about destinations.

N3.28 Although it is not possible to give a single, precise definition of development, however long, it is possible to describe in general terms the potential scope of development to be controlled, from an examination of *bestemmingsplannen* and legislation. The rest of this section does this, drawing on the *bestemmingsplannen* summarised in Appendix (b). They are respectively for an inner urban area and an urban fringe/rural area, thus spanning the range of situations (see Plates 27 and 28). Obviously, the definitions in any particular plan will vary, and the precise standards to be adopted will also vary. But the description gives a broad indication of the all-embracing character of control. It is described under six headings, namely

   (i) basic principles

   (ii) the urban built environment, including the non-agricultural activities in rural areas

   (iii) the rural, non-urban environment including agriculture, forestry and recreation

   (iv) infrastructure

   (v) 'associated construction not being building'

   (vi) the question of elaboration.

## Basic Principles

N3.29  The basic principle is that all uses and buildings contrary to the plan are illegal, including existing uses which do not conform to the destination in the plan unless covered by the transitional arrangements for the introduction of the plan. This means that except where new development or redevelopment is intended, the plan tends to be a description of the existing pattern of land uses and built form. Any divergence from the plan requires a permit which will be given only for very small departures, such as exceeding the specified height of buildings by a very small margin, or for very small non-residential buildings. However, there may be a provision that the municipal executive may grant an exemption from the regulations if a strict interpretation would prohibit without good cause the most efficient use of the site or buildings.

N3.30  The new section 18a in the Physical Planning Act 1985 however offers scope for relaxation. It gives a municipal executive power to grant exemption from the requirement to comply with the plan. Article 21 of the Physical Planning Decree 1985 spells out the details of exempted development. They cover, with a size limit for each in terms of height, or area, or plot cover etc

   (i) outbuildings in the curtilage of a dwelling, for the 'greater enjoyment' of that building, including garages, garden sheds, greenhouses etc

   (ii) extensions to the side or rear wall of a dwelling

   (iii) roof shelters, verandahs, dormer windows, and other minor changes to buildings which do not imply an extension of the built area, or a change of use

   (iv) garden furniture, pergolas, swings, sundials etc

   (v) aerials, fences

   (vi) certain constructions serving traffic, energy supply, telecommunications etc.

N3.31  The significance of this list is that it indicates the degree of control that is possible through a *bestemmingsplan*. Yet, even now, section 18a is not mandatory. It merely gives municipalities the ability to relax control by limiting the definition of development if they wish.

## Urban Built Environment

N3.32  The list of definitions covers the fairly typical list of urban land and building uses, though with a capacity for variety.

   (i) Residential refers specifically to the dwelling itself, within front and rear building lines shown on the map. A distinction may be made between single-family and multiple family dwellings; dwellings with and without attics; and the regulations stipulate height, width, floor area, etc. There are separate definitions and regulations for

      (a) that part of the plot (the *erf* or yard) within which various extensions to the dwelling, or the construction of sheds etc, may be permitted within defined limits of height, plot cover, and distance between buildings and plot boundaries. This is the area within which section 18a will have most effect

      (b) the rest of the plot (the *tuin* or garden) within which virtually no construction other than hedges or fences is permitted

      (c) the inner courtyard in multi-storey blocks (the *binnenterrein*) within which no construction is permitted.

   (ii) Retail is separately defined, though with a possible distinction between that solely in retail use and retail with residential above. The regulations are likely to cover the siting on the plot, building lines, height restrictions, etc. Formerly, in some plans retail was included under the general heading, commercial (including industrial), thus enabling supermarkets to be developed in industrial areas outside town centres. However, retail is now separately identified to ensure strict control (Borchert, 1979).

   (iii) Commercial uses, including offices, wholesale distribution and manufacturing, storage, commercial garages, etc may be grouped either as a single category or more usually as separate sub-categories. The regulations cover building lines, height, floorspace, plot cover etc and may permit one dwelling for use by the firm.

Commercial uses are also defined in terms of their environmental impact based on the severity of the environmental nuisance at source including noise, pollution, scale, etc; the spread of that impact; and its effect on dwellings. The Association of Netherlands

353

Municipalities has classified hundreds of separate types of commercial enterprises into six categories depending on their impact on residential areas. The effect is for the *bestemmingsplan* to define categories 1 and 2, for instance, as being permissible in particular areas; categories 3 and 4 being permissible, with qualifications and conditions; and categories 5 and 6 to be prohibited other than in specific areas (Ministry of Housing, 1986a).

The system has been endorsed by the Ministry which intends to recommend it to all municipalities "as a central instrument for the management of (industrial and commercial) activities in a *bestemmingsplan*". The list prepared by the Association provides a uniform starting point although municipalities may vary it to meet local circumstances (Ministerie VROM, 1986).

(iv) 'Special purposes' covers a wide variety of uses such as old persons' homes, schools, sports halls, post offices, welfare uses etc separately identified, if necessary, with regulations covering building lines, height etc.

(v) Parks and green spaces are also separately defined, the latter referring to a mixture of shelter-belts, roadside verges and incidental open spaces on which any development other than that incidental to its use and enjoyment (such as litter bins, benches etc) is prohibited.

(vi) 'Mixed uses' is shown as a single category on the inner city example (Plate 28). It refers to groups of buildings where the mixture of uses on different floors is so complex as to defeat cartographic skill. The regulations therefore define in a diagram the permitted uses on each floor in each building. The actual uses are those defined previously.

*Rural Environment*
N3.33 The rural example of a *bestemmingsplan* (Plate 27) shows that virtually all of the urban list of definitions above applies. But, in addition, there is a list of more purely rural definitions which are subject to control.

(i) The use of land for crops, animal husbandry, market gardens on open ground, orchards, forestry etc in areas allocated for these uses in general does not require permission, though the definition for Albrandswaard explicitly excludes intensive livestock farming under cover, and greenhouses more than 1 m high, a major land use in the Netherlands. The only constraint on use is in areas also designated as being of archaeological, landscape or natural ecological value (see below). However, within the areas allocated for agricultural enterprise, there are strict controls on agricultural buildings, defining limits governing their location in relation to roads, height, floor area, width, space between buildings, etc such that

(a) only one agricultural dwelling is permitted for each enterprise, although the municipal executive does have power to permit a second one under certain circumstances; and to permit a change of use from an agricultural to a non-agricultural dwelling, under strict conditions

(b) all of the agricultural buildings must be located on a single plot not more than 1 ha in size, and not more than 100 m from a road. Additional buildings may be located elsewhere, depending on the circumstances, and subject to conditions

(c) elsewhere, there is an absolute prohibition on any buildings of any form.

(ii) A number of other related enterprises are separately defined in the regulations and on the map, each with limits on the number, size and location of buildings, plot cover, etc. They include auxiliary agricultural enterprises, agricultural storage enterprises and commercial nursery gardens.

(iii) The map and regulations define two other categories of area where there are additional controls or limits on development including agriculture. Any development within these areas requires the appropriate building or construction permits from the municipal executive, on the advice of the relevant experts. They are

(a) areas of archaeological value in which any digging other than normal ploughing, ditching or piling is prohibited without permission

(b) areas of landscape and natural (ie ecological) value where roads, paths and car parks, earth moving, tree planting or felling, and drainage are among the activities which are prohibited without permission.

(iv) There is also one example of a country estate where the controls are particularly stringent, in order to ensure its preservation and conservation.

*Infrastructure*
N3.34 Infrastructure is a defined use in terms of a *bestemmingsplan*, with regulations governing the development permitted in those areas, confined usually to that associated directly with the particular infrastructure. The uses, or definitions, fall into two broad categories.

(i) The physical use of land for traffic and transportation, defining separately, for instance, residential streets, other streets, major traffic roads; footpaths, cycle paths; bus/tram lanes; railways; canals; and dykes or embankments.

(ii) The lines of overhead high-tension electricity and various underground pipelines including a very wide corridor for major pipelines serving Europoort. Each has its own defined width within which other specified uses may be permitted subject to the conditions necessary for safeguarding the infrastructure.

## Associated Construction Not Being Building

N3.35 Particularly in rural areas, there is a wide variety of development which strictly does not involve building construction, and therefore cannot be controlled only by building permits, but which does need to be controlled in order to safeguard the ultimate destination of land according to the *bestemmingsplan* (Brussaard, 1979). The Physical Planning Act therefore provides that in areas to be defined in the *bestemmingsplan*, a construction permit has to be obtained from the municipal executive for the development specified in the regulations. The list is potentially very long and can include (Ministerie VROM, 1985)

(i) roads, airfields, hard surfaces

(ii) piling, timbering, quays, sea wall protection

(iii) mineral excavation, surface mining

(iv) underground pipelines, cables etc

(v) altering ground levels, land reclamation, grading

(vi) planting and felling of trees and other vegetation

(vii) deep ploughing, land drainage

(viii) use of fertilisers, pesticides.

## The Elaboration of Bestemmingsplannen

N3.36 The point about elaboration is to give the municipal executive the authority to approve the working out in detail of the general principles set down in global fashion in the original plan. Thus it forms an important element in at least three types of situation, namely urban renewal; development of extensions to towns and villages; and the landscape treatment etc of rural areas. By this means the definition of development can be greatly increased. Two examples illustrate the point, further matters for increasing flexibility being discussed in paragraph N5.43 as they do not affect the definition of development as such.

(i) For urban renewal, and indeed new urban development, the global definitions in the plan can specify

  (a) the types and sizes of dwellings in terms of house price (and therefore eligibility for subsidy under the Housing Act), single/multiple family dwellings, and number of rooms

  (b) the range of proportions of dwellings in each category

  (c) the maximum and minimum floor areas for specified uses such as residential, retail, commercial etc

  (d) schematic, or precise, indications of the pattern of land uses, densities, access and vehicular and pedestrian circulation.

(ii) For landscape and forestry areas where the purpose is to conserve or rehabilitate areas of special landscape value, or the reafforestation of other areas, the global regulations again demonstrate the potential range of control, covering

  (a) the maximum and minimum areas for tree planting, outdoor recreation and agricultural uses

  (b) the maximum and minimum areas of floorspace for recreation, maintenance and management, with standards of construction

  (c) the layout and construction of roads, parks, car parks, shelters etc.

N3.37 In summary, therefore, the potential range of development which can be controlled through the *bestemmingsplan* is very wide. It not only covers the use of land and buildings in fine detail but also their layout and design in terms of height, width, space between buildings etc. It overlaps and is complemented by the definitions under the building regulations. However, the plan itself in effect controls only the massing, bulk and location of buildings. Aesthetic control, in terms of the outward appearance of buildings, is controlled through the building regulations rather than the plan.

N3.38 The scope of control can be extended to include a number of matters not covered by the building regulations, notably those affecting rural areas such as minerals, excavations and even certain aspects of agriculture. Tree felling and planting too can be controlled, if the *bestemmingsplan* so authorises, whether in urban or rural areas, using the device of construction permits.

## Listed Buildings and Conservation Areas

N3.39 A further definition of development subject to control covers the alteration, extension or demolition of buildings of architectural or historic interest, and conservation areas (van Voorden, 1981). It had its origins in 1918, and the appointment of a National Committee for the Preservation of Historic Buildings (Amsterdam, 1983) and is currently established under the Monuments and Historic Buildings Act 1961 (Ministry of Cultural Affairs, undated). The responsible minister is the Minister of Welfare, Health and Cultural Affairs, formerly the Ministry of Cultural Affairs, Recreation and Social Welfare (Department of Monuments, undated).

(i) A register or list of buildings to be subject to control is prepared by the National Monuments Agency. The list includes most buildings from before 1850 and is currently being revised to 1900; the current total is about 40,000. Listing has the effect of prohibiting the demolition or use of a building in such a way as to endanger or disfigure it; and any alteration requires permission.

(ii) A register of town and village conservation areas designated by the National Monuments Agency is also maintained. These are groups or concentrations of listed historic buildings and their immediate surroundings. The areas are based on a typology of urban and rural settlements which identifies relatively intact historic structures, but boundaries tend to be drawn very tightly. The details of the area and the buildings for conservation are contained in a preservation order and plan with explanatory memorandum. Once approved, the area is subject to interim control until it has been incorporated into a *bestemmingsplan*. By 1979, 69 areas had been thus designated and protected for more than ten years although only 35 had been incorporated into a plan (van Voorden, 1981).

N3.40   In addition, municipalities may adapt their own local regulations for the preservation of historic buildings, or conservation areas. Thus the City of Amsterdam passed an ordinance in 1980, giving additional powers for preparing their own lists, and exercising control (Amsterdam, 1983).

## Urban Renewal Areas

N3.41   The larger municipalities have been working towards ways of improving their capacity to plan and control the redevelopment of their older areas, especially their older residential areas built during the nineteenth century. Amsterdam had 12 urban renewal areas by 1972, each with its own project group (Amsterdam, 1983). It became a priority for Rotterdam in 1974, and by 1984 the city had 20 urban renewal areas in those parts of its inner city not destroyed during the war, and in some of the older suburbs in the outer areas (Rotterdam, 1984). The available procedures were very complex, partly because of the financial issues, and the dependence on national government; partly because of the difficulty of applying planning instruments, such as the *bestemmingsplan*, which were designed for controlling new development rather than redevelopment; partly because of the need to involve local residents in the process. The whole issue was given greater priority in the Third National Report on Physical Planning. In fact, the first memorandum at the national level about the planning, financial and procedural problem of urban renewal came in 1969; the first, draft urban renewal bill in 1974; and the eventual Town and Village Renewal Act in 1984 (Amsterdam, 1983).

N3.42   Under this Act, the definition of development for control once more depends on the content of the various policy plans or regulations which may be prepared by a municipality, all of which have the force of legally binding plans (de Vries-Heijnis, 1985). In summary, they are

(i)   *stadsvernieuwingsplannen* (urban renewal plans), providing the full regulations and procedures for renewal. The definitions of development for control are similar to those in the normal *bestemmingsplan*. The differences lie in the procedures and financial arrangements for implementation

(ii)   *leefmilieuverordeningen* (residential environment regulations) for areas without an urban renewal plan or *bestemmingsplan*, and threatened with urban

decline by the physical decay and dereliction of buildings, or the incursion of alien uses such as major offices or industrial development. The areas are meant to be mainly residential, though with mixed small businesses and commercial uses. In Amsterdam, for instance, they are the inner ring of canals (the *grachten*) and the areas immediately beyond of low income housing, none of which in general were covered by *bestemmingsplannen* or the earlier *uitbreidingsplannen*. The aim is to arrest the process of decay by a mixture of positive environmental improvement and strong controls to prevent unwelcome uses. The regulations therefore do not specify positive destinations, as in the *bestemmingsplan*. Rather, they define the uses and building which are not permitted, with particularly strong controls over demolition, stronger than in the normal building regulations. The regulations may cover buildings, uses and other construction not being building. They have a limited life of five years, with the possibility of a further five years.

## Other Environmental Controls

N3.43   Finally, reference must be made to environmental matters in identifying the scope and definitions of development subject to control. The Netherlands has a number of acts, or bills in preparation, dealing with such matters. They are a response on the one hand to the country's geography: its low-lying character, its situation at the mouths of three of Europe's largest rivers, its high degree of industrialisation and urbanisation, and its intensive agriculture. Land use and development must have regard to environmental impact. At the same time, they reflect the growth of the environmental lobby, acknowledged in 1982 by adding the word 'environment' to the Ministry's title, with responsibility for general coordination of government's policies and responsibilities in the field (Ministry of Housing, 1983f).

N3.44   The list is long. It includes (Ministry of Housing, 1985a)

Nuisance Act, pre-1969
Nuclear Energy Act, pre-1969
Surface Water Pollution Act, 1970
Air Pollution Act, 1972
Sea Water Pollution Act, 1977
Chemical Waste Act, 1977
General Environmental Provisions Act, 1980
Water Substances Act, 1981
Noise Nuisance Act, 1983
Soil Sanitation Act 1983
Soil Protection Bill
Extremely Dangerous Substances Bill.
Environmental Impact Statement Bill.

N3.45   In each case, the legislation defines activities for which a licence is required, or for which mandatory standards have to be met. Three in particular relate back to the definitions and scope of physical planning, to be controlled through building or construction permits.

(i) A *bestemmingsplan* may define areas within which residential and other sensitive buildings or uses in general are prohibited, or permitted only with conditions, according to standards defined in the Noise Nuisance Act.

(ii) A *bestemmingsplan* may also define areas within which specified commercial and manufacturing uses are prohibited, or are permitted only with conditions, according to the classification based on a synthesis of environmental issues and developed by the Association of Netherlands Municipalities for the guidance of its members (Ministry of Housing, 1986a).

(iii) A *bestemmingsplan* may also define other uses and buildings for which advice is needed from environmental agencies such as the Central Environmental Agency for Rijnmond before a building permit may be granted. This relates mainly to atmospheric pollution (Rijnmond, 1979, 1984). But the proposed soil protection bill, for instance, may provide a basis for control of the building materials to be used in development (Ministry of Housing, 1986a).

## Conclusions

N3.46   It will be clear from this chapter that any attempt to draw up a definition of development in the Dutch planning system is extremely difficult. It is not just that there is no single, universal definition nor that authority for defining development is left to the individual municipality. It is that the concept of development as in the English system appears to have no place in the Netherlands.

N3.47   For physical planning, development may be viewed as a process of change from one state of the environment to another, whether it involves physical construction or simply a change of use. One method of establishing control therefore is by reference to a single, non-flexible but comprehensive definition of development as a process. The future product of the process is judged on each occasion at the actual time of change, for each piece of development, on its merits by reference to the general purposes of the planning legislation. A particular plan, or even a statement of national policy, is there only to provide guidance, not to determine the outcome. This means that the definition of development is crucial as it provides the legal basis for control.

N3.48   The alternative method, actually used in the Dutch system, controls the process of development by reference to a very precise, legally certain, criterion contained in a particular, legally binding plan or regulation, usually the *bestemmingsplan* and the building regulations in combination. Together, they define what is permitted, the 'realised destination' of planning policy and control. The system therefore is plan-led, the definitions being specific to each plan, itself a piece of legislation. Thus it does not, and does not have to, rely on a general definition of development to provide the legal basis of control.

N3.49   The control of development under the Dutch system in principle is therefore very clear, certain and precise up to the time that the *bestemmingsplan* is fully realised, namely that the reality of the built environment corresponds to the plan. Thereafter, any further change requires alteration of the plan. Otherwise the process of development in theory has come to an end. Admittedly, the plans and regulations can allow for some relaxation, by defining the grounds for a possible exemption in the case of minor alterations and extensions to dwellings and other buildings or some changes of use.

N3.50   One consequence of this is that the definition and scope of development depends very much on the procedures for the preparation and approval of plans at every level of government, as well as the instruments for its control.

N4.1 Development, whether building construction, demolition or change of use is controlled by one of two methods under the Housing, Physical Planning, Monuments and Historic Buildings, and Town and Village Renewal Acts.

(i) Formally, it is obligatory under all circumstances to apply for and be granted a *bouwvergunning* (building permit), supplemented, if necessary, by an *aanlegvergunning* (construction permit) or demolition permit.

(ii) However, in effect control is more often exercised through the procedures under the Physical Planning Act for the elaboration of a *globaal bestemmingsplan* (section 11), or in anticipation of the declared preparation or review of a *bestemmingsplan* (section 19). In either case, the building or construction permit is issued as part of the process of plan elaboration, preparation and review.

## Building, Construction and Demolition Permits

N4.2 There are three basic instruments of control.

(i) The building permit, issued by the municipal executive, is required for all building construction. It can be refused only if the proposal is not in compliance with

   (a) building regulations

   (b) an approved *bestemmingsplan*, urban renewal plan or residential environment regulations

   (c) permission under the Monuments and Historic Buildings Act.

(ii) The construction permit is also required under special circumstances for construction not being building, and other action such as ploughing or tree-planting, where such development could prevent a piece of land becoming less suitable for its intended use in a *bestemmingsplan*, or is needed to maintain and protect an existing use in the plan. The circumstances and areas within which this applies have to be specified in the *bestemmingsplan*, or the decree for the preparation of a *bestemmingsplan*.

(iii) A demolition permit can be required under building regulations, urban renewal plans or residential environment regulations.

*Building Permits*
N4.3 A building permit must be granted within the

definitions laid down in the regulations or plans, with or without conditions, unless there are specific grounds for refusal, or in some circumstances unless there is objection from the provincial executive. The decision must be made within two months (with a possibility for an extension of time for a further two months). A refusal or conditions must have stated, explicit reasons and can be appealed to the municipal council and, if unsuccessful, to the Judicial Department of the Council of State.

N4.4 The normal procedure for a building permit is as follows, a fee of 0.9% of the cost of construction being payable to the municipality (see Figures N3 and N4).

(i) The application for a building permit under section 46 of the Housing Act is submitted to the building inspectorate of the municipality. Everybody wishing to undertake building works, including the municipality itself and all government departments, must apply (see Figure N5 for an illustration of an application form).

(ii) The building inspectorate checks the application

   (a) with other municipal departments and agencies as required by the plan or regulations. These can include housing departments and the regional housing inspectorate of the Ministry if the application relates to the conversion of non-residential to residential use, or the sub-division of a dwelling into apartments, both of which require a permit under the Housing Act in certain municipalities

   (b) with the agencies and committees responsible for agricultural, landscape or archaeological advice, or environmental licensing and control; or the public utilities, transport and infrastructure organisations

   (c) with the *welstandscommissie* (aesthetic commission), which makes its own recommendation direct to the municipal executive. If this is not accepted by the building inspectorate, the inspectorate has to obtain further, independent professional advice

   (d) with the National Monuments Agency. If the municipality is otherwise minded to grant a building permit, the decision must be deferred until the Agency has been notified with a period of one month for it to object
   - in the case of one of its listed buildings, if a

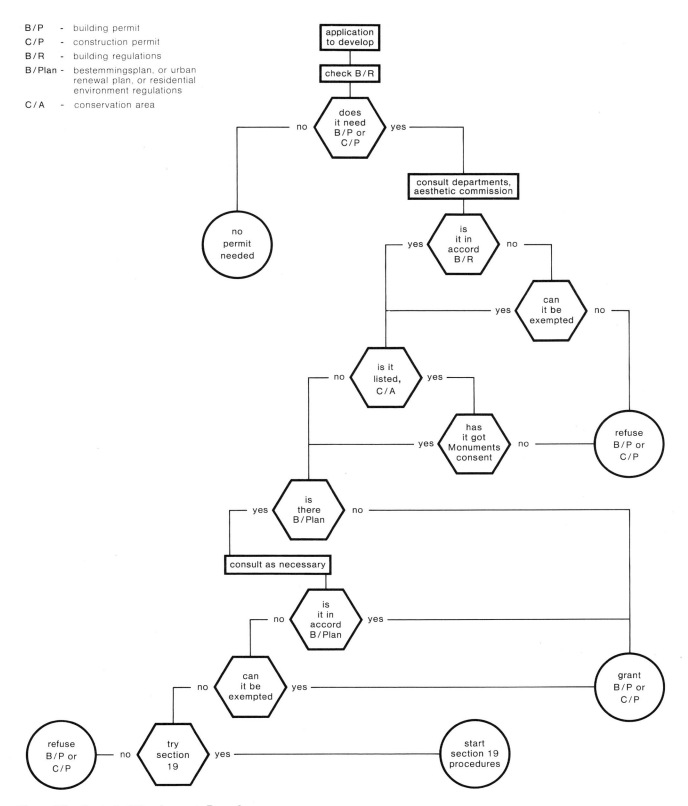

B/P   -   building permit
C/P   -   construction permit
B/R   -   building regulations
B/Plan -  bestemmingsplan, or urban
           renewal plan, or residential
           environment regulations
C/A   -   conservation area

**Figure N3   Control of Development: Procedures**

Figure N3 shows the procedures for the grant or refusal of a building permit, or a construction permit for associated construction (including certain agricultural operations etc.) in circumstances specified in a *bestemmingsplan*. In principle, the permit can be granted only if the proposal is in accord with the building regulations and, if there is one, with the

*bestemmingsplan*. There are limited circumstances under which exemptions can be granted for very minor variations. Otherwise, a building permit for a proposal in conflict with a plan can be granted only in anticipation of a formal review of the plan under section 19 of the Physical Planning Act (see Figure N6).

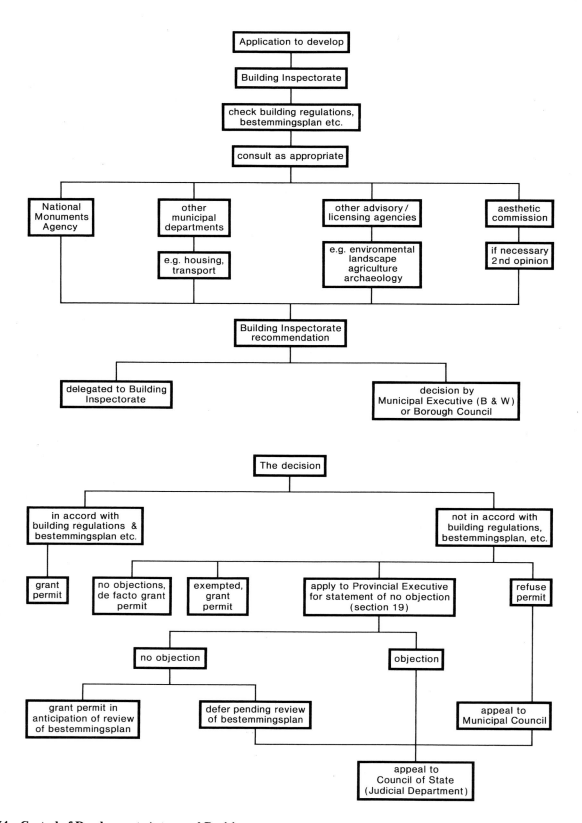

**Figure N4    Control of Development: Actors and Decisions**

Figure N4 shows, in the top half, the various consultations which the Building Inspectorate of a municipality has to carry out when dealing with an application for a building permit, and the making of the decision. The bottom part shows the different possible decisions which can be made, including appeals against a decision by the municipality.

permit has not been obtained from the Agency for its demolition, alteration or change of use
- in the case of an application in a designated conservation area not yet included in a *bestemmingsplan*.

(iii) The building inspectorate makes a recommendation to the municipal executive. Formally, the decision on a permit is made by the municipal executive or one of the aldermen deputed to act in this way. If the municipality has set up a system of elected borough or district councils, some decisions may be delegated to these, and, in some circumstances, some classes of decision may be delegated to the building inspectorate.

    (a) If the application is in accord with the building regulations and, if there is one, with the *bestemmingsplan*, urban renewal plan or residential environment regulations, then the municipal executive issues the permit.

    (b) If the application does not conform to either the regulations, or the plans, or both, either the permit is refused; or, the applicant negotiates for an exemption from the regulations and/or plans after which, if agreed, the permit is issued, with or without conditions. But the circumstances under which exemption can be granted are strictly defined within the regulations, as discussed below, under the heading of flexibility and control.

    (c) If the application does not conform to the *bestemmingsplan*, the applicant may apply for a permit in anticipation of a review of the plan under section 19 (see below).

    (d) In some cases, however, a building permit may be issued even if the application does not conform to the regulations or the plan provided it is for a minor development to which there has been no objection from neighbours, the aesthetic commission, or any other agency or department. This strictly is illegal but a blind eye is turned on grounds of common sense and pragmatism.

(iv) The applicant is notified of the decision, with any conditions, and issued with a permit. The work then may proceed but the building cannot be occupied or the new use commenced until the completion of works has also been certified.

(v) If the permit is refused, or granted with conditions, or not decided within the specified time, the applicant may appeal to the municipal council and, if still not satisfied, to the Judicial Department of the Council of State. In addition, aggrieved people such as neighbours can object to the grant of a building permit in the first instance to the municipal council and then the Council of State.

## Construction Permits

N4.5   The application for a construction permit under section 14 of the Physical Planning Act is made to the building inspectorate of the municipality and is issued by the municipal executive within the criteria specified in the *bestemmingsplan* or the decree for preparation of the plan. In addition, however, under some circumstances as laid down in the plans, the permit may be granted only if there is no objection from either the provincial executive or the government's physical planning inspectorate for the region. The decision must be made within one month. If refused, or granted subject to conditions, or if the decision is not made within one month, the applicant may appeal to the municipal council and, eventually, to the Judicial Department of the Council of State.

## Demolition Permits

N4.6   In general, building regulations may require a permit for the demolition, wholly or in part, of any building. The permit must be decided in the same way as for a building permit.

N4.7   However, the more significant, compulsory, control is exercised under the Town and Village Renewal Act. In an urban renewal plan, and in an area covered by residential environment regulations, the conditions are strict, the decisions being made by the municipal executive.

    (i) Applications for permission to demolish buildings must be refused if they involve listed buildings or conservation areas; or deferred until a building permit has been granted for a building in place of the demolished one.

    (ii) The building permit must be accompanied by a bankers' guarantee for 20 per cent of the value of the replacement building, and this is forfeited to the municipality if construction is not started within a specified period, and the building permit is cancelled.

    (iii) The application for the demolition permit must be decided within two months, with a possible extension of two months, or be deferred until the replacement building permit has been granted.

    (iv) There are opportunities for an appeal against a refusal of a demolition permit.

## Other Authorisations

N4.8   The grant of a permit is not, in itself, sufficient authority for development. Apart from those relating to demolition

    (i) a building authorised by a building permit may start to be used only after completion of the building works has been certified

    (ii) construction of other works authorised by a construction permit must not be started for one month, to give an opportunity for objection by the government's physical planning inspectorate for the region.

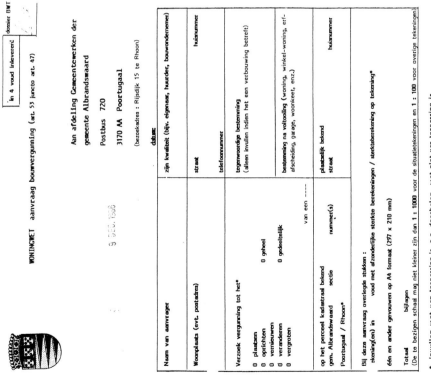

**Figure N5    Application Form for a Building Permit**

Figure N5 reproduces the application form for a building permit under the Housing Act, using the example of the municipality of Albrandswaard.

The first page requests basic information about the applicant; the type of application sought (ie new construction, renewal etc.); the present and proposed use; the land registration reference number of the site; and a list of accompanying drawings etc.

The second page covers further details about who are responsible for design and construction; detailed measurements and costs of site preparation and construction; and the reasons for any temporary uses.

Invulling is nodig voor zover de hier gevraagde gegevens op het plan van toepassing zijn en niet op de tekening kunnen worden aangegeven.

Wie is de eigenaar van het bouwperceel?

Staan er bomen of struiken op het bouwterrein? — ja / neen*

Is het door de eigenaar gekocht van de gemeente? — ja / neen*

Heeft de aanvrager het in huur of in erfpacht — ja / neen* in huur / erfpacht*

Ligt op het bouwperceel een erfdienstbaarheid de van invloed kan zijn op het plan?

Wat wordt de hoogteligging van het onbebouwd blijvende gedeelte van het bouwperceel ten opzichte van de aangrenzende weg?

Waardoor worden de erfafscheidingen gevormd?

Welke materialen zullen worden gebruikt voor

de fundering?
de trasramen?
de dragende buitenwanden?
de niet dragende buitenwanden?
de dragende binnenwanden?
de niet-dragende binnenwanden?
de dakbeschieting?
de dakbedekking?
de vloeren?
de plafonds?
, trappen?
de bodemafsluiting?

de schoorstenen?
de gas/voetkanalen?
de ventilatiekanalen?

de standleidingen?
de verzamelleidingen?
de grondleidingen c.p. huisaansluitingen?

*Doorhalen wat niet van toepassing is.

hoogte

beg.gr.     verd.

binnen het gebouw     buiten het gebouw

Abrexkward 886/6     -1.778.511

---

Op welke wijze zal worden voorzien in:
drink- en houdhoudwater?
bedrijfswater?
bluswater?
gas?
electriciteit?
verwarming?

Welke voorzieningen worden getroffen voor:
antennecontructies?
reclameconstructies?
wasen van de ramen aan de buitenzijde?
binnenbrengen van huisraad?

Op welke wijze zal worden voorzien in parkeergelegenheid en/of stallingsruimte voor motorrijtuigen, bromfietsen en fietsen?

Hoeveel personen zullen naar schatting ten hoogte in het gebouw verblijven?     overdag ............ personen     des nachts ............ personen

Wat zullen de kleuren en/of materialen zijn van:
het buitenwerk?
het buitenverfwerk?
het buitenmetselwerk?
de dakbedekking?

volgens welke regeling zal/zullen de woningen worden gefinancierd?     0 ongesubsidieerd     0 gesubsidieerd, volgens .........

(Alleen beantwoorden wanneer de aanvraag betrekking heeft op het bouwen van één of meer woningen.)

Indien de aanvraag een vernieuwing, verandering of vergroting betreft, dienen bovendien nog de volgende vragen te worden beantwoord

Zijn de bestaande privaten van een spoelinrichting voorzien?

Hoe is de bestaande afvoer:
a van de faecaliën?
  is deze afvoer gemeenschappelijk?
  zo ja, voor hoeveel gebouwen?
b van het hemel-, huishoud- en bedrijfswater?
c is deze afvoer gemeenschappelijk?
  zo ja, voor hoeveel gebouwen?

VOOR EVENTUELE BIJZONDERHEDEN EN TOELICHTING: zie het aanvraagformulier

Plaatsnaam en datum     Handtekening gemachtigde     Handtekening aanvrager

---

The third page requests details about ownership, tenure and leases for the site; existing landscape; and details of construction.

The fourth page covers, where relevant, supply of utilities and services, advertisements, car parking; number of persons by day and by night; external colour; and details of housing subsidies.

## Public Participation in Control

N4.9　Strictly, there is no statutory provision for public participation, no requirement that the general public must be notified and consulted about applications for any one of the permits. The theory is that the plans and regulations, once approved, are legally binding as much on the municipality as on the applicants. Thus the only ground for objection can be that a decision to grant or to refuse a permit was illegal, in which case the procedures discussed in paragraph N6.16 come into play.

N4.10　The only formal exception is when neighbours have to be consulted according to building regulations or the *bestemmingsplan*, in those instances when the municipal executive may issue a permit in exemption from the regulations. They also have to be consulted in the course of a section 19 application.

N4.11　In practice, it would seem that the opinions of the general public, and especially those of neighbours, are often widely canvassed and taken into account very seriously whenever the plans or regulations permit. This is particularly true, it is alleged, where decisions are delegated to the local borough councils. The situation may change however if the promised legislation on public participation materialises.

## Flexibility in Control

N4.12　Building permits and construction permits provide the basic instruments of control, supplemented where necessary by the procedures for demolition control. The procedures seem straightforward and rigid, giving the legal certainty required by Dutch law and administration. But in practice the system is far less rigid, with much more flexibility than is at first apparent and much room for negotiation.

N4.13　The relationship between plans and regulations, and the role of the different actors in the process is illustrated in Figures N3 and N4. Figure N3 emphasises the parts played by the building regulations and the *bestemmingsplan* (or, alternatively, the urban renewal plan or the residential environment regulation) in the process of issuing a building or construction permit. Figure N4 shows the process of consultation and decision-making. Both, but especially the latter, show that of the various possible outcomes only two in fact result in a straightforward decision either to refuse, or to grant immediately, a building or construction permit.

N4.14　The building regulations may specify the area or circumstances under which the municipal executive may grant exemption from particular requirements. In general, they refer only to very small, non-residential buildings or slight divergences from say the permitted height of a building (see paragraph N3.15) and in any case a permit is nevertheless required.

N4.15　The amount of flexibility that is possible in the absence of a *bestemmingsplan* can be illustrated by the following example from Leiden (Thomas et al, 1983). It involves the case of a burnt-out factory in the inner city where the factory owner wished to move to another site, and the municipal executive wanted to see flats for the elderly, despite this involving a change of use contrary to the building regulations, and despite the site having been bought by a property developer.

(i)　The municipal executive discussed the proposal with the developer, several housing associations and the urban design department as a result of which provisional, informal agreement was reached.

(ii)　The developer applied for a building permit, as a result of which the use of the site for housing was approved informally and conditions imposed on design, car parking etc.

(iii)　The municipal executive and the other actors held detailed discussions with the regional housing inspectorate of the Ministry regarding layout, subsidies etc, and with the aesthetic commission on design.

(iv)　The housing association applied for an annual subsidy from the Ministry; and loan guarantees from the municipal council.

(v)　The developer applied for a formal, detailed building permit using a modified design to satisfy the aesthetic commission.

(vi)　Demolition work started on the site, and the new building was in use before the final certificate of completion of construction had been issued by the municipal executive.

N4.16　In the case of the *bestemmingsplan*, the Physical Planning Act contains a number of sections which give opportunities for flexibility although, in each case, the plan has to define the areas and circumstances within which the flexibility can be exercised, and the limits of discretion involved. Thus a *bestemmingsplan* may

(i)　stipulate that a particular proposal or destination be either

(a)　provisional, applying only for a limited period which may be extended with the permission of the provincial executive (section 12);
or

(b)　deferred until after a specified date, in effect a postponement of the coming into operation of the plan (section 13)

(ii)　specify that permission be given for temporary uses and buildings for a period of up to five years unless expressly prohibited by the plan (section 17)

(iii)　specify the areas and circumstances in which the municipal executive may either stipulate additional requirements beyond those in the plan, or grant exemption from the regulations in the plan provided there is no objection from the provincial executive (section 15). This refers for instance to the discretion given in some plans to permit, with conditions, manufacturing uses in category 3 of the industrial classification described in paragraph N3.32.

N4.17 The possibilities for exemption have in principle been greatly extended by the new Physical Planning Act though it is too early to assess how far in practice they will be used.

(i) Section 10 of the original 1962 Act placed a duty on municipalities to prepare and adopt a *bestemmingsplan* and this remains unchanged. The form and content of the plan is laid down in article 12 of the Physical Planning Decree, 1985. It now allows for the destinations of the plan to be described 'in headlines', giving more room for discretion in what is permitted. Thus for instance, instead of separately identifying the garden, the actual building, and the use of each separately (eg a village street with several shops), the plan may make a general statement which would permit changes of use within stipulated limits and a defined area (see Figure N8). This provision is different from that contained in section 11 of the Act and discussed below, as under article 12 the formal *bestemmingsplan* remains the original document.

This is a new idea, introduced by the 1985 Decree as part of the process of deregulation, so that, if required, the plan can be "more global, programmatic in character" (Staatsblad, 1986). Previously this kind of description was confined to the explanatory memorandum which accompanies the plan, rather than the formally approved plan itself. However, it is still possible to use the old form, and keep the plan very detailed and precise.

(ii) Section 18a considerably extended the very limited, general scope in the original section 18 for granting exemption from the plan. The original section referred only to greenhouses and other industrial buildings. By the new section, the Minister was given authority to make regulations for granting exemption. These are now contained in article 21 of the 1985 Decree, giving a long list of very minor extensions and small buildings for which the municipal executive may grant exemption from a *bestemmingsplan* without having given further details in the plan itself (see paragraph N3.30). Any such application for an exemption must however be open for objection for two weeks and then decided within one month.

N4.18 Two things are clear from this. The actual amount of flexibility in either the building regulations or a *bestemmingsplan* even now is still quite small, with the possible exception of the opportunities created by the new article 12 of the Decree for the so-called *globaal-eind* plan. The discretion is limited to very small divergences, or very minor developments and, even so, in the majority of cases a building or construction permit is still required.

## Control and the *Bestemmingsplan*

N4.19 In the view of at least some commentators, however, the chief means of ensuring flexibility is to switch the instrument of control away from the building permit issued after approval of a *bestemmingsplan*, to the procedures for the preparation of the plan itself. The argument is that in the former case, the issuing of a building permit becomes a formal, statutory process which "churns along in the wake of development" whereas, in the latter case, it can become an "informal, flexible, opportunistic, pragmatic process responsible for development actually taking place" (Thomas et al, 1983).

*Elaboration of a* Bestemmingsplan *(Section 11)*
N4.20 An approved *bestemmingsplan* can specify areas within which the final details of the plan can be worked out and approved by the municipal executive when development is imminent. This procedure, under section 11 of the Act, is a crucial means of implementing the plan, in effect creating a building action plan as an elaboration of the original *bestemmingsplan*. Once approved, the building action plan becomes the formal *bestemmingsplan*, replacing the original global plan. In this it differs from the idea in the 1985 Decree, whereby the plan can remain global, thus retaining its flexibility.

N4.21 Under this procedure

(i) the municipal executive is required to consult with the various interests involved, although this does not involve the opportunities for formal objection by the general public which had already occurred when the original global *bestemmingsplan* was prepared. But the building action plan must remain within the limits and area laid down in the global plan

(ii) the building action plan has to be approved by the provincial executive within two months (or three, if necessary and agreed) with an opportunity for comment by the government's regional housing inspectorate if housing is involved; otherwise it is deemed to have approval. But, if the provincial executive refuses to grant approval, the municipal executive can appeal to the Judicial Department of the Council of State within one month.

N4.22 The municipality must also enact land development regulations for the implementation of a *bestemmingsplan* where, as is usual, it has acquired the land for development. These add a further element of control which however will be subject to negotiation between the municipal executive and a developer for the land. The regulations cover the acquisition of land by the municipality for public purposes, its subsequent preparation and disposal to a developer, and the share of the costs to be apportioned between the interests involved in the development. The provincial executive has to approve any such land development regulations, after consultation with the government's housing inspectorate for the region. It may also grant exemption from the need for such a regulation.

N4.23    Examples from a study in Eemland demonstrate the process at work, illustrating the sort of time scale which can be involved (Masser et al, 1978). Thus, one example for a major extension to a medium sized town on a site of 3.5 hectares shows

(i) 1965-67, preliminary studies by the municipality for the preparation of a global *bestemmingsplan* to meet its own housing needs and priorities, and its eventual submission to the provincial executive for approval

(ii) 1967-70, negotiations between the municipality and the province about a possible compromise between national, provincial and municipal housing and planning policies, including the number, type and tenure of dwellings, and the level of subsidy; and eventual approval of the global plan

(iii) 1970-72, compulsory purchase proceedings started, but growing public opposition to the high rise content of the original global plan

(iv) 1972-74, negotiations between the municipality and potential developers, taking into account the objections to the housing type, to achieve a scheme financially acceptable to the municipality and the chosen developer within the constraints imposed by the possible mix of subsidised housing and free market dwellings

(v) 1974-76, modifications to the global *bestemmingsplan* in the light of the final, agreed housing mix, and its detailed implementation through elaboration of the plan and final approval by the provincial executive for building permits to be issued and construction to start.

*In Anticipation of a* Bestemmingsplan *(Section 19)*
N4.24    The other, highly significant method of control through the procedures for preparing a *bestemmingsplan* lies in the powers given by section 19 of the Act. By this, the municipal executive may grant exemption from the provisions of an existing plan provided (a) that the provincial executive has no objection, having consulted with the government's physical planning inspector in order to ensure that regional and national interests are protected; and (b) that the municipality has adopted a resolution under section 21 that a *bestemmingsplan* is in preparation.

N4.25    This use of a decree or resolution that a *bestemmingsplan* is in preparation is very important as it has the dual effect of freezing the situation, including development in accord with the original plan and the one in preparation, yet at the same time allowing development to proceed under control. A planning decree under section 21 has a limited life, within which the draft *bestemmingsplan* must be placed on deposit for public inspection, or the decree lapses. If adopted by the municipality on its own, the decree has a life of one year.

This is extended to two years, with the possibility of a third, if there is an approved structure plan; or if (rarely) the provincial executive has issued the decree thus imposing a duty on the municipality to prepare a *bestemmingsplan*. However, one practice is for a municipality to renew annually its section 21 declarations in order to ensure that it is in a position to respond quickly to applications under section 19.

N4.26    If a preparatory decree is in force, any building or construction permit which could not be refused under the normal rules must be deferred during the life of the decree.

(i) The municipal executive however may grant a building or construction permit in anticipation of the new *bestemmingsplan*, and in exemption from the existing plan, subject to the approval of the provincial executive in consultation with the physical planning inspectorate, and after the proposal has been open for public inspection and objection for a period of 14 days.

(ii) The provincial executive gives its approval either (a) having issued a general declaration of no objection in anticipation of the plan for specified types of development in the area of the plan; or (b) by issuing a specific declaration of no objection in relation to the particular proposed development (Zuid Holland, 1983, 1986). The declaration must be issued or refused within two months, or it is then deemed to be refused.

N4.27    This method has been used extensively, 14,000 times in 1973 (Brussaard, 1979). It is a way of enabling development to go ahead which strictly is either out-of-accord with an approved *bestemmingsplan* or where there is not an approved plan but which, in the view of the municipal executive, it would be reasonable to permit on general planning grounds or in anticipation of a new plan. Its use is demonstrated in the case of Leiden, for instance (Thomas et al, 1983)

(i) for an increase in the permitted number of subsidised flats from 340 to 360 in a residential development, the building permit being given in anticipation of an alteration to the *bestemmingsplan*

(ii) for an office block which would exceed the permitted height and encroach over the building line in an approved *bestemmingsplan*, but to which the aesthetic commission and the building inspectorate broadly had no objection. In this case, the municipal executive gave its approval in principle and the section 19 procedures were invoked. Construction actually started before the building permit was issued, and the building was occupied before all the final details of the permit had been approved

(iii) for the conversion of a garage to a surgery in conflict with the *bestemmingsplan*. This would have been an example of a 'postage stamp' plan, covering a single, small site with a single small use. They are much used, though in this particular case the procedure

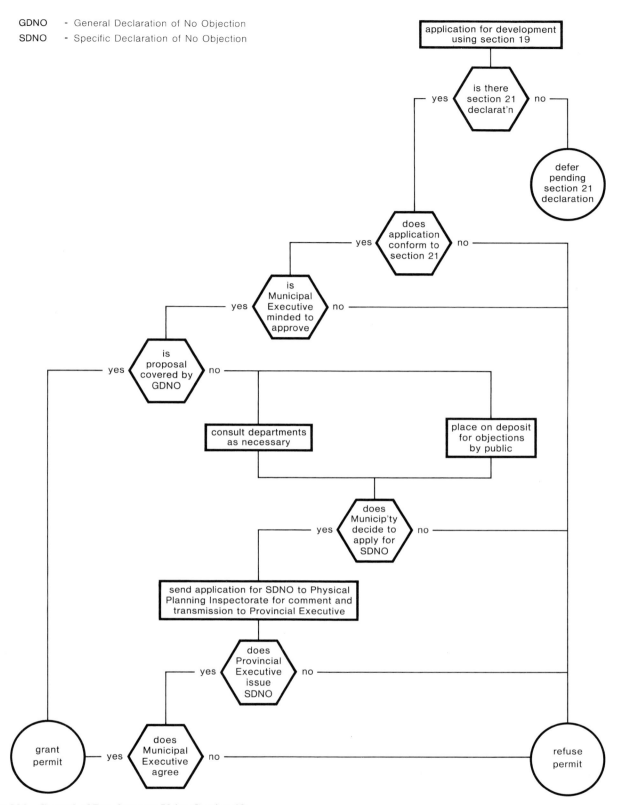

GDNO — General Declaration of No Objection
SDNO — Specific Declaration of No Objection

application for development using section 19

is there section 21 declarat'n

defer pending section 21 declaration

does application conform to section 21

is Municipal Executive minded to approve

is proposal covered by GDNO

consult departments as necessary

place on deposit for objections by public

does Municip'ty decide to apply for SDNO

send application for SDNO to Physical Planning Inspectorate for comment and transmission to Provincial Executive

does Provincial Executive issue SDNO

grant permit

does Municipal Executive agree

refuse permit

**Figure N6   Control of Development Using Section 19**

Figure N6 shows the procedure for dealing with an application for a building permit in anticipation of a review of a *bestemmingsplan* under section 19 of the Physical Planning Act. This is a widely used procedure which involves the grant of a general or a specific declaration of no objection by the provincial executive before the permit can be granted by the municipality.

367

was not used and the municipal executive issued a building permit irrespectively as there had been no objection from either the aesthetic commission or the neighbours.

N4.28    Other more recent examples include

(i) the elaboration of part of a global *bestemmingsplan* for a major suburban extension of Rotterdam to permit use of the land for the development of a major office development. The plan had specified uses appropriate to a residential area in very broad terms. The offices thus were, in effect, a departure from the original plan

(ii) the refurbishment of an older, 1960s planned shopping centre in Capelle aan de IJessel including its partial redevelopment, and some changes of use.

N4.29    The actual procedure for invoking section 19 is complex and can be lengthy, taking in theory up to about eight months though in practice it can take much longer as some of the examples show. The cause of this is the amount of negotiation which may be necessary before all parties can agree, for instance on the financial arrangements and share of costs. A summary of the process is given in Figure N6.

N4.30    As befits a procedure which is used so much, it has accumulated a lot of legal decisions and interpretation. A circular letter to all municipalities from the provincial executive of Zuid Holland summarises the key points (Zuid Holland, 1983).

(i) The application for a section 19 exemption in anticipation of a plan can only be entertained if it is urgent; of a size and character or in a location where its use is warranted; and if it will not jeopardise the future *bestemmingsplan* in preparation. Otherwise, the application must be deferred until the new plan has been approved.

(ii) The general declaration of no objection can cover only changes to the interior of a building, a small increase in the volume of an existing building with no change of use, or a very small new building.

(iii) The special declaration of no objection otherwise must be used for the particular application.

## Enforcement

N4.31    The Housing Act prohibits building construction without a permit in all cases, and prohibits action which is stipulated as needing a construction permit by the regulations in a *bestemmingsplan* or planning decree under section 21 or otherwise. The Physical Planning Act prohibits any transgression of a *bestemmingsplan* provided the grounds are stated explicitly in the regulations with the plan.

N4.32    The transgressions are indictable offences under the criminal code, punishable by imprisonment or a fine. Police powers of investigation are in the hands of the government's regional inspectors of housing and physical planning; provincial officials; and the housing and building inspectorate of the municipality.

N4.33    Infringements of the provisions of a *bestemmingsplan* may also be countered by administrative action in which case the remedial action to make good the breach of regulations is undertaken by the municipality and costs recovered from the transgressor under the civil code of justice.

## The Development Control Workload

N4.34    No statistics about the volume of work, or the time taken, were available or apparently are regularly collected by the municipal, provincial or national governments. Thus Table N4.1 gives only some very broad indications based on ranges and estimates given by officials. The permits fall into three categories, namely

(i) those issued strictly in accordance with the plan, including the limited amount of discretion permitted by the plan under sections 15 and 18

(ii) those issued through the procedures for elaboration or in anticipation of the plan under sections 11 or 19

(iii) those which, at least in the large municipality, were granted using a strictly unofficial discretion, in effect turning a blind eye to possible breaches of control (the so-called 'section 5').

N4.35    The variation in the number of building permits in relation to population size is wide, but its pattern reflects the expected pressures for development, being highest in a rapidly developing planned growth centre, and lowest in the central municipality of a large conurbation.

N4.36    The high proportion of permits issued or refused either in accordance with the plan, or through procedures for which the amount of discretion is defined explicitly in the plan, regulations or legislation, indicates the extent to which control is plan-led. Nevertheless, section 11 and 19 procedures are clearly important. Informal estimates for Zuid Holland show that they are used for between a quarter and a third of all building permits, which is roughly in line with the three figures in Table N4.1, ranging from about 11 per cent to 40 per cent. In discussion, it was alleged that section 19 is used more frequently than section 11 on the argument that it was quicker. A more likely explanation however is that section 19 is used for the unforeseen variation from a plan, or when there is no plan, whereas section 11 is used for the foreseen development where options have deliberately been kept open until it is imminent.

**Table N4.1  The Control of Development: Workload***

| Type of municipality | A small semi-rural municipality | A medium-size, growth centre municipality | A large urban municipality |
|---|---|---|---|
| Approximate number of building permits | | | |
| (i) Plan-led procedures granted/refused in accord | n/e | n/e | 600 |
| Exempted, under sections 15/18 | n/e | 200 | 300 |
| Sub-total | 90 | 800 | 900 |
| (ii) Elaboration/ revision of plan, section 11/19 | 60 | 100 | 300 |
| (iii)Granted discretion, 'section 5' | n/e | n/e | 700 |
| Total | 150 | 900 | 1,900 |
| Number of refusals (% of total) | 5 | n/e | n/e |
| Number of appeals to the municipal council (% of total) | 2 | n/e | 10 |
| Number of appeals to the Council of State (% of total) | 1 | n/e | 1 |
| Approximate length of time for decision (months) | | | |
| Ordinary | 1–4 | 1–2 | 2–6 |
| Section 19 | 2–6 | n/e | 8–18 |
| Delegated to building Inspectorate (%) | 0 | 0 | 50 |

* No statistics apparently are collected or were obtainable. The figures are based on informal estimates by officials in meetings, all of which were actually given as ranges or as very broad indications, in round figures.
n/e no estimate given.

N4.37   The only real, uncontrolled discretion is through the so-called 'section 5' procedure. It would appear that in some municipalities a building permit may be issued even if contrary to the *bestemmingsplan*, and for which no formal exemption could be granted. This applies for minor developments to which neither neighbours, nor the aesthetic commission, nor any other departments, raise an objection, and for which there are no good grounds for refusal.

N4.38   Strictly speaking in such cases, section 19 should be invoked, as it involves a departure and therefore needs a formal variation from the *bestemmingplan*. But the procedure is cumbersome, especially if there is no basic objection, so the permit is issued anyway. Examples include the conversion of a small garage at the side of a house into a doctor's surgery (Thomas et al, 1983); and the rebuilding in replica of a listed building in a designated conservation area, with a change from its destined use as a hostel to that of an office. In the latter case, the National Monuments Agency, the aesthetic commission and the neighbours raised no objection; and, in the former, the neighbours were invited to comment, but did not object.

N4.39   The number of objections or appeals, especially to the Crown, is small, and this is confirmed in paragraph N6.16. Furthermore, the discussions suggested that the objections and appeals are very seldom successful. This would reflect the extent to which the plans are legal documents which have already gone through due process. Objections and appeals in this context are on the legality of the decision rather than on matters of policy, as expressed in the plan or regulations.

N4.40   The time limits for making a decision are laid down in the legislation. In practice, the discussions suggest that most of the smaller applications are dealt with in the specified period of two months, with larger ones taking longer. The figures for obtaining a permit through sections 11 or 19 clearly are much longer, the range being up to about 18 months from the time of the original application, though this may shorten under the 1985 Decree. Discussions with developers broadly confirmed this, figures of six months being quoted for a fairly straightforward office development, and up to about 18 months for a larger, more complex shopping development with a further two years for construction. It should be noted however that, in practice, the final building permit might be issued long after construction has started, or even after it is completed, informal permission having been given at an earlier, outline stage (Thomas et al, 1983).

## Conclusions

N4.41   A number of points emerge from this review of the planning control of development in the Netherlands. The system

(i) links together building construction and planning issues in the one instrument of control, the building or construction permit, and can provide control of demolition

(ii) binds together policy guidance and control into legally binding plans and regulations

(iii) creates a very powerful, but rigid instrument of control in the shape of the *bestemmingsplan*, one so rigid that many different statutory and informal procedures have been devised to allow flexibility in the implementation of the plan, thus reducing the amount of certainty in the system

(iv) leaves a potentially more vague situation in the existing built-up areas not covered by a *bestemmingsplan* or its predecessor (the *uitbreidingsplan*), though this is being clarified by the new urban renewal plans and residential environment regulations under the Town and Village Renewal Act.

N4.42   This once more confirms that the Dutch system in theory is a plan-led system. Thus the procedures for the preparation and approval of the *bestemmingsplan* and similar legally binding plans and regulations are as much a machinery for the control of development as is the formal instrument, the building or construction permit. This is likely to become even more crucial as deregulation progresses and the arrangements for greater flexibility in plans come into effect.

N5.1   The basic, legally binding, plan is the *bestemmingsplan*, prepared by the municipality. It may include whatever the municipality deems necessary for the physical planning of its area, but it is constrained by the policies of the higher levels of government, chiefly to ensure that national and regional interests are protected and promoted. Thus the *bestemmingsplan* as an instrument of policy must be placed in the context of the higher levels of government. But the decentralised decision-making in the preparation of these plans and the control of development means that national and regional interests can be delayed, or even blocked, by municipalities.

## National Planning Policies

*Physical Planning and Urban Renewal*
N5.2   There is no longer any expectation or requirement for preparing a national physical plan. Nevertheless, there are national physical planning policies in a very real, positive sense. Their purpose is, *inter alia*, to ensure horizontal coordination of sectoral (ministry or departmental) plans affecting physical planning, in the end through cabinet; and the vertical coordination between levels of government such that national policies and priorities are expressed in lower tier plans and eventually in *bestemmingsplannen*.

N5.3   Ministers can make regulations and decrees for the implementation of legislation, provided the authority is given by the legislation. The Physical Planning Decree 1985 lists the various instruments for national policies. They have evolved since 1960, and include

(i)   the annual report or memorandum required by the legislation to be prepared for the States-General by the Minister to accompany the budget, and subject to debate

(ii)  the major reports to the States-General on physical planning, prepared by the Ministry of Housing and Physical Planning and Environment with extensive consultation and cabinet approval. They have defined the main directions of policy (Hazelhoff, 1981; van der Cammen, 1984) namely

   (a)   the First Report (1960), with emphasis on dispersal to the declining regions

   (b)   the Second Report (1966), a long range blue print for regional growth poles; and concentrated decentralisation to growth centres and cities

   (c)   the Third Report (1975-78), in three parts on general orientation, urbanisation, and rural areas, with emphasis on conservation in the general sense and revitalisation of the existing urban areas in particular.

(iii) Structural Outline Sketches on urbanisation and rural areas in amplification and review of the Third Report (Ministry of Housing, 1983d, 1984a).

(iv)  Structural Outline Schemes for particular sectors (eg electricity supply, traffic and transport, housing, military training areas, outdoor recreation, etc.), to be reviewed every five years

(v)   special reports of the interdepartmental national physical planning council on particular areas, such as the Waddenzee.

N5.4   The two structural outline sketches are thus the nearest to a national, spatial policy for physical planning. The original 1979 version showed, *inter alia*,

(i)   for urbanisation, the location and projected size of growth towns and centres; urban regions based on public transport; buffer zones between the main urban centres in Randstadt Holland; and the areas to be kept open including the so-called "green heart" of Randstadt Holland (see Plate 29).

(ii)  for rural areas, areas where agriculture, or the natural environment, or a mixture of the two, were to receive priority; and the areas with a restrictive policy to limit growth, including those in the urban areas structural stretch for the two are coordinated (see Plate 30).

N5.5   The urban areas structural sketch was revised in 1983, with different target figures and both outline sketches are now national key planning decisions with the status that implies. As noted below, this means that the sketches provide the authority for government directives to the provinces. If necessary, they also afford through the procedures for the preparation and approval of regional plans and *bestemmingsplannen*, the basis for challenge to those plans by the government's regional physical planning inspectorate.

N5.6   Since 1972, and a report on publicity in the preparation of physical planning policy, a formal procedure has been devised for the approval of *rijksplanologische kernbeslissigen* (national physical planning key decisions). All the reports, sketches and schemes starting with the Third Report have gone through this process, giving them extra authority and legitimacy. The procedure for their preparation is now a statutory requirement by the 1985 Act and Decree. The procedure is

(i) preparation of a draft cabinet resolution

(ii) publication, administrative consultation, and, if the issue is potentially contentious, democratic public participation organised through the Advisory Physical Planning Council, for a period of one year

(iii) review of the resolution, with reasons for modifications or rejection of comment during consultation and participation

(iv) laid before States-General for six months

(v) if no objection, or when approved by the States-General, the decisions become formally effective and are implemented.

N5.7　The key planning decisions therefore are formal statements of national policy. Although they are not legal instruments as such, they are the basis of directives by the Minister to the provincial and municipal governments, and their implementation is monitored by the Ministry's regional physical planning inspectorate. In particular, the directives may

(i) order a provincial council to prepare or review a regional plan, with a reserve power for the Ministry itself to prepare the plan if necessary

(ii) direct a provincial council (a) on the content of a regional plan; and (b) on provincial directives to municipal councils about the content of their plans

(iii) grant exemption or deferment from *bestemmingsplan* regulations

(iv) be the policy basis for ministerial decisions on appeals to the Crown covering objections to draft *bestemmingsplannen* etc.

N5.8　The most direct instrument of national policy, however, is the power of central government to allocate funds through the provincial governments to the municipalities. This is especially important in central control of national policies for housing and urban renewal, for which the *bestemmingsplannen* and urban renewal plans provide the local basis for the implementation of policy.

*Listed Buildings and Conservation Areas*
N5.9　The Ministry of Welfare, Health and Cultural Affairs is responsible for policy regarding listed buildings and conservation, working through the National Monuments Agency (Department of Monuments, undated).

N5.10　Listed monuments cover most pre-1850 buildings whose preservation is in the public interest on the grounds of their scientific, historical and cultural interest, together with their surroundings. The list is prepared,

published and kept up-to-date by the Agency and is now being reviewed and carried forward to 1900.

(i) Buildings are selected for listing by the Agency, or may be proposed by owners or pressure groups such as the various local and national amenity and historical or architectural societies. Spot listing is also possible, within the rules of 'good government'.

(ii) The entire building has to be listed as, in law, it is an 'immovable monument'.

(iii) Owners are notified, with an opportunity for objection, and there are funds for preservation.

N5.11　Conservation areas in towns and villages containing a number of listed buildings are identified and designated jointly by the Ministries of Cultural Affairs and Physical Planning on the recommendation of the Agency, in the form of a preservation order and plan with explanatory memorandum. The criteria for designation are the scientific, historic and cultural interest of the area. They are not designated according to their beauty as such, as aesthetic criteria are the responsibility of the local aesthetic commissions and municipalities.

N5.12　The conservation area plan is a legally binding plan but one with a limited life during which it gives interim control of development pending its incorporation in a *bestemmingsplan*. The procedure is

(i) the Monuments Agency prepares a designation proposal in consultation with the National Physical Planning Agency and the National Housing Directorate

(ii) there is a two month period for objection and appeal to the Crown

(iii) once approved by the Ministers, the plan is entered into a register and becomes operative in theory for no more than one year, although in practice the interim control continues as long as necessary.

## Provincial Planning Policies

N5.13　Regional or provincial planning has many aspects involving not only housing and physical planning but also matters coming under other national ministries such as economic affairs and industrial development, transportation and the interests of the very powerful *Waterstaat* concerned with land drainage and reclamation. The province thus becomes the level at which broad directives from different ministries are coordinated, integrated and translated into orders and directives for implementation by the municipalities.

N5.14　The key planning instrument for this coordination, communication and control is the *streekplan* (regional plan). Although it is not legally binding on citizens in the way that *bestemmingsplannen* are, and therefore does not generally appear directly as an instrument for the control of development, the regional plan is administratively binding on all municipalities and other public sector agencies within its area. It provides the yardstick for the evaluation by the province of *bestemmingsplannen* and other plans, and provides the authority for provincial orders and directives.

*Regional Plans*

N5.15 A regional plan or plans has to be prepared or reviewed for all or part of a province either on its own initiative or on direction from the Ministry which may also issue directions on the content of regional plans. To this end, the provincial executive has the duty to survey continuously all matters affecting physical planning in the provinces including demographic, economic, social and cultural trends; the natural environment; and the possibilities for and desirability of development. The results are published.

N5.16 The procedure for preparing and approving a regional plan is as follows (Brussaard, 1979)

(i) the provincial executive prepares the plan, with extensive consultations including

(a) the provincial physical planning committee, which includes, *ex officio*, the government's regional inspector of physical planning to ensure compliance with national policy

(b) other government departments and municipalities

(ii) notice is published that the draft plan is to be made available for inspection and the draft plan is placed on deposit in a designated way

(iii) objections to the plan are made to the provincial council within two months

(iv) the provincial council must approve the plan within four months (plus one extension of two months if necessary), giving reasons for rejecting objections or modifying the draft plan

(v) the approved plan is published and placed on deposit, for public information

(vi) the provincial council notifies the Ministry, although there is no requirement for formal approval by government, nor is there opportunity for appeal to the Crown as the plan is not legally binding

(vii) the plan immediately takes effect, and must be reviewed within not more than ten years.

N5.17 The regional plan provides a general statement about the future direction of physical planning. Typically, it covers the form and purpose of the policies governing future development and the ways in which the provincial executive is given authority for its elaboration in greater detail without having to go through the full procedures for approval. It usually comprises

(i) the *toelichten* (explanatory memorandum), including target numbers of dwellings etc

(ii) the *beschrijving* (regulations) which, after approval by the provincial council, are binding on the provincial executive in its dealings with municipalities such as for the approval of a *bestemmingsplan*. It covers

(a) *beleidsstrategie*, the broad strategy of the plan

(b) *kernpunten*, the structural elements for each sub-region within the plan showing specific, or general, locations for new residential or industrial development, or new directions for growth; planned new roads; new recreation areas etc together with a requirement that a *bestemmingsplan* be prepared or revised to cover each new development

(c) the *richtlijnen* (directions) for the preparation and content of *bestemmingsplannen* including the general contents of such plans, that they must for instance cover noise levels and policies; infrastructure; open spaces; and the built environment

(iii) the map, usually though not necessarily on a topographic base.

N5.18 The entire country is covered by a total of about 35 approved regional plans (van der Cammen, 1984). The province of Zuid Holland has four, covering the far south, the west around The Hague, the east central and Rijnmond. The plan for the Rijnmond area covering Rotterdam and Europoort is illustrated in Plate 31 (Zuid Holland, 1986).

*Other Plans and Policy Instruments*

N5.19 A second collection of policy documents are the *ad hoc* reports on physical planning matters ranging from statements of goals to policies on specific topics such as out-of-town shopping, caravan sites or environmental standards. These planning reports, once approved, are binding on the provincial executive, finding expression either in regional plans, or in provincial ordinances. The procedure for such reports covers

(i) preparation by the provincial executive

(ii) consultation, including public participation

(iii) approval by the provincial council.

N5.20 Finally the provincial executive can

(i) prepare provincial regulations or ordinances for approval by the provincial council covering specific topics such as landscape conservation

(ii) issue orders to a municipality to prepare, and give directions on the contents of, structure plans and *bestemmingsplannen*

(iii) object to the grant of, or to an exemption from the need for, a building or construction permit under certain circumstances

(iv) hear objections to a draft plan, or appeals against the grant or refusal of a building or construction permit.

## Municipal Physical Planning

N5.21 The municipality relies on the two main policy instruments for physical planning provided for in the Physical Planning Act, namely the structure plan and the *bestemmingsplan*. In addition, it relies on the building regulations, under the Housing Act, and the various instruments under the Town and Village Renewal Act.

N5.22 Formerly, although there were opportunities for objection to a draft plan of either kind before its approval, there were none for public participation at earlier stages. However, the 1985 Act requires municipalities to adopt regulations for mandatory public participation during the preparation of plans.

*Structure Plans*
N5.23 A structure plan is not obligatory. It can be prepared for all or part of a municipality, or jointly with other municipalities. The provincial executive can also direct that a structure plan be prepared and issue directions about its content. The actual form and content comprises a description of desirable long-term development, with phasing, plus maps in a specified form, with an explanatory memorandum.

N5.24 The procedure for a structure plan is as follows

(i) the municipal executive prepares the draft plan based on a specified range of surveys similar to the regional plan; consultations with adjoining municipalities and provincial and central government; and public participation

(ii) notice is given of the intention to publish the plan, and the plan is placed on deposit for objections to be made to the municipal council within one month

(iii) the municipal council approves the plan, and

(a) once more giving notice, places it on deposit for public inspection

(b) notifies the provincial executive and the regional inspector of physical planning though, as it is not legally binding, it does not require provincial or national approval, nor is there appeal to the Crown.

(iv) the plan must be reviewed within ten years.

N5.25 Structure plans have not been used as widely as expected, in part because of the procedures. Instead, municipalities have used, and have been encouraged by provinces to use, informal structure plans as the technical planning framework for *bestemmingsplannen* (Staatsblad, 1986).

N5.26 A structure plan is not a set of regulations and it is not binding on *bestemmingsplannen*; but it does

safeguard future *bestemmingsplannen*, extending from one to three years the life of a planning decree in a built-up area to give interim development control pending the preparation of the plan. Furthermore, there have been cases where the Crown has disallowed a *bestemmingsplan* on appeal in the absence of a structure plan, notably in a plan for part of a town centre with traffic implications (Crince le Roy, 1975). Zuid Holland, for instance, usually requires any new or revised *bestemmingsplan* to be set in the context of a structure plan at least for its immediate surroundings, if not the entire municipality.

N5.27 Most of the larger municipalities take the initiative and prepare a structure plan for their entire area. Amsterdam for instance prepared an outline structure plan in 1974, later extending it to include amendments on housing in the older residential areas, and employment. Ideas of further amendments, to cover traffic and public facilities, however, were dropped as the perceived need was for a more up-to-date comprehensive structure plan. Accordingly a draft plan was prepared and published for public comment in 1984 (Amsterdam, 1983). The typical form of a plan is illustrated in Plate 32, using the draft structure plan for the Hague as an example ('S Gravenhage, 1981).

*Bestemmingsplannen*
N5.28 *Bestemmingsplannen* are obligatory for all parts of a municipality outside the built-up area, unless the provincial executive waives the requirement. They may also be prepared for built-up areas and may also cover a single topic, such as sites for gypsy caravans in an entire municipality provided they take into account surrounding areas (Staatsblad, 1986). The plan may either be global or the full detailed plan. In either case, it comprises

(i) a detailed, topographic plan, with a formal description, showing the destined uses of land and buildings, including the possibility of mixed land uses, and the continuation of existing uses; and the form of physical development, including the demolition and replacement of buildings

(ii) regulations about the use of land and buildings including

(a) the uses permitted by the plan, with criteria and standards for their design and layout, in effect the definitions of development in section N3

(b) procedures to be used through section 11 of the Act for the elaboration of specified parts of the plan by giving authority for negotiation and approval to the municipal executive, within specified limits, subject to approval within two months by the provincial executive in consultation with the physical planning inspectorate for the region. The plan is then elaborated when development is imminent through the preparation and approval of a building action plan

(c) prohibition of certain specified uses and construction without a construction permit

(d) specified ways in which the regulations may be relaxed for the control of development including provisional, deferred and temporary uses, and exemption from the regulations using sections 18 and 18a of the 1985 Act.

(iii) an explanatory memorandum, explaining the logic and context of the plan.

N5.29 The *bestemmingsplan* is accompanied by an economic feasibility study which is compulsory, under the Act, and has to be approved by the municipal council. The study contains a financial appraisal of the development, the *exploitatie opzet*, and a statement of the conditions under which the municipal executive will cooperate in the development of land, covering arrangements for the donation of land and the allocation of costs of provisions made in the public interest. Together they form the basis for the agreement, the *exploitatie overeenkomst*, between the municipality and the developer, whether housing association or private sector organisation, for the acquisition, disposal and development of the land. They are subject to approval by the provincial executive which may however exempt a municipality from having to go through this procedure, especially if the municipality does not intend to acquire the land for site preparation and development.

N5.30 In outline, a *bestemmingsplan*
(i) is prepared by the city planning department or planning consultants for the municipal executive during which time the executive can exercise interim development control under a resolution or decree by the municipal council that the plan is in preparation

(ii) is adopted in principle by the municipal council

(iii) is finally approved by the provincial executive on the advice of the provincial physical planning committee.

N5.31 Embedded in this administrative and legislative process is the quasi-judicial process giving opportunities for objection to the plan at each level to the municipal council, the provincial executive and, if necessary and ultimately, the Crown. The appeal process is discussed in section N6.

N5.32 In more detail the procedure starts when the municipal council on its own initiative, or if directed by the provincial executive, passes a resolution or decree under section 21 of the Act that a *bestemmingsplan* is under review or in preparation.

(i) The decree must define the area for the proposed plan and specify any construction not normally covered by building regulations which is to be controlled by a construction permit, to enable the municipality to prevent land becoming less suitable for its intended use in the new plan.

(ii) The decree must be published and placed on deposit for objection to the municipal council within one

month against the terms of any prohibition under (i) above.

(iii) Under the decree, applications for building and construction permits may be refused or deferred, giving interim development control, or granted in anticipation of the new plan (see section N4).

(iv) The decree expires in one year if the new draft *bestemmingsplan* has not been placed on deposit for public objection, with an extension for up to a further two years with provincial approval if the decree includes built-up areas covered by a structure plan.

N5.33 Once resolved, the preparation of the plan at the municipal level then involves the following stages.

(i) The municipal executive, and its officials or consultants,

(a) prepares the draft plan, by carrying out surveys and studies of the financial feasibility of imminent changes of use and development; consulting adjoining municipalities and provincial and central government; and undertaking public participation, a new requirement under the 1985 Act

(b) places the draft plan on deposit for public inspection, with prior notice, to give an opportunity for objections to the plan within one month to the municipal council.

(ii) The municipal council

(a) hears objections to the draft plan at a public enquiry conducted by a committee of the council

(b) approves the plan within three (or if necessary six) months giving reasons for dealing with objections, and, if necessary, modifying the plan.

N5.34 The draft plan then goes for final approval to the provincial executive. If there were no objections at the municipal level, or if the original objectors have been satisfied by the modifications to the plan, and if there is no further objection to the modifications, that is virtually the end of the story. The provincial executive

(i) consults the provincial physical planning committee (which includes the government's physical planning inspector) and approves the plan in whole or in part within six (or, if necessary twelve) months; a failure to do so in the time period means that the plan is deemed to be approved

(ii) notifies the municipal council and the national physical planning inspectorate for the region

(iii) publishes the plan.

N5.35 If however the provincial executive withholds approval in whole or in part, the municipal council must approve a new plan within 12 (or 18) months, in accord with the provincial executive's decision. But, if there are

(a)

(b)

mogelijke verkaveling kepplerstraat e.a. - minimum rekenmodel -

**Figure N7 Regentesse Valkenboskwartier, Den Haag: Area for Elaboration Under Section 11 of the Physical Planning Act**

Figure N7 shows a section to the north-east of the area illustrated in Plate 28 and covered by the *bestemmingsplan*. The extract from the *bestemmingsplan* (a) defines the area of existing buildings to be comprehensively redeveloped. When development is imminent, the plan will be worked up in detail – within the limits prescribed in the accompanying regulations accompanying the plan and approved under Section 11 of the Physical Planning Act. The regulations describe the maximum and/or minimum numbers of dwellings of different kinds and the other uses to be accommodated in the area. The sketch plan (b) illustrates one possible layout for redevelopment.

still objections at the provincial level, either because the original objectors were not satisfied, or there are new objections to the modifications, or by the municipal council to the adverse decision by the provincial executive, the process becomes quasi-judicial, involving appeals to the provincial executive and eventually to the Crown.

N5.36 The process can be long and complex. Even a straightforward *bestemmingsplan*, not involving the withholding of approval by the province or appeal to the Crown, can take one or two years. In more complex cases the process can take five or six years.

N5.37 *Bestemmingsplannen* are enforceable in law for any breach of the regulations provided that it is expressly specified as such in the regulations. The methods are

(i) administrative action, in which remedial action is taken by the municipality to make good the breach

(ii) the municipality recovering the costs of making good the breach of regulations from the transgressor through the civil courts

(iii) imprisonment or fine as they are infringements of the criminal code, detected by the government's inspectors of physical planning, or officials designated by the provincial governor or the burgomaster.

N5.38 *Bestemmingsplannen* also make provision for compensation if a person suffers, or will suffer, damage as a result of the regulations, subject to a number of conditions.

N5.39 Virtually the entire country is covered by *bestemmingsplannen*, providing strict control albeit made more flexible by the various methods for exemption or variation described in section N4. Thus Rotterdam is almost entirely covered by plans, including the older inner city where plans were necessary for the redevelopment of the blitzed areas and the renewal of the older housing areas, as well as the surrounding rural areas where development is prohibited, and the newer suburbs built since 1962. Plate 33 illustrates the global *bestemmingsplan* for a large extension of Rotterdam built during the 1970s, and subsequently elaborated using section 11.

N5.40 A small, semi-rural municipality, for instance Albrandswaard, is covered by a number of plans, the only exceptions being a few very old villages and some small, open rural areas under no threat of development. The plans include

(i) one for the western third of the municipality, with only one 'agricultural' destination, to prevent any urban development, apart from some infrastructure uses

(ii) several to conserve in existing use small areas of scattered dwellings and fields

(iii) a larger plan covering a village in detail, including an area for 570 dwellings to have been elaborated under

section 11, although in practice section 19 was used to vary the plan as it was thought to be a quicker procedure

(iv) the northern area where a contentious issue concerned the extension of an existing container storage terminal (see Plate 27 and Appendix (b)). The municipality is opposed to the terminal but Rotterdam, in which the terminal was located before a boundary change on the reorganisation of local government in the area, and Zuid Holland province want the terminal as part of the Europoort complex. The outcome so far is that the province has not approved the draft plan; has changed its regional plan to include the terminal; and has ordered the municipality to prepare a revised plan and directed that it include the terminal. In the meanwhile the municipal council, against the advice of its executive, has appealed to the Crown against the province's refusal to approve the draft *bestemmingsplan*.

N5.41 Capelle aan/de IJssel was one of the growth centres shown in the Second Report on Physical Planning for attracting growth in the Rijnmond area, and therefore was given priority in the allocation of housing and local government finance. It therefore has a high proportion of social housing, much of it high-rise built in the 1960s and 1970s. It has about 50 *bestemmingsplannen* covering each stage of its town development, including housing neighbourhoods of between 400 and 8,000 dwellings; industrial areas; open spaces; and a shopping centre. They include at least one plan covering just one or two buildings which raised very contentious issues. Planning has now started on the latest stage of its development, for an area of housing, industry and open space.

(i) An informal structure plan is in preparation, to resolve the key issues and to provide the framework for the eventual *bestemmingsplannen*. The issues are

(a) safety and drainage for what will be an area below sea-level, involving major investment by government

(b) prevention of pollution from dredging, and from motorway noise

(c) negotiations about the numbers of dwellings, and therefore the level of government subsidy, in the context of the targets in the regional plan

(d) relation to existing development.

(ii) The procedure continues with preparation of the first industrial *bestemmingsplan* for which a section 19 procedure will be used to approve site preparation and infrastructure to enable a quick start without having to wait for formal approval of either the structure plan or the *bestemmingsplan*.

N5.42 Ideas about *bestemmingsplannen* have been evolving and a memorandum is being prepared by the National Physical Planning Agency, to give guidance to municipalities following the 1985 Act and Decree

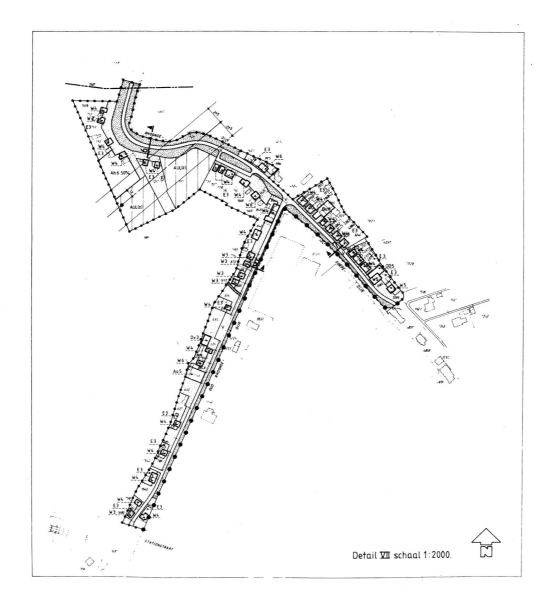

Detail VII schaal 1:2000.

**Figure N8    Albrandswaard-Noord Bestemmingsplan: inset plan VIII showing detail of village**

Figure N8 shows in great detail the planned destination of each building or piece of land within one of the villages covered by the *bestemmingsplan* illustrated in Plate 27. The zoning distinguishes for instance between a single house (W, with its permitted height) and its curtilage or *erf* (E, with strict controls over what can be permitted); and between a road (V) and the adjacent open spaces (light stripple). Particular buildings are shown for retail (DD), or permitted industrial or business uses (B).

A village of this kind could, in principle, be covered by a new, more flexible form of *bestemmingsplan*. Article 12 of the

Physical Planning Decree 1985 allows the destinations in a plan to be described 'in headlines', that is in more general terms, thus giving a municipality more discretion in deciding what can be permitted. It could for instance specify the number of shops to be allowed in a village street without necessarily identifying the individual buildings. Article 18a of the Decree also gives a municipality more discretion in deciding whether or not to grant an exemption from the requirements of a *bestemmingsplan* for a long list of minor types of development especially those for instance covered by the zoning for the curtilage of a dwelling (E).

(Ministerie VROM, 1986). The ideas are a mixture of deregulation and harmonisation. Deregulation is seen in the move towards encouraging municipalities to give greater flexibility in their plans. One method was mentioned in paragraph N4.17, namely the greater use of exemptions for minor development, under section 18a of the Act.

N5.43    The more important idea is to extend the use of

global plans. Formerly, the global *bestemmingsplan* defined the area and gave the framework for detailed elaboration. The detailed scheme, prepared under section 11 (or 19) would then become the actual *bestemmingsplan*. Figure N7 illustrates the principle at work. Figure N7(a) shows a part of the *bestemmingsplan* for a section of the inner city of The Hague, identifying areas for elaboration. Figure N7(b) shows one of the ways for elaborating the global plan, with a suggested layout for redevelopment which, when approved, becomes part of the *bestemmingsplan*.

N5.44 The new idea is that the global plan simply gives the 'headlines' of the development permitted in the defined area, but does not require elaboration. The plan remains in its original global form, and only a building or construction permit is required. One use could be, for instance, to define a village street containing a few shops and houses as a single global area with a specified list of permitted developments instead of, as now, identifying each separate use in detail. Figure N8 shows an inset from the Albrandswaard plan illustrated in Plate 27 where this new flexibility could apply.

N5.45 The other trend is towards harmonisation imposing a unified set of definitions on municipalities. Progress towards a new decree giving a unified, comprehensive set of building regulations was noted in paragraph N3.7. The new memorandum is intended to do the same for the definitions in *bestemmingsplannen*.

### Building Regulations

N5.46 These are mandatory, issued by the municipal council and approved by the provincial executive, with power to grant exemptions by the municipal executive. They cover the construction, use and demolition of buildings etc.

### Urban Renewal Plans

N5.47 Finally, a municipality is empowered to prepare various types of plan under the Town and Village Renewal Act 1984. They are

(i) *stadsvernieuwingsplannen* (urban renewal plans), for the comprehensive renewal of an existing built-up area

(ii) *leefmilieuverordening* (residential environment regulations), for areas without an urban renewal plan or *bestemmingsplan*, for a limited period to arrest decay, dereliction and change of use. Their impact on the definition of development and the exercise of control were described in paragraphs N3.42 and N4.7.

N5.48 Urban renewal plans are similar to a *bestemmingsplan* in terms of policy for the control of development and procedures for their preparation and approval, except for the greater involvement of the housing arm of the Ministry, especially on their housing mix and therefore financial implications. The chief difference is the addition of a mandatory, and administratively binding, executive or action plan (*uitvoeringsschema*). This shows how the plan is to be implemented, including redevelopment proposals, temporary re-housing, phasing, and financial arrangements. As part of the urban renewal plan, the action plan is subject to public participation, objection and, finally, approval by the provincial executive.

N5.49 The residential environment regulations are different from an urban renewal plan or *bestemmingsplan* in two respects. They do not define what is permitted, simply what is prohibited without permission; and they have a life of only five years, with one possible extension for a further five years. In addition, they contain the municipality's positive proposals for environmental improvements. Their preparation and approval involves

(i) placing the draft regulations for public inspection and objection for one month

(ii) the municipal council, after hearing objections, approves the draft regulations and forwards them to the provincial executive

(iii) after allowing a further one month for objections, the provincial executive being compelled to make a decision on the regulations within three months, in consultation with the provincial urban renewal committee. Failure to do so in the time period means that they are deemed approved.

## Conclusions

N5.50 The key policy documents in the control of development therefore are the *bestemmingsplannen*, and these now cover virtually the entire country. They are designed to give legal certainty to the future use and development of land, by the concept of a 'realised destination'. And the procedures enshrine ideas of democratic accountability. Originally these were limited to a right to object to a *bestemmingsplan* and ultimately to appeal to the Crown. Since the 1985 Act they have been extended to include a right to public participation during the preparation of the plan though this is only now starting to be put into practice.

N5.51 However, the *bestemmingsplan* (and its near neighbour, the urban renewal plan) is explicitly constrained by the policies of the higher levels of government. Not only is there the *post hoc* approval of plans by the province and, if there has been objection, the Minister. More importantly, the higher levels of government give directions on policy to the lower levels, to ensure protection of their national or provincial interests, and the implementation of their policies. Furthermore, these national and provincial policies have to go through formal, statutory procedures for their preparation and approval.

N5.52 The result is that policy for the control of development can be seen as a highly formal, complex and rigid framework. In practice, however, rather than being seen in these terms, a more accurate description is that policy and control involve a continuous process of negotiation between the different levels of government and with the private sector. While the various policy instruments do exercise control, there is enough built-in, or informal, flexibility to ensure that development takes place within an acceptable and predictable time frame in any particular case.

## Administrative Law and the Council of State

N6.1   Dutch law and the constitution derive from Roman law and the *Code Napoléon*. Built into this during the nineteenth century was the concept of *rechtszekerheid* (legal certainty). Individual liberty could be guaranteed only by the principle of *rechtstaat* (the constitutional state) through legislation, judicial control and police enforcement in which a framework of rules afforded citizens protection from infringement of their rights by other individuals (van Gunsteren, 1976). For this to work, the essential requirements were a constitution, basic human rights, and a separation of powers between the legislature, the judiciary and the executive or administration.

N6.2   Dutch law is found in the statutes and decrees, and in the civil, criminal and penal codes, rather than made by the courts. But the separation of powers means that the courts cannot judge whether an act is constitutional: "the statutes are inviolable" (Koopmans, 1971). It means that court decisions are not legally a source of law, and there is no doctrine of a binding precedent.

N6.3   The rights of property are found in the civil code, enumerating the different rights and interests in property including ownership and acquisition. Those rights themselves are subordinate to the legislation and, during the twentieth century, there has been an ever increasing volume of administrative law which increasingly has seemed "to give the administration a free hand to act 'creatively' rather than to provide certainty for, and to protect the liberty of, the individual" (van Gunsteren, 1976, p. 87). This creates a tension between the declared aim of the legislation, including the *bestemmingsplan*, to give certainty about the rights in property which it guarantees; and the actual practice of administration which increasingly gives discretion to the executive in implementing policy.

N6.4   For planning and property, as for other aspects of society, Dutch law therefore falls into two categories, namely

(i)  administrative, public or constitutional law, concerning the processes of state administration and their relationships with the citizen, for which the Council of State has responsibility

(ii)  civil, criminal and penal law for which the ordinary courts up to and including the Supreme Court have responsibility.

N6.5   With very rare exceptions, chiefly concerned with enforcement, planning and building matters come under the administrative law. They cover two fairly distinct areas of law, each of which has its own procedures. They are

(i)  hearing objections to draft plans or regulations on the grounds of their legality and effectiveness before their final, irrevocable approval and passage into law. The rules for these are in the relevant legislation such as the Physical Planning Act, and they cover for instance the approval of *bestemmingsplannen*

(ii)  hearing appeals against administrative decisions of any of the three levels of government on the grounds of their lawfulness as they affect the interests of individual citizens and organisations. The rights and procedures for this are found in *Wet Administratieve Rechtspraak Overheidsbeschillingen* 1976 - AROB (Administrative Tribunal on the Decisions of Public Authorities Act) 1976. For planning, the decisions are chiefly those on building or construction permits, listed buildings, and section 19 applications.

N6.6   The distinction between the two sections is crucial, even though for both the Council of State is the responsible organ of state. For the first category, the responsible section of the Council of State is the *Afdeling Geschillen* (Disputes Department) acting only in a quasi-judicial capacity. Its decision is not final. It makes a recommendation to the responsible minister who then makes a decision on the grounds of policy and the public interest, rather than legality as such. Details of the process are given below.

N6.7   By contrast, for the second category of decision under AROB, the responsible section of the Council of State, *Afdeling Rechtspraak* (the Judicial Department), acts precisely in a judicial capacity. It is the first and last court of appeal and its decision is made entirely on its interpretation of the law (van Ittersum, unpublished).

N6.8   A recent successful appeal to the European Court of Human Rights in 1984 has brought into question the principle underlying the Disputes Department. The case was an environmental issue involving planning matters for which the legislation laid down the procedures for appeal under which the Crown's decision was made. The argument was that an appeal to the Crown in this case is an administrative, not a judicial, process since the final decision is made by a minister, on the basis of 'law and expediency' thus denying an individual his or her constitutional rights of appeal to the courts under the European Convention on Human Rights.

N6.9 The European Court found in favour of the plaintiff. The eventual outcome however is likely to be fresh legislation to clarify the distinction between the two procedures. Decisions affecting only one individual, including, for instance, those on a *bestemmingsplan* which is a 'postage stamp' plan, will go to the Disputes Department for final decision with no further action by a minister. Decisions affecting people in general, such as an ordinary *bestemmingsplan*, will continue to be heard by the Disputes Department as now with the final decision being made by the responsible minister (van Ittersum, unpublished).

## Objections to Plans and Regulations

N6.10 For planning purposes, these are the objections to draft plans. In the case of the two types of non-legally binding plans, the only opportunity for objection is to the draft plan when it has been placed on deposit, through a designated procedure in which the elected council hears objections to the draft plan prepared by the executive. These plans are

(i) regional plans, placed on deposit by the provincial executive for two months, after which the provincial council considers any objections and comes to a decision within four (or six) months, giving reasons for its acceptance or rejection of the objections and modification of the plan

(ii) structure plans, placed on deposit by the municipal executive for one month after which the municipal council considers any objections and comes to a decision.

N6.11 The procedure for a *bestemmingsplan* or urban renewal plan is more complex, ultimately involving an appeal to the Crown as the plan has the force of law. Residential environment regulations follow broadly a similar process except that there is no final right of objection to the Crown. The *bestemmingsplan* procedure is as follows.

(i) The plan is placed on deposit by the municipal executive for one month, after which the municipal council conducts a public inquiry at which objectors are heard, and the council then makes its decision within three (or if necessary six) months, giving reasons for dealing with objections and, if necessary, modifying the plan.

(ii) The plan is placed on deposit by the municipal executive for a further period of one month during which the original objectors, if still unsatisfied, and any objectors to the subsequent modifications, can object to the provincial executive; the provincial executive then approves the plan, wholly or in part, or withholds approval within six (or if necessary twelve) months.

(iii) The plan is placed on deposit by the provincial executive yet again for one month to give an opportunity for an appeal to the Crown by the municipal council, the government's physical planning inspector, still unsatisfied objectors from the first two rounds, or anyone objecting if the provincial executive has withheld approval.

N6.12 Thus, even up to this stage, before an appeal to the Crown, the procedure can have taken anything up to 12 months from being first placed on deposit, or up to about 21 months if the necessary extensions of time are agreed. The only safeguard is that failure to decide in time means that the plan is approved.

N6.13 The appeal then reaches the Crown.

(i) The Disputes Department receives the objections and passes them to the *bureau van de adviseur t.b.v. raad van state* (advisory bureau) in the Ministry of Housing, Physical Planning and Environment for a technical investigation of the objections as such rather than the plan itself. The staff of the bureau

   (a) invite further written evidence from objectors with a legal interest (ie who have objected according to the procedures, as there is no other limitation)

   (b) make a site visit

   (c) prepare a technical report with recommendations.

(ii) The Disputes Department

   (a) publishes all documents including the advisory bureau's report

   (b) invites written comment from objectors and the relevant administrative bodies

   (c) holds an inquisitorial hearing into the cases brought by objectors

   (d) writes its recommendation in the form of a draft royal decree.

(iii) The Minister receives the Council of State's recommended draft decree and either accepts the recommendation or refers it back to the Council of State for reconsideration on one occasion. The Minister then, finally, makes a decision which can be either to approve the plan, or withhold approval, in which case it is referred back to the municipality for revision within one year.

N6.14 The whole process, from receipt of the appeal to the Minister's final decision is intended to take between 18 and 21 months but there are cases of up to five years. And there were, in 1985, about 400 plans to which objections were raised (Raad van State, 1985).

N6.15 The most interesting point in this process, once it reaches the Council of State, is the sharp separation of functions in professional, administrative and legal terms. The Ministry's advisory bureau is staffed by professional planners whose task is to make a technical assessment of the plan in relation to the objections. The Council of State is staffed by professional lawyers. Its task is to assess the merits of the objections in legal terms, taking into account the technical evidence. Finally, it is for the Minister to make the decision on a mixture of 'law and expediency', that is, in terms of public policy. Whether this is a real, as opposed to a formal, distinction is uncertain. In fact it is apparently very rare for the Minister not to accept the recommendation of the Council of State.

## Appeals Against Administrative Decisions

N6.16 For planning purposes, these are the appeals against decisions about building and construction permits and section 19 applications. In 1985 there were about 1,289 decisions on appeals of this kind to the Council of State, roughly one in seven of all such appeals (Raad van State, 1985). They included appeals against

(i) a refusal by the municipal or provincial executive to grant the relevant permit, by the aggrieved applicant

(ii) a grant by the municipal or provincial executive of a permit, by aggrieved neighbours or anyone else.

N6.17 The grounds for appeal are strictly limited, namely that (Raad van State, 1987)

(i) the decision was contrary to a generally binding regulation or plan

(ii) the municipality or province used its powers for a purpose other than that specified in the legislation

(iii) the decision was unreasonable in the light of the interests involved

(iv) the municipality or province acted contrary to the 'principles of good government'.

N6.18 The only decisions which are excluded from the AROB procedure are those of general significance rather than affecting individual interested persons; those for which there is a formal opportunity for appeal to the Crown (as in a *bestemmingsplan*); those covered by the civil codes (ie not involving administrative departments acting within the law); and cases explicitly excluded by the AROB law.

N6.19 The procedure is that appeals against decisions by a municipality or province are heard by the municipal council or the AROB committee of the provincial council, with a further appeal to the Judicial Department of the Council of State (Zuid Holland, undated). There is no time limit for deciding these appeals but a person who is appealing against a grant of permission can apply to the Chairman of the Judicial Department for an injunction to defer construction pending the outcome of the appeal.

N6.20 At each level, the committee or department holds a public inquiry into the appeal, and makes its decision according to the 'principles of good government'. These derive from the *Code Napoléon* and mean that the decision was made correctly within the given rules of the legislation or regulations; that the legislation or regulations were used for their intended purpose; that there was no discrimination; and that the decision was reasonable.

N6.21 The relative significance of this level of appeal to Judicial Department of the Council of State cannot be definitely established. In total, there were 1,289 appeals in 1985 under the Housing and Planning Acts (Raad van State, 1985). Figures from Zuid Holland suggest that about 12-16,000 building and construction permits are issued annually. If this estimate is correct, it suggests *pro rata* a total for the country of between 54,000 and 72,000 permits in which case the proportion appealed is comparatively low, at between 1.7 per cent and 2.4 per cent; especially as it contains an unknown proportion of appeals against the grant, as well as the refusal, of a permit.

## The Civil and Criminal Courts

N6.22 These are involved only either when a municipality is taking enforcement action against an infringement of building regulations or a *bestemmingsplan*; or conversely when an aggrieved citizen is taking legal action against a municipality or province for acting unlawfully, for instance for having demolished a building without having gone through the correct procedures and issued written notice.

## Introduction

N7.1 The Netherlands (population, 14 million) is a decentralised unitary state with a written constitution under which government is exercised at three levels; the national; the 12 provinces (population, 180,000 to 3,200,000); and more than 700 municipalities (fewer than 2,000, up to 700,000). At each level, the higher tier of government has a supervisory power over the lower; but the lower levels are independent and can pursue their own policies provided they do not conflict with those of the higher.

N7.2 Thus for planning, the primary authority for the control of development is delegated by national legislation to the municipalities. The main role of the provinces is to ensure vertical and horizontal coordination of policies affecting physical planning. The national government formulates outline planning policies, the responsible ministry being Housing, Physical Planning and Environment.

## Planning Legislation and Instruments

N7.3 The first planning legislation was in the 1901 Housing Act. Today

  (i) the 1962 Physical Planning Act (amended 1985) creates the administrative framework for planning at all three levels and defines the statutory procedures for the main planning instruments

 (ii) the 1962 Housing Act (amended 1982) places a duty on all municipalities to adopt building regulations and to issue building permits for all construction

(iii) the 1961 Monuments and Historic Buildings Act gives the Ministry of Cultural Affairs power to list buildings of historic interest and to designate conservation areas

 (iv) the 1984 Town and Village Renewal Act gives municipalities power, if they wish, to adopt urban renewal plans and residential environmental regulations

  (v) a mass of environmental legislation covers pollution etc, some of which is linked to planning control (eg noise).

## The Control of Development

N7.4 The basic system of control derives from the principle of legal certainty which underlies Dutch law and administration.

  (i) The *bestemmingsplan* is a legal document prepared by the municipality showing the 'destination' (ie the intended use and built form) of every piece of land. The plans, created after 1962, and their predecessors from 1901, are mandatory for the entire area of municipalities outside the original built-up area. By now, they cover virtually the entire country including the built-up areas in the larger cities.

 (ii) The purpose of planning control is to safeguard, and eventually to 'realise' (ie implement), the destinations shown in the plan.

(iii) Many plans are action-oriented, creating opportunities for development which in many cases involve the municipality in acquiring, preparing and servicing land for development in partnership with private developers and housing associations etc, to whom the land is then leased or sold in accordance with the plan. The origin for this state involvement lies in the country's history and geography, reclaiming land from the sea, or land with very difficult ground conditions.

 (iv) Many other plans however are designed to safeguard green areas and agricultural land, and areas for conservation, from any unauthorised development.

N7.5 The method of control is by building permit and construction permits designed to safeguard destinations in the *bestemmingsplans* from inappropriate development other than building.

  (i) The municipality's building inspectorate receives the application and

   (a) consults other departments; the aesthetic commission (a local panel of architects and others who advise on aesthetic matters); any other special agencies (eg on landscape, agricultural or environmental matters); if necessary, the National Monuments Agency

   (b) checks building regulations, and the *bestemmingsplan*.

(ii) The decision is made, and the permit issued or refused, with or without conditions, by the municipal executive (the burgomaster and aldermen for the municipality) unless it has been delegated to a locally elected sub-municipal council or the building inspectorate.

    (a) If in accord with plan and regulations, a permit cannot be refused.

    (b) If not in accord, a permit cannot be granted.

(iii) The decision must be made within two months, plus one possible extension of two months after which it is deemed to be refused.

(iv) An applicant refused a permit, or a person aggrieved by the grant of a permit, can appeal in the first instance to the municipal council, and thereafter to the Judicial Department of the Council of State on the legality of the decision.

(v) The Town and Village Renewal Act provides additional controls for dealing with problems of urban renewal. As well as normal control, they enable strict control over demolition. Permission to demolish can be granted only when a building permit has been issued for redevelopment, and a financial guarantee, lodged with the municipality, is forfeited if construction is not started in due time.

## The Definition of Development and the Scope of Control

N7.6 Development requiring permission is defined uniquely for each municipality in its regulations and plans. There is no single, comprehensive, universal definition although the Association of Netherlands Municipalities has model building regulations, and the Ministry is currently working on a further unification of definitions for building regulations and *bestemmingsplannen* as part of the policy of deregulation. But, broadly, the control of development can cover

(i) all building construction, including site planning and layout; the height and mass of buildings; and aesthetic control matters including the outward appearance of buildings

(ii) demolition of buildings

(iii) all agricultural buildings, but in principle excluding agricultural land use other than in areas of special landscape, ecological or archaeological interest

(iv) all 'construction not being building' in specified areas of a *bestemmingsplan*, including tree-planting and felling; mineral workings; land grading and reclamation; piling; deep ploughing, fertilisers and pesticides etc

(v) open spaces, parks, recreation areas and green areas (eg shelter belts, roadside verges etc)

(vi) certain changes of use in areas not covered by a *bestemmingsplan* and all changes of use in areas covered by *bestemmingsplannen* although the use classes tend to be fairly wide. The Association of Netherlands Municipalities has developed a very detailed classification of industrial and commercial activities in terms of their distance from, and impact on, residential areas. There are also special controls over changes of use involving residential property, including the conversion of single family dwellings and control over vacant dwellings

(vii) car-parking and advertisements

(viii) all infrastructure and utilities including different categories of roads, footpaths, cycleways, railways and 'bus-lanes; areas for the protection of overhead and underground pipelines and cables; canals and dykes.

N7.7 The degree of detail varies depending on the required level of control. At its most extreme, every floor in every building in an inner city area is separately identified; or there are separate regulations in a residential area for the actual dwelling, the area of the plot on which extensions, sheds etc may be permitted (*erf*), and the garden on which fences only may be permitted (*tuin*). The plan however can prescribe only what can or cannot be allowed. It does not place any obligation on what must be done.

N7.8 The only exemptions from the prohibition on development must be specified in the regulations or plan, and they still have to receive permission under sections 15 or 18 of the Act. They have recently been extended by the 1985 amendment to the Physical Planning Act and Decree and now include certain short-term, and/or temporary uses; minor construction and extensions to buildings; small, non-residential buildings etc.

## Flexibility in Control

N7.9 However, the system does provide room for flexibility and negotiation, using three principal methods. First, section 11 of the Physical Planning Act enables a *bestemmingsplan* to be expressed in general terms, for further elaboration into its final, legal form when development is imminent. For instance, a plan for an inner city area would identify sites for redevelopment, giving a range of figures for the numbers of dwellings, commercial floorspace etc; or areas for new development would show a table of accommodation and other mandatory requirements for access, circulation etc. These are then worked up in detail in negotiations between the municipality and the developer.

N7.10 Second, article 12 of the Physical Planning Decree was amended in 1985 to enable a *bestemmingplan* to remain 'global', with the regulations expressed only in 'headlines'. It is similar to the section 11 procedure except that the general description remains the actual plan, and any development may be given a permit so long as it remains within the specification. It could be used, for instance, for an inner city area with very mixed land uses; or a village street with shops and houses. In each case, the general description would specify, for instance, the permitted uses, or the number of dwellings. However, this is new, and its use and impact have not yet been tested.

N7.11 Third, section 19 of the Act is the most widely used method. Provided the municipality has declared its intention to prepare, or review, a *bestemmingsplan* then, for a limited period, it can grant permits in anticipation of the new plan provided the province (which has formally to approve the *bestemmingsplan*) issues a general, or a specific, declaration of no objection.

N7.12 Strictly, the section 19 procedure should be used for any development which diverges from the original *bestemmingsplan* beyond the limits for a possible exemption. In practice, provided there is no objection from neighbours, the aesthetic commission, or other departments, it is possible that permission would be given anyway for a comparatively minor development, in effect turning a blind eye (the so-called 'section 5') to the infringement.

## Control Workload

N7.13 There are no generally available statistics but the situation in one large city is roughly as follows.

| | |
|---|---|
| number of permits per annum granted/refused approximately | 1,900 |
| straightforward, in accord with plan | 32% |
| within limits of discretion permitted by the plan under sections 15/18 | 16% |
| by elaboration or revision of the plan under sections 11/19 | 15% |
| sub-total, plan-led | 63% |
| minor developments not in accord with the plan but granted anyway, no objection ("section 5") | 37% |
| total | 100% |
| of which, appealed to the municipal council | 10% |
| appealed to the crown | 1% |

Provincial estimates suggest that a more usual situation would have a higher proportion (30 per cent to 40 per cent) of section 11/19 permits and a lower proportion of the so-called 'section 5' permits.

N7.14 The time taken is supposed to be not more than four months maximum. In practice, ordinary applications take between two and six months; those needing section 11/19 take between eight and eighteen months although the recent changes should result in quicker decisions.

## Policy and the Preparation of Plans

N7.15 The basic plan is the *bestemmingsplan* which, as the one legally binding document, has complex procedures for its preparation, approval and challenge which ultimately can, though do not necessarily, involve the Crown. The plan is

(i) prepared for the municipal executive by the city planning department with consultations and, since 1985, public participation

(ii) placed on deposit for objections from anyone to the municipal council

(iii) the subject of a public inquiry by the municipal council into objections and the plan is approved with or without objections

(iv) placed on deposit for repeated objections, or new objections to modifications, to the province

(v) finally approved by the provincial executive after consultation with the government's regional planning inspector and the provincial planning committee (an advisory committee with members from the provincial offices of government departments and lay experts).

N7.16 The entire process up to this stage, which completes the normal procedure for approval, can take between one and two years for a straightforward plan. If however there is a further round of objections, to the Crown, the process can take five or six years (see below).

N7.17 The *bestemmingsplan* is the only legally binding plan and is therefore the only one which involves a hierarchy of objection and appeal. All other plans are policy instruments which have to go through their own statutory procedures but, not being legal orders or regulations, cannot be challenged by appeal to the Crown, and do not require approval by higher levels of government. They are

(i) national key planning decisions, stating government policy for sectors, such as transport, or physical planning issues such as policy for rural areas

(ii) regional plans prepared by provinces and now covering the entire country

(iii) structure plans, prepared by municipalities as a framework for *bestemmingsplannen*.

N7.18 All three types of instrument have to go through public participation during their preparation. In addition, they are considered by the national or provincial advisory planning committees before final approval by the States-General, provincial or municipal council respectively.

N7.19 These plans are administrative policy instruments. They are not a necessary precondition for the preparation of a *bestemmingsplan* although they are used when evaluating a *bestemmingsplan* for approval. But they can be the authority for directions by national government for the preparation and content of regional plans which are administratively binding; and similarly for directions by the provinces for the preparation of *bestemmingsplannen*.

## Objections and Appeals

N7.20 Dutch law is divided into public, or administrative, law which comes under the Council of State; and private, or civil and criminal, law which comes under the Supreme Court. In both, the function of the courts is to interpret the law by reference to the legislation and the constitution.

N7.21 Objections and appeals, as part of administrative law, come under the Council of State. Objectors to a *bestemmingsplan* not satisfied at the provincial level, or municipalities and the government's regional planning inspector aggrieved by a provincial decision, can appeal to the Crown. This involves the Disputes Department of the Council of State.

(i) A technical planning report on the objections is prepared by the Advisory Bureau of the Ministry.

(ii) The Disputes Department publishes the report and all other documents, holds a public inquiry and makes its recommendation based on the law, in the form of a draft royal decree approving, or withholding approval, of the plan.

(iii) The Minister either accepts the recommendation and publishes the decree, or refers it back for reconsideration on one more occasion.

N7.22 The process is lengthy and can take up to between 18 and 21 months at the minimum, and five years is not unknown, with 400 plans being appealed during 1985. It is a quasi-judicial procedure in that the Crown's (in effect, the Minister's) decision is made on grounds of 'law and expediency' as a matter of government policy. There is no further appeal to the courts.

N7.23 Appeals against administrative decisions such as on the grant or refusal of a building permit or a section 19 application follow a different procedure. They are made under the so-called AROB law of 1976. The appeal is to the Judicial Department of the Council of State in its capacity as a final court of appeal; and the decision is made purely on its interpretation of the law and 'the principles of good government': that the decision was made correctly within the legislation and the given rules. As an appeal can be made against the grant of a permit, there is also a means for obtaining an injunction, pending a decision on the appeal. There were 1,289 appeals of this kind in 1985.

N7.24 Finally, the civil and criminal courts generally are involved only when enforcing building or planning control; or, rarely, if a municipality or government department acts illegally, for instance in failing to comply with proper procedures in enforcement.

## Conclusions

N7.25 In conclusion, the chief characteristics of the system of control in the Netherlands are that it

(i) links building and planning control in the one instrument, and allows for the control of demolition

(ii) provides ample opportunity for objections to plans, and appeal against decisions both for the grant and refusal of the permission, and now also enables public participation in the preparation of plans

(iii) tries to provide legal certainty for developers and their neighbours through the *bestemmingsplan*

(iv) allows a measure of local, municipal discretion in deciding what to control and a more limited measure of flexibility in exercising that control

(v) provides a policy framework at both national and provincial levels.

N7.26 Thus it is reasonable to describe the Dutch planning system as being plan-led. Virtually the entire country likely to experience development is covered by legally binding plans, and permits have to be granted or refused by reference to the mandatory requirements of the plan. The only room for discretion is that expressly laid down in the plan, including the limits of that discretion, which are narrowly defined. The only alternative is for the municipality to start the procedures for preparing or amending a plan, and then possibly decide to grant permission.

N7.27 The system as a consequence is open to the criticisms that it

(i) involves potentially very lengthy procedures for the preparation and revision of plans

(ii) is overly detailed and inflexible

(iii) is more appropriate for positive planning and the control of major new development such as the planned extension to existing built-up areas, and the prevention of urban sprawl in rural areas; but is less appropriate for control of the incremental adaptation and renewal of existing built-up areas.

N7.28 The control of development therefore in practice results in a lot of uncertainty as ways are found for overcoming the rigidities, either by the section 19 procedures for interim control, or by the so-called section 5 method of in effect ignoring the plans, although it is not clear how widespread is the latter practice.

N7.29 Some of these criticisms are being addressed by government, by unification of definitions; by simplification of procedures; and, above all, by deregulation or at least enabling a wider degree of flexibility in the system. It is too soon to assess the likely success of these measures which still have as their aim the maintenance of legal certainty and decentralisation of control.

# GLOSSARY

**aanlegvergunningen**
construction permit required under the Physical
Planning Act for any construction and other works not
being building, specified in a *bestemmingsplan*, which
would otherwise jeopardise the plan

**Afdeling Geschillen**
Disputes Department of the Council of State,
responsible for hearing objections to
*bestemmingsplannen* etc

**Afdeling Rechtspraak**
Judicial Department of the Council of State,
responsible for hearing appeals against administrative
decisions, including the grant or refusal of building
permits

**agrarische adviescommissie bouw-en aanlegvergunning**
independent committee of agricultural experts
appointed by a municipality to comment on
applications for building and construction permits

**beleidsrategie**
the broad strategy of a regional plan

**beschrijving**
administratively binding regulations in a regional plan

**Besluit Ruimtelijke Ordening**
ministerial decree on the form and content of local
plans

**bestemmingsplan**
literally, destination plan (also land allocation plan;
zoning plan; development plan); the legally binding
plan prepared by a municipality under the 1962
Physical Planning Act for its non built-up area, and
now, with its predecessor, the *uitbreidingsplan*, covering
most areas built up since 1901

**binnenterrein**
inner courtyard of a block of apartments, likely to be
covered by separate regulations in a *bestemmingsplan*

**bouw-en woningtoezicht**
the building and housing inspectorate, or department,
of a municipality, responsible for processing building
permits

**bouwvergunning**
building permit required for all building construction
under the Housing Act

**bouwverordening**
building regulations prepared by all municipalities
under the Housing Act, containing planning, as well as
building, matters

**Bureau van de Adviseeur t.b.v. de Raad van State**
Advisory Bureau in the Ministry of Physical Planning
responsible for the technical planning report on
objections to a *bestemmingsplan*

**burgemeester**
burgomaster, the chief official of a municipality,
appointed by the Crown

**college van burgemeester et wethouders**
municipal executive, the committee of aldermen
(*wethouder*) chaired by the burgomaster, and
responsible for all executive decisions under the
municipal council

**Dienst Centraal Milieubeheer Rijnmond**
Central Environmental Control Agency for Rijnmond,
responsible for monitoring, and issuing licences relating
to, environmental pollution or nuisance

**dienst ruimte en groen**
the newly created provincial department responsible for
physical planning, replacing the former *provinciale
planologische dienst*

**erf**
the garden of a house, usually subject to separate
regulations in a *bestemmingsplan*

**exploitatie-opzet**
economic feasibility study of the development proposed
in a *bestemmingsplan*

**exploitatie overeenkomst**
financial agreement between a municipality and
developer, for implementation of a *bestemmingsplan*

**facet plan**
national policy for specific topics, produced by
government

**gedeputeerde staten**
provincial executive, the committee of provincial
deputies, chaired by the provincial governor, and
responsible for all executive decisions under the
provincial council

**gemeente raad**
the elected municipal council

**globaal**
a term used to describe a generalised *bestemmingsplan* requiring further elaboration under section 11 before permits can be issued

**Hoge Raad**
the Supreme Court, under the civil and criminal codes, but not for administrative law

**hoofdlijnen**
literally, headlines, meaning a new form of *bestemmingsplan* under which the regulations may be expressed in more general terms, without needing further elaboration

**inspecteur ruimtelijke ordening**
government inspectors of physical planning, appointed by the Minister of Physical Planning for five regions in the country, *ex officio* members of the *provinciale planologische commissie*

**kernpunten**
the structural elements for each sub-region of a regional plan, showing the location of future development etc

**komplan**
a detailed plan and regulations for an existing built-up area, formerly prepared under 1931 legislation

**landelijke gebieden en kwaliteizord**
independent committee of experts appointed by a municipality to comment on applications for building permits affecting areas of landscape quality and the natural environment in a *bestemmingsplan*

**leefmilieuverordeningen**
residential environment regulations prepared under the Town and Village Renewal Act to control changes of use and demolition

**Ministerie van Volkshuisvesting en Ruimtelijke Ordening**
Ministry of Housing, Physical Planning and Environment

**Ministerie van Welzijn, Volksgezondheid en Cultuur**
Ministry of Welfare, Health and Culture, responsible *inter alia* for listed buildings and conservation

**modelbouwverordening**
model building regulations prepared by the Association of Netherlands Municipalities

**planologische kernbeslissigen**
administratively binding key planning decisions by the national government, approved by the States-General

**premie**
partially subsidised housing for rent, intermediate between unsubsidised free housing and subsidised social housing

**provinciale planologische commissie**
provincial physical planning committee, including the provincial heads of government departments, lay experts and the government's inspector of physical planning, responsible for independent advice to the provincial executive on all planning matters

**provinciale planologische dienst**
provincial planning agency/department, now replaced in most provinces by the *dienst ruimte en groen*

**provinciale stateen**
the elected provincial council, chaired by a provincial governor appointed by the Crown

**Raad van Advies voor de Ruimtelijke Ordening**
Advisory Physical Planning Council, a committee of representatives of different public interest and sectoral organisations and lay experts appointed under the Physical Planning Act to advise the Minister of Physical Planning

**Raad van State**
Council of State, the highest advisory organ of state and responsible for administrative justice and law

**Raad voor Ruimtelijke Ordening en Milieuhygiene**
cabinet committee for national physical planning and environment, the chief coordinating body of ministers

**raadsleden**
elected municipal councillor

**rechtstaat**
literally, the constitutional state, the basic principle underlying Dutch law and administration (see van Gunsteren, 1976)

**rechtszerheid**
the idea of legal certainty

**richtlijnen**
directions in a regional plan for the content of a *bestemmingsplan*

**Rijksdienst Oudheidkundig Bodemonderzoek**
National Agency for Archaeological Surveys, responsible for archaeological sites

**Rijksdienst voor de Monumentenzorg**
National Monuments Agency, responsible for listed buildings and conservation areas

**Rijksplanologische Commissie**
National Physical Planning Committee, the interdepartmental committee of officials, established by the Physical Planning Act for coordinating the policies of ministries as they affect physical planning

**Rijksplanologische Dienst**
National Physical Planning Agency, established by the Physical Planning Act to provide technical advice and support to the Minister of Physical Planning

**stadsvernieuwingsplan**
urban renewal plan prepared under the Town and Village Renewal Act

**Staten-Generaal**
States-General (Parliament)

**streekplan**
administratively binding regional plan, prepared in a province for all or part of its area

**struktuurplan**
administratively binding structure plan, prepared by a
municipality for all or part of its area

**toelichten**
the explanatory memorandum with a regional or
structure plan or a *bestemmingsplan*

**tuin**
the grass verges and other open land alongside roads,
usually subject to separate regulations in a
*bestemmingsplan*

**uitbreidingsplan**
legally binding town extension plan, prepared under the
1901 Housing Act for all new development, replaced by
the *bestemmingsplan*

**uitvoeringsschema**
administratively binding plan for the implementation of
an urban renewal plan

**Verening van Nederlandse Gemeenten**
Association of Netherlands Municipalities

**voorschriften**
regulations in a *bestemmingsplan*, giving precise details
of permitted uses and building

**welstandscommissie**
aesthetic committee of architects and lay people
appointed by a municipality under the building
regulations to comment on the outward appearance of
all proposals for development

**Wet Administratieve Rechtspraak
Overheidsbeschikkingen - AROB**
Administrative Tribunal on the Decisions of Public
Authorities Act 1976, establishing basis for appeal
against grant or refusal of permits etc

**Wet Ruimtelijke Ordening**
Physical Planning Act 1962, 1985

**Wet Stads-en Dorpsvernieuwing**
Town and Village Renewal Act 1984

**wethouder**
alderman, member of a municipal council elected by the
council to be a member of the municipal executive, with
responsibility for particular functions

**Woningwet**
Housing Act 1901, 1962, 1982

# REFERENCES

ALBRANDSWAARD (1986), **Ontwerp-Bestemmingsplan Albrandswaard-Noord:** Voorschriften (Draft Plan: Regulations and Explanation), Gemeente Albrandswaard

AMSTERDAM PHYSICAL PLANNING DEPARTMENT, (1983), **Amsterdam: Planning and Development,** Amsterdam: City of Amsterdam

BEKKER, R et al (1981) **Growth Centres and Growth Cities: The Planned Development of Existing Communities in the Netherlands,** The Hague; Ministry of Housing, Physical Planning and Environment

BOMNER J (1967), **Housing and Planning Legislation in the Netherlands,** Rotterdam: Bouwcentrum, in co-operation with the Ministry of Housing and Physical Planning

BORCHERT, J (1979), "Retail Planning in the Netherlands", in DAVIES, R L (editor), **Retail Planning in the European Community,** Farnborough, Hants: Saxon House, pp 81–98

BORCHERT, J (1985), "Dutch Retail Planning Reviewed", **The Planner,** vol 71, no 5, pp 16–18

BRUSSAARD, W (1979), **The Rules of Physical Planning,** The Hague: Ministry of Housing and Physical Planning

BRUSSAARD, W (1986), "Physical Planning Legislation in the Netherlands", in GARNER, J F and GRAVELLS, N P (editors), **Planning Law in Western Europe,** Oxford: Elsevier Science Publishers B. V. (North Holland), 2nd edition

CARRICK, R J and WRATHALL, J E (1983), "Urban Renewal in the Netherlands", **The Planner,** vol 69, no 3, pp 88–90

CRINCE LE ROY, R (1975), "The Netherlands: The Dutch Planning Act", in GARNER, J F (editor), **Planning Law in Western Europe,** Oxford: North Holland Publishing Co, 1st edition

DEKKER, F et al (1978), "A Multi-level Application of Strategic Choice at the Sub-regional Level (Eemland)", **Town Planning Review,** vol 49, no 2, pp 149–162

DE KLERK, L A (1986), **Rotterdam: Urban Crisis and Urban Policy** (conference paper for the City, The Engine behind Economic Recovery), Rotterdam: City of Rotterdam

DEPARTMENT (for the preservation) OF MONUMENTS AND HISTORIC BUILDINGS (undated), **The Preservation of Historic Buildings: the Role of Government,** Zeist: Rijksdienst voor de Monumentenzorg

DE SCHMIDT, M (1985), "National Development and Economic Policy", in DUTT, A K and COSTA, F J, (editors) (1985), **Public Planning in the Netherlands: Perspectives and Change Since the Second World War,** Oxford: Oxford University Press

DE VRIES, G W (1986), **Krachtens de Bouwverordening: Bouw-en-Woningtoezicht Rotterdam 1861–96** (Rotterdam Building and Housing Inspection 1861–96, English Summary), Rotterdam: Uitgeverij Ad Donker

DE VRIES-HEIJNIS, G E en VAN WOERKOM, H C M (1985), **Wet op de Stads-en Dorpsvernieuwing,** Editie Schuurman &

Jordens, Zwolle: W E J Tjeent Willink

DUTT, A K and COSTA, F J (editors) (1985a), **Public Planning in the Netherlands: Perspectives and Change since the Second World War,** Oxford: Oxford University Press

DUTT, A K and COSTA F J (1985b), "Introduction", in DUTT and COSTA, **op cit**

FALUDI, A and DE RUIJTER, P (1985), "No Match for the Present Crisis? The Theoretical and Institutional Framework for Dutch Planning", in DUTT and COSTA, **op cit**

FASSBINDER, H and KALLE, E (1982), **Comparative Study of Urban Renewal Policy,** The Hague: Ministry of Housing and Physical Planning

FLOOR, J W G, (1981), **Rents, Subsidies and Dynamic Cost,** The Hague: Ministry of Housing and Physical Planning

FRIELING, D H (1986), "A Garden in Europe: the Case of the Ijsselmeer District of the Netherlands", **Town Planning Review,** vol 57, no 1, pp 35–50

HABERER, P, et al (1980), **The Neighbourhood Approach: Improvement of Old Neighbourhoods by and on Behalf of their Inhabitants,** The Hague: Ministry of Housing and Physical Planning

HABERER, P. (1982), **Urban Renewal in the Netherlands,** The Hague: Ministry of Housing and Physical Planning

HAMNETT, S (1975), "Dutch Planning: a Reappraisal", **The Planner,** vol 61, no 3, pp 102–105

HAMNETT, S (1982), "The Netherlands: Planning and the Politics of Accommodation", in McKAY D H (editor), **Planning and Politics in Western Europe,** London: Macmillan

HAMNETT, S (1985), "Political Framework and Development Objectives of Post-war Dutch Planning", in DUTT and COSTA, **op cit**

HAZELHOFF, D (1981), **Introduction to Physical Planning at the National Level,** The Hague: Ministry of Housing and Physical Planning

HAZELHOFF, D (1981), **Government Reports on Physical Planning,** The Hague: Ministry of Housing and Physical Planning

HETZEL, O J (1983), **A Perspective on Governmental Housing Policies in the Netherlands,** The Hague: Ministry of Housing, Physical Planning and Environment

HETZEL, O J (1985), "Government Housing Policies in the Netherlands since 1945", in DUTT and COSTA, **op cit**

HEYWOOD, P (1970), "Regional planning in the Netherlands and England and Wales", **Journal, Royal Town Planning Institute,** vol 56, no 10, pp 428–434

JONES, M G (1984), "The Evolving EIA Procedure in the Netherlands", in CLARK, B D, (editor), **Perspectives on Environmental Impact Assessment,** Dordrecht: D Reidel Publishing Co

KITS NIEUWENKAMP, J (1985), "National Physical Planning:

Origins, Evaluation and Current Objectives," in DUTT and COSTA, **op cit**

KOOPMANS, T (1971), "Netherlands", in KNAPP, V (editor), **International Encyclopaedia of Comparative Law, Volume 1, National Reports,** The Hague: J C B Mohr (Paul Siebeek)

KREUKELS T, MASSER, I and PUTS, E (1978), "Spatial Planning in Eight Eemland Municipalities: Research Perspectives", **Town Planning Review,** vol 49, no 2, pp 119–126

KUNZMAN, K R (1985), "Educating Planners in Europe", **Town Planning Review,** vol 56, no 4

MACMULLEN, A L (1979), "Netherlands", in RIDLEY, F F (editor), **Government and Administration in Western Europe,** Oxford: Martin Robertson

MCCLINTOCK, H and FOX, M (1971), "The Bijlmermeer Development in the Expansion of Amsterdam", **Journal, Royal Town Planning Institute,** vol 57, no 7 pp 313–316

MARKLAND, J and WILKINS, C. (1978), "Professional Planning Agencies in the EEC", **The Planner,** vol 64, No. 3

MASSER, I et al (1978) "The Dynamics of Develoment Processes: Two Case Studies (Eemland)", **Town Planning Review,** vol 49, no 2, pp 127–148

MINETT, J (1978), "Local Planning in the Netherlands and England and Wales", **Journal, Royal Town Planning Institute,** vol 56, no 10, pp 428–434

MINISTRY OF CULTURAL AFFAIRS, RECREATION AND SOCIAL WELFARE (undated), **The Dutch Monuments and Historic Buildings Act 1961,** The Hague: Ministry of Cultural Affairs

MINISTRY OF HOUSING AND PLANNING (and ENVIRONMENT, after 1982), English language reports on housing and planning, The Hague: Ministry of Housing, Physical Planning and Environment

(1981a), **Main Characteristics of the Land-Use Policy in the Netherlands**

(1981b), **Government Control over Housing in the Netherlands,** Factsheet

(1981c), **The Municipality and Housing in the Netherlands,** Factsheet

(1981d), **Compulsory Purchase Act,** Factsheet

(1981e), **The Muncipalities' Right of Pre-emption Act,** Factsheet

(1981f), **The Agrarian Land Transactions Act,** Factsheet

(c 1981g), **Empty Property Act,** Factsheet

(1982a), **National Physical Planning Key Decisions**

(1982b), **The Housing Act in the Netherlands,** Factsheet

(1982c), **Physical Planning Act,** Factsheet

(1983a), **The Future of Randstad Holland: A Netherlands Scenario Study**

(1983b), **Multiyear Plan for Urban Renewal in the Netherlands**

(1983c), **Some Reflexions on the Interaction Between Settlement Patterns and Economic Performance in the Netherlands**

(1983d), **Structural Outline Sketch for the Urban Areas,** Factsheet

(1983e), **Non-Profit Housing Associations,** Factsheet

(1983f), **Environmental Protection in the Netherlands**

(1984a), **Rural Areas Report,** Factsheet

(1984b), **The Chemical Waste Act,** Factsheet

(1985a), **Environmental Program of the Netherlands 1985–1989**

(1985b), **Summary of the Netherlands Urban and Village Renewal Act 1985,** Factsheet

(1985c), **The Financing of Housing and Housing Subsidies in the Netherlands**

(1985d), **Individual Rent Allowance,** Factsheet

(1985e), **Decentralization of Housing Policy,** Factsheet

(1986a), **Environmental Hygiene and Urban and Village Renewal**

(1986b), **Outline of the Netherlands' Noise-Nuisance Act,** Factsheet

(1986c), **Industrial Hazardous Waste Management in the Netherlands,** Factsheet

(1986d), **Government Assistance in Housing in the Netherlands,** Factsheet

(1986e), **Randstad Country-Planning Outline,** Factsheet

MINISTERIE VAN VOLKSHUISVESTING RUIMTELIJKE ORDENING EN MILIEUBEHEER (1985) **Wegwijzer Vergunningen in de Ruimtelijko Ordening** (Guide to Permits in Physical Planning), Den Haag: VROM

MINISTERIE VAN VOLKSHUISVESTING RUIMTELIJKE ORDENING EN MILIEUBEHEER (1986), **Bestemmen met Beleid: Voorlopig Standpunt** (Destinations with Prudence: interim statement), Den Haag: VROM

RAAD VAN STATE (1985), **Jaarverslag 1985,** Den Haag: Raad van State

RAAD VAN STATE (1987), **Raad van State** (in English), Den Haag: Raad van State

RIJKSPLANOLOGISCHE, **Annual Reports,** chapter 1 (in English), The Hague: National Physical Planning Agency

(1979), "Physical Planning Developments in Europe" (1978 Report)

(1981), "Aspects of the North Sea" (1979 Report)

(1983). "Energy and Physical Planning" (1981 Report)

RIJNMOND (1979), **Central Environmental Control Agency: Annual Report, 1979,** Schiedam

RIJNMOND (1984), **Dienst Centraal Milieubeheer: Jaarverslag 1984**

ROTTERDAM (1981), **A Systematic Approach to Urban Renewal in Rotterdam,** Rotterdam: City of Rotterdam

ROTTERDAM (1984), **Rotterdam in Focus** (Factpack, including local government, city services, city planning, urban renewal, housing etc), Rotterdam: City of Rotterdam

ROTTERDAM (1986), **Bouwverordening der Gemeente Rotterdam** (Building Regulations), Rotterdam: Stadzimmerhuis, Marconiplein

ROTTERDAM (1979), **Bestemmingsplan Centrum Oost,** Rotterdam: Stadzimmerhuis, Marconiplein

SCHOLTEN, H (1984), "Planning for Housing Construction and Population Distribution in the Netherlands: The Use of Forecasting Models", **Town Planning Review,** vol 55, no 4, pp 405–419

'S-GRAVENHAGE (1985), **Bestemmingsplan Regentesse-/Valkenboskwartier Zuid: Toelichten, en Voorsdiriften** (Plan: Explanation and Regulations), Den Haag: Gemeentelijke Dienst Stadsontwikkeling-Grondzaken

'S-GRAVENHAGE (1981) **Ontwerp Structuurplan 's-Gravenhage-eerste stap** (Draft Structure Plan), Den Haag: College van Burgemeester en Wethouders

STAATSBLAD VAN HET KONINKRIJK DER NEDERLANDEN (1985), **Wet en Besluit op de Ruimtelijke Ordening** (Physical Planning Act and Decree, with explanatory note), Staatsblad 626, 627, Den Haag: Staatsuitgeverij

THOMAS D et al (1983), **Flexibility and Commitment in Planning: A Comparative Study of Local Planning and Development in the Netherlands and England,** The Hague: Martinus Nijhof, Publishers

VAN DELFT, A and KWAAK, A (1985), "National Development and Economic Policy" in DUTT and COSTA, **op cit**

VAN DEN BERG, L et al (1982), **Urban Development and Policy Response in the Netherlands,** Rotterdam: Netherlands Economic Institute

VAN DER CAMMEN, H (1984), "Netherlands" in WILLIAMS, R H (editor), **Planning in Europe,** London: George Allen & Unwin

VAN DER HOUT, C and SEGOUD VON BOUCHET, G (1985), **Area-Based Approach to Urban Renewal in the framework of an urban policy,** paper for a United Nations Working Party on Urban and Regional Planning, (unpublished)

VAN DER SLUYS, F and VAN EVERT, G (1985), "Housing Renewal in the Hague", in DUTT and COSTA, **op cit**

VAN GUNSTEREN, H (1976), **The Quest for Control: a critique of the rational-central rule approach in public affairs,** London: John Wiley

VAN HEZIK, M J M and VERBEIJEN, L (1983), **The Netherlands in Brief,** The Hague: Ministry of Foreign Affairs

VAN ITTERSUM (c 1986), **The Council of State,** The Hague: Raad van State (unpublished)

VAN VOORDEN, F W (1981), "The Preservation of Monuments and Historic Townscapes in the Netherlands", **Town Planning Review,** vol 52, no 4, pp 433–453

VERENING VAN NEDERLANDSE GEMEENTEN (1986), **22e Serie Wijzigingen (Model-) Bouwverordening,** (22nd Series of Modifications to the Model Building Regulations), Den Haag: VNG

VOOGD, H (1982), "Issues and Tendencies in Dutch Regional Planning", in HUDSON R and LEWIS J (editors), **Regional Planning in Europe,** London: Pion (1979)

WOOD, C and LEE, N (1978), **Physical Planning in the Member States of the EEC,** Occasional Paper No. 2, Department of Town and Country Planning, Manchester: University of Manchester

ZUID HOLLAND (undated), **de Wet-AROB en de Provincie,** (Administrative Disputes, Law and the Province), Den Haag: Provinciehuis, Koningskade 1

ZUID HOLLAND (1983, 1986), Circular letter from the Provincial Executive to Burgomasters and Alderman on Section 19 of the Physical Planning Act, Den Haag: Privinciehuis, Koningskade 1

ZUID HOLLAND (1986), **Streekplan Rijnmond** (Regional Plan for Rijnmond), Den Haag: Provinciehuis, Koningskade 1

ZUID HOLLAND (1987), **De Nieuwe Organisatie** (New Organisation), Den Haag: Provinciehuis, Koningskade 1

# APPENDIX A  THE SCOPE AND DEFINITION OF DEVELOPMENT:

## (a) BUILDING REGULATIONS

1.  Every municipality is required, by section 2(1) of the Housing Act 1962, to adopt legally binding building regulations. The regulations have to give the general definitions of building which requires a building permit; an elaboration of criteria and standards; building which is exempted from needing a permit; and building for which the municipal executive may give a permit notwithstanding the regulations. In addition, the regulations give the procedures to be followed in dealing with an application for a permit, and in enforcement.

2.  Virtually every municipality now bases its regulations on the model building regulations first prepared by the Association of Netherlands Municipalities in 1965, and kept up-to-date. In addition, municipalities may add their own regulations to meet local circumstances.

3.  The following notes summarise the contents of building regulations, giving greater emphasis to matters relating to physical planning. They are drawn from two sources.

(i)  The City of Rotterdam Building Regulations, 1986 (Rotterdam, 1986) as an illustration of the regulations adopted by a large, independent municipality with a history of building regulation going back to 1857 (de Vries, 1986). They are very largely based on the model regulations.

(ii)  The most recent, 22nd, series of comment on, and modifications to, the model building regulations prepared by the Association and incorporating legal decisions of the administrative and other courts (Verening van Nederlandse Gemeenten, 1986).

| | |
|---|---|
| **Chapter 1:** | **Introduction and Definitions** |
| article 1 | defines building as "any construction of sizeable proportions made of wood, stone, metal or other material which is directly or indirectly linked with or based on the ground in its location" |
| **Chapter 2:** | **Administration** |
| article 14 | gives a general requirement for a building permit for any building |

| | |
|---|---|
| (1) | grants the possibility of certain exemptions |
| (a) | for minor constructions in which aesthetic considerations are insignificant, if the municipal executive agrees |
| (b), (c), (d), | for certain fences, aerials and very small non-residential buildings |
| article 31 | requires a special permit, under section 56A of the Housing Act, for the conversion of a dwelling into flats/apartments. This applies only in certain cities with housing problems, including Rotterdam |
| **Chapter 3:** | **Technical Regulations** |
| **Part 3A:** | **The Building and its Environment** |
| para 1 | the relation of a building with its surroundings |
| articles 32, 33 | a building must not interfere with existing or planned surrounding buildings and their compliance with regulations, and *vice versa* |
| article 34 | a building etc must "in itself and in conjunction with the surrounding existing or planned development ... (comply with) ... reasonable aesthetic requirements". This refers explicitly to the "outward appearance" of a new building, rather than its use. It can cover matters such as house extensions, dormer windows, the colour of external painting, aerials, and large, illuminated advertisements and billboards, on all of which the legal principles have been tested in the administrative courts. Aesthetic matters have to be considered, and there is a formal procedure for securing advice |
| para 2 | buildings in areas not covered by a *bestemmingsplan* |
| article 36 | prohibits all building in rural areas, with certain exemptions, namely buildings for public utilities, greenhouses |

393

| | |
|---|---|
| para 3 | buildings on road frontages |
| article 37 | prohibits building other than on sites with a frontage to certain classes of road |
| paras 4-6 | lay down standards and criteria for building lines, front and rear; space about, and between, buildings; and height of buildings |
| **Part 3B:** | **Internal Planning of Buildings** |
| | this covers the internal space requirements; daylighting, ventilation and sound-proofing; sanitation, water and electricity |
| **Part 3C** | **Construction of Buildings** |
| | this covers foundations, structure, etc |
| **Part 3D** | **Special Regulations** |
| para 1 | fire and safety |
| para 2 | special requirements |
| article 258 | requires the municipal executive to make further rules and standards for on-site car parking. The Rotterdam rules divide the city into three types of area and define parking standards for each type of use in each area |
| article 263 | gives further control over the interior of buildings of historic or cultural importance but not listed under the Monuments Act |
| **Chapter 4** | **Roads and Open Spaces, On-Site** |
| **Part 4** | **Construction of Roads** |
| article 287 | prohibits road construction on-site without a permit, though roads built by specified public authorities are exempted |
| **Part 4B** | **Layout and Maintenance of Open Spaces** |
| article 294 | layout of open space, on-site |
| **Chapter 5** | **Demolition of Buildings** |
| article 304(1) | prohibits the demolition of buildings without the permission of the municipal executive. This is a strong control which is found in most large towns and cities such as Rotterdam, but not necessarily in smaller municipalities. The only exemptions in Rotterdam are if the municipal executive has ordered demolition; if the demolition is also covered by a building permit for a new building; or, with permission, for minor demolitions |
| **Chapter 6** | **Existing Buildings** |
| | this covers their general maintenance, especially for sanitation, water supply, gas and electricity |
| **Chapter 7** | **Use of Buildings and Curtilage** |
| para 1 | residential use |
| article 346 | requires a special permit under section 55(1) of the Housing Act for the conversion of non-residential to residential uses, the only exemption being if the previous non-residential use was temporary |

| | |
|---|---|
| para 2 | unfit buildings |
| para 3 | use of buildings, open spaces and sites according to their *bestemmings* (destination) |
| article 352(1) | prohibits the use of a building or open space contrary to the destination in an extension plan prepared under the Housing Act 1901, or to the nature of a building implied by its construction unless there are specific regulations in a subsequent *bestemmingsplan*. |
| | This is very far reaching and covers, for instance, "associated construction not being building", ie car parks etc. It also covers non-agricultural uses and the use of agricultural buildings in rural areas although most modern *bestemmingsplannen* will specify these matters in detail. The main exemptions, according to the administrative courts, are for temporary, short-term uses such as a cycle racing track for one week, with the permission of the municipal executive; and for uses the prohibition of which would be unconstitutional (eg freedom of worship). |
| article 352(2)(a) | prohibits changes of use of buildings and open spaces contrary to the destination which is apparent from their construction, styling and/or layout in specified areas of the city, mainly the older areas of the inner city. |
| | In general however the Association of Netherlands Municipalities notes that a distinction can be made between the use of a building implied by its construction and that laid down in a *bestemmingsplan*, the *verwezenlijkte* (realised destination). Thus the courts have found that certain changes of use are acceptable provided no external construction is involved, for instance the conversion of a garage for a supermarket, or even the use of a factory as a supermarket shown as an industrial area in a *bestemmingsplan*. |
| article 352(2)(b) | explicitly prohibits the use of vacant sites for industrial purposes |
| article 352(3) | prohibits the use of buildings in rural areas for uses other than those specified in the *bestemmingsplan*, or vacant sites for uses other than agriculture |
| article 352(4) | states that articles (1), (2) and (3) do not apply to existing uses provided that they were lawful at the time when the regulations were approved, and that they continue unchanged. |

However, the Association states that the

existing use may not be used as a reason to prevent the realisation of a destination in a *bestemmingsplan*. Furthermore there are circumstances in which a municipality either may, or must, grant exemption from the prohibitions in article 352(1). These include where prohibition would prevent the most efficient use of a building or open space without good reason; where a building cannot sensibly be used according to its original destination; or if a change of use would not cause a far-reaching, irreversible effect on the original destination. Examples tested in the courts include the continued use of a factory in operation, to prevent it becoming vacant; or the use of a disused barn for car repair. But the administrative courts have upheld a refusal to allow a surgery in a predominantly residential street to continue in use when the municipality wished to retain the residential character of the street

| article 352(5) | provides for a limited period of exemption to articles 352(1)-(3) |
| article 353 | permits the use of space within the curtilage for parking, loading and unloading |
| para 4 | covers safety and welfare |
| para 5 | covers overcrowding of houses |

**Chapter 8** **General Executive Regulations**
these cover procedures for inspection etc

| **Chapter 9** | **Other Regulations** |
| para 1 | covers road safety |
| para 2 | covers advertisements |
| article 388a | prohibits all advertisements, except as provided for by special, local regulations. In Rotterdam, these define different kinds of areas/streets, and define standards for each. The only general exemption is for small, non-illuminated signs |

**Chapter 10** **General Rules for Appeals**

**Chapter 11** **Sanctions, Transitional and Title**

# (b) THE BESTEMMINGSPLAN

1   A *bestemmingsplan* is a legally binding map, with regulations and explanatory memorandum, indicating the intended pattern of development for its area. That is, the buildings and uses specified in the regulations and on the map are the 'destination' of planning policy ultimately to be 'realised' in precise detail. Any divergence from that destination, including the current existing use when the plan was finally approved, strictly is illegal unless the municipal executive is empowered by the regulations to grant permission within specified limits for the divergence, or there is a general exemption for what will be a very minor divergence.

2   At present, each *bestemmingsplan* is prepared by, or for, a municipality to meet its local circumstances, and within the policies laid down by the province. Each plan therefore is unique and incorporates its own definitions. The following notes summarise the definitions used in two *bestemmingsplannen*, describing different types of area, concentrating purely on the scope and definition of development subject to planning control.

## Albrandswaard-Noord
(see Figure N8 and Plate 27)

3   Albrandswaard is a small municipality with a population of 13,600 which was created in 1985 following the amalgamation of two smaller municipalities. The villages of Poortugaal and Rhoon lie about 2 km south of the Nieuwe Maas river and about 7 km south-west of the centre of Rotterdam.

4   The area covered by the Albrandswaard-Noord *bestemmingsplan* is about 4 x 1.5 km, lying between the villages and the northern boundary of the municipality (Albrandswaard, 1986). It is mainly in agricultural use, with a scattering of isolated houses and small groups of houses, and some industry. It is crossed by various underground pipelines and overhead lines, and is bounded on the north by the main east-west motorway serving Europoort.

5   The plan was prepared for the municipality by a firm of planning consultants in April 1986 and is currently going through the process of approval. It contains one major, contentious proposal, for the afforestation and landscape conservation of an area which the provincial council will probably wish to see continue and expand its existing temporary use as a container storage terminal but to which the municipality is opposed. However, the plan does illustrate the scope and definitions of development in a rural area, where the proposed policy essentially is to prevent further urban development.

6   There are 37 regulations covering every detail of the plan; the plan itself, to a scale of 1/5,000, defining buildings and uses with the precision and detail of a topographical base map; inset maps to a scale of 1/2,000 for the built up areas; road profiles; and a supplementary map for the noise regulations.

7   The general regulations regarding the use of land and buildings are

(i)   any use contrary to the plan is prohibited

(ii)  certain uses of land (eg deposit of rubble or dredging materials, storage/dumping of boxes, and in certain areas, camping/holiday homes, motor racing etc) are specifically prohibited

(iii) existing uses of land and buildings at the time of approval of the plan may continue in use, even if contrary to the plan, with a number of provisos regarding an intensification of the use, or replacement or alteration of the building, which would then require permission from the municipal executive.

8   The predominantly urban uses, shown mainly on the inset plans, are as follows.

(i)   The map shows the following specific destinations

(a)   residential, meaning specifically the actual dwelling house in the location shown on the map, for single family dwellings only

(b)   the area of the curtilage (*erf*) within which garages and other outbuildings, and other construction not being buildings (such as driveways, or paved yards etc), may be constructed

(c)   the rest of the curtilage (*tuin*) within which only fences and hedges are permitted, in effect the space between the front building line and the road or public open space

(d)   retail, covering shops and related uses such as hotels, restaurants and cafés (public houses), including the buildings and associated construction, but excluding residential

(e)   commercial enterprises with separate sub-divisions for manufacturing; wholesale; utilities;

storage; and transport, covering the buildings and associated construction but excluding retail

(f) public and special uses which, in this plan, is confined to a couple of clubhouses for recreation activities.

(ii) The regulations for each destination or zone specify standards or criteria including

(a) for most buildings, the building lines (front and rear); minimum width, maximum height, minimum floor area; distance between outbuildings, the main building, and plot boundaries

(b) for commercial uses
- one associated or tied dwelling is permitted, within specified size limits
- the maximum percentage of the site to be covered by buildings and construction is specified
- the list of permitted uses within each of the five principal commercial uses, being those within categories 1 and 2 of the very detailed industrial classification prepared by the Association of Netherlands Municipalities (ie those with very little environmental impact on residential areas).

(iii) The regulations contain only two opportunities for an exemption to be allowed by the municipal executive, though still requiring a building permit, namely

(a) for dormer windows and roof extensions on dwellings, within certain limits

(b) for industrial uses in category 3 of the industrial classification, subject to consultation.

(iv) Finally, the regulations stipulate that where any of these destinations coincides with those for certain specified infrastructure zones, the latter predominates.

9 The uses in the predominantly rural areas, which cover most of the map, accompanied by detailed regulations, are as follows.

(i) Agricultural enterprises in general, excluding intensive livestock farming and market gardening under glass, as defined

(a) the use of land for crops and animals in general does not require permission

(b) the following may be given permission
- one farmhouse, within specified limits of height, width, floor area and location
- non-residential agricultural buildings, located on a single plot for each enterprise, with limits on the height, etc of the buildings and their location in relation to each other and the plot boundary
- other associated construction not being buildings, within the plot and subject to limits etc

(c) two sub-categories are separately identified on the plan

- auxiliary agricultural enterprises, subject to limits of height and plot cover
- agricultural storage enterprises, subject to limits of height

(d) in addition, the municipal executive may permit the following exemptions from the regulations, under strict rules about the exemption, and the size etc
- a second farm worker's house
- non-residential agricultural buildings outside the specified plot
- extensions to existing greenhouses
- replacement of buildings destroyed by fire etc
- conversion of an unwanted farmhouse to residential.

(ii) Two more restrictive zonings overlap agricultural enterprises, and other rural destinations, namely

(a) areas of archaeological value within which any digging, ploughing (other than normal farming), drainage or piling is prohibited without permission and subject to conditions and advice from the relevant, independent committee

(b) areas of landscape and natural (ie ecological) value, where a long list of activities is prohibited without permission, mainly through construction permits, and subject to conditions and advice. The list covers roads, paths and car parks; reclamation of land, earth moving, grading etc; tree planting and felling of trees, woods and other vegetation; filling in or digging of ditches; overhead or underground lines.

(iii) Nursery gardens, within which the use does not require permission; one tied house and associated construction not being buildings is permitted, subject to conditions.

(iv) Country estates, referring in this case to the Castle of Rhoon, with very strict controls over the siting and design of buildings and other constructions and the preservation of areas of landscape and natural value etc, in addition to controls over the castle itself as a listed building under the Monuments Act.

(v) Areas for more detailed elaboration, for which the regulations give only the broad criteria within which subsequent development must occur. Some of this is the area in dispute, currently used for a container terminal but which the municipality wishes to restore to rural uses.

(a) Landscape area, for forestry, nature, and recreation, with associated roads, paths, shelters, car-parks etc. The regulations state that the aim is restoration of the landscape and open air recreation if necessary in combination with agriculture. They state
- maximum floorspace and height of buildings for recreation, maintenance and management
- maximum height for any other construction
- maximum area for outdoor recreation

- minimum and maximum area for afforestation and nature.

(b) Forest purposes, including footpaths, cycle paths, car parks, water areas and associated construction. The regulations state the limits, namely
- minimum area for tree planting
- maximum size of car parks
- maximum area for existing agriculture, to remain
- a "zone for visual relation" between the forest and the village, defined on the map and to be kept open in character.

10   The rest of the map covers destinations associated with infrastructure and utilities.

(i) The map shows the following

(a) traffic, including roads and, as a separate category, footpaths and cycle paths, with permitted road profiles illustrated

(b) water and canals

(c) dykes and hard shoulders of roads

(d) a wide, major zone for underground pipelines

(e) overhead high tension lines

(f) underground pipelines

(g) aerials.

(ii) In each case the regulations contain strict limits on what may be built, with permission, within these zones. Where the uses overlap with other uses, the infrastructure allocation has priority.

(iii) The map also shows green areas which include open spaces, roadside verges and shelter belts, and which may overlap with landscape, nature and archaeological areas. In effect, the only constructions which may be permitted are, for instance, benches or litter bins.

11   Finally, a supplementary map shows noise contours from motorways, railways and industrial areas within which a building permit is required under the Noise Nuisance Act for certain classes of buildings.


## Regentesse Valkenboskwartier, 's-Gravenhage
(see Figure N7 and Plate 28)

12   The area covered by this *bestemmingsplan* is part of the nineteenth century inner city of The Hague, about 2 km west of the central area ('s Gravenhage, 1981). It is a typical mixture of old, densely packed terrace houses, many with shops etc on the ground floor, intermingled with small workshops and businesses.

13   The purpose of the plan is to provide a framework for the urban renewal of the older, more decayed parts of the area whilst stabilising land uses in the rest of it. It forms part of the urban renewal programme which includes the nearby Schilderswijk district, the Painters' Quarter (van der Sluys & van Evert, 1985).

14   The plan covers an area of about 1.5 x 0.6 km. There are 41 regulations and several schedules and tables covering every detail of the plan. The plan itself is to a scale of 1/1,000, showing precise building lines and land uses, and the precise boundaries of the areas for urban renewal. The plans for the urban renewal areas are scheduled for elaboration under section 11 of the Physical Planning Act. The relevant regulations indicate the permitted range of development within urban renewal areas, and the explanatory memorandum to the *bestemmingsplan* contains illustrative plans showing alternative maximum and minimum proposals for their redevelopment.

15   The plan itself shows the following uses, in each case identifying precisely the specific buildings or terraces to which the use refers, and in effect therefore the front and rear building lines. Apart from the renewal areas, it is a map of existing uses. The list is

(i) residential areas

(a) distinguishing between single and multiple dwellings; dwellings with attics; annexes including extensions, outhouses, garages

(b) identifying streets or terraces within which a specified range and number of non-residential uses is permitted at specific locations or addresses, namely shops, offices, manufacturing industry and workshops (within specified categories), and other uses which are specifically prohibited

(ii) retail, distinguishing between a single use, and shops with residential above

(iii) commercial, including offices, workshops, manufacturing and associated residential, storage, parking, distinguishing between a single use and commercial with residential above

(iv) commercial garages

(v) special purposes, distinguishing between old persons' homes; welfare; schools; covered sports hall; post office; and various combinations, in some cases with residential above

(vi) areas for mixed uses, including residential, offices, retail and catering (hotels, restaurants and cafés), with a schedule for each area showing the use of each floor in each building

(vii) other open space uses, distinguishing between parks; green spaces such as verges, incidental open spaces etc; the courtyards or open spaces associated with blocks of flats (*binnenterrein*)

(viii) traffic, circulation and infrastructure uses, distinguishing between parking lots; footpaths; cycle paths; bus/tram lanes; residential streets; other streets; major traffic roads; canals; energy supply; and sewerage.

16 The regulations accompanying the plan define precisely the requirements or standards for each separate use. In particular

(i) for all buildings, the regulations specify the maximum heights of the buildings and of annexes; the minimum width of individual dwellings; the length of terraces etc

(ii) for groups of mixed uses, the range of permitted uses, and of specifically prohibited uses such as use for retail or carparking other than as shown in the plan, and the use of sites for dumping of waste

(iii) for commercial enterprises, the only permitted uses are those in categories 1 and 2 of the classification devised by the Association of Netherlands Municipalities, that is uses which are compatible in terms of environmental impact with residential areas.

17 However, the regulations do permit a certain amount of flexibility.

(i) In general, sites and buildings can be used only in conformity with the plan. Any other use is prohibited. But the municipal executive does have power to grant exemption if a strict interpretation would prohibit without good reason the most efficient use of the site or buildings; and for the continuation of existing uses.

(ii) The municipal executive may also grant an exemption for very small, non-residential buildings; for exceeding the permitted height of buildings by a small margin (1 m); or for permitting manufacturing uses in categories 3 and 4 of the Association of Netherlands Municipalities' classification under exceptional circumstances and with conditions.

(iii) More generally, the municipal executive may grant exemption on specified minor details in residential areas, such as extending the balcony of a dwelling beyond the building line shown on the map; or exceeding the permitted size of small professional offices or surgeries; or garages for parking.

18 Finally, the map shows the areas for urban renewal, in which the *bestemmingsplan* showing the eventual form and use of land and buildings is to be elaborated and approved using section 11 of the Physical Planning Act. The areas are in three categories, each with a different specification in the regulation. Two extend over several streets, whereas the third comprises parts of three terraces, one containing only six dwellings.

(i) The general regulations indicate that the site cover should not exceed 60 per cent; the houses should be in the price range to qualify as subsidised Housing Act dwellings; and the proportions of different sizes of dwellings should be within specified limits.

(ii) The more detailed regulations for each category of area cover, in the case of the largest

(a) the permitted uses: residential; retail; commercial, including manufacturing and storage in categories 1 and 2; and welfare

(b) the maximum proportion of single-family dwellings

(c) the maximum height of buildings and annexes

(d) the permitted range of residential floorspace; retail floorspace; commercial floorspace; carparking spaces

(e) vehicular access points to the site.

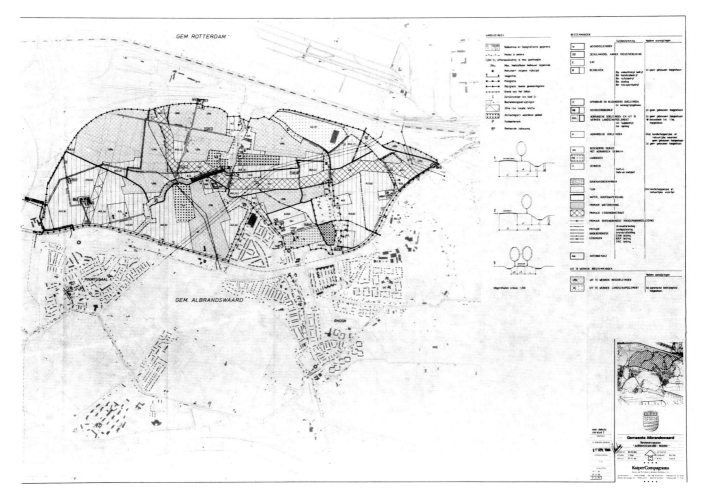

**Plate 27   Albrandswaard-Noord: Bestemmingsplan**

Plate 27 shows an example of a *bestemmingsplan* for the
northern part of a small, semi-rural municipality adjoining
Rotterdam. The main purposes of the plan are to restrict
urban development, to retain the generally open character of
the area, and to protect various infrastructures including the
main underground pipelines serving Europoort.

The notation shows the destined land uses, or restrictions
on land use, together with road sections etc. Further details
are given in the Regulations (see Appendix (b)). The chief
rural notations including land to be restricted for agricultural
uses of various kinds (A); protected zones including landscape
and ecological areas (1nz), archaeological areas (bA), and
country estates (NI), within which the restrictions cover

matters such as certain agricultural operations as well as the
use of land and buildings; and areas for more detailed
elaboration for future use as landscape/recreation (UL) and
forestry (UNb).

The plan shows the alignment of the numerous pipelines
and overhead lines crossing the area, the roads and the dykes
for all of which the accompanying regulations give details of
restrictions on development.

The plan shows the boundaries of built-up areas (small
villages etc.) for which inset maps give more detailed zoning
(see Figure N8).

An accompanying plan shows noise contours within which
further restrictions on development apply.

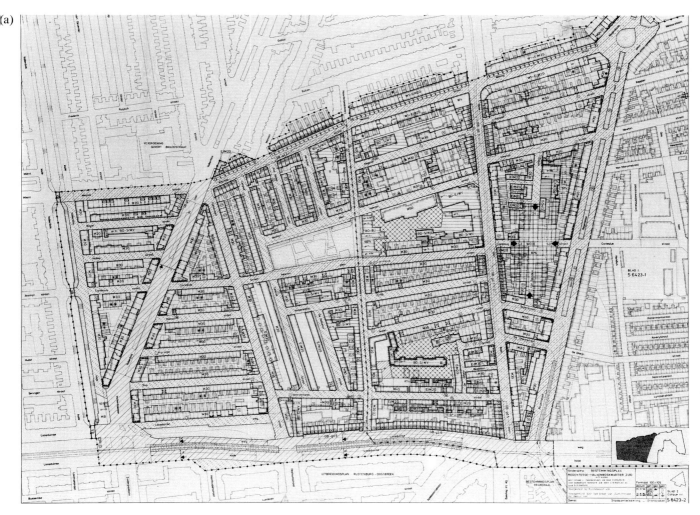

(a)

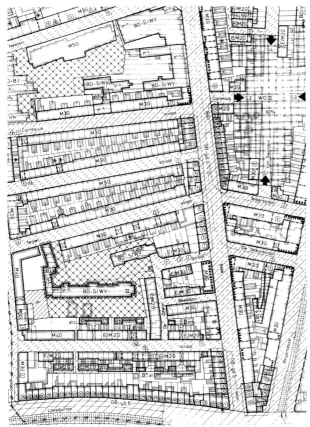

b)

## Plate 28   Regentesse Valkenboskwartier, Den Haag: extract from the Bestemmingsplan

Plate 28 shows an example of part of a *bestemmingsplan* prepared in the municipality for part of the inner city of The Hague. (a) shows a general area (the full plan extends further to the east, towards the central area of the city); (b) an enlargement of the bottom right-hand part of the plan; and (c) the notation, including road sections.

The notation shows the destined land use of every individual building and piece of land, the definitions being given in detail in the regulations with the plan (see Appendix (b)). Thus E, M are residential in single or multiple occupation; W is shops; BD is various special uses; GB is mixed uses, specified precisely in the Regulations. The plan shows the number of storeys, and building lines; the courtyards (*binnenterrein*) of apartment blocks for which there is a separate zoning, shown by cross-hatching; and the zoning for various classes of roads including public transport, etc.

The general aim is to stabilise existing land uses and the built environment for this part of the inner city, together with the areas for comprehensive redevelopment (indicated by WOI, etc.) for which detailed plans are to be elaborated using section 11 of the Physical Planning Act (see Figure N7).

(c)

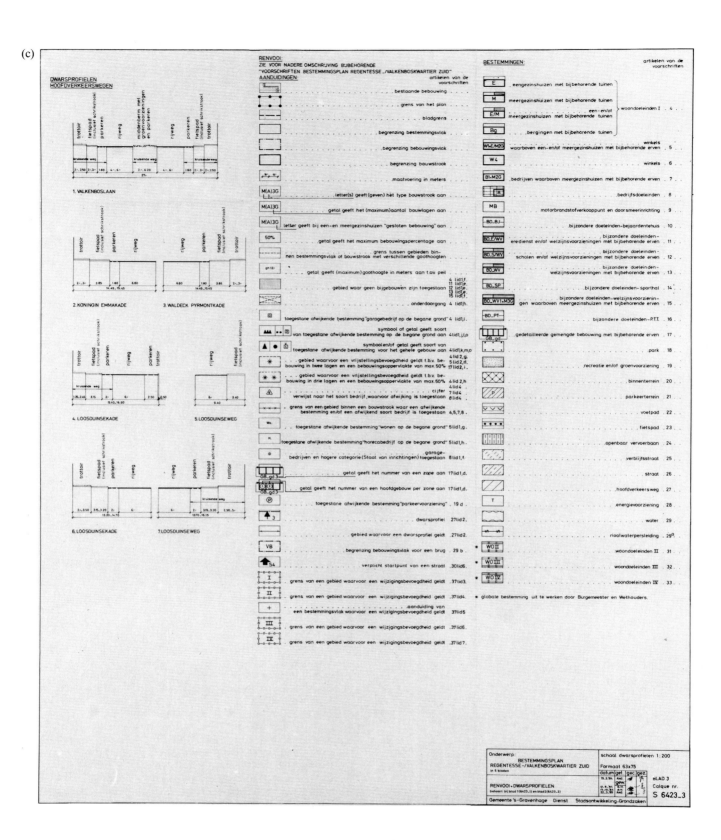

402

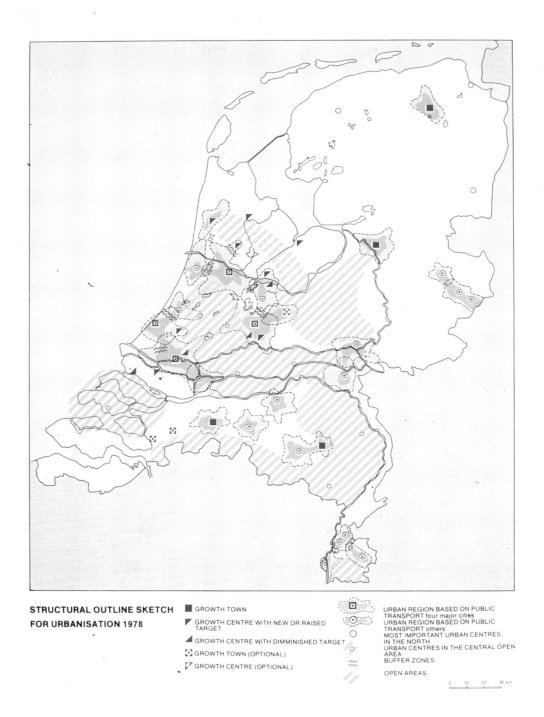

STRUCTURAL OUTLINE SKETCH
FOR URBANISATION 1978

■ GROWTH TOWN

◤ GROWTH CENTRE WITH NEW OR RAISED TARGET

◢ GROWTH CENTRE WITH DIMMINISHED TARGET

GROWTH TOWN (OPTIONAL)

GROWTH CENTRE (OPTIONAL)

URBAN REGION BASED ON PUBLIC TRANSPORT four major cities

URBAN REGION BASED ON PUBLIC TRANSPORT others

○ MOST IMPORTANT URBAN CENTRES IN THE NORTH

URBAN CENTRES IN THE CENTRAL OPEN AREA

BUFFER ZONES

OPEN AREAS

0   10   20   30 km

**Plate 29    Structural Outline Sketch for Urbanisation**

Plates 29 and 30 shows the two maps illustrating the structural outline sketches for urbanisation and for the rural areas, prepared by the National Physical Planning Agency, and formally approved as national physical planning key decisions. The maps indicate the general spatial implications of national physical planning policies, to be incorporated in regional plans prepared by the provinces, and in directives to the municipalities. The Urbanisation plan was subsequently reviewed in 1983, giving greater emphasis to urban renewal in Randstadt Holland.

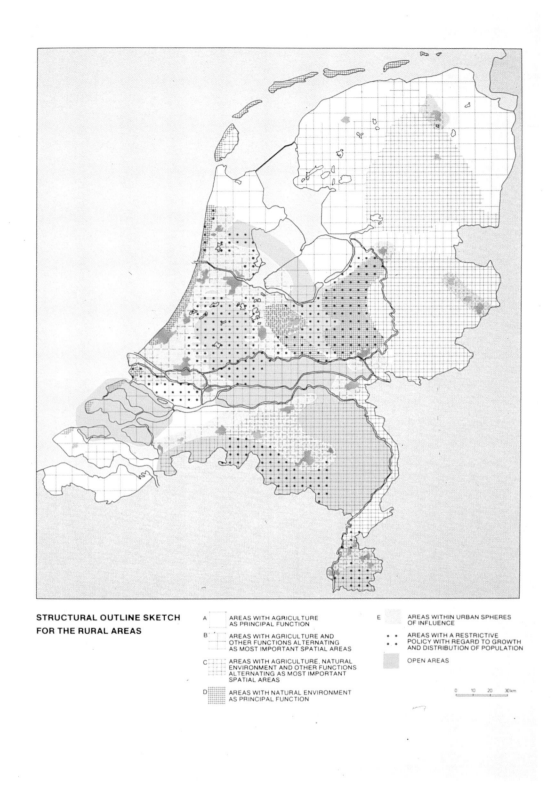

STRUCTURAL OUTLINE SKETCH
FOR THE RURAL AREAS

A — AREAS WITH AGRICULTURE AS PRINCIPAL FUNCTION

B — AREAS WITH AGRICULTURE AND OTHER FUNCTIONS ALTERNATING AS MOST IMPORTANT SPATIAL AREAS

C — AREAS WITH AGRICULTURE, NATURAL ENVIRONMENT AND OTHER FUNCTIONS ALTERNATING AS MOST IMPORTANT SPATIAL AREAS

D — AREAS WITH NATURAL ENVIRONMENT AS PRINCIPAL FUNCTION

E — AREAS WITHIN URBAN SPHERES OF INFLUENCE

•• — AREAS WITH A RESTRICTIVE POLICY WITH REGARD TO GROWTH AND DISTRIBUTION OF POPULATION

OPEN AREAS

0  10  20  30 km

**Plate 30    Structural Outline Sketch for the Rural Areas**

404

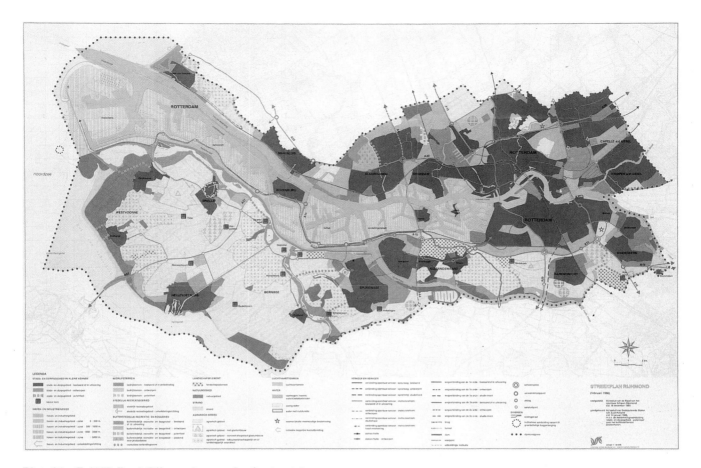

**Plate 31     Zuid Holland, Rijnmond: example of a streetplan
(Regional Plan)**

Plate 31 shows the proposed regional structure for the
Rijnmond area, surrounding Rotterdam and Europoort. The
plan was prepared by the South Holland province and, with
its accompanying regulations, provides a detailed,
administratively-binding framework for the preparation of
*bestemmingsplannen* by the municipalities. The key features

include the provision of land for the future development of
Europoort; the location of land for new housing etc; land to
be retained as open space, or for nature preservation, or for
agriculture, nurseries and forestry; and the communications
system for the region.

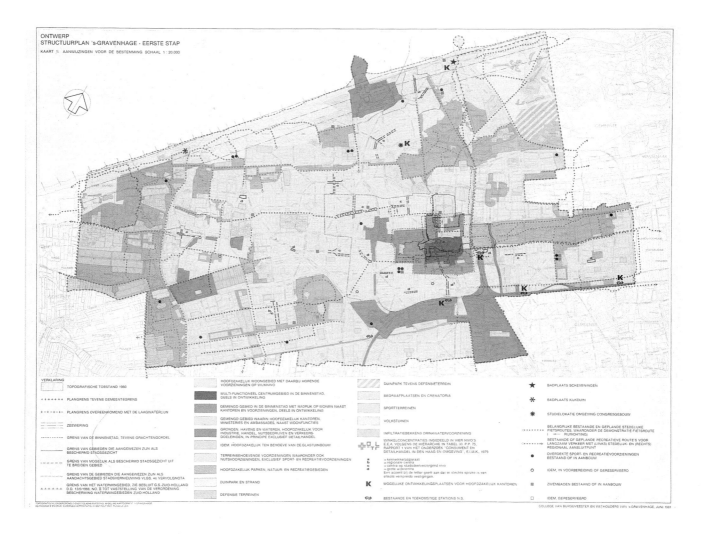

**Plate 32    Den Haag: Structuurplan (Structure Plan)**

Plate 32 shows the advisory structure plan prepared by the municipality as a framework for future planning. The city, which includes the resort area of Scheveningen, is virtually fully built-up. The plan identifies the basic urban structure in terms of land uses, cycle routes and railway stations. Other communications are shown separately.

The chief land uses include the central area and other shopping centres or streets; residential and industrial areas; areas of special interest such as protected townscapes or water catchment areas; and the sites of major social and recreational facilities. Possible sites for new development (K) are identified, and future townscape areas and areas for urban renewal.

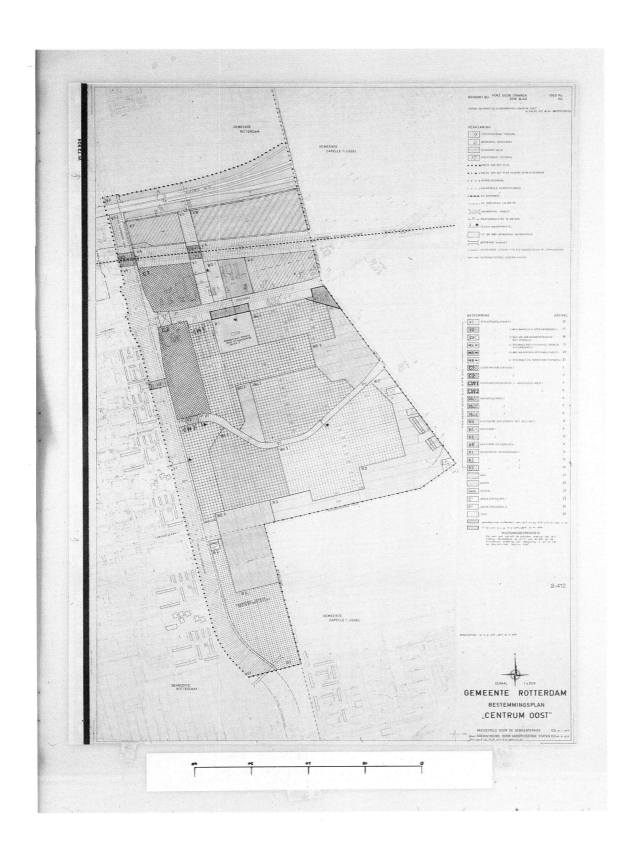

**Plate 33   Rotterdam Centrum Oost: Example of a Globaal Bestemmingsplan**

Plate 33 illustrates an example of a global *bestemmingsplan* laying down the basic planning principles for a large area for new, suburban development. The notation shows the zoning for various major traffic routes (V), commercial areas (C), residential (W), offices (K) and recreation (R). The accompanying regulations contain schedules of the numbers of dwellings, floorspace etc. to be provided. The details were then worked out during the 1970s within the framework provided by the plan and regulations, under section 11 of the Physical Planning Act.

# COMPARATIVE STUDY

HWE Davies
D Edwards
AJ Hooper
JV Punter

# C1 INTRODUCTION

C1.1 The object of the study was to examine the operation in practice of the planning systems in Denmark, France, West Germany and the Netherlands and to compare them with that in England. The focus of the research was on obtaining an accurate description of the methods and procedures of development control and the relationship between them and any development plan-making function.

C1.2 Earlier sections in the report provide the detailed description. This final report attempts the comparison. It puts the four systems and the English system into a common framework. However, it must be emphasised that this endeavour presents problems. Any comparison inevitably risks generalisation and simplification. More fundamentally, development control as it is practised in England differs crucially from control in the other four countries which, although differing amongst themselves in many ways, share a broadly similar type of system. Three distinctions in particular are crucial.

(i) In England there are separate permissions for the control of development under planning legislation and for the control of building construction; in the other four countries the two are combined in a single permit.

(ii) The relationships between development control and the development plan are quite different. In England, development control has to have regard to the development plan, but also to any other material considerations. In the other four countries the local plan and building regulations are legally binding on all citizens, and therefore on all proposals for development.

(iii) Each of the other four countries has what the Danes call a system of 'framework management and control', that is, a formal system of plans at different levels of government in which those of the higher levels are administratively binding on the lower plans. Thus local plans have to comply with, or at least not conflict with, regional plans, and so on.

C1.3 The effect is that planning in the overseas countries in principle is plan-led to a greater extent than in England. Control in principle is more a purely administrative process working within the procedural rules and substantive policies contained in the legislation and the legally binding plans and regulations. Conversely, control in England to a greater extent is a flexible response to the pressure for, or constraints on, development working

within the procedural rules laid down in the legislation, but in which the substantive policies do not have formal, legal expression.

C1.4 This distinction means that any attempt to place the various systems into a common framework is not only difficult but can be positively misleading unless done with care, and read with circumspection. In addition, despite the broad similarities identified in this final report, there are many more detailed differences between the various overseas systems. It is the national reports which contain the factual descriptions of planning and control in those countries, not this report which, in a sense, forces the five systems into a single framework to facilitate comparison.

## The Origins and Context of Planning and Control

### Local Issues and Control

C1.5 The control of development in all five countries is primarily a function of local government, albeit within a framework of national legislation. Its origins were similar in all five countries: it was largely a continuation of government concern about the housing and urban conditions of the nineteenth century and the rapidly growing suburban expansion of towns. By 1909 in England, and somewhat later in the other countries except the Netherlands, this concern found expression in legislation to plan and control suburban development. The measures were to bring forward properly laid out and serviced land for the extension of towns; and ensure that the housing and other construction met the rising standards of health, safety, convenience and amenity. The chief instrument was the local planning scheme, with building controls. Later, planning was extended to a wider scale with attempts to limit the uncontrolled spread of towns into open countryside and smaller towns and villages.

C1.6 The actual legislative forms taken by the control of development in the different countries largely reflects their constitutional structure. The chief contrast between England and the rest is that the four overseas countries have a written constitution and a legal system based on the *Code Napoléon* or, in Denmark, the Scandinavian traditions of law, whereas England has no written constitution and has a legal tradition based on the common law. This has four main consequences which directly affect the control of development.

(i) The written constitutions establish and define the principal elected levels of government. Thus provincial and local governments and, in Germany's federal system, the *Länder* are autonomous bodies, protected by the constitution and capable of independent policies and action provided they comply with, or at any rate do not conflict with, those of higher levels of government.

(ii) The written constitutions also explicitly protect the rights of citizens (including, in Germany and Denmark, property rights) and in particular reserve a right to challenge the administrative decisions of government through the courts.

(iii) The legal system in the four countries in general is one in which the law is found in the enacted statutes and decrees, and the civil, criminal, penal and other codes, rather than being continuously made by the courts as in England, through their interpretation of statute law. However, France also has to some extent an administrative tradition of judge-made as well as statute law.

(iv) There is also a distinction between public (or administrative) and private law, each with its own hierarchy of courts. Planning and control, as public law, thus come under the administrative courts in France, Germany and the Netherlands, and the civil or criminal courts are usually involved only in certain aspects of enforcement. However in Denmark, as in England, there is just the one, all-purpose hierarchy of courts within which the distinction between public and private law is strongly maintained.

C1.7    The most important principle, deriving from and underlying the four characteristic features in all four countries, is that of legal certainty. Administrative decisions such as issuing a building permit are made by reference to binding legal instruments such as plans and regulations. There is markedly less administrative discretion beyond that specifically laid down in the legislation, plans or regulations, thus guaranteeing legal certainty. Development control as such is thus more purely an administrative function, separated from policy-making which is found in the preparation of the plan. Furthermore, appeals against administrative decisions, ultimately to the administrative courts, can be made in general only on their legality, including both procedural and substantive matters but not policy. Nevertheless, all administrative systems require discretion to make them work, either in whether to apply the rules or how to interpret them. Certainty thus becomes an ideal to be striven for as far as possible, rather than something which in all instances characterises the overseas systems.

C1.8    Even as an ideal, the certainty in overseas systems contrasts with the English system in which control is characterised by a high degree of administrative discretion, and the appeals are made to the responsible minister who is also able to exercise discretion in the interpretation and application of policy. However, there is also an opportunity for challenge to the courts in England on the legality of the decision in terms of its definitions and procedures.

*National and Regional Issues and Policy*

C1.9    Although the control of development is essentially a matter for legislation and local government, it is nevertheless affected directly by wider issues, notably structural changes in employment and economic activity and interregional shifts in population, with their consequences for differential levels of unemployment, rising land prices and housing shortages, and new spatial patterns of property investment. These are the subject matter of national and regional policies which constrain, or at least attempt to influence, local policies and control.

C1.10    The five countries are of differing size and characteristics (see Figure C1). Thus Germany and the United Kingdom have similar sizes of area and population but very different urban structures, with the latter dominated by London but with no single dominant city in Germany as Berlin is an exceptional case. France and Denmark have a similar density, which is only half that in the United Kingdom, and both are dominated by their capital cities, but their sizes are very different. The Netherlands is unique in being the smallest and most densely populated, with much of its territory reclaimed from the sea with implications for the level of state activity in planning and development.

C1.11    The regional issues take different forms. In France, it is the dominance of Paris, the decline of the older mining regions of the north east and, more recently, the rapid growth of urban populations especially in the south. In Germany, following the massive post-war reconstruction of the urban areas, the chief issues have included the decline of the Ruhr and Saar, the problem of the peripheral or border regions in the east, and the rapid growth in the south. In England, too, very similar problems are characterised by the current phrase, a 'north-south divide' and the decline of the larger conurbations, with ex-urban growth in much of the south and also on a smaller scale in parts of the north. Many of the cities of the south in all three countries are still facing problems of growth whereas those of the north are more concerned about decline, or revitalisation. Thus national and regional policies for tackling these problems, as well as social and market forces, interact strongly with policies for local planning and control.

C1.12    The problems of the two smaller countries are equally difficult. In the Netherlands they have been a combination of housing shortages, ex-urban expansion, land reclamation and pollution in Randstad Holland, and of low economic growth in the north east and the former mining areas of the south east. Even in Denmark, the more remote areas of Jutland suffer from relatively high unemployment and out-migration while the peripherality of the Copenhagen metropolitan area and its continuing

# Figure C1 Political Structure of Planning: Elected Levels of Government

Notes: + parishes in England and Denmark are in mainly rural areas; arrondissements in France, boroughs in Netherlands and stadtebezirke in Germany are only in a limited number of large cities;
o non-elected, ad hoc groupings of communes, in France
* non-elected administrative districts of Länder in Germany
population figures are for c.1980-85

| Country | England | Denmark | France | West Germany | Netherlands |
|---|---|---|---|---|---|
| area (sq km) | (UK) 244,780 | 43,080 | 551,208 | 248,706 | 33,612 |
| population (000s) | (UK) 56,400 | 5,100 | 54,300 | 61,200 | 14,100 |
| **national** legislature | Parliament | Folketing (Parliament) | L'Assemblée Nationale (National Assembly) | Bundestag & Bundesrat | Staaten-Generaal (States General) |
| planning ministry | Department of the Environment | Miljøministeriet (Ministry of the Environment) | Ministère de l'Urbanisme, du Logement et de l'Equipement. Direction de l'Architecture et de l'Urbanisme | Ministerium für Raumordnung, Bauwesen und Städtebau (Regional Planning, Building and Urban Development) | Ministerie van Volkshuisvesting, Ruimtelijke Ordening en Milieubeheer (Housing, Physical Planning and Environment) |
| administrative court | = | = | Conseil d'Etat (Council of State) | Bundesverwaltungsgericht (Federal Administrative Court) | Raad van Staat (Council of State) |
| formal advisory body | = | = | = | Ministerial Conference on Spatial Planning | Rijksplanologische Commissie (National Physical Planning Committee) |
| **state** number | = | = | = | Land 11 | = |
| population (000s) | | | | 1,000-16,800 | |
| average | | | | 5,610 | |
| **province/region/intermediate** | = | = | Région (region) | *Regierungsbezirk (administrative district) | Provincie (province) |
| number | | | 22 | 26 (in 6 Länder) | 12 |
| population (000s) | | | 1,000-11,000 | = | 180-3,200 |
| average | | | 2,500 | 2,374 *Regionalverband (regional association) in other Länder | 1,167 |
| **metropolitan (ad hoc)** | = | Greater Copenhagen Council | oCommunauté Urbaines   oAgence d'Urbanisme | Landschaftsverbande (planning association) | = Rijnmond Council (abolished 1986) |
| number | | 1 | 9   32 | 3 (in Nordrhein'Westfalen) | 1 |
| population (000s) | | 1,723 | 100-1,100 | = | |
| **county/department** | "shire" County   Metropolitan County (abolished 1986) | Amt | Département | Landkreis   Kreisfreie Stadt | = |
| number | 39   7 | 14 | 96 | 237   91 | |
| population (000s) | 121-1,511   1,140-6,696 | 47-612 | 100-2,500 | 49-630   34-1,889 | |
| average | 746   1,949 (excl London) | 324 | 580 | 169   237 | |
| **district/municipality** | "shire" District   Metropolitan District/London Borough | Kommune | Commune | Gemeinde   Kreisfreie Stadt is also a Gemeinde | Gemeente |
| number | 296   69 | 273 | 36,433 | 8,505 | 809 |
| population (000s) | 34-281   132-1,003 | 0.1-479 | 0.1-1,000 | 1-10 | 0.3-712 |
| average | 98   256 | 19 | 1.5 | 7 | 18 |
| **submunicipal +** | Parish | Sogn (parish) | Arrondissements | Stadtbezirk (city district) | Borough |

413

rapid growth and dominance combine to create different types of problem.

C1.13  The regional and structural contrasts are found not only between and within countries but also over time. In the first two decades after world war II, the main priority was coping with growth and reconstruction. This was most dramatic in the rebuilding of German cities, and the rapid urbanisation of France. In the Netherlands, it was the economic recovery and housing shortages of Randstad Holland, and in England, planning and containing suburban development.

C1.14  However, since the energy crisis of 1973, there has been a convergence of planning issues in all five countries, in the wake of relatively low economic and demographic growth. Attention has turned to the problems of the older, existing built-up areas, and those of more distant, ex-urban growth. The problems are common: the changing forms and location of residential, commercial and industrial development; the financing and implementation of housing redevelopment and rehabilitation; and the growing importance of conservation in town and country.

C1.15  Two points emerge from this brief review. The first is that in all five countries the control of development has been primarily a local matter concerned with the traditional issues of land use planning and development, originally the extension of towns but more recently their conservation, rehabilitation and redevelopment. The problems and policies have varied depending on the different stages of urban development both within, and between, the five countries. But in all five, the form of control has been based on, and reflected, constitutional relationships and the nature of the legal systems which, to a considerable extent, have resulted in inertia. Thus only in the 1970s did the systems of local planning and control start to be reviewed in the four overseas countries to give greater decentralisation and flexibility in local control. Even so, the basic relationship between a binding local plan and a building permit remains unaltered. In England, similar pressures led to a concern about the efficiency and effectiveness of planning and control. In the 1980s, in all countries, there was a trend towards deregulation, or a slow but definite relaxation of planning control.

C1.16  The second point is that the control of development cannot be divorced from the wider planning system. It forms one method by which governments at every level and in every country have sought to achieve their aims. This becomes much clearer when reviewing the control of development in the four overseas countries because of the much stronger, formal and explicit relationship between the various levels of plans, and between the local plans and control than in England. But the relationship is there also in England; it is just that the relationships are less formal and structured in the absence of a written constitution.

## Outline of the Report

C1.17  The rest of this section contains a comparative framework for understanding the primary control of development. It reviews separately the political structure of government and, in more detail, the local administration of planning; the various plans and regulations involved in the control of development, with particular emphasis on the status, form, preparation and approval of local plans; and the scope of development control, its administrative procedures and the opportunities for appeal. In each, the different facets of planning and control are reduced as far as possible to a simple form of matrix based on the English system, using a common language. The aim is to facilitate comparisons even though this involves the risk of over-simplification.

C1.18  A short section, C5, then notes various supplementary and related systems of control, and the report concludes with a discussion of the more general considerations about the control of development emerging from the comparative study.

## The Political Structure of Planning and Control

C2.1   The political structure of planning is shown in Figure C1, identifying the various elected levels of government. The actual structure of government differs considerably between the five countries. England, a unitary state in which all power is centralised in government and parliament, saw a complete reorganisation of its local government in 1974. The formerly independent county boroughs were abolished and a two-tier system created incorporating county councils in the upper tier, and districts in the lower. The counties were fairly similar in size and area to the pre-1974, historic counties, but the new districts were created by amalgamating the old, smaller urban and rural districts and municipal boroughs. For planning, the same two-tier system applied over the entire country, including the main conurbations. The system was further altered in 1986 when the upper tier of metropolitan counties was abolished, leaving the conurbations to be administered by groups of independent district or borough councils.

C2.2   Germany has a federal structure in which most domestic legislation is enacted at the state level, and each state, or *Land*, has its own system of local government. In general there is a two-tier system of autonomous counties and districts or municipalities outside the main towns, and a single authority combining the functions of both levels of government in the towns. The number of municipalities was substantially reduced by amalgamation between 1968 and 1978. In addition, several of the *Länder* have created a subordinate, non-elected level of administrative districts or regional associations with considerable, delegated powers, and a more loose system of planning associations with only limited powers.

C2.3   Denmark and the Netherlands have roughly similar two-tier systems of local or provincial government in both of which the smallest units of local government have been, and are currently being, amalgamated into slightly larger areas, with an intended minimum population of 10,000 or 15,000. The first difference between the two is that the provinces in the Netherlands in general are bigger in population than the counties in Denmark. Partly as a result, the introduction of metropolitan government in the two countries has taken a different form. The Greater Copenhagen council is superior in function to the counties. The metropolitan councils in the Netherlands would have been intermediate between municipality and province. In fact only one was

created, for the Rotterdam conurbation, and it was abolished in 1986.

C2.4   France has had for many years, a complex, four-tier system of government, including national government, region, *département* and *commune* although the regional government became elected only in 1983. Formerly it was a highly centralised system but under the most recent reform, a considerable amount of authority has been delegated to the most local level of government. In addition Figure C1 draws attention to the *ad hoc* administrative groupings of *communes*, the *Communaute* and *Agences d'Urbanisme*, which have been created especially for a number of larger towns and urban regions. In addition, it is common for *communes* to come together into syndicates to prepare local plans. Despite the sweeping nature of the changes under decentralisation, there is much debate over the extent to which the central state has relinquished its control.

C2.5   One implication of this variety is the very different size (in population) of the municipalities at which local planning and control is located. Local government in all four countries is decentralised to many more small units of population than has been the case in England since its local government reorganisation in 1974. Thus the average size of the *communes* in France is less than 2,000, or in Germany, about 7,000; and even in Denmark and the Netherlands, it is about 18,000; whereas in England the average, even outside the metropolitan districts, is 98,000.

  (i)  Decentralisation of local planning and control has gone furthest, to the smallest units, in Denmark and Netherlands. The municipalities have been given both functions irrespective of the size of their population which can be as small as 1,000.

  (ii)  The responsibility for local planning and control may be divided between different levels of government in Germany and France. In principle, control as an administrative function is at the upper level, the *Kreis* or county in Germany with a population of at least 49,000; and the *département* in France, with at least 100,000. And, in principle, local plans are prepared by the lower tier, the *Gemeinden* or *communes* with a population which can be less than 1,000.

However, in practice, the two functions may either be combined, or local plans are not in fact prepared. The former is the case in Germany, where there is a single local government for the larger towns, combining both functions and both levels of government. It is the case too in France where many of the larger, urban *communes* have prepared local plans either on their own, or in groups, and by so doing have acquired development control powers. Local plans are not always prepared by the smaller, more rural areas in France, or Germany, in which case control remains with the upper tier of larger authorities.

(iii) Thus only in England as a matter of course are both development control and local plans located in fairly large local authorities with a minimum size of 34,000. In practice however it is also the case in Germany where the urban authority provides a single tier of government for the larger towns.

C2.6    Two other points are noteworthy. The first is that in every country there is a form of sub-municipal, elected governments, most of which have at least an advisory or consultative role in planning and control. In England and Denmark, they are elected parish (or town) councils found chiefly in the former, pre-1974 rural districts. In the Netherlands, Germany and France, they are elected borough councils, districts or *arrondissements* respectively found only in the largest cities.

C2.7    A second noteworthy feature is the extent to which the Netherlands has built in through its planning, and other, legislation advisory committees with a formal role in the preparation of planning policies at the national and provincial levels of government.

C2.8    Finally, in three countries, France, Germany and the Netherlands, there is a separate system of administrative courts. In each, these are very powerful organisations with the twin roles of advice on all pending legislation, and adjudication on cases in administrative law. The courts themselves in France and the Netherlands are very old established, pre-dating the *Code Napoléon* and the growth of administrative law.

## Administrative Structure for Local Planning and Control

C2.9    There is even more variety in all five countries in the ways in which local planning and control is administered at the local level (see Figure C2). In fact, the only common feature is that, in all five, government at the local level is by an elected council, though even that similarity breaks down at the next highest level in Germany, where the *Regierungsbezirke* is an administrative district of the *Land*.

C2.10    England is unique in having a system under which the final decisions on both plans and control are, in principle, made by the elected council. In practice, the decisions may be delegated to a committee of council or

even to the planning officers for routine planning applications. The council is advised by a planning department staffed by professional planning officers responsible for the administration of development control and the preparation of plans including consultations and negotiations with other departments and organisations, developers and applicants for planning permission. But the function of officers strictly is to advise, not to make decisions.

C2.11    In no other country does this tight relationship apply. Instead three sets of relationships below the elected council are significant for planning and control.

(i) The official leader, mayor or burgomaster may be either a political chairman, or a chief executive, or someone who combines both roles; he or she may be directly elected, appointed by the council, or even (in the Netherlands) appointed by a minister of the Crown. Thus in France the mayor is simultaneously an elected official and a representative of government. He or she may also hold office in other levels of government, helping to reinforce the close relationships between different levels of government.

(ii) The chief decision-making body (except where the decisions are delegated to professional officers) may be the council, or a committee of council; or the mayor or chief executive; or (in the Netherlands and sometimes in Germany), a municipal executive comprising the burgomaster and a few councillors elected by the council but functioning not as a committee of council but an independent, decision-making body. The distinction is between those countries (England, Denmark) where the decision on both plan-making and control in the end is a political decision by the elected representatives either in council or committee; and the other countries where the decision on the plan is political, by the council, and that on control is administrative, by the mayor, burgomaster, municipal executive or even, as in Germany, by the professional officers.

(iii) The professional officials responsible for the technical work fall into three categories. As noted above, in England planning and control are the responsibility of a single planning department staffed by local government officers. In the other four countries, the two functions are located in different departments, reflecting the distinction between the policy function of plan-making, and the more technical, administration function of control. In all four, too, the smaller municipalities usually do not have a planning department, relying instead on consultants or the planning officers of other levels of government. In particular France relies heavily on the *Directions Départementale de l'Equipement* (DDE) which are composed of civil servants appointed by central government supported by contractual staff. Only the largest *communes*, or groups of *communes*, have their own technical staffs, including those which have access to one of the *Agences d'Urbanisme*.

## Figure C2 Administrative Structure for Local Planning and the Control of Development

Notes: 1. Official bodies concerned with building/planning permits and the preparation and approval of local plans at the local/county/provincial level
2. Each Land in Germany has its own structure of local administration; this refers to Nordrhein-Westfalen

| | England | Denmark | France | West Germany | Netherlands |
|---|---|---|---|---|---|
| **primary control at the local level:** | | | | | |
| principal authority | District | Kommune | Commune | Landkreis/Kreisfreie Stadt/ Gemeinde county/municipal council | Gemeente |
| elected body | district council | kommune council | council | Bürgermeister (elected/ appointed by council) | municipal council |
| chief executive/leader | leader/chairman of council (elected by council) | mayor (elected by council) | mayor (elected) | | burgemeester (appointed by government) |
| executive decision making: local approval of local plans | district council | municipal council | mayor, or intercommunal public establishment, and council | Gemeinde/Kreisfreie Stadt municipal council | municipal council |
| approval of permits | district planning committee, (councillors) or delegated to officers | technical and environment committee, or delegated to officers | mayor (if approved local plan), or Commissaire de la République (formerly Préfet) (if no plan) | Landkreis/Kreisfreie Stadt (building department) officials | burgemeester & wethouder (councillors nominated by council), or delegated to submunicipal councils, or delegated to officers |
| technical officials: local plan preparation | district planning department | technical services (planning) department or consultants | Directions Départementale de l'Equipment (DDE - technical officers of central government at departmental level, or Agence d'Urbanisme for larger cities/ad hoc intercommunal plans, or consultants | planning department or consultants | city planning department (larger municipalities), or consultants (smaller municipalities) |
| evaluation of permits | district planning department | technical services (building) department | technical officials of commune, or DDE or both | building authority officials | municipal housing & building inspectorate |
| **higher level of government involved directly:** | | | | | |
| name of authority | "shire" County, outside metropolitan areas | Amt (county) | Département | Regierungsbezirk | Provincie |
| elected body | county council | county council | council | = | provincial council |
| chief executive/leader | leader/chairman of council (elected) | county mayor (elected by council) | Commissaire de la République (appointed by government) | Regierungspräsident (appointed by Land minister) | provincial governor (appointed by government) |
| executive decision making: | county planning committee (councillors) | county technical & environment committee | Commissaire de la République | as above | provincial executive (governor and councillors nominated by council) |
| technical officials | county planning department | technical services (planning) department | DDE (as above) | Bezirksregierung (district administrative office) | provincial planning agency |
| official advisory bodies with formal role | = | = | = | = | provincial physical planning committee |

417

## The Hierarchy of Plans

C3.1   If the federal level in Germany is excluded, all five countries have, or have had, a system involving four levels of plan corresponding to national (or *Land*, Germany); regional; structure; and local (see Figure C3). The terms are used for convenience, though the overseas plans are not strictly comparable with the English equivalents. However, there are strong contrasts, chiefly between England and the rest, and in Germany the federal government provides an additional tier, the broad national framework for planning in the *Länder*.

C3.2   First, in every country but England, the most local level of plan, as has been noted, is a form of local plan which is legally binding on all citizens and government agencies. It is the primary instrument for the control of development in those countries and, with the building regulations, is the only plan used in this way. Its importance is such that it is described in more detail in the next two sections, together for comparison with the local plan in England (Figures C4 and C5).

C3.3   Second, each of the five countries has, or had, a hierarchy of three levels of plans and policies above that of the local plan.

(i)   At the national level (and in Germany, the *Land*), there are ministerial directives and statements on particular issues or sectors, and advisory circulars, apart from the actual legislation and decrees. In addition Denmark, Germany and the Netherlands require annual or other regular reports on planning policies and issues to be submitted to parliament.

(ii)   At the regional or intermediate level, or, in Denmark, the county level, regional plans or strategies are prepared. Their form varies. In France, the new *contrat de plan* is mainly a public sector investment programme supplementing the regional plan prepared as part of the national policy. In the other three countries, the plan is usually a combination of a broad, spatial activity plan and a set of structural, or sectoral policies. Broadly similar regional strategies were prepared by the national government in England during the 1960s and 1970s but they have not since been updated in comparable detail.

(iii)   At the local level, structure plans (or roughly similar documents) provide the more detailed framework for local planning, usually on a comprehensive basis

covering the entire area of a county or district except in the Netherlands where they are optional. The German document however is more strictly a preparatory land use plan, with a narrower, but more detailed, land use form than the structure plans of the other countries. Instead, it is the regional development plan in Germany which corresponds most closely in form and function to a structure plan.

C3.4   The main distinction once again is between England and the others. In England, the only plan required by law is the structure plan but its status is only that of a material consideration to be taken into account in control. Except in France, the plans and programmes at each level above the local plan in the overseas countries are required by law. However, in all four countries preparation has to go through a formal procedure, and they have in principle a powerful status. Each of the plans or programmes is administratively binding on all government departments and public sector agencies, and provides an equally binding framework on all lower tier plans and programmes. Although the precise requirement varies, lower tier plans in general must either not conflict with, or in some countries must be in accord with, higher tier plans.

C3.5   The main purpose of these higher level plans is to facilitate vertical and especially horizontal coordination of resources and public sector programmes within government at the higher levels, widened to include land use in the more local levels. The procedures for their preparation include extensive consultation and negotiation within the public sector and, in some countries, notably Denmark and the Netherlands, every type of plan has to be subjected to public participation. The legislation in the Netherlands also establishes formal advisory committees, including representatives of non-governmental organisations and experts, at the national and provincial levels of government.

C3.6   In none of these overseas countries can the plans and programmes be challenged either in the higher levels of administration or the courts in the way that there can be objection to local plans. The point is that although they are administratively binding, they are not in themselves an administrative decision similar in status to the issue of a building permit; nor are they binding and enforceable in law. Rather, they provide the framework for subsequent administrative decisions such as the approval of a plan, a directive about the content of a plan, or the allocation of resources.

**Figure C3  Plans and Regulations for Planning and the Control of Development**

Notes: All plans etc except (*) are administratively binding on government departments and municipalities; all plans and regulations marked (°) are legally binding on citizens 100% - full territorial coverage is required for local governments

| level of government | England | Denmark | France | West Germany | Netherlands |
|---|---|---|---|---|---|
| **national** | * circulars<br>* regional strategies | landsplandirektiv (directives)<br>°bygningsreglement (building regulations)<br>°bygningsreglement for småhuse (small buildings)<br>*landsplanredegørelse (annual statement)<br>=circulars | °Code de l'Urbanisme (planning law)<br>°Code de la Construction et de l'Habitation (building regulations)<br>*circulars | Bundesbaugesetz (national framework for land use planning and control)<br>Bundesraumordnungesetz (national framework for regional development planning)<br>*federal regional planning programme (statement of principle)<br>*4-yearly statement | planologische kernbeslissige (key planning decision)<br>*annual report |
| **state** | = | = | = | Landesentwicklungsplan (comprehensive development plan)<br>Landesentwicklungsprogramm (development programme)<br>°Landesbauordnung (state building orders)<br>=circulars | = |
| **regional/provincial or intermediate** | = | = | contrat de plan (state/region infrastructure expenditure plan) | Gebietsentwicklungsplan (regional development plan)<br>regionalplan (regional plan) | streekplan (regional plans) |
| **metropolitan/county/department or supramunicipal** | structure plan - 100% | regionalplanskitse (regional plan) | schéma directeur (structure plan for urban regions) | Regionalplan fur Landschafts verband (regional plan for planning association) | = |
| **municipality** | local plan | kommuneplan (structure plan) - 100%<br>°lokalplan (local plan) | °plan d'occupation des sols | Flächennutzungsplan (preparatory land use plan - 100%)<br>°Bebauungsplan (binding land use plan) | strukturplan (structure plan)<br>°bestemmingsplan (destination plan)<br>°bouwverordening (building regulations) |

419

C3.7 Despite the very general similarities, there are differences between the four countries, and also trends in the development of plans. At the national level there has been a clear shift away from the types of national plans and policies which characterised the 1960s, which increasingly are regarded with scepticism. The shift has been to more sharply focused directives on specific sectoral or spatial policies and programmes.

C3.8 The regional plans in Denmark, Germany and the Netherlands are regarded as relevant and useful instruments, performing a key role especially in public sector planning. In effect each entire country is now covered even though they are not mandatory other than in Germany and Denmark. The French *contrat de plan* is more problematic as, so far, only a few regions have prepared a fully developed programme, but its value is not questioned.

C3.9 The structure plans (or, in Germany the preparatory land use plans as the structural elements are more properly covered in the higher level, regional plans) are especially important. As in England, these plans are mandatory in Denmark and Germany and have to cover the entire territory of a county or municipality, providing the comprehensive framework for local plans. Even in the Netherlands, where the plans are not mandatory, current practice is for the provinces to require a form of structure plan to provide a context for their approval of local plans. Only in France is the situation less firm. *Schéma directeur* were prepared in the 1970s for the main urban agglomerations by the *départmental* planning staffs (DDE), in theory as the framework for the local plans. However, the plans were based on long-term forecasts resulting in unrealistic and inappropriate land use allocations. They have not been systematically revised, have become rapidly out-of-date, and have had to be amended every time a local plan is approved. Revision and approval of these plans however seems less likely to occur in future with the decentralisation of planning to the local *communes*.

C3.10 The third main category of documents, the building regulations, are legally binding on all citizens in all five countries. But the overseas regulations include a number of considerations which in England are the province of the planning legislation: considerations such as layout and site planning, the outward appearance of buildings, carparking, advertisements, and, to some extent, the use of land and buildings. Building regulations are required to cover the entire country whether prepared by central or local government. In Germany, they are of equal status to the local plan, covering different aspects of control. Elsewhere, the regulations are legally inferior to the binding local plans in the other three overseas countries where they either supplement, or even (in Denmark) are supplanted by, the plan.

C3.11 Authority for the building regulations is given by the national legislation. In France, they are contained in the *Code de la Construction*; in Denmark, in the national regulations, with a separate set for small buildings; and in Germany, in the *Land* building orders. However, in the Netherlands the legislation merely identifies the broad categories of construction, and the actual regulations are a local responsibility, subject to approval by the province. Even there, the Association of Netherlands Municipalities has prepared model building regulations, which are widely used, and monitors decisions of the Council of State.

## The Local Plan: Status and Function

C3.12 The status and function of local plans is summarised in Figure C4. The chief contrast is between the legally binding character of the four overseas plans, and the English local plan which is there to provide only a guideline for development control. The other differences mainly flow from this fundamental contrast.

C3.13 In none of the overseas countries are local plans required under all circumstances for the entire territory. As in England, the general principle is that they are to be prepared only where necessary. A local plan is specifically needed for all significant or major developments, as well as in more general circumstances in Denmark. In the Netherlands a local plan has been required for all non-built up areas since the first legislation in 1901, repeated in 1962 and 1985, which in turn has meant frequent amendments and alterations as pressures for development not in accord with the original plan had to be accommodated. Although there is no mandatory requirement for a local plan, much the same has happened in Germany, so that in all three countries, the phenomenon of the 'postage stamp' plan covering a very small area is common.

3.14 The situation in France is different. Local plans are optional but by now they cover much of the territory of the most highly urbanised *communes* But there are significant government restrictions on development if no plan is prepared, and no devolution of development control powers. Thus, in the smaller, more rural *communes*, there are few if any local plans and consequently there is a general prohibition on urban development outside the built-up areas. Within those built-up areas without a plan, control is exercised through the national rules of urban planning, as part of the *Code de l'Urbanisme*.

C3.15 Each of the four countries offers a slightly different explanation of the function of the local plan. In Germany it is to implement the preparatory land use plan; in Denmark to ensure the implementation of the municipal structure plan and to control development proposals; in France, to provide a comprehensive statement of planning restrictions and opportunities; in the Netherlands to define the planned 'destination' (ie use and built form) of its territory to be 'realised' through development and control. But, because of their comprehensive coverage of all non built-up areas, local plans in Netherlands also have the function of preventing unacceptable development especially in rural areas where the existing environment is to be protected.

C3.16   In every case, above all, the local plan enshrines existing and future development rights. Historically the function of the plan was to lay down the pattern of development in the suburban extension of towns and, by definition, thus to prevent any further development beyond the planned area either, as in the Netherlands, by having a local plan which specifically excluded such development; or, as in France since 1983, by not allowing development in areas not covered by the plan. Only later did the plan need to be used for the renewal or redevelopment of existing built-up areas, or their adaptation to new uses, whether or not there was an earlier local plan.

C3.17   Of course certain exceptions to this very rigid framework were permitted, provided the authority was there explicitly in the legislation, plans or regulations. But the intention was to provide a precise framework, in conformity with the principle of legal certainty. Accordingly the basic character of the local plan is that of a comprehensive land use plan, with regulations specifying the permitted dimensions of buildings and the space between buildings, the use of land and buildings, servitudes and easements for utilities, and any other restrictions that might be needed (eg for noise) in whatever detail was required. The plans could range from 1/500 to 1/5,000.

C3.18   Local plans in the overseas countries are thus intended chiefly as a stage in progressing development. Their link with the building permit is precise and absolute, making the local plan in those countries a primary instrument of control. But they also have other functions, such as providing authority for compulsory purchase of land; for subdivision and as a record for land registration; and, for instance in the Netherlands through the land development regulation which may have to accompany the plan, as a device for allocating the costs of urban development (including land acquisition, site preparation and servicing) between government, the municipality, land owners and developers.

C3.19   In all of these respects, local plans could not differ more greatly from the current English model with its much more flexible character and, in particular, its lack of any requirement that it be an explicit, comprehensive, land use plan. The English local plan is intended to carry forward the policies of the structure plan, to coordinate land use and development, and to provide a framework for development control. The key document is the proposals map, showing policy constraints such as conservation areas, and future opportunities and proposals for development. But it does not in any way enshrine either existing or future development rights. Nor does it generally provide a comprehensive guide to the considerations to be taken into account in development

control. Many of these considerations may be found in other, non-statutory documents such as the development control policy notes of central and local government, or design guides and development briefs, or even in informal local plans.

## The Local Plan: Preparation and Approval

C3.20   The actual process of preparing a local plan is broadly similar in all five countries, the divergences appearing at the stage of possible objection or challenge to a plan, and its final approval. Briefly, the process is as follows, noting significant variations (see Figure C5).

(i)   The preparation of a local plan is initiated by the lowest tier in the political structure, the municipality or district, albeit in England with the approval of the next tier, the county, which might also prepare its own local plan. Higher tiers may also direct a municipality to prepare a plan. Also in France, several municipalities may combine to form an intercommunal organisation for the purpose.

(ii)   The preparation is initiated formally, the consequence being to defer the grant of all building permits until the plan has been published. But in France, the Netherlands and, to some extent, Denmark, a formal declaration that a plan is in preparation triggers the possibility of granting a building permit in anticipation of the plan, subject to provincial approval.

(iii)   Preparation in every country involves surveys, consultations with government departments and other organisations, public participation and, if necessary, detailed negotiations with the intended developers of the land. The work is usually done by the municipal or city planning department. But, as the municipalities may be very small in the overseas countries, consultants are used extensively in the Netherlands and applicants in Denmark may be asked to assist. In Germany the work may be done for the municipality by the county, on an agency basis. In France the DDE provide technical planning staff appointed by government; or a permanent *Agence d'Urbanisme* of planners may be established by a group of *communes* in an urban agglomeration.

C3.21   Once the draft plan has been prepared, in every country it is placed formally on public deposit for a period of one or two months, to give an opportunity for objections. In addition in France there will have been a commission of conciliation to resolve conflicts between members of the inter-departmental working group involved in preparation of the plan. In theory too the *département* in France should also certify that the plan conforms to the structure plan if there is one, but this is treated less rigorously as there are few such plans and they tend to be out-of-date. At this stage the plan will also have been certified by the county in England and Denmark as being in conformity with their structure plan. This step is

## Figure C4  The Local Plan for the Primary Control of Development: Status and Function

| | England | Denmark | France | West Germany | Netherlands |
|---|---|---|---|---|---|
| **name of plan** | local plan | lokalplan | plan d'occupation des sols | Bebauungsplan | bestemmingsplan |
| **coverage** <br> by law | optional, as required | optional, as required for areas/for every major development if no previous plan | optional, as required | optional, as required, mainly for town extensions and new housing | mandatory for non-built up areas as at 1962, replacing the earlier uitbreidingsplan which, from 1901, was also mandatory for non-built up areas |
| in practice | c. 25% of country, with very few district-wide plans, but extending | wide coverage | c. 80% of urbanised communes c. 36% of smaller communes poor coverage in smallest, rural communes | wide coverage c. 25% of average Gemeinde (but 100% in some cases, eg Hamburg) | virtually 100% except for a few older villages and unchanging, pre-1901 built-up areas |
| **legal status** | material consideration for control, but not binding | legally binding on all citizens and government agencies, in conjunction with building regulations which they may supercede | legally binding on all land owners; and devolves decision-making in control from the state to the commune | legally binding on all public and private sectors | legally binding on all development, in conjunction with building regulations to which it is superior |
| **basic function** | guideline for development control, and for the coordination of land uses | to ensure implementation of structure plan and to control development <br><br> record for land registration | comprehensive statement of all planning restrictions, densities, public utility servitudes etc <br><br> record for land registration | binding land use plan prescribing full details on building construction and use | statement of the planned 'destination' of all land and buildings to be 'realised' through development and safeguarded by control, including restrictions on development <br><br> record for land registration |
| **basic character** | topographic base map with proposals for specific developments, greenbelts, conservation areas etc, usually 1/2500 or 1/10000; written statement of policies | land use plan (1/500-1/2000) with detailed, specific land use and other regulations including legal, administrative and procedural matters | land use zoning map (1/2000-1/10000) (with supplementary thematic and detailed maps); detailed, specific land use and other regulations for each zone | land use plan (1/1000-1/5000) of permitted land uses and densities, building mass, noise zones etc | land use map (1/1000-1/5000) with a written description of policies and land use/built form; detailed land use and building regulations: permitted uses, restrictions, exemptions, noise zones etc; procedures for elaboration of outline plans; areas where a general plan is sufficient |
| **format and content** | local discretion, within general guidelines, <br> (a) general district plans <br> (b) subject plan, eg for minerals <br> (c) action area plan, for comprehensive development | non-standard form and content, local discretion | standardised format and minimum content defined by Code de l'Urbanisme, though level of detail and precision can vary enormously | national standardised form and content | non-standard form and content, subject to provincial approval |
| **implementation** | planning permission | building permit | building, demolition permits subdivision approval land clearance and other authorisations | building permits | building permits construction permits, for development other than building as specified in the plan land development regulations, covering allocation of financial costs of development, land acquisition etc |

not necessary in Germany or the Netherlands as the same local authority prepares both plans.

C3.22 It is from here on that significant variations appear.

(i) Objections to the plan are assessed in different ways: by an inspector appointed by the ministry in England; by a commission appointed by the local administrative tribunal in France; or by a committee of the municipal council in Germany or the Netherlands. In every case there is the possibility of a public hearing, though in France this is limited to property owners affected by the plan. But there is no public inquiry in Denmark as the previous participation and consultations are considered sufficient.

(ii) The municipal council then approves the plan with or without modifications, with a further round of objections to the modifications.

(iii) Approval by the municipal council is the final stage in England and France unless, in England, the minister has 'called in' the plan for his approval. But there is a further round of approvals elsewhere. In Netherlands and Germany, final approval is by the province or the intermediate regional administration; in Denmark, by the minister if the county has sought to veto the plan, and in France by the *Commissaire de la République* of the *département*.

C3.23 However, in most countries there is opportunity for a further round of challenge to the plan through the courts. In England and Germany, the challenge is strictly on the legality of the procedures and content of the plan, in the former to the High Court and thereafter to the Court of Appeal and House of Lords (very rarely used); and in the latter to the administrative courts if the appellant expects to suffer disadvantage as a result of the plan. In France too there is opportunity for challenge in the administrative tribunal. The Netherlands has the most complex and lengthy procedure involving appeal to the administrative courts which recommends a decision to the minister which may be to approve the plan, or withhold approval in which case the municipality has to start again. But there is no further challenge either administratively or through the courts in Denmark; that is left to the subsequent decisions on building permits.

C3.24 The process of preparing and gaining approval of local plans in every case can be protracted, with a minimum period of about one or two years for a straightforward plan with few objections, or up to five or six years if the full process of appeal has to be gone through.

C3.25 Local plans are a vital part of the system of control in the overseas countries. The procedures for the alteration or review of plans thus become important if the plans are to retain their relevance and credibility. In Denmark, Germany and the Netherlands the full process has to be gone through, the only saving being that the amendment can be to a part of a plan, rather than the entire plan, and that at least in the Netherlands and Denmark, the process does not need to halt development as a building permit can be given in anticipation of approval of the plan. The problem is that this can result in a plethora of small plans which, again in the Netherlands, has led the courts to direct that a much amended plan be replaced by a new plan rather than to continue to approve further amendments. In France, however, there is a simplified procedure for the preparation and approval of alterations to local plans.

C3.26 More importantly, the complexities of the process and the rigidity of the plans have led several countries to seek other ways of controlling development. Denmark, the Netherlands and to some extent France have gone down the path of making plans more flexible, for specified areas and within specified limits: what the Danes call 'empty' plans, the Dutch, planning 'in headlines', and the French, the 'simplified' plan. Germany has gone a different path, by trying to do without the local plan altogether in particular circumstances, as is shown in the next chapter.

C3.27 The problem does not arise to the same extent in England. The local plans are already 'empty' by overseas standards. The chosen path instead has been to make new types of plan, for enterprise and simplified planning zones. These could be even more similar to the 'empty' overseas plans as they will specify a precise range of permitted development. The difference is that a building permit is still required in the overseas countries, presumably at least in part to comply with building regulations. In England, planning permission is granted automatically for development which complies with the planning scheme for the zone.

**Figure C5   The Local Plan for the Primary Control of Development: Preparation and Approval**

| | England | Denmark |
|---|---|---|
| **name of plan** | local plan | lokalplan |
| **who initiates** | district council, in agreement with county council | municipality, or a development proposal |
| **initiation/authorisation** | inclusion in official county development plan scheme | s.21 proposal that plan in preparation, enables permits to be issued if in conformity prior to formal approval of plan |
| **who is responsible for preparation** | district/county planning committee with advice from officials | municipal council |
| **who prepares the plan** | district/county planning department | planning department or consultants, assisted by initiating developer when required |
| **technical preparation of draft plan** | surveys<br>consultations with other departments, county, DOE etc<br>formal public participation on alternative plans etc | surveys<br>alternative proposals may be published<br>consultations with other departments, county, ministry etc<br>extensive public participation negotiations with developers |
| **initial approval of draft plan** | approval of draft plan by district council<br>certification in accord with structure plan by county council<br>on deposit for 6 weeks for objections | approval of draft plan by municipal council<br>certified by county, in accord with regional plan and municipal structure plan<br>on deposit for 2 months for objection/veto by county |
| **approval, 1st level** | planning inspector (DOE appointed) holds public inquiry, writes report and recommendations<br>district council receives report, approves/modifies plan<br>if modified, on deposit for 6 weeks for objections to modifications<br>repeat 1st level, if objections, and final approval | municipal council approves/modifies plan after consideration of objections (no public inquiry)/negotiation and agreement with county about veto |
| **approval, 2nd level (if necessary)** | = | if county veto not withdrawn dispute referred to minister for decision |
| **approval, 3rd level (if necessary), or legal challenge** | appeal to the High Court etc on legality of procedures, definitions etc | = |
| **time scale for preparation and approval** | 4-5 years average | about 1-1½ years average |

424

| France | West Germany | Netherlands |
|---|---|---|
| Plan d'Occupations des Sols | Bebauungsplan | bestemmingsplan |
| mayor or president of ad hoc intercommunal public establishment set up for the purpose (EPCI) | municipality (Gemeinde/ Kreisfreie Stadt) | municipality, or directive from province |
| formal resolution of council/EPCI to prepare plan | formal resolution of council | s.21 preparatory decree that plan in preparation enables interim control to defer application pending municipal approval of plan; or permit in anticipation of plan, subject to no objection by province |
| mayor or president of EPCI, with advice from DDE officials | municipal council | municipal executive (burgemeester & wethouder) |
| Direction Départementale de l'Equipement, Agence d'Urbanisme or consultants | building/planning department, or Kreis or consultants (small municipalities) | city planning department (large municipalities), or consultants (small municipalities) |
| working group of all government levels, chamber of commerce etc informal discussion with mayor preliminary public consultations resolution of conflict by commission of conciliation | surveys consultations with other departments, county etc formal public participation negotiation with developers | surveys consultations with other departments, province etc formal public participation negotiations with specific developers if necessary |
| approval of plan by mayor or president, and municipal council | municipal council approves draft plan | approval of draft plan by municipal executive |
| on deposit for 2 months for objections by property owners | on deposit for 1 month for objections | on deposit for 1 month for objections |
| public enquiry by a commission appointed by Tribunal Administratif; with report and recommendations mayor or president accepts report, approves/modifies plan if modified, on deposit for 2 months and repeat, then final approval | municipal council approves/ modifies plan (no public inquiry) if modified, on deposit for 1 month for objection to modifications final approval by municipal council | municipal council hears objections, approves/modifies plan within 3 months on deposit for 1 month for objections to modifications etc |
| | approval by Regierungs- bezirk (after approval by municipal council) | provincial executive hears objections, consults provin- cial planning committee approves/modifies within 6/12 weeks (otherwise deemed approval) on deposit for 1 month for objections to Crown by aggrieved citizens, municipal council |
| challenge in Tribunal Administratif possible | appeal to higher administrative courts on strictly legal grounds, by anyone expected to suffer disadvantage by the plan, within 1 year | Council of State (disputes department) hears objections with an independent published technical report on the objections, recommends a decision to the minister minister decides to approve, or withhold approval (in which case municipal council has to revise plan within 12 months) |
| 1½-2 years average 3 years maximum from initiation, or invalid | 1-3 years, or longer | if straightforward, by province, 1-2 years if complex and/or objection to Crown, 5-6 years |

425

## Planning Considerations and the Scope of Control

C4.1   The primary instrument of control is a form of permit, or permission, which has to be obtained from the appropriate local authority, and this applies in all five countries. But there is a very substantial distinction to be drawn between England and the overseas countries. It is a distinction which has a very strong legal basis and implications, and can, though not necessarily does, have major consequences for the day-to-day technical work of the development controllers.

C4.2   Briefly, development in the English planning system is a process affecting the use and development of land for which there is an all-embracing national definition in the Town and Country Planning Act, 1971. There, in section 22, development is defined as

"(i)   the carrying out of building, engineering, mining and other operations in, on, over or under land; or

(ii)   the making of any material change in the use of any buildings or other land."

The definition is further amplified in statutory instruments, namely the General Development Order and the Use Classes Order. The process of control then means that permission has to be sought for any development, and the consent or refusal is given having regard to the development plan and any other material considerations.

C4.3   Control in England therefore is based not only on a fixed definition of development but also on a concept of planning considerations which may be taken into account provided (a) they are proper considerations in terms of the purposes of planning as enacted in the legislation and interpreted by the courts; and (b) they are material (in the legal sense) to the circumstances of a particular application for permission to develop.

C4.4   Planning considerations cover a wide and continually evolving range of material. Depending on the particular circumstances of time and place, the list could well include any of the following broad categories

(i)   amenity, such as the design, visual quality of development and its relation to its surroundings etc

(ii)   the arrangement of development including site planning and layout, highways and infrastructure, provision of open spaces, etc

(iii)   the efficiency of the development in terms of use of resources, condition of buildings etc

(iv)   the phasing of development and its coordination with other land uses etc

(v)   the quantity, character and location of development etc

(vi)   other factors, including the possibility of planning gain, precedent etc.

They are found in the basic legislation and the various policies and plans of every level of government, but also in local custom and practice, and in the professional training and ideology of planners. Their authority is continually challenged in, and legitimised by, the courts in their interpretation of the legislation, decisions which create legally binding precedents.

C4.5   Thus the potential scope of control in any one locality, at any one time, is an amalgam of the current state of case law about proper planning considerations and the test for materiality, and the policy stance, custom and practice of the local authority, together with the policies and advice of government.

C4.6   In the strict legal sense, there is nothing in the overseas countries comparable to an evolving list of potentially material planning considerations. To a greater or lesser degree, their legislation provides a broad definition of development and then establishes the legal instruments within which the definition is amplified, be they national building regulations or legally binding local plans. Control then becomes a matter of testing compliance with those definitions which, in the case of the local plans, in effect define a specific pattern of land use and development to be promoted, achieved and safeguarded. This means that there can be no question of other material considerations. The only considerations are those specifically identified in the legislation, plan and regulations.

C4.7   The scope of control in all four countries is thus defined precisely in the plans and regulations and is specific to a particular area covered by that plan and set of regulations or, in the absence of the plan, by the regulations on their own. It can change only when the plan or regulations are themselves changed. And the challenges to the administrative decisions in control are not against their legitimacy in general planning terms but specifically whether they have correctly applied the definitions in the plans and regulations.

C4.8   The overseas plans and regulations cover the same

broad categories of considerations as was mentioned previously for England. Nevertheless, in general, the scope of planning control can be much wider, and development be defined much more sharply than in England.

C4.9 England and the overseas countries also differ in one further, crucial respect. Planning and building control are separate functions in England requiring separate permissions. In the overseas countries they come under a single system of control, with a single building permit. The local plans and the building regulations overseas each contain what in England would be described separately as planning considerations and building matters of safety, stability and construction.

C4.10 However, in more practical matters of the *de facto* definition of development, and the day-to-day procedures for control, the five countries can be placed in a single comparative framework, provided the crucial distinctions are not lost sight of.

## The Definition of Planning Control

C4.11 The key features in defining planning control are summarised in Figure C6. The basic scope of the definitions in all five countries covers all building construction, mining and engineering operations including site planning and the outward appearance of buildings; and all changes in the use of land and buildings. The most significant differences are that control in all four overseas countries is much more detailed (partly because it also includes building control) and extends to the demolition of buildings; and, in general, changes of use are controlled only in those cases where there is a local plan. However, changes of use outside built-up areas involving construction are strictly controlled in France, Germany and Denmark even in the absence of a plan.

C4.12 The precise definition and scope of control depends on the sources. In England the definitions are found in the national legislation, in the General Development and Use Classes orders, and in court decisions. In France they are amalgamated in the *Code de l'Urbanisme* including the national rules of planning, and in the local plans. Elsewhere they are in the building orders or regulations and local plans. The difference then lies between the highly decentralised system in the Netherlands, in which local plans and building regulations are both a local responsibility; the more centralised system in Germany in which Federal legislation specifies the permitted content of local plans and the *Land* provides the building regulations; and the mixed system in Denmark with national building regulations and autonomous local plans. But in every case, the operational instruments are the actual building regulations and the specific local plan, with the latter in every case superior to the former other than in Germany where the two are complementary and of equal status as they cover different matters.

C4.13 Within the basic scope and definitions there are a number of exceptions, or exemptions which apply more or less in all five countries.

(i) The use of land for agriculture and forestry everywhere is excluded from planning control, although under certain circumstances specified in particular local plans tree felling and planting, alterations to vegetation, land drainage and reclamation, and agricultural methods including deep ploughing and use of pesticides can be brought within planning control. Agricultural buildings in principle are everywhere subject to control unless specifically excluded.

(ii) Existing uses in general are exempted from control, and their continued use or replacement after fire etc may be permitted, even if they are in conflict with an approved plan, depending on the circumstances.

(iii) Certain very insignificant constructions, such as aerials, fences, small non-residential buildings, underground pipelines etc and some forms of temporary building usually will not require a building permit unless so specified in a local plan or building regulations in an overseas country. But the list of exemptions is nowhere as long as that exempted from requiring planning permission in England by the General Development and Use Classes Orders. However, even in England, the exemptions can be cancelled under special circumstances, for instance in Areas of Outstanding Natural Beauty.

(iv) Certain other developments may be exempted from having to comply with the local plan or building regulations though still subject to control through a building permit. These include minor deviations within specified limits for instance in minimum plot area, or maximum height of buildings, or small extensions to buildings, or (in Denmark) garages. In the same category are the changes in use of buildings depending on the degree of specificity in the local plan which can vary between an extremely narrow set of definitions, or very broad categories. There is no comparable category in England, because of the different legal status of the plan.

## Procedures for the Primary Control of Development

C4.14 The most significant features of the procedures for control are shown in Figure C7. As with other aspects of control, the chief differences are between England and the rest, all of them stemming from the basic difference in the legal status of the plan and regulations. The contrast starts with who must apply for permission. In all four overseas countries the plan and regulations are binding and therefore all citizens and all government departments and agencies must apply. However in Germany, as in England, federal and *Länder* government departments

**Figure C6   The Scope and Definition of Planning Control**

Notes: 1. This refers only to the primary system of planning control, through planning permission or building permits.  It excludes related systems such as listed buildings/conservation, urban renewal, rural areas, subdivision, etc.
2. The definitions, exceptions, permitted exemptions, etc are not comprehensive but illustrate general principles

| | England | Denmark |
|---|---|---|
| general, legislative source of control | Town & Country Planning Act 1971 etc | Municipal Planning Act<br>Building Act |
| basis of control and definition of development | universally applicable, national definition in the legislation | within the general framework of national policy and legislation, national building regulations and where necessary a local plan |
| operational instruments for primary local control | Town & Country Planning Act<br>General Development Order<br>Use Classes Order | national building regulations<br>national building regulations for small buildings<br><br>lokalplan |
| basic legal definition | universally, all building etc operations; all changes of use of land and buildings | all construction & demolition, including parking and access, site planning/aesthetic, advertisements, change of use, land grading etc |
| general exclusions from control | use of land for agriculture and forestry;<br>continuation of existing use and minor increases in volume<br>development by government departments (quasi-control) | use of land for agriculture and forestry<br>minor development for utilities and certain very small buildings<br>continuation of existing uses |
| development usually not requiring a building permit or planning permission, or at most a lesser form of permission | minor developments and extensions (unless exemption suspended by article 4 direction<br>temporary buildings<br>agriculture and forestry buildings in some cases<br>changes of use within a use class | minor building and demolition work<br><br>= |
| development requiring a permit but exempted from need to comply with plans/regulations | not applicable | minor deviations from minimum plot area/ratio or height<br>certain other minor developments including some leisure/ agricultural buildings, garages etc (unless required by plan)<br>larger deviations from building regulations with approval of minister<br>minor deviations from plan/ regulations which do not change the special character of an area |

428

| France | West Germany | Netherlands |
|---|---|---|
| Code de l'Urbanisme | Federal Building Law<br>Land Building Order | Housing Act 1982<br>Physical Planning Act 1985<br>Physical Planning Decree 1985 |
| universally applicable, national definition in the Code de l'Urbanisme, including provision for a local plan where necessary | within the general framework of federal/land policy, land building regulations and local plans where necessary | within the general framework of national policy and legislation, local building regulations and where necessary a local plan |
| Code de l'Urbanisme (Règles Nationales d'Urbanisme) | Federal Land Utilization Order<br>Land building regulations | local building regulations |
| plan d'occupation des sols | Bebauungsplan<br>Flächennutzungsplan (for s.34/35 procedures if no Bebauungsplan) | bestemmingsplan |
| all construction/demolition including work on existing buildings which changes use, external appearance or volume; applies to public services and agencies | all construction, alteration change in use, demolition or removal of building, including any change in use of land | local definitions, in general covering all building etc operations, including demolition; site planning/ aesthetic; car parking, advertisements<br>(especially if there is a plan) use of land/buildings; and other 'construction' not being building, as specified in a plan (eg tree planting/ felling, land reclamation) |
| use of land for agriculture some minutiae of development | use of land for agriculture, forestry | use of land for agriculture, forestry<br>continuation of legal existing uses |
| underground pipelines etc transport and utility operational works street furniture works of art small, pylons etc | very small buildings for non-residential use | with approval, a few very insignificant constructions aerials, fences, small non-residential buildings in certain cases |
| minor developments including small extensions, swimming pools, small leisure buildings etc | as specified in Bebauungsplan or federal Land Utilisation Order | minor deviations (eg from specified height, balconies, use), certain temporary, short term uses, changes of use not requiring construction<br>agricultural dwellings/ buildings, as specified in the plan<br>development whose prohibition would prevent without good reason the use of land |

## Figure C7  Primary Control of Development: Summary of Procedures

Note:  This refers only to the primary system of planning control, excluding listed buildings/ conservation, urban renewal, special rural areas, subdivision, etc.

| | England | Denmark |
|---|---|---|
| **basic instrument** | planning permission | byggetilladelse (building permit) |
| **legally binding criteria** | material considerations | lokalplan and/or national building regulations |
| **who must apply** | all citizens (special procedure for government departments) | all citizens, government departments |
| **to whom is the application made** | district council | municipal building departments |
| **form of application** | outline permission, with reserved matters<br>full permission | building permit |
| **procedure** | receive/register application<br>notify neighbours, if required<br>consult other departments, refer to local and structure plans, circulars<br>notify DOE of departure from structure plan | receive/register application<br>neighbours 'learn'<br>consult other departments<br>refer to plans/regulations<br>negotiate with applicant |
| **who makes decision** | district council planning committee unless delegated to officers | technical & environment committee unless delegated to officers |
| **what is the decision** | grant/refuse, in outline/ full, with/without conditions, according to material considerations<br>deemed refusal if no decision in 2 months | grant/refuse, in compliance/ not complying with binding criteria, with conditions |
| **post-decision procedure** | decision notice issued to applicant | decision notice issued to applicant |
| **discretion to grant exemption from binding criteria** | not applicable | within limits specified in plan/regulations |
| **flexibility to interpret the plan/regulations** | wide, as other material considerations apply | limited to minor matters which do not affect the special character, and with their consent from regulations on distance from neighbours<br>new style "empty" plans give more discretion within limits |
| **flexibility to vary the plan** | DOE notified of proposed 'major' departures from development plan | none, except by full procedure |
| **permission to occupy** | = | permit to use required (for most developments) |
| **enforcement** | enforcement/stop notice, for breach of planning control, through the courts, to recover costs/impose fine, if no successful appeal | through the courts to recover costs/impose fines for non-compliance with plan/ regulations or permit |
| **time taken for a decision**<br>    legal requirement | 8 weeks | probably, no time limit |
|     actual | 67% less than 8 weeks<br>11% more than 13 weeks | 2-4 weeks average |
| **refusals** | 15% average | less than 1% |
| **appeals**<br>    to council/courts | = | not available |
|     to crown/council of state petition committee of Land parliament | 3% average | |

430

| France | West Germany | Netherlands |
|---|---|---|
| permis de construire | Baugenehmigung (building permit) | bouwvergunning (building permit) aanlegvergunning (construction permit) |
| plan d'occupation des sols or national rules in Code de l'Urbanisme | Bebauungsplan and/or state building order | bestemmingsplan and/or local building regulations |
| all citizens, government departments and agencies | all citizens, government departments and agencies | all citizens and government departments and agencies |
| commune (mayor), if approved POS Département (Commissaire de la République), if no POS | the competent authority (Kreis/Kreisfreie Stadt) | municipality (building inspectorate) |
| outline permission full permission | outline permission interim building permit building permit | building/construction permit, and (a) development in elaboration of outline plan (s.11), or (b) development needing amendment of local plan (s.19) |
| receive/register application advertise application consult other departments etc check conformity with plan and national rules | receive/register application notify neighbours (if required) consult other departments refer to plans/regulations negotiate with applicant | receive/register application neighbours usually 'learn' consult other departments consult aesthetic commission refer to building regulations/ local plan negotiate with applicant/ consult province etc if proposed elaboration of outline plan/amendment of approved plan |
| mayor/Commissaire de la République, as above | chief planning officer in Kreis/Kreisfreie Stadt | municipal executive unless delegated to officers or submunicipal council |
| grant/refuse, in compliance/ not complying with binding criteria, with/without conditions deemed permission if no decision in 2 months | grant/refuse, in compliance/ not complying with binding criteria appeal to administrative courts for a decision after 3 months | grant/refuse, in compliance/not complying with binding criteria deemed refusal if no decision in 2/4 months |
| decision notice issued to applicant site notice posted, to give opportunity for objection | decision notice issued to applicant site notice posted, for information, including public authorities | decision notice issued to applicant |
| within limits specified in plan or code, with some scope for interpreting complex rules | within limits specified in plan/regulations (ie reuse, minor) | within limits specified in plan/regulations (ie precise, minor) unofficial discretion is used for minor deviations if no objection from neighbours, aesthetic commission etc |
| only to regularise minor deviations, unless specified in plan or national rules | none, unless specified in plan/regulations | s.11 procedure for elaboration outline plan, within specified limits and subject to provincial approval s.12a(new) style of outline plan, giving discretion within limits |
| none, except by modification procedure | none, except by formal adjustment of preparatory land use & binding plan with Regierungsbezirke approval | s.19 procedure to grant permit in anticipation of a declared intention to amend/prepare a plan, subject to a declaration of no objection by province |
| certificate of conformity required | certificate of completion required | certificate of completion required |
| 'verbal process' system to hear infringements, stop works etc severe fines and penalties (prison/demolition/ restoration), with recourse to civil courts for damages | strict, by the competent authority, with demolition/stop notices/ fines | through the courts to recover costs/impose fines for non-compliance with plan/ regulations or permit |
| 2 months | in effect 3 months | 2 months (or extension to 4) |
| 85% in 2 months 95% in 5 months | 12 weeks | most within 2 months 6-18 months if section 11/19 decision |
| 5% | 0.4% | approx 5% |
| 0.5% | not available | approx 2-10% |
| 0.5% | | approx 1% |

receive their permission by a different method, from government, although current practice means that the local authority is consulted.

C4.15 Applications are made to the municipality, that is the lowest general tier of government, except in Germany where the county is responsible in the more rural areas, and France where the *département* is responsible unless there is a plan which has been prepared by the municipality. In general, the application is for a building permit except in England, France and Germany where a preliminary application may also be made for a form of outline permission, to establish the basic right to develop. The actual processing of the application is similar in every country, with consultations, negotiations with applicants etc. However in none of the five is there a general legal right of prior notification to neighbours about an application before the decision is made although in most countries it is possible for people to find out about applications. In the overseas countries, this reflects the administrative nature of the decision. There are local variations. Thus in the Netherlands each municipality must appoint, and consult, an independent aesthetic commission whose advice cannot easily be overruled.

C4.16 The first significant difference is who actually makes the decision on whether or not to grant permission. In England and Denmark it is by a committee of council (though in practice it may well be delegated to the officials); in the Netherlands, the municipal executive, that is, not the elected body; in France, the mayor or, if there is no plan, the *Commissaire de la République* of the *département* unless it has been delegated to the departmental officials; and in Germany, the decision is nearly always made by officials unless it is a very major proposal. Thus three of the four overseas countries follow the administrative route, regarding the decision as purely administrative, and this stems from the basic nature of the decision, namely whether the application complies with, or is in conflict with the plan and regulations. In principle there is relatively little capacity for the exercise of discretion actually in the decision on the permit. Any discretion is more likely to be in negotiations about amending a plan to fit, as far as possible, the applicant's wishes.

C4.17 Nevertheless, discretion is exercised in the control of development in all five countries to a greater or lesser extent. In England there is very wide discretion, although strictly any departure from an approved plan is intended to be gone into more carefully, possibly involving the minister who approved the plan. Elsewhere, three levels of discretion can be identified.

(i) Discretion legally to grant exemption from the binding requirements of the plan and regulations is confined to very narrow criteria specified in the plans. However, in practice in the Netherlands, there is fairly widespread use of discretion even in this degree with a permit being issued if the development is insignificant and there are no objections from neighbours, the aesthetic commission or any other departments.

(ii) Discretion to interpret the plan or regulations is more varied. In principle there is none in Germany. The French national rules of planning, which apply if there is no local plan, do allow discretion through the use of words such as 'may' or 'can'. In addition France, Denmark and the Netherlands have created instruments which allow interpretation within specific limits: the so-called 'empty', generalised plans in the three countries; the Dutch procedure for elaboration of an approved outline plan; and the Danish possibility of deciding that minor developments which do not affect the special character of an area may be permitted.

(iii) Discretion actually to vary the plan or regulations is possible in general only by going through the formal procedures for amending a plan. As previously noted, this can mean in the Netherlands and, to a lesser extent, Denmark the possibility of an early decision on the permit, in anticipation of approval of the plan. This is a very widely used procedure. However, Germany has provided in its legislation two other methods for issuing a permit in the absence of a plan. One covers applications in conformity with adjacent development in an existing built-up area; the other allows, more rarely, for larger applications for development in peripheral areas around towns, subject to conditions regarding proper planning and the public interest.

C4.18 The actual decision is straightforward: a grant of permission or the issue of a building permit, with conditions if that is possible within the discretion allowed by the plan or regulations; or refusal. In most countries, failure to come to a decision in the specified time limit means that the applicant can assume permission to have been refused although in France the opposite is true, and a tacit permission can be assumed. But that is not the end of the story. At least in France and Germany, notice of the decision has to be posted on site, to give people an opportunity for objection in France, or to provide information about where the approved plans can be inspected in Germany. Then, when building construction is completed the control process in the overseas countries always includes a requirement for a certificate of completion before the new development can be brought into use. Such a requirement applies in England only under building control. In all five countries there are procedures for the enforcement of planning control.

C4.19 The contrast between England and the rest is seen in the comparative statistics although these must be treated with caution. Several countries do not collect statistics, and others may not be comparable. However, in every country a decision is intended to be made within about two or three months, and, as in England, the majority of decisions are made within the time limit in the overseas countries provided there is no need for an amendment to the plan. The refusal rate tends to be much lower in those countries than in England, and the rate of appeal against the decision is also probably much lower.

C4.20   Both of these latter statistics reflect very largely the greater certainty in the overseas systems, and the more public knowledge about the amount of discretion and therefore room for negotiation. It must be emphasised however that the big contrast is that the way of obtaining a permission on a major proposal not covered by the plan in England is to negotiate approval of a planning application or, failing that, to appeal to the minister; in the other countries it is much more likely by negotiating an amendment to the plan as any appeal in general can only be on grounds of legality.

## Appeals in the Control of Development

C4.21   Following on from the last paragraph, the purpose of appeals in England and the other countries is quite different, as is shown in Figure C8. The first difference is that the appeal in England is to the minister against a refusal by a district council on grounds of planning policy; in the others it is against any administrative decision, whether it be the grant or refusal of a permit or conditions, and the appeal is on the grounds that the decision violated the proper legal procedures or the legal definitions in the plan or regulations. This is a crucial difference, although there is also opportunity for appeal in England on grounds of procedural legality but that is by a separate route, to the courts.

C4.22   Since appeals in the overseas countries are on grounds of legality, the appeal route itself is more complex than in England. Thus a first appeal against the decision can be made to the elected municipal council in the Netherlands and thereafter to the provincial committee and then the Judicial Department of the Council of State. At these two subsequent appeals, the action is thus heard by the courts whose decision is final. There is no appeal to the minister. The situation in France and Germany is roughly similar, except that in Germany there is also a first appeal to a higher administrative authority to reconsider the decision whereas in France the appeal is immediately to the administrative courts and then on to the higher courts as necessary. The system is similar in Denmark except that the final appeal, still on grounds of legality, is to the minister, with a first appeal to the county.

C4.23   Appeals against a refusal of a permit are made by the applicant or, in some circumstances, by a municipality or government department. Those against the grant of a permit or conditions are more restricted and only in Denmark and the Netherlands can anyone appeal. In France and Germany, such an appeal is open only to those citizens whose property interests are directly affected, or whose constitutional rights are violated, by the proposed development or, in France, to recognised environmental groups. In every country there is an opportunity to apply for operation of the permit to be suspended until the appeal has been heard.

C4.24   The proportion of applications going to final appeal in the courts or to the minister in those overseas countries where statistics are available is probably a lot smaller than in England, for reasons already mentioned. The point is made in France and the Netherlands for instance that access to the courts is comparatively simple. But the available information suggests that the process can be very lengthy, up to two or three years in France, whereas for the vast majority of appeals in England that are decided by an inspector rather than the minister, the process is much quicker, on average about five or six months.

# Figure C8  Appeals in the Primary Control of Development

| | England | Denmark | France | West Germany | Netherlands |
|---|---|---|---|---|---|
| **who can appeal** | any applicant aggrieved by a refusal of planning permission (including deemed refusal after 8 weeks), or conditions | any applicant aggrieved by the grant/refusal/conditions for a building permit | any applicant aggrieved by refusal of a building permit, or conditions any citizen directly affected or any environmental group aggrieved, by grant of a permit/tacit approval | any citizen whose constitutional rights (property interests) are violated by grant/refusal/conditions for a building permit | any citizen aggrieved by grant/refusal of a building/construction permit by the municipal executive (including deemed refusal after 2 months), or conditions any citizen/municipality aggrieved by grant/refusal of a declaration of no objection (S19) by the provincial executive |
| **the basis of the appeal** | the decision (a) was unacceptable to the applicant on substantive terms; or (b) violated the procedural rules and definitions under the act | the decision was illegal (ie went against procedures/legal definitions in plan/ regulations) the decision is of general interest/has substantial consequences for the appellant | the decision violated proper procedures/or the substantive rules of urban planning (including the plans/regulations) | the decision was illegal (ie violated procedures/ definitions in the plan/ regulations) | the decision was illegal (ie went against procedures, or definitions in the plan/ regulations) |
| **to whom is the appeal made initially** | for (a), the minister appoints an inspector who takes written representations/holds a public inquiry; and either is delegated to make the decision or recommends a decision to the minister for (b) the High Court | the county council (or, in Copenhagen/Frederiksburg, the minister) | Tribunal Administratif | higher administrative authority (ie Kreis or Regierungsbezirk) for reconsideration of the decision | municipal council, against a decision of the municipal executive with a public hearing of the appellant's case provincial committee of the council of state, against a decision by the provincial executive, with a public hearing of the appellant's case |
| **to whom is a subsequent appeal made, if still aggrieved** | for (a), no further appeal for (b), the Court of Appeal and, subsequently, the House of Lords whose decisions constitute a legally binding precedent | the minister, on the strict legal interpretation of the act/regulations | Council of State, whose decision is final and constitutes a legally binding precedent | administration courts, strictly on legal process and constitutional rights | Council of State (Judicial Department) with a public hearing of the appellant's case, and whose decision is final but does not constitute a binding precedent in law, though is in effect treated as such |
| **suspension etc** | none | not available | application can be made for the suspension of a permission pending the hearing of an appeal, with a decision within 24 hours | application can be made for suspension of permission pending hearing of appeal | application can be made to the council of state for suspension of the grant of a permit pending hearing of the appeal, with a decision within 24 hours |
| **time taken for an appeal** | 20 weeks for written representation/inspector 55 weeks for public inquiry | not available | up to 3 years | no time limit, no data available | no time limit, no data available |

# C5 SUPPLEMENTARY AND RELATED SYSTEMS OF CONTROL

C5.1 Obtaining a planning permission or building permit is the major step in the planning control of development. But there are two other kinds of control which may have to be gone through before final approval and occupation of the new development.

(i) There are the related systems in which the primary system of planning control is modified to include other requirements or consultations before the permission can be granted. An example of this in every country concerns the alteration or demolition of listed buildings, for which prior approval from some other agency has to be obtained before a building permit can be issued.

(ii) There are the supplementary systems of control which have to be satisfied before development can go ahead whether or not planning permission has been granted. A unique example in England is the requirement for building control independently of planning control.

C5.2 The following notes identify five different areas in which, at least in some countries, additional authorisations are required. They are topics which cover many of the issues and considerations of concern for planning, or demonstrate techniques which are relevant to control. However the list is not consistent, nor is it comprehensive for any one of the five countries. The precise relation between planning control and these other systems varies between the different countries. More importantly, in any case the main purpose of these other systems may not be regulatory control as such, but rather the implementation of some policy for which other instruments may be more important such as the payments of grants or taxes, or the registration of land ownership.

## Listed Buildings and Urban Conservation

C5.3 These two areas are closely related, although the regulation is more strict in the case of listed buildings. Briefly, for listed buildings, the situation in all five countries is as follows.

(i) There is special legislation governing buildings of historic or architectural significance enshrined in the case of France in the *Code de l'Environnement* as well as in the *Code de l'Urbanisme*.

(ii) There is a special agency of the national government or, in Germany, of the *Länder*. The aim of control is to prevent the alteration or demolition of listed buildings although in every case an additional aim is to assist in their preservation by fiscal measures.

(iv) The agency which prepares the list has to be consulted and, depending on the classification of the listed buildings, has to give its authorisation before a building permit or planning permission can be granted.

C5.4 The situation with regard to conservation areas is more varied and complex, although in most countries there are special procedures. The situation is least specialised in England and Germany where conservation areas are designated by the municipality and control is exercised through the normal planning permission or building permit. Control will usually be more rigorous in designated conservation areas in England and Germany, with an extension of control in the former over any proposal to demolish buildings in conservation areas. In Germany, the conservation area is simply part of a binding local plan which may give the requisite degree of control.

C5.5 Denmark and the Netherlands both have two classes of conservation area. One type is designated by the national agency and is little more than a group of listed buildings and their environs. Any development in these areas, including alterations and demolitions, requires the approval of, or at least no objection by, the national agency. Eventually such areas are intended in the Netherlands to be included in a local plan but in the meanwhile full control can be exercised. In addition, municipalities may designate their own conservation areas within which control is exercised through the normal building permit.

C5.6 France has a more complete set of arrangements under the *Code de l'Environnement* in which there are two classes of conservation area both of which are designated by the national agency for historic buildings. In the higher class, any proposals for demolition and development have to be approved by the national agency, with a full public inquiry. There is no local control, and the areas are not included in a local plan. In the second class of conservation areas, control is exercised by the departmental *Commissaire de la République*. These areas include the wider environs of the first class of conservation areas giving a much stronger, more comprehensive and much wider centralised control than in the other four countries. However, the control is highly selective compared with, say, England as there are comparatively few conservation areas or listed buildings.

## Rural Conservation and Protection of the Countryside

C5.7    Most of the planning control discussed in this report is about urban development. The aim of planning policy for the countryside beyond the built-up area as far as possible is that it be 'protected' from intrusive building. Planning for rural conservation is really a separate, and very complex, field with its own legislation and procedures in every country. However, roughly, two levels of planning can be identified which interact with planning control.

C5.8    The first level of policy involves designating national or regional parks, and areas of special landscape quality or ecological or scientific interest. All five countries have many such designations, which have the effect of focusing or intensifying planning control. Thus designated areas have to be integrated with regional plans in Germany, Denmark and the Netherlands. For instance

(i) Denmark has separate, legally binding conservation plans for such areas, prepared by the county as part of its regional planning, which then exercises planning control in which any exemptions have to be authorised by the government agency responsible for identifying the areas

(ii) the Netherlands includes the areas in the legally binding local plans with planning control exercised by the municipality subject to approval, or no objection, by the appropriate national or provincial agencies or local, independent expert committees.

In both countries, planning control in these areas can be extended by the plans to include farming methods such as deep ploughing or use of pesticides, land drainage, planting or removal of vegetation.

C5.9    In France, the methods of control for the so-called sensitive perimeters, including natural areas and woodlands, are through the preparation of binding local plans, although control remains with the *département* rather than being delegated to the *commune*. But the chief interest in France lies in the even more intensive level of control in coastal and mountain areas which have been the subject of national planning directives. Their aim is not only protection of the landscape, but also of the traditional economy, agricultural and fishing practices and tourist development. These directives have the status of laws, with control exercised by the *départements* through a national set of rules in the *Code de l'Environnement*.

C5.10    The degree of control in the various special kinds of area in England in theory is less restrictive than in the other countries. Nevertheless the policies are relatively strictly enforced within the limits of planning legislation. National parks, areas of outstanding natural beauty, heritage coasts and various types of nature reserve are designated by the appropriate national government organisation which in some nature reserves also owns the land. In addition local authorities identify areas of special landscape quality in their structure or local plans. Control over all these types of area remains at the local, district level or, in the case of national parks, at the county level or national park authority, and is exercised through the normal planning procedures. There is no obligation to consult the designating organisation except in the case of nature reserves.

C5.11    At the second more general level beyond the various specially designated areas, any construction (other than some agricultural works) is prohibited in the countryside outside built-up areas in France through the *Code de l'Urbanisme*, and in rural zones in Denmark through the Urban and Rural Zones Act. This prohibition applies universally in both countries unless there is a local plan, or, in exceptional circumstances in Denmark, if permission is granted for a 'zone dispensation', in which case normal planning controls apply. In effect, the same applies in the Netherlands except control is through the binding local plans which have to cover the entire country outside built-up areas, where necessary designating areas exclusively for agriculture or forestry. Germany has a roughly similar prohibition under the Federal Building Law, although there is the possibility of exemption under special circumstances on the edge of towns.

C5.12    In England each proposal is decided on its merits as there is no general, binding prohibition of development in the countryside. Nevertheless there are strict policies affecting such development. Many structure plans contain policies preventing isolated dwellings in the countryside. Green belt policies nationally and in structure and local plans restrict certain classes of development especially residential, industrial and commercial.

## Urban Renewal

C5.13    Urban renewal is an area of great and increasing importance in all five countries although emphasis has been shifting since the early 1970s away from slum clearance and comprehensive redevelopment towards rehabilitation. For the most part, it has been carried on through housing or urban renewal legislation providing for the positive implementation of policies: the acquisition, transfer and disposal of land; the clearance of slum and other dwellings and the removal of non-conforming uses; above all the finance for municipal housing or, more usually, in the overseas countries, housing associations, and grants, subsidies and tax relief for the private sector.

C5.14    Urban renewal involves development and is therefore subject to the normal controls through a planning permission or building permit and, in the overseas countries, the binding local plan. However in Denmark and the Netherlands legally binding urban renewal plans have been substituted for, or co-exist with, the local plans though, for planning control, they have virtually the same scope. The only variation is in the Netherlands where, in an urban renewal plan, a demolition permit can be issued only when approval has also been given for a building permit, and the developer has lodged a financial guarantee with the municipality that redevelopment will occur in a specified time period or the guarantee is forfeited.

C5.15 The Netherlands has also sought to cope with the problem of changes of use in built-up areas threatened with urban decay, but for which control has to rely exclusively on the building regulations if there is no local plan or urban renewal plan. Under these residential environment regulations, control can be exercised over changes of use in a specified area for a period of ten years, in the hope that either the decay can be arrested or a proper urban renewal or local plan can be approved, giving full control. Some municipalities had previously exercised similar control by amendment to their building regulations.

C5.16 This problem in principle does not arise in England as planning control can be exercised fully over all changes of use whether or not there is a local plan.

## Environmental Protection

C5.17 Environmental protection is also a matter of growing significance and will become even more so after 1988 when the European Commission directive comes into force, making an environmental impact analysis mandatory at least for a list of major classes of development.

C5.18 So far, however, France is the only one of the five countries to have had, since 1976, a formal system of environmental impact analysis. The *Code de l'Environnement* contains a list of 400 classes of development for which an environmental impact analysis has to be prepared, covering the proposal, its environmental impact, the reasons for going ahead and the measures to be taken for mitigating the environmental consequences. If the proposal is of major significance, a public inquiry has to be held by an inspector appointed by the *Direction Interdépartementale de l'Industrie* (DII). The *Commissaire de la République* for the *département* may then issue a decree authorising the development in principle, after which normal planning control takes over, through the procedure for a building permit.

C5.19 Elsewhere the field of environmental protection is covered by a very wide range of legislation under which separate authorisations may be required, for instance for nuclear installations, or potential pollution of the soil, atmosphere or water, etc. In some cases, notably for noise, the control may be exercised through the normal planning procedures. Thus local plans in Germany and the Netherlands must define areas within which development is restricted because of noise from industrial areas or motorways.

C5.20 In addition, the Association of Netherlands Municipalities has classified roughly a thousand types of industrial and commercial establishments into six categories depending on their environmental impact on, and distance from, residential areas. These are included as land use zoning regulations in local plans, with a limited amount of discretion in their application.

## Land Registration

C5.21 It is important to stress that planning control in the overseas countries is closely linked to questions of land ownership. As was previously noted, the binding local plan, in addition to its other functions, establishes or records existing and future development rights. The plans therefore contain information about land registration and ownership.

C5.22 In France and Denmark, for instance, there is legislation requiring a separate authorisation for the subdivision of open land before a building permit can be issued. In Denmark, this is found in the Urban and Rural Zones Act which defines rural areas where the permission of the county is required, with a local plan if major development is proposed.

C5.23 The system of *lotissement* in France is even more significant for planning control, as demand increases for single-family detached dwellings. Before any building permits can be issued, an application for subdivision of land must be made to the competent authority, the *commune* if there is a local plan, or the *département* if there is no local plan and the national rules for planning apply. The application covers many of the considerations which in England would be the subject of an outline planning permission, conditions and reserved matters, together with a section 52 or similar agreement: considerations such as the planning of development, servicing and infrastructure as well as the subdivision of land for ownership, which in effect means a site layout. This can be a lengthy process, especially as it has to be followed by the normal procedures for a building permit.

## Planning Gain and Finance

C5.24 Finally, mention must be made of what in England is called 'planning gain', for there are roughly similar objectives for planning control in the other countries. Strictly, in England planning gain is not a method of control. The agreement between a developer and local authority, say under section 52 of the Town and Country Planning Act, is not a necessary or prior condition for obtaining a planning permission. Nevertheless apart from any other factors, a planning agreement might cover financial arrangements under which the developer agrees to make a contribution to the cost of roads or infrastructure, or donate land for public uses.

C5.25 The arrangements in the other countries tend to

be more formal, explicitly part of the process by which a building permit is obtained, or a consequence to be taken into account. In Netherlands for instance a high proportion of the new land for urban development is literally new, being reclaimed from the sea. The usual procedure in these and often in more ordinary circumstances is for the municipality, working with the appropriate national or provincial agencies, to acquire the land after reclamation, to prepare it for development, including services and infrastructure, and then to dispose of the land to housing associations or private sector developers. The local plan is the instrument through which this is coordinated, with a special financial appraisal of the proposed development and an agreement embodying the financial arrangements, including the share of the costs to be borne by the different parties in the development, the rent or sale price for the land, and the future ownership. Thus the preparation and approval of the local plan, and the issuing of building permits, involves negotiations going far beyond those followed in England, involving all three levels of government and the intending developers.

C5.26   The relationship between planning gain and the control of development is even more varied and complex in France. The law provides for a number of different kinds of taxes to be levied, or contributions assessed, before a building permit can be issued. Some of the taxes are mandatory, other contributions are discretionary or at least the level of payment can be negotiated. The payments can be related most directly for instance to the cost of infrastructure or public services within the development, or they can be a levy for permission to exceed the plot ratio in a local plan. With decentralisation and the increasing fiscal problems of local government, development taxes and planning requirements loom ever larger in the negotiations over development control. A particular area of dispute can be the negotiations over subdivision or *lotissement*.

C5.27   In Germany, the arrangements are more formal, and less open to negotiations. The landowner is explicitly required to pay a fixed proportion (90 per cent) of the costs of providing streets, parking places and other public facilities, and the supply of gas, water, electricity and sewerage.

# C6 CONCLUSIONS

C6.1 The last of the objectives for the study set out in the terms of reference was to comment on the efficiency and effectiveness of the planning systems as seen by their operators and users. This is a very difficult task to undertake in a comparative study. Efficiency and effectiveness are complex ideas to handle in relation to an administrative process such as the control of development seen from the perspective of its users and operators. Each will have their own ideas depending on what they are expecting of control, how they relate it to planning in general and the part it plays in their own operations as citizens, developers or administrators. And those expectations vary not only depending on who is answering the question, but in which country they are located.

C6.2 The detailed comparative studies in the report show that the systems for planning control in the four overseas countries differ amongst themselves, despite sharing a common reliance on legally binding plans and building regulations; and in this last, common element, they differ sharply from the system in England. In principle, the English system is characterised by flexibility and administrative discretion; the overseas systems by rigidity and legal certainty. In practice, the overseas systems also incorporate ways of achieving flexibility through negotiation within government and with the private sector. But in all four the basic instruments of a legally binding plan and a building permit are generally accepted as giving a necessary and desirable level of certainty.

C6.3 In none of the five countries are the general aims for the control of development spelt out in detail. They are obviously closely related to the more general aims for planning but those are not precisely defined either. However, the systems of control fundamentally are constrained by constitutional and legal structures which in the overseas countries derive from, or have affinity with, the *Code Napoléon* rather than the English system of common law. This distinction is fundamental to any comparison between the systems, but despite the roughly common legal basis, each of the four countries has its own, unique system of control responding to its particular social, political and geographical context.

C6.4 Furthermore the systems of control in all five countries are currently evolving in response to a number of factors including

(i) geographical, economic, social and technological trends in patterns of urbanisation and economic activity, the scale and character of development, environmental issues and attitudes to rural areas

(ii) political and administrative issues including deregulation, decentralisation, public participation and fiscal stress in local authorities.

C6.5 Thus, to a considerable extent, any strict comparison and evaluation of efficiency and effectiveness in five such different countries and control systems is highly problematic. Even if comparison between the five countries is possible, comparability does not necessarily imply transferability. With that warning, a number of themes emerge from the study based on a consideration of efficiency and effectiveness. They are, first of all, the more general issues about the managerial efficiency of control systems and their effectiveness in achieving their objectives. Secondly, they relate to other operational criteria, notably predictability of outcome; democratic accountability; and credibility or acceptability.

## Efficiency and Effectiveness in Control

C6.6 One measure of efficiency which has been widely used in England is the time taken to prepare plans and to process planning applications and appeals. Statistics are compiled and many developers and users of the system regard what is described as delay as a significant factor affecting the costs of development.

C6.7 Comparative statistics by this measure of efficiency are not readily available as, in general, this is not regarded as a matter of major importance by either local authorities or developers in the overseas countries, the real aim being to achieve an acceptable permit or local plan. However, the indications are as follows.

(i) On permits, overall there is probably not much difference between England and overseas, the shorter time possible for proposals in conformity with legally binding plans being offset by the longer time likely to be needed for a permit based on preparation of, or amendment to, a plan, and the combination of building and planning control.

(ii) On local plans, the legal status of overseas plans, and the opportunities for challenge in the courts before approval (except in Denmark), means that in general a longer time is needed overseas than in England.

C6.8 The effectiveness of control is an even more

difficult concept. At its simplest, it can mean the extent to which control is capable of achieving the objectives set in the local plans and regulations. On this score, the overseas systems are probably more effective than that in England; after all, permits can be granted only if in accord with the plan and developers can negotiate to get their proposals expressed in the local plan. But since plan amendment is a direct part of the control system, the argument is circular. It can be argued that in the long run the close legal relationship between plan and control, and the legal status of the plan, ensures that the overseas systems are reasonably effective in achieving local objectives and are less likely to be thrown off course, as, for instance, local policies in England may be affected by appeal decisions.

C6.9 The situation at this local level of planning in England is much more complex and not strictly comparable with the overseas countries. The flexible relationship between the development plan and development control introduces the possibility of there being planning objectives for control which are not necessarily explicitly contained in the plan. Indeed, it could be argued that the purpose of control in England is to provide a forum in which a democratically accountable decision can be made openly and fairly, rather than to stick rigidly to some policy expressed in a plan.

C6.10 A further measure of effectiveness could be the degree to which local control, as part of the planning system, contributes to the objectives of national planning. That becomes extremely difficult to measure. None of the five countries has a national physical plan, although several have flirted with the idea, and found it impractical. However, compared with those in England, overseas countries in general have more explicit, and accountable, national planning policies which then provide the guidelines for regional plans and, in turn, the local plans. Thus one question could be, to what extent does a decentralised system of local control found in all countries assist in the achievement of those national policies?

C6.11 The answer is probably that administrative control as such is not the issue. Municipalities in the overseas countries can and do affect or, at any rate, delay the implementation of approved national policies, but through the plan-making procedures rather than the power of control. National policies, for instance for the regional distribution of population and employment, usually are intended to be the positive guidelines for public and private investment and thus depend for their achievement on the availability of resources rather than negative systems of control. But, given the legal nature of the local plans, those national policies tend to be more specific, more capable of being expressed in a set of regulations, than is the case with the government circulars in England.

C6.12 At a much higher level, the question follows whether national or local planning policies and control are effective in furthering wider national objectives. But this raises more fundamental questions about the relationship between physical and other forms of planning, and market forces, questions for which no clear answers are forthcoming in any of the countries.

## Certainty and Flexibility: the Predictability of Outcome

C6.13 A further measure of efficiency and effectiveness can be the extent to which the outcome of control is predictable, especially for developers and citizens; in other words the balance between certainty and discretion in control which in turn implies that between rigidity and flexibility in plans. In theory there is a complete contrast between the flexible, discretionary system in England, with a relatively unpredictable outcome, and the rigid, certain system overseas, with a predictable outcome capable of being altered only after a lengthy process of plan preparation or amendment.

C6.14 In practice, the contrast is not so great. The comparative difficulty of reliable forecasting, the complex nature of the central/local relationships, and the changing forms of development all make some form of flexibility essential, and procedures have been devised to allow for this. The question then becomes whether or not there is merit in starting from the relatively certain position of a legally binding local plan, and having to go through formal procedures and negotiations for elaboration or amendment of the plan. In the case of town extensions and control of urban development in non-built up areas, the answer is probably yes, given the complex negotiations needed to ensure servicing and infrastructure. That certainly was the view expressed overseas. The approach instead has been to retain, but modify, the legally binding local plan, either by developing new forms such as the more general outline plans, or by adopting easier procedures for the elaboration or alteration of plans.

C6.15 However, in existing built-up areas which have either never had a plan, or in which the plan has been fully implemented, the answer is more problematic, in that proposals in such areas are likely either to be minor developments and changes of use in which the direction of change and its cumulative impact is uncertain; or to be major redevelopment with wider consequences for the surrounding areas. Neither the rigidity of the binding local plan, nor the flexibility of an English local plan, nor the absence of any local plan, provides an effective answer to the problem of managing change through control.

## Democratic Control and Local Accountability

C6.16 This raises two related issues. The first is the opportunity for objection and participation in decision-making about control. The opportunities are widespread, the main difference being the right overseas to object to any administrative decision, whether for the grant or the refusal of a permit. This is limited however in that the objection must be to the legality of the decision, rather than its expediency or policy and the rights may be confined to certain classes of person. The more clear, concise and certain the policy, and the more rigid the relationship between plan and control, the less opportunity there is for objection.

C6.17 The more significant point probably is that, given the nature of the relationship between plan and control, and the necessity for preparation or amendment of a plan before a permit can be issued, the opportunity for challenge comes through the rights of public participation or consultation, and objection in plan-making.

C6.18 The other issue concerns the locus of authority for taking the control decision. This goes back to the relationship between the different levels of government, and between government and the courts. Once a plan or regulations have been approved, the responsibility overseas is usually that of the municipality or county, and objections to control are made ultimately not to the province or minister on a matter of policy, but to the courts on a matter of law. In effect, this gives the municipality greater autonomy than in England where the appeal system provides a basis for intervention by government. The chief way of direction by higher levels of government overseas is through the procedures for plan-making and approval. In that too the municipality has considerable autonomy, losing it only insofar as it tries to approve policies in conflict with those of higher levels of government. Underlying this is the more general autonomy of local government under the constitution, and on the whole the increasing decentralisation of government to local levels.

## Credibility and Support for the Planning System

C6.19 The last theme concerns the extent to which planning and control is credible and has widespread support from politicians, developers and citizens. The first point concerns the balance between acceptability to developers, and to citizens in general. It was claimed overseas that both parties in general benefit from both the certainty in control and the formality of the procedures for gaining a permit, whether directly or by amendment of a plan. The time taken to complete the procedures is accepted as the price to be paid.

C6.20 Obviously, developers wish for greater speed in decision-making, and less concern with detail. Nevertheless, slow, detailed procedures are accepted as necessary adjuncts to the certainty eventually given by the system although in every country there is a trend towards deregulation, and a search for greater efficiency. Developers in most countries criticised the growing influence of anti-development lobbies, fuelled by the right of objection to the grant of permission and in the Netherlands to the delegation of decision-making to sub-municipal councils.

C6.21 Another general criticism, especially in France and Germany, concerned the complexity and opacity of the system, to such an extent that many accepted control as it were by default. Like death and taxes, it is always with us. All that can be done is some simplification and relaxation of control, provided it does not too seriously create uncertainty. A more specific criticism voiced in the Netherlands is the number of levels through which objections to a plan can go before a final decision.

C6.22 However, it is important to stress the fundamental stability of the system of control which underpins the degree to which apparently it has very general support overseas. That stability derives in part from the care and effort which goes into drafting the original legislation over periods of years if the time for preliminary investigation and consensus building is included.

C6.23 But that leads to a last impression, namely, what has been described as the distinction between administration and government. In principle, the task of local government in the overseas countries tends to be regarded more as the administration of duly enacted laws, plans and regulations, especially in the regulatory control and implementation of policy. Thus there is less questioning of control, as such, than is the case in England. There is less propensity to see it as something almost independent of plan-making for, in reality, the two are much more directly interrelated, with control being the administrative implementation of policy.

## Conclusions: Transferability or Comparability?

C6.24 Any attempt to draw conclusions about the different systems of control has to recognise that their aims, and indeed the concepts of planning and control, differ. Strictly, the systems should be assessed only in relation to their own aims, within their own legal and constitutional contexts. It is not possible to say that one country's system is better than another's. And, despite similarities, every system is unique to that country.

C6.25 The aims of planning and control are not laid down with any precision in any of the countries. In very general terms however they are broadly the same in each, namely, ensuring a proper, balanced pattern of urban settlement and conservation of resources, and providing a good quality of environment for living and working. The differences lie in the relationship between policy and control, and the status and function of policy. For the overseas systems, control is policy-led, with policy being legally established in plans and regulations. This still leaves room for discretion in control, and recent political and economic trends are tending to widen the amount of discretion by increasing flexibility in policy.

C6.26 Policy is made explicit at each level of government, with formal procedures for its preparation and adoption, and its enforcement by higher levels of government on lower levels. National policies are intended, through regional policies, to provide a framework for local plans. Financial and administrative mechanisms facilitate and ensure implementation of national policies at the local level, in the end through legally binding plans and regulations. But, at that local level, implementation is seen as a partnership, the municipality and the developer negotiating to achieve the best results in the national and the local interest.

C6.27 This, of course, is an idealised picture. In reality, national policies can be, and often are, thwarted by uncooperative municipalities and unwilling developers. But the framework of policies does give clarity and a starting point for negotiation; and the procedures for preparing and approving those policies formally incorporate consultation between different levels of government and participation of non-governmental interests. The point is that the formality of the procedures and the need for clarity in the policies derive from the constitutional powers of the different levels of government, the constitutional rights of citizens and the principles of legal certainty.

C6.28 The contrast in England is that the control is much more flexible, more capable of accommodating change relatively quickly and easily in administrative terms. What is lost, is the comparative certainty of the overseas systems, the greater predictability of outcome at least of the minor, routine proposals for development, or indeed of major developments in policy areas where there is a strong, legally enforceable consensus such as in much of the countryside in every country or, more narrowly for instance in the policy for preventing out-of-town regional shopping centres in the Netherlands.

C6.29 Comparisons are therefore highly problematic. Nevertheless, by looking at the operation and context of planning control in the overseas countries, fresh insight can be gained into the relationship between development plans and development control in England. It shows that any simplistic transfer of basic ideas would be very difficult as it would raise major issues about the role and status of central and local governments, the relationship between administration and the courts, the nature of public/private partnership in development, or land policy.

C6.30 Thus the English system of discretionary development control would probably be as unacceptable for the overseas countries as would their legally binding local plans be for England. The tendency rather, in both, is to retain the basic relationships but, in the overseas countries, to make the plans more flexible whilst retaining their legally binding status; and in England to reduce or limit the amount of discretion in control in particular circumstances, say, through the use of enterprise or simplified planning zones or more generally through circulars issued by government.

Printed in the United Kingdom for
Her Majesty's Stationery Office
Dd. 289794   C10   2/89   36145